CAMBRIDGE LIBRARY COLLECTION

Books of enduring scholarly value

Physical Sciences

From ancient times, humans have tried to understand the workings of the world around them. The roots of modern physical science go back to the very earliest mechanical devices such as levers and rollers, the mixing of paints and dyes, and the importance of the heavenly bodies in early religious observance and navigation. The physical sciences as we know them today began to emerge as independent academic subjects during the early modern period, in the work of Newton and other 'natural philosophers', and numerous sub-disciplines developed during the centuries that followed. This part of the Cambridge Library Collection is devoted to landmark publications in this area which will be of interest to historians of science concerned with individual scientists, particular discoveries, and advances in scientific method, or with the establishment and development of scientific institutions around the world.

The Scientific Papers of the Honourable Henry Cavendish

Henry Cavendish (1731-1810) was an English scientist whose published work was mostly concerned with electricity. He was elected a Fellow of the Royal Society in 1760. Cavendish was a prolific scientific investigator, performing experiments on not only electricity but also magnetism, thermometry, gases, heat potential and the chemical composition of water. Although he published some of his research, including his discovery of hydrogen, the majority of his work remained unpublished until 1879, when James Clerk Maxwell published a collection of Cavendish's electrical experiments. These papers showed that Cavendish had discovered many important electrical concepts which had since been credited to other researches, including the concept of electric potential. First published in 1921, these volumes are a collection of Cavendish's results from his many experiments. Volume 2 contains previously unpublished papers showing the results of Cavendish's chemical, magnetic and thermometry experiments.

Cambridge University Press has long been a pioneer in the reissuing of out-of-print titles from its own backlist, producing digital reprints of books that are still sought after by scholars and students but could not be reprinted economically using traditional technology. The Cambridge Library Collection extends this activity to a wider range of books which are still of importance to researchers and professionals, either for the source material they contain, or as landmarks in the history of their academic discipline.

Drawing from the world-renowned collections in the Cambridge University Library, and guided by the advice of experts in each subject area, Cambridge University Press is using state-of-the-art scanning machines in its own Printing House to capture the content of each book selected for inclusion. The files are processed to give a consistently clear, crisp image, and the books finished to the high quality standard for which the Press is recognised around the world. The latest print-on-demand technology ensures that the books will remain available indefinitely, and that orders for single or multiple copies can quickly be supplied.

The Cambridge Library Collection will bring back to life books of enduring scholarly value (including out-of-copyright works originally issued by other publishers) across a wide range of disciplines in the humanities and social sciences and in science and technology.

The Scientific Papers of the Honourable Henry Cavendish

Volume 2

Edited by Sir Edward Thorpe

CAMBRIDGE UNIVERSITY PRESS

Cambridge, New York, Melbourne, Madrid, Cape Town, Singapore,
São Paolo, Delhi, Dubai, Tokyo, Mexico City

Published in the United States of America by Cambridge University Press, New York

www.cambridge.org
Information on this title: www.cambridge.org/9781108018227

This edition first published 1921
This digitally printed version 2010

ISBN 978-1-108-01822-7 Paperback

THE SCIENTIFIC PAPERS OF THE HONOURABLE HENRY CAVENDISH, F.R.S.

CAMBRIDGE UNIVERSITY PRESS

C. F. CLAY, MANAGER

LONDON : FETTER LANE, E.C. 4

NEW YORK : THE MACMILLAN CO.
BOMBAY, CALCUTTA, MADRAS } MACMILLAN AND CO., LTD.
TORONTO : THE MACMILLAN CO. OF CANADA, LTD.
TOKYO : MARUZEN-KABUSHIKI-KAISHA

CAVENDISH'S HOUSE AT CLAPHAM

THE SCIENTIFIC PAPERS OF THE HONOURABLE HENRY CAVENDISH, F.R.S.

VOLUME II

CHEMICAL AND DYNAMICAL

Edited from the Published Papers, and the Cavendish Manuscripts in the possession of
HIS GRACE THE DUKE OF DEVONSHIRE, K.G., F.R.S.

by

SIR EDWARD THORPE, F.R.S.

with contributions by

DR CHARLES CHREE, F.R.S.
SIR FRANK WATSON DYSON, F.R.S.
SIR ARCHIBALD GEIKIE, O.M., F.R.S.
SIR JOSEPH LARMOR, F.R.S.

CAMBRIDGE
AT THE UNIVERSITY PRESS
1921

PREFACE

In pursuance of the policy of the Cambridge University Press to render available in collected form the occasional writings of men eminent in science and learning who have been associated with the University, a complete edition of the works of Henry Cavendish has long been contemplated. This project was naturally suggested by the position of Cavendish as one of the greatest of British men of science, and by the fact that his manuscripts were known to contain very important contributions to knowledge which had never been published to the world: in justice to his fame, and in the interests of the history of science, it was desirable that they should see the light. It was, moreover, an obligation which had very special claims on Cambridge on account of the intimate connection of the House of Cavendish with the University, maintained now through many generations.

As regards what was considered the more pressing part, the Electrical Researches, this duty was discharged with signal efficiency by the late Professor Clerk Maxwell, in a volume published by the University Press in 1879, which now, in revised form, constitutes Vol. I of this complete edition.

With the concurrence of the Duke of Devonshire, who again allowed the use of Henry Cavendish's manuscripts preserved at Chatsworth, the task of preparing for press the researches, published and unpublished, other than electrical, was begun some years ago; but the printing has been greatly delayed owing to circumstances arising out of the war. This portion of the work now forms Vol. II of the complete edition. It is largely, although by no means exclusively, concerned with chemical subjects. In the history of science it has been the custom to regard Cavendish mainly as a chemist. It is true that his chemical discoveries are among the greatest of his scientific achievements. But, as these volumes abundantly prove, to consider him simply as a chemist is to take a very partial and incomplete view of his scientific activities. In truth he was a Natural Philosopher of the broadest possible type, who occupied himself in turn with every branch of physical science known in his time, and who impressed the marks of his genius and the extraordinary penetrative force of his intellect on them all.

In its general plan the present volume is substantially similar to the volume of Electrical Researches edited by Professor Clerk Maxwell. It gives a brief account of Cavendish's personal history and characteristics, followed by a short commentary on his investigations, as published in part in the *Philosophical Transactions* of the Royal Society, where, as a matter of fact, everything that he chose to print is to be found. This commentary attempts to show the relation of his work to the knowledge of the time, to point out wherein it marks a new departure and how it permanently influenced the progress and development of science. Such advances as the recognition of the individuality of the gases, the discovery of the real nature of atmospheric air, and of the compound nature of water and its quantitative composition, were, it need hardly be stated, epoch-making. The twenty-two years that Cavendish devoted to chemical inquiry constitute indeed one of the most brilliant periods in the history of that science. In his work on thermometry, on cryoscopy, and on terrestrial magnetism he often breaks new ground and anticipates discoveries which have been attributed to much later observers.

This section is followed by a reprint of Cavendish's papers, other than those relating to electricity, which, as already stated, form the subject of Vol. I, with reproductions of the original illustrations. The papers are arranged in chronological order, and certain obvious errors and necessary typographical corrections have been indicated.

The last section deals with the unpublished manuscripts other than those already dealt with by Professor Clerk Maxwell, and is based upon the material in the possession of the Duke of Devonshire. Some small portion of these has already been made known. The Rev. William Vernon Harcourt examined the papers in connection with the famous Water Controversy, which he made the main subject of his Presidential Address to the British Association in 1839; he eloquently and convincingly advocated the just claims of Cavendish as the first and true discoverer of the compound nature of water, and printed in *facsimile* some excerpts from the manuscripts. Mr Harcourt appears to have gone through the whole of the papers and to have arranged them according to their subject matter. He was so impressed with their character and with the light they threw on the nature and range of Cavendish's intellectual activities that, in the course of his address, he strongly urged that they should be critically examined and arranged for publication. The present edition of Cavendish's works may be regarded as a compliance, however belated, with a desire that has often been urged.

The unpublished papers relating to chemistry show that Cavendish had anticipated Scheele in the discovery of arsenic acid, which he prepared by the method now in use. They also appear to indicate that if he did not actually anticipate Scheele in the isolation of tartaric acid, he was an independent discoverer of the true nature of "tartar," and of the relation of cream of tartar to "soluble tartar" or normal potassium tartrate. He seems to have been familiar with certain of the general principles underlying the phenomena of gaseous diffusion and to have experimentally verified them. He was perhaps the first to attempt to investigate quantitatively the phenomena of gaseous explosions. His work on Heat was original and independent; and had it been published at about the time it was performed, it would have placed him on a par with Black in regard to the historical development of the subject and have established his priority to Irvine, Crawford, Wilcke and Charles. His long and patient work on the tension of aqueous vapour was remarkably accurate and compares favourably even with that of Regnault. It was thus far superior to the measurements of Dalton, made nearly 30 years subsequently, and long regarded, despite the criticisms of Biot, as authoritative, especially in this country.

With respect to his other physical investigations it can hardly be matter for surprise to those who have studied the Electrical Researches as edited and expounded by Maxwell, that their author should have possessed views on these subjects far in advance of his age. The manuscripts reveal him as an indefatigable student of physical science of every kind, remarkable both for originality of ideas and for skill in their scrutiny and development by the rough direct methods of Newtonian fluxional calculation. There can be no doubt that Cavendish was the directing force in the project to determine the mean density of the Earth, whether from observations of the attraction of a mountain, or, twenty years later, by the vibrational method which he had inherited from his friend Michell. A preliminary discussion, which must date from some earlier time, of the results of the French geodetic expedition to Peru, had this object mainly in view: it dealt in a practical way with all the questions of geodetic compensation and isostasy which emerged later in more precise fashion in the great Indian Survey, and are now so prominent. The error in observations due to change of refraction on a heated slope was not overlooked, and was discussed (p. 393) in a masterly manner. The dynamical variation of latitude, now an important correction to refined observation, which had only been mentioned in passing by Euler, is fully grasped, and concisely expounded, though its actual amount was of course then unknown.

But perhaps most striking of all were his reasoned views on the Conservation of Energy, including a precise introduction of the idea of Potential Energy, and the recital of the causes of the degradation of Energy into the form of Heat. To obtain a parallel to the lines of his argument, cramped in expression as it naturally was at that time, we have to come down to Helmholtz's famous Essay of 1847. An effort has been made by Sir Joseph Larmor, in brief footnotes, to correlate his views with the pronouncements of Newton and Daniel Bernoulli, and other writers of the eighteenth century on mathematical physics, to whom Helmholtz has recorded his obligations in connection with the doctrine of Energy. Especially acute is Cavendish's expression of his conviction that the degradation of any likely amounts of kinetic and potential energies of moving particles, as then known, would prove insufficient to provide the large amounts of heat that are found to be involved in chemical changes; the next word on this subject belongs to Faraday and Maxwell and the Electric Molecular Theory.

He knew, as the Astronomer Royal points out, about the tidal retardation of the diurnal rotation of the Earth, and understood (p. 437) the principles of rotational torque and exchange of energy that are involved, apparently in a more direct and complete way than did Kant, recalling in fact the modern developments by Lord Kelvin and Sir George Darwin. The amount of deviation of a ray of light passing near the Sun, on the Newtonian corpuscular theory as extended by Michell to include gravitation of the corpuscles, did not escape (p. 437) his universal scrutiny of physical Nature. In most of his work the trend of thought seems to have been straight towards the course of the subsequent progress of science.

From the very outset of his career as a scientific observer, Cavendish seems to have occupied himself with the problems of terrestrial magnetism; and his interest in that subject continued unabated so long as he lived. Much of the material preserved at Chatsworth, and which is incidentally referred to in a postscript to Clerk Maxwell's Introduction to Vol. I, relates to this subject. Although it did not fall within the scope of the work on which he was engaged, Maxwell gave a brief summary of the character of the magnetic papers, and concluded that "they may supply important materials for the magnetic history of the Earth, and are in all respects excellent specimens of Cavendish's scientific procedure."

These papers have now been carefully examined by Dr Chree, the superintendent of the Kew Observatory, who has furnished a very complete account of their contents. He shows how Cavendish anticipated the ideas and procedure of subsequent observers in determining, for example, the best form of dip needles, in tracing the influence of "bending" of the needle on the observed value of its inclination, and as regards other sources of error.

Cavendish continued to make observations on magnetic dip and declination for nearly 40 years, and accumulated a considerable mass of material. This has been thoroughly sifted and discussed by Dr Chree; thus his summary of the results constitutes an important contribution to the magnetic history of the Earth, and its known close relation to the solar cycles, during the last third of the eighteenth century for which our information is so scanty and uncertain.

Acknowledgments are due to Sir Joseph Larmor for the active interest he has taken in the production of the work, and in particular for his contribution concerning the mathematical and dynamical papers and memoranda to be found among the Chatsworth manuscripts, and for certain general remarks on Cavendish's merits as an original thinker and observer which have been incorporated in the Preface. Thanks also are due to Sir Archibald Geikie, to the Astronomer Royal, and to Dr Chree for their several communications on such of Cavendish's unpublished papers as come within their special scope and interests.

Sir Horace Darwin was good enough to afford a critical estimate of Cavendish's published paper on his manner of dividing Astronomical Instruments; and Dr Burnett, of the British Museum, gave a similar estimate of the value of Cavendish's paper in the *Philosophical Transactions* on the Civil Year of the Hindoos.

Finally it is a pleasure to acknowledge the very efficient assistance rendered in every way by the staff of the Cambridge University Press. It cannot be doubted that the enterprise of the University Press, in now at length giving to the world an adequate account of the researches of one of the greatest of the intellectual sons of Cambridge, will meet at the hands of all men of science with the appreciation it deserves.

T. E. THORPE.

July 1920.

CONTENTS

INTRODUCTION BY THE EDITOR

Personal history of Henry Cavendish. His birth, Oct. 10th, 1731 (O.S.). Early education. Enters Peterhouse, Cambridge. Rejoins his father, Lord Charles Cavendish, in London. His laboratory in Great Marlborough Street. His wealth. His various residences. His death, Feb. 24th, 1810. His characteristics. Sketch of his scientific work. His chemical researches. Paper on *Factitious Air* (1766), Inflammable Air (hydrogen), Fixed Air (carbon dioxide). Recognition of individuality of the "Airs." On the *Rathbone-Place Water* (1767). Cause of temporary hardness of water. Solubility of calcium carbonate in aqueous solution of carbonic acid. Lime process of water softening. *A New Eudiometer* (1783). Uniformity of composition of Atmospheric Air established. Analysis of the upper air collected during a balloon ascent. *Experiments on Air* (1784). Discovery of compound nature of water. Nature and proportion of its constituents. The Water Controversy. Priestley, Watt, Lavoisier. Controversy with Kirwan. Discovery of *Composition of Nitric Acid* (1785). *Account of the Royal Society's Meteorological Instruments* (1776). Cavendish on Thermometry. Determination of fixed points in graduating thermometers. Correction for emergent column. *Freezing point of Mercury* (1783). Work on *Freezing Mixtures* (1786). Isolates and determines *Freezing points of hydrated mineral acids.* Implicitly recognises laws of chemical combination. Cavendish's physical and astronomical papers. *Height of Aurora* (1784). *Civil Year of the Hindoos* (1792). *Mean Density of the Earth* (1798). *Dividing Astronomical Instruments* (1809) 1–74

REPRINT OF PAPERS COMMUNICATED BY CAVENDISH TO THE ROYAL SOCIETY AND PUBLISHED IN THE *PHILOSOPHICAL TRANSACTIONS*: ARRANGED IN CHRONOLOGICAL ORDER

Three papers, containing Experiments on Factitious Air. *Phil. Trans.*, vol. 56, 1766, 141, with a Plate 77–101

Experiments on Rathbone-Place Water. *Phil. Trans.*, vol. 57, 1767, 92 102–111

An Account of the Meteorological Instruments used at the Royal Society's House. *Phil. Trans.*, vol. 66, 1776, 375, with a Plate . 112–126

An Account of a new Eudiometer. *Phil. Trans.*, vol. 73, 1783, 106, with a Plate 127–144

Observations on Mr Hutchins's Experiments for determining the Degree of Cold at which Quicksilver freezes. *Phil. Trans.*, vol. 73, 1783, 303 145–160

Experiments on Air. *Phil. Trans.*, vol. 74, 1784, 119 161–181

Answer to Mr Kirwan's Remarks upon the Experiments on Air. *Phil. Trans.*, vol. 74, 1784, 170 182–186

UNPUBLISHED PAPERS FROM THE ORIGINAL MANUSCRIPTS IN THE POSSESSION OF THE DUKE OF DEVONSHIRE, K.G., LL.D., F.R.S.

With explanatory notes by the Editor

LIST OF PLATES*

*These plates are available for download from www.cambridge.org/9781108018227

INTRODUCTION

COMPARATIVELY little is known concerning the personal history of the author of these memoirs. Nor is there much hope now that more may be gleaned. It may be doubted, indeed, whether there is much more to learn, for, apart from his scientific achievements, his life was singularly uneventful. He lived a solitary, secluded existence, and, despite his rank, and, in his later years, his great wealth, he deliberately refrained from any attempts to exercise the slightest social influence. He left no personal records, and few of his letters seem to have been preserved, possibly because few were written. Such as are known relate almost exclusively to matters of science and are otherwise of very slight human interest. All the knowledge of him we possess is based upon the fragmentary notices of a few contemporaries, principally Thomas Young, Thomas Thomson, of Glasgow, Sir Humphry Davy, and Lord Brougham. Their accounts, together with the reminiscences of others who had a certain small measure of personal acquaintance with him, or were able to communicate hearsay information concerning his character, habits and mode of life, have been brought together by the late Dr George Wilson, of Edinburgh, whose *Life of the Honble Henry Cavendish,* written at the request of the Cavendish Society, and published in 1851, still remains the only authoritative biography of the philosopher.

The following brief summary of his life and scientific achievements, which it seems desirable to prefix to this collection of his memoirs and papers, is almost wholly based upon that work.

The Honourable Henry Cavendish was born on October 10th, 1731, at Nice, where his mother was residing at that time for the sake of her health. His father, Lord Charles Cavendish, was the fifth son[1] of the second Duke

[1] In Wilson's *Life*, and also in the Introduction to Professor Clerk Maxwell's account of Cavendish's *Electrical Researches* (now vol. I of Scientific Papers), it is stated that Lord Charles Cavendish was the *third* son of the second Duke of Devonshire. The Rev. Walter H. Green, Vicar of All Saints' Church, Derby, to whom I am indebted for information concerning Cavendish's place of burial, and the non-existence of any memorial to him in the church, informs me, on the authority of Cox and St John Hope, who wrote a *History of All Saints' Church*, and who give copies of all the inscriptions on the coffin plates in the vault beneath the Devonshire Chapel, that the third son of the second Duke was Lord James Cavendish, who was buried in 1741. Lord Charles Cavendish who died on April 28th, and was buried in the Devonshire vault on May 7th, 1783, is stated, on his coffin-plate, to have been the *fifth* son of the second Duke.

of Devonshire; his mother, formerly Lady Anne Grey, was the fourth daughter of the Duke of Kent. She died when her son Henry was about two years old, shortly after the birth, in England, of a second son, Frederick.

Very little has been recorded concerning Henry Cavendish's earliest years. When eleven years old he was sent to Dr Newcome's school at Hackney, together with his brother Frederick and other members of the Cavendish family. This seminary is described by Lord Campbell in his *Lives of the Chancellors* as "a most excellent school," and the master as "a sound classical scholar, and a strict disciplinarian," but we have no information concerning the courses of instruction or of the degree of proficiency which Cavendish reached in them. He remained at school until 1749 when he was entered at Peterhouse, Cambridge. No particulars of his life at the University have come down to us. He left in 1753 without taking a degree, it is surmised, because he objected to the tests, which at that time were very stringent. Cavendish, in fact, was not a member of any religious body, and seems at no time to have professed any religious faith and never to have attended a place of worship. His brother Frederick, who came up to Peterhouse in 1751, also left without taking a degree. After leaving Cambridge the brothers would appear to have made a journey together on the Continent, but no particulars of their tour have been recorded beyond the gruesome story of their having seen a corpse in their hotel at Calais, laid out for burial in a room adjoining that which they occupied, to the absolute unconcern of the elder brother.

On leaving the University, Cavendish took up his residence in his father's house in London, where, according to Thomas Thomson, a set of stables were fitted up for his accommodation. It is probable that the stables were simply his laboratory and workshop, for, as his early writings show, it was at about this time that he entered upon the mathematical, chemical and physical studies which led to his investigations and discoveries. There is reason to suppose that his first experimental work was in connection with his father's scientific labours. Among his MSS. papers is a quarto sheet in his handwriting, headed "Table to reduce divisions on nonius of father's thermometer to degrees on new plate"; and this is followed by "Trials of father's th. by father, April 12, 1757." Lord Charles Cavendish had joined the Royal Society in 1727. He paid considerable attention to thermometry, and in 1757, when a Vice-President, contributed to the *Philosophical Transactions* descriptions of two forms of maximum and minimum thermometers for which he received the Copley Medal, or, as it is expressed in the terms of the award, "for his curious invention of making thermometers shewing respectively the greatest degree of heat and cold which have happened at any time during the absence of the observer." There is a paper by him in the Journal-Book of the Royal Society on Canton's experiments on the compressibility of water (*Phil. Trans.* 52, 1765, p. 640). As his son informs us in a paper which will be referred to subsequently, he also made accurate observations of the depression of the mercurial column in barometer tubes of various bores, which first established the relation between the depression and the internal diameter of the tube. Occasional allusions are to be met with in Cavendish's papers to work done in concert with his father, and to instruments which they used in common. These facts serve to throw light upon the relations of father and son. It has been stated that the family were disappointed that Cavendish should have declined to enter public life, and that his father treated him with niggardliness in consequence. There is no real evidence to support this supposition, which would seem to rest mainly on the doubtful authority of Lord Brougham. Cavendish, with his nervous embarrassed manner, his extraordinary shyness, his thin shrill voice, and hesitation of speech, was singularly unfitted for a public career, and it is unlikely that he was ever pressed to embark upon it. It is true that for the first forty years of his life he was a comparatively poor man, and stories have been related of the parsimonious habits he thereby contracted. But Lord Charles Cavendish was not rich, and, according to the botanist, Robert Brown, who had good means of knowing, he allowed his son as much as he could afford; and it is added he fully appreciated his son's abilities and never treated him unkindly. As regards money, he had probably as much as his habits and simple tastes required.

At some period subsequent to 1780 he became wealthy—how is not exactly known, but probably through bequests from relatives. Although he was, as Biot expressed it, "le plus riche de tous les savans et le plus savant de tous les riches," he was singularly indifferent to money and had little interest or concern in spending it. There is a well-known story of his threatening to remove his money if his bankers, who were concerned at the amount lying idle in their hands, continued to bother him about it. He could be liberal at times, almost to the point of extravagance, when some one pressed a worthy object on his notice, but an impulsive generosity was wholly foreign to his disposition, and he made little use of his wealth, which steadily accumulated, until, at his death, he was found to be the largest holder of bank-stock in England and to possess upwards of a million in different public funds—in addition to £50,000 in the hands of his bankers, a freehold estate of £8000 a year besides canal and other personal property.

Cavendish, who was never married, would appear to have resided in his father's house until he was near fifty years of age. Lord Charles Cavendish died in 1783, when, or possibly shortly before, his son moved to a house at Hampstead, and to a town-house close to the British Museum, at the corner of Montague Place and Gower St. We are told that few visitors crossed the threshold of either place, but those who were admitted found that its chief furniture consisted of books and apparatus. But the greater part of his library was contained in a separate house in Dean St., Soho, then a fashionable residential neighbourhood. Here he had brought together a large collection of works on science which he freely allowed all engaged in research to consult, and to which on occasion he himself went, signing a formal receipt for such books as he took away. For a time he lived at 4, Bedford Square.

During his later years he resided on Clapham Common, in a low white building surrounded by a garden. Very little in the house was set apart for personal comfort. What was intended to be the drawing room was converted into a laboratory. A forge stood in an adjoining room. The upper apartments constituted an astronomical observatory. A large registering thermometer of its owner's design (see p. 395) formed a sort of land-mark to the house; on the lawn was a wooden stage affording access to the top of a large tree to which in the course of his meteorological or electrical researches he would occasionally ascend. He lived most abstemiously and seldom saw company. We are told that if anyone dined with him he was invariably treated to a leg of mutton and nothing else. It was said that when on one occasion three or four scientific men were to dine with him the housekeeper remarked that one leg would not suffice. "Then get two," was the reply.

This solitary eventless life came to an end, after a very short illness, on February 24th, 1810. He died unattended, and was buried in All Saints' Church, Derby, in the Devonshire vault, near the splendid tomb which his ancestress, the redoubtable Elizabeth Hardwicke, the founder of his family, had built for herself[1]. No slab or monument of any kind marks the place of his sepulture.

The accounts of Cavendish which have been furnished by his contemporaries are singularly uniform in the impression they convey. They represent him as a man almost morbidly shy, nervous and embarrassed, extremely taciturn and reserved, who hated to be addressed, and who had a horror of a strange face; when he could be induced to talk he spoke in a thin shrill voice, and as if he had difficulty in articulation. He was described as of fair complexion, with small and not marked features, awkward in manner, and walking, with one hand behind his back, with a peculiar slouching gait. His dress was that of the preceding half-century and was never varied—a faded violet suit, frilled shirt-wrists, high coat

[1] Wilson speaks of a "funeral tablet," but according to Professor Clerk Maxwell, on the authority of Mr Cooling, a former Churchwarden of All Saints', there is nothing of the kind in the church. This is confirmed by the present vicar, Rev. Walter H. Green, who, moreover, thinks it doubtful if there ever was one. Several historians of Derby and its churches make mention of Henry Cavendish, but say nothing of any tablet or memorial of him. The vault in which he was buried is now permanently built up. It appears to have been last opened by Cox and St John Hope in 1879 for the purpose of making a full list of all interred there. They carefully copied all the coffin plates, 44 in number. Amongst them is No. XXVIII—that of Henry Cavendish who is stated to have died on February 24th and to have been buried on March 12th, 1810.

collar, knocker tailed periwig and cocked hat. He was a confirmed misogynist. He would never see a female servant, and if an unfortunate maid showed herself she was promptly dismissed. On one occasion he met one of his domestics with a broom and a pail on the stairs, and was so annoyed that he immediately ordered a back staircase to be built. There is, however, a story to the effect that, during one of his solitary walks, he so far overcame his bashfulness as to rescue a lady from the attacks of an infuriated cow—a circumstance which made a great sensation at the time at Clapham, where he was regarded with a certain amount of awe as a woman-hating wizard.

As to his character as it appeared to his biographer Wilson:

He was almost passionless....His brain seems to have been but a calculating machine....His Theory of the Universe seems to have been, that it consisted *solely* of a multitude of objects which could be weighed, numbered and measured; and the vocation to which he considered himself called was, to weigh, number and measure as many of those objects as his allotted three-score years and ten would permit. This conviction biassed all his doings, alike his great scientific enterprises and the petty details of his daily life....Throughout his long life he never transgressed the laws under which he seems to have instinctively acted.... It seems, indeed, to have been impossible for Cavendish to investigate any question otherwise than quantitatively....

Whatever, accordingly, we may think of the ideal which Cavendish set before him, we must acknowledge that he acted up to it with undeviating consistency; and that he realised it to a far greater extent than most men realise the more lofty ideals which they set before them. The pursuit of truth was with him a necessity, not a passion. In all his researches he displayed the greatest caution, not from hesitation or timidity, but from his recognition of the difficulties which attend the investigation of nature; from his delight in reducing everything to numerical rule, and his hatred of error as a transgression of law. *Cavendo tutus* was the motto of his family, and seems ever to have been before him.

On the other hand, Davy, in the course of an eloquent tribute to the character and achievements of Cavendish as a man of science, delivered in the theatre of the Royal Institution, a few weeks after his death, stated with perfect justice,

it ought to be mentioned in estimating the character of Mr Cavendish that his grand stimulus to exertion was evidently the love of truth and of knowledge:—unambitious, unassuming, it was often with difficulty that he was persuaded to bring forward his important discoveries. He disliked notoriety; he was, as it were, fearful of the voice of fame[1]. His labours consequently are recorded with the greatest dignity and simplicity and in the fewest possible words, without

[1] It was said of him that "he was peevishly impatient of the inconveniences of eminence, detested flattery, and was uneasy under merited praise." Brand's Preface to Suppl. to *Encycl. Brit.*

parade or apology; and it seemed as if in publication he was performing not what was a duty to himself, but what was a duty to the public.

Indeed, as regards the style in which his memoirs are put together, Cavendish would seem to have been ever faithful to the injunction laid upon him by the original Statutes of the Royal Society and which remained in force long after he had passed away, viz. that "in all reports of experiments to be brought into the Society, the matter of fact shall be barely stated, without any prefaces, apologies, or rhetorical flourishes."

Cavendish is usually regarded in the personal history of science as a chemist. Although his chemical discoveries are among his greatest scientific achievements, it may be questioned whether he himself would so reckon them. In reality he was a natural philosopher on a very broad gauge. Almost every department of the physical science of his time appealed to him with equal force and he pursued all with equal zeal and success. Nevertheless, compared with some of his contemporaries, the bulk of his published work is not large. Most of his investigations seem to have been made to satisfy his own spirit of inquiry, and he pursued them as opportunity and his mood impelled him. He cared little for the judgement and opinion of his fellows and was wholly indifferent to scientific fame. At no time was he in a hurry to give the results of his labour to the world, and it not seldom happened that these either remained unpublished, or were withheld for years after they were actually obtained. It may be, as many of his drafts would seem to imply, that literary composition was irksome to him. This was unfortunate no less for science than for his own reputation. The earlier publication of the cardinal discovery of the compound nature of water, for example, would, possibly, although by no means certainly, have effected the speedier downfall of phlogistonism; it would at least have spared us what is known in the history of science as the Water Controversy, with its regrettable and unfounded aspersions on Cavendish's scientific character and moral worth.

In his dislike of notoriety, his taciturnity, and disinclination to publish, Cavendish, as in many other respects, affords a striking contrast to his contemporary Priestley. It was characteristic of Priestley to write:

> When, for the sake of a little more reputation, men can keep brooding over a new fact, in the discovery of which they might, possibly, have very little real merit, till they think they can astonish the world with a system as complete as it is new, and give mankind a prodigious idea of their judgment and penetration; they are justly punished for their ingratitude to the fountain of all knowledge, and for their want of a genuine love of science and of mankind, in finding their boasted discoveries anticipated, and the field of honest fame pre-occupied by men, who from a natural ardour of mind, engage in philosophical pursuits, and with an ingenuous simplicity immediately communicate to others whatever occurs to them in their inquiries. (Preface to *Experiments and Observations on Different Kinds of Air*, Second Edition, 1775.)

This, we may be reasonably sure, must have been read by Cavendish, who was on good terms with Priestley and occasionally corresponded with him, but it is no less certain that he was entirely uninfluenced by it.

Cavendish became a Fellow of the Royal Society in 1760 and its meetings, in Crane Court, Fleet St., together with those of the Royal Society Club, which at that period dined at the "Mitre" or the "Crown and Anchor Inn," the Sunday conversaziones at Sir Joseph Banks's residence in Soho Square, and an occasional christening at Devonshire or Burlington House, were the only forms of social relaxation he seems to have allowed himself.

In spite of his shyness and his dislike of publicity, he was an active member of the Society, served on its Council, took a leading part in a number of its Committees, and was a loyal and zealous supporter of Sir Joseph Banks throughout the long and occasionally turbulent reign of that masterful President.

In commenting on Cavendish's work, other than his electrical researches which have been edited, mainly from the unpublished manuscripts in the possession of the Duke of Devonshire, by Professor Clerk Maxwell, and form vol. I of this collection, it may be desirable to deal first with his published chemical memoirs, upon which his fame mainly rests; and then with the rest of his published papers, which are not so conveniently grouped.

His published chemical researches are described in seven papers contributed at intervals between 1766 and 1788 to the *Philosophical Transactions*, where, in fact, all his published work of whatever kind first appeared. The seven papers referred to deal mainly with pneumatic chemistry.

They may be said, in a certain sense, to form a series in so far as they are concerned with a particular class of phenomena, but strictly speaking their subjects are independent.

The first communication, published in 1766, is entitled, "Three Papers, containing Experiments on factitious Air." The title is significant having regard to current doctrine concerning the nature of "air." "By factitious air," he says, "I mean in general any kind of air which is contained in other bodies in an unelastic state, and is produced from thence by art." It was in this sense that the term was first used by Boyle. Cavendish then proceeds to give, by way of introduction, a general account of his methods of manipulating the several varieties of factitious air he describes. His arrangements were not very dissimilar to, nor hardly an improvement upon, those of Stephen Hales, the real father of pneumatic chemistry, with whose work he was, no doubt, familiar. Like his predecessor, he collected the "airs" over water, suspending the collecting vessels by means of strings. The idea of the shelf, which constituted the essential feature of the pneumatic trough contrived by Priestley, the "tub," as it was called in the terminology of the time, may be said to have originated with Dr William Brownrigg and is described in a paper on "An Experimental Enquiry into the Mineral Elastic Spirit, or Air, contained in Spa Water; as well as into the Mephitic Qualities of this Spirit," printed in the *Phil. Trans.* for 1765, p. 218. Brownrigg's shelf, however, was a sort of rack fixed above the level of the water and perforated with holes in which the cylinders used for collecting the "air" were inserted and held in position by wedges. This was certainly a more convenient arrangement than the method of suspending the vessels by strings, adopted by Hales and Cavendish. It was a very simple step to place the shelf below the level of the water, but it was left to Priestley's nimble wit, some five or six years after the publication of Cavendish's paper, to make it. Cavendish, in fact, showed no great amount of ingenuity in the construction of new apparatus or in the modification of that already existing. He introduced no new or permanent contrivance into operative chemistry but was content to make the best use possible to him of such as lay ready to his hand. Whatever alterations he may have devised were directed to making them capable of affording quantitative results. He cared little for their appearance as regards elegance of form: his main concern was that they should be efficient.

The published paper on factitious airs is divided into three parts. Part I consists of a description of experiments on inflammable air from metals (hydrogen). Cavendish says he knew of only three metallic substances, namely zinc, iron and tin, that generate inflammable air on solution in acids; and those only by solution in the diluted vitriolic acid (sulphuric acid) or spirit of salt (hydrochloric acid). He made observations on the comparative rates of solution of the several metals in the different acids, studied the influence of dilution and temperature, and ascertained the volume of "air" evolved. He found that 1 ounce (480 grains) of zinc produced 356 ounce measures of "air," the quantity being "the same which-soever acids of these it is dissolved in." The same quantity of iron wire produced about 412 ounce measures of air; the quantity was "just the same, whether the oil of vitriol was diluted with 1½ or 7 times its weight of water." One ounce of tinfoil by solution in strong spirit of salt yielded 202 ounce measures of inflammable air. The volumes of air were measured "when the thermometer was at 50° and the barometer at 30 inches." These results are in substantial accordance with those demanded by the respective atomic weights of the metals.

He next studied the action of "nitrous acid" (nitric acid) and of undiluted oil of vitriol when heated upon the three metals, but found that the "air" which was generated in each case was "not at all inflammable." He made no further study of the "airs" produced in these cases but evidently considered them as essentially the same as in the former case but "modified" in their nature by the action of the acid used in their production.

In explanation of the difference in the character of the "airs" he says:

It seems likely from hence that when either of the above-mentioned metallic substances are dissolved in spirit of salt, or the diluted vitriolic acid, their phlogiston flies off, without having its nature changed by the acid, and forms the inflammable air; but that, when they are dissolved in the nitrous acid, or united by heat to the vitriolic acid, their phlogiston unites to part of the acid used for their solution, and flies off with it in fumes, the phlogiston losing its inflammable property by the union. The volatile sulphureous fumes, [sulphur dioxide] produced by uniting these metallic substances by heat to the undiluted vitriolic acid, shew plainly, that in this case their phlogiston unites to the acid; for it is well known, that the vitriolic sulphureous acid consists of the plain vitriolic acid united to phlogiston[1].

As to the inflammable air, produced by dissolving these substances in spirit of salt or the diluted vitriolic acid, there is great reason to think that it does not contain any of the acid in its composition; not only because it seems to be just the same whichever of these acids it is produced by; but also because there is an inflammable air, seemingly much of the same kind as this, produced from animal substances in putrefaction, and from vegetable substances in distillation, as will be shown hereafter; though there can be no reason to suppose that this kind of inflammable air owes its production to any acid.

We have here a striking illustration of what has been styled "the prejudice of that epoch" which was "not to regard compound substances as simple, but to consider undecompounded substances as compound." It was no doubt a prejudice, which as Mr Harcourt asserts, "infected the whole of chemistry" at the time, and Cavendish was not insensible to it. But in spite of the special pleading of the learned President of the British Association there can be no question that Cavendish remained fettered by the complexity of the phlogistic theory to the end; he never at any time sat "loose" to the favourite hypothesis: on the contrary he clung to it with as much persevering tenacity as any one of his countrymen.

Cavendish, it will be gathered, imagined that the inflammable air in all cases came from the metals as they were dissolved, and not from the diluted acids. This of course was in conformity with current doctrine. He was at first led to believe that the inflammable air unmodified by the acid was in fact pure phlogiston, but although he afterwards changed this opinion, the identity of phlogiston with hydrogen was, from this time forth, held to be established by one section at least of the Phlogiston School, notably by Priestley, Kirwan and Watt.

Cavendish then proceeded to study the properties of hydrogen. He

[1] Footnote in original paper. "Sulphur is allowed by chymists, to consist of the plain vitriolic acid united to phlogiston. The volatile sulphureous acid appears to consist of the same acid united to a less proportion of phlogiston than what is required to form sulphur. A circumstance which I think shows the truth of this, is that if oil of vitriol be distilled from sulphur, the liquor, which comes over, will be the volatile sulphureous acid."

found that it had no tendency to lose its elasticity by keeping, and that it was not absorbed by water or by fixed or volatile alkalis. Its explosibility when mixed with common air had, he says, been observed by others, and he sought to determine how the effect varies with the proportions in which the "airs" are mixed, by noting the loudness of the sound when a piece of lighted paper is applied to the mouth of the bottle. He offers no opinion as to the real cause or significance of the explosion, nor, inasmuch as the mixture was made over water, is it at all surprising that he failed to notice the appearance of moisture as a result of the combustion of the inflammable air.

He compared the hydrogen obtained by the use of the different metals and by the action of both sulphuric and hydrochloric acids, but was unable to perceive any difference. His general conclusion is that

it appears from these experiments that this [inflammable] air, like other inflammable substances, cannot burn without the assistance of common air. It seems too, that, unless the mixture contains more common than inflammable air, the common air therein is not sufficient to consume the whole of the inflammable air; whereby part of the inflammable air remains, and burns by means of the common air, which rushes into the bottle after the explosion.

He next attempted to ascertain the relative density of the inflammable air from metals as compared with common air by ascertaining the weight of the same bladder when filled successively with the two "airs." The experiments were repeated with hydrogen from the various sources but no certain difference could be perceived: "the small difference observed in these trials is in all probability less than what may arise from the unavoidable errors of the experiment." The average of the trials showed that 80 ounce measures of inflammable air weigh 41 grains less than an equal bulk of common air. On the assumption that water is 800 times denser than common air, this air would be seven times heavier than inflammable air: if Hauksbee's value of 850 be adopted, then common air will be eleven times heavier than hydrogen. Both values are, of course, erroneous, hydrogen being more than fourteen times lighter than air. Cavendish's method was faulty in principle. He was aware of certain sources of error due to leakage and admixture with common air and the presence of water-vapour. He made an attempt to estimate the influence of water-vapour by passing a known volume of the gas over dried pearl-ashes, and noting the increase in weight of the drying material. To check his results he made another series by a second method based on a different principle. He sought to determine the weight of hydrogen evolved on the solution of a known weight of metal, taking the precaution to pass the hydrogen over dry pearl-ash which he found by direct experiment not to absorb any sensible quantity of inflammable air. Knowing from his previous experiments the volume of hydrogen at a known temperature and pressure produced by the solution of a known weight of the metal he was in a position to calculate the specific gravity of hydrogen as compared with water and thence with common air.

The experiments, as in the first series, were made with hydrogen produced by the action of sulphuric and hydrochloric acids upon the different metals but, as in that case, no certain difference could be ascertained. "By a medium of the experiments, inflammable air comes out 8760 times lighter than water, or eleven times lighter than common air."

Cavendish concludes this section of his paper by an account of the result of an attempt to ascertain if any inflammable air could be obtained from copper by solution in spirit of salt. He could not obtain it by this means but the phenomena seemed "remarkable enough to deserve mentioning." In the cold there was no action;

> but, with the assistance of a heat almost sufficient to make the acid boil, it made a considerable effervescence,...when the water [from the pneumatic trough] rushed violently through the bent tube...and filled it almost intirely full.

He then varied the experiment so that the end of the tube of the generating vessel was exposed to the air.

> As soon as the effervescence began, the vapours issued visibly out of the bent tube; but they were not at all inflammable, as appeared by applying a piece of lighted paper to the end of the tube. A small empty phial was then inverted over the end of the bent tube, so that the mouth of the phial was immersed in the water, the end of the tube being within the body of the phial and out of the water. The common air was by degrees expelled out of the phial, and its room occupied by the vapours; after which, having chanced to shake the inverted phial a little, the water suddenly rushed in and filled it almost full; from thence it passed through the bent tube into the bottle and filled it quite full.

Had Cavendish followed up this observation he would in all probability have discovered the nature of the "elastic fluid which," as he says, "immediately loses its elasticity, as soon as it comes in contact with the water." As it was, the discovery of the "Marine Acid Air" was left to Priestley, who in 1772, on repeating Cavendish's experiment, found that the copper was not really concerned in the generation of the "air" which came from the acid when heated.

The second part of the paper is entitled: *Containing Experiments on Fixed Air, or that Species of Factitious Air, which is produced from Alkaline Substances, by Solution in Acids or by Calcination.*

The particular form of the title calls for a few words of comment. The term Fixed Air although used prior to Black to denote air which entered into the composition of substances and became thereby latitant was restricted by Black to the gas we now term carbon dioxide. Strictly speaking it was of course equally applicable to any form of "factitious air," as this was defined by Boyle and Cavendish. The fact that it was not sufficiently descriptive was perfectly obvious to Cavendish, as he implies in the opening paragraph of his paper. But inasmuch as the term was current in the chemical literature of the period, he uses it in the sense to which it was first restricted by Black in or about 1754 or 1755. In his famous paper *Experiments upon Magnesia Alba* Black gives no detailed account of the properties of Fixed Air in the free state; his main object was to show its influence in determining the "mildness" of alkalis. What he actually stated was that

quicklime...is capable of being joined to one particular species [of air] only, which is dispersed through the atmosphere, either in the shape of an exceedingly subtle powder, or more probably in that of an elastic fluid. To this I have given the name of *fixed air*, and perhaps very improperly; but I thought it better to use a word [air] already familiar in philosophy, than to invent a new name, before we be fully acquainted with the nature and properties of this substance, which will probably be the subject of my further inquiry.

The results of this further inquiry, if made, were never published by him.

Black was clearly conscious that his fixed air was distinct from common air and, as stated, may possibly have illustrated the distinction in his lectures at Glasgow and Edinburgh, but no published account of these appeared until some years after the date of Cavendish's paper.

Cavendish prepared the fixed air of Black by the action of spirit of salt upon marble, collecting the gas in the usual way over the pneumatic trough. He found, as had been already observed by Macbride, as indeed he remarks, that the air was absorbed by water and that the solution precipitated "the earth from lime-water; a sure sign that it had absorbed fixed air." He noticed that it could be preserved, apparently indefinitely, over mercury. By collecting a known volume of the gas over the mercurial pneumatic trough in a graduated cylinder in which was introduced a known volume of water "well purged of air by boiling," it was found that water, when the thermometer is about 55° F., will absorb rather more than an equal bulk of the gas, more being absorbed in cold weather than in warm. "Water heated to the boiling point is so far from absorbing air that it parts with what it has already absorbed." After being so heated it was found, on cooling, "not to make any precipitate, or to become in the least cloudy on mixing it with lime water." Exposed to the open air in a saucer for a few days the solution also parts with the fixed air, as the lime-water no longer renders it cloudy. Spirits of wine at the heat of 46° F. absorbs near $2\frac{1}{4}$ times its bulk of fixed air, and olive oil rather more than an equal bulk. Cavendish was disposed to think, however, that "fixed air contained in marble consists of substances of different natures, part of it being more soluble in water than the rest," an opinion from which Black dissented, but which, curiously enough, has recently been revived in a modified form in Russia.

A determination by Cavendish of its specific gravity by means of a bladder showed that fixed air is 1·57 times heavier than common air—a fair approximation to the truth. "Fixed air has no power of keeping fire alive, as common air has; but, on the contrary, that property of common air is very much diminished by the mixture of a small quantity of fixed air." Air containing about one-ninth of its bulk of fixed air will not support the flame of a candle.

He next made a series of determinations of the amount of fixed air in various "alcaline substances" by observing the loss of weight which known amounts suffered by solution in hydrochloric or sulphuric acid. The apparatus was identical with that employed in ascertaining the weight of hydrogen evolved by the action of acids on metals, except that the drying tube "was filled with shreds of filtering paper instead of dry pearl ashes; for pearl ashes would have absorbed the fixed air that passed through them." The substances examined were marble, volatile sal ammoniac (ammonium carbonate), pearl ashes, and acid potassium carbonate which Cavendish prepared by exposing "salt of tartar" (potassium carbonate) to an atmosphere of carbonic acid. The quantitative results are only rough approximations to the truth, but they are historically interesting as being the first fairly successful attempts to analyse carbonates.

Part III contains the results of "Experiments on the Air produced by Fermentation and Putrefaction."

Macbride had shown that vegetable and animal substances yield fixed air by fermentation and putrefaction. Cavendish found that this was the case in the alcoholic fermentation of sugar and apple-juice, and that the fixed air formed was identical with that produced from marble. Nor was the nature of common air affected by the fermenting liquid, since

> a small phial being filled with one part of this air and two of inflammable air; the mixture went off with a bounce, on applying a piece of lighted paper to the mouth, with exactly the same appearances, as far I could perceive, as when the phial was filled with the same quantities of common and inflammable air.

The "air" produced from gravy broth and from raw meat by putrefaction was found to consist of a mixture of fixed air and inflammable air, which could be separated by soap-leys. He determined the explosibility of a mixture of this inflammable air and common air: "it went off with a gentle bounce on applying the lighted paper; but I think not so loud as when the phial was filled with the last mentioned quantities of air from zinc and common air." A determination of its specific gravity showed "that this factitious air should seem to be rather heavier than air from zinc; but the quantity tried was too small to afford any great degree of certainty." The conclusion was that "on the whole it seems that this sort of inflammable air is nearly of the same kind as that produced from metals. It should seem however, either to be not exactly the same, or else to be mixed with some air heavier than it, and which has in some degree the property of extinguishing flame, like fixed air."

It would appear that this paper was originally intended to consist of four parts. Part IV, although completed and ready for press, for some reason was withdrawn. It has already been made public by the Rev. William Harcourt in the Postscript to his Presidential Address to the British Association in 1839[1]. It is to be found among the MSS. preserved at Chatsworth and will be reprinted in this volume and referred to subsequently.

The papers by Cavendish on "Factitious Airs" and that of Black, published a dozen years previously, together constitute an epoch in the history of chemistry. The early alchemists appear to have had some slight knowledge of an inflammable air produced by the action of acids upon metals. Van Helmont, to whom we owe the term *gas*, was aware of the existence of various forms of "air," as were Boyle and Mayow. That hydrogen was combustible was clearly recognised by Turquet de Mayerne in the seventeenth century and Lémery in 1700 noticed the detonating property of a mixture of this gas with common air. That an inflammable air was occasionally present in coal-workings was also known, and a distinction was drawn between fulminating- or fire-damp and choke-damp. Stephen Hales in his *Vegetable Staticks* (1727) and in his *Hæmastaticks* (1732) had shown that "airs" of different characters might be obtained by a variety of operations. But, like his predecessors, he attached no very definite importance to the diversities of colour, smell, solubility in water, inflammability, etc., which he noticed; and his imperfect attempts to determine the relative densities of the various airs only served to confirm him in his belief that they were essentially identical, but "infected" or "tainted" with extraneous "fumes," "vapours" and "sulphurous spirits." Hales's Essays had a considerable vogue in their day but whilst they may be said in one sense to have laid the foundations of pneumatic chemistry, in another they probably retarded its development, by serving to strengthen the belief, as a relic of mediæval scholasticism, in one universal and elementary air. It was only after the publication of Cavendish's paper when he conclusively established that the constant differences in the characters of "airs," to some of which he was able to assign quantitative values, were the real indications of their individuality, that Hales's Essays acquired their full importance. The properties which Hales noted, but of which he failed to perceive the significance, were then recognised as possible clues to the existence of a variety of elastic fluids differing essentially from common air.

This, then, is the great merit of Cavendish's paper: it gave a final and decisive blow to the conception of a universal air, elementary and primordial. That its true significance was everywhere clearly perceived is abun-

[1] *Report of the Ninth Meeting of the British Association for the Advancement of Science*, London, 1840, p. 58.

dantly proved by the literature of the period. The Royal Society showed their appreciation of its merit by awarding its author a Copley Medal. It gave a great impetus to the study of pneumatic chemistry and no doubt influenced Priestley in choosing this particular branch of chemical inquiry for his first essays in original investigation, thereby leading to a great extension of our knowledge of gases, and eventually to a complete revolution in chemical doctrine.

It seems to have been Cavendish's practice to keep several investigations running concurrently, although these were not invariably experimental, or concerned with a single department of science. During the time that he was engaged on his inquiry into "factitious airs" he was also occupied with an examination of certain natural waters, made primarily with a view of throwing light upon the cause of the "suspension [solution] of the calcarious earth which is separated from them by boiling." The results of this inquiry were read to the Royal Society on February 19th, 1767, and are published in the *Phil. Trans.* for that year.

Dr Brownrigg in a paper entitled "An Experimental Enquiry into the Mineral Elastic Spirit, or Air, contained in Spa Water; as well as into the Mephitic Qualities of this Spirit," published in the *Phil. Trans.* for 1765, had already thrown some light upon this matter and had shown that this "spirit was" identical with that

> most subtile and active exhalation, which, in many places perspires from springs and lakes, and other openings of the earth; or arises in pits and mines, where it is discovered by extinguishing flame; and from its pernicious effects, in killing all animals that breathe therein, is known to our miners by the name of choak-damp.

Shaw had previously collected "air" from Scarborough water, as also had Home from the chalybeate water of Dunse, in Scotland, but they appear to have regarded it as identical with common air. Brownrigg drew attention to the fact "that in proportion as this mineral air is separated by heat in the same proportion the more gross earthy parts of the water seem also to separate from it," but he offers no sufficient explanation of the phenomenon. The examination of the Rathbone Place water by Cavendish resulted in clearing up the matter. At that period most districts in London were supplied by wells or springs; the water of Rathbone Place was formerly raised by an engine and served the immediate neighbourhood.

Cavendish incidentally made an attempt to gain some information concerning the amount and nature of the saline ingredients in the water which he found to contain 17·5 grains of soluble matter to the pint, of which about half was observed to consist of calcium carbonate and a little magnesium carbonate, the rest being sodium chloride with small quantities of the sulphates of lime and magnesia. An unsuccessful search was made for nitrates, which were found in other London well waters, but a notable amount of "volatile alcali" was detected, which seems to reflect somewhat upon the purity of the supply. The details of the inquiry are interesting as an early example of water-analysis. The examination of the gaseous contents of the water appears to have chiefly interested Cavendish, and the greater part of the paper is devoted to a description of the method of procuring and analysing them. He found that the greater quantity was fixed air, and that the rest was common air, which he tested by his explosion method with hydrogen, and by determining its specific gravity. With characteristic caution he satisfied himself that the fixed air was not generated by boiling, but that it pre-existed in the water. As regards the effect of fixed air in determining the solubility of the earthy carbonates his experiments led him to conclude that:

> It seems likely from hence, that the suspension [solution] of the earth in the Rathbone place water is owing merely to its being united to more than its natural proportion of fixed air; as we have shown that this earth is actually united to more than double its natural proportion of fixed air, and also that it is immediately precipitated, either by driving off the superfluous fixed air by heat, or absorbing it by the addition of a proper quantity of lime-water.

The experiments, in fact, plainly foreshadow the lime-process of water-softening. Cavendish comments upon the paradoxical fact that calcareous earths should be

> rendered soluble in water, by furnishing them with more than their natural proportion of fixed air, i.e. that they should be rendered soluble, both by depriving them of their fixed air, and by furnishing them with more than their natural quantity of it,

and he multiplies experiments in proof of it.

Cavendish's third chemical memoir was not published until 1783 when he communicated to the Royal Society a paper on "A New Eudiometer" printed in the *Phil. Trans.* for that year.

During the years that followed the publication of his paper on the Rathbone Place water he was occupied with other than chemical work, notably with the electrical researches which form the subject of the volume to which this is a companion, as well as with other inquiries which will be referred to later.

The seventeen years which elapsed between the publication of his second and third chemical memoirs were singularly fruitful in discoveries in pneumatic chemistry, thanks more especially to the labours of Priestley and Scheele. The existence of a large number of gaseous substances, some of which, it was subsequently found, had actually been prepared, but not clearly recognised by Hales, was now definitely established. Owing, however, to the limited means of analytical chemistry at this period the composition and true relationships of these gases were very imperfectly understood. The phenomena they exhibited were explained, whenever possible, in terms of the phlogistic hypothesis, and the explanation, so far from elucidating the facts, frequently wholly obscured them. Until late in the eighteenth century the various "airs" were still roughly grouped as *fixed*, *mephitic* and *inflammable* although more than one observer had pointed out the inadequacy of such a classification, many "airs" being both inflammable and mephitic, whilst the term "fixed" was, strictly speaking, applicable to all forms of "factitious air."

Even after the existence of a variety of "airs" had been made known, it was years before the fact that they were individual and independent substances and not necessarily co-related was clearly and definitely recognised—so ingrained was the general belief of the schoolmen in "air" as an elementary and primordial substance. The specific differences in mode of origin and characters of the various "inflammable airs" were, it might be thought, sufficient to indicate that, as Keir wrote to Priestley so late as 1790, "there is not one inflammable but many inflammables, which opinion you now think as heterodox as the Athanasian system." Priestley's inability to recognise the difference between carbonic oxide and hydrogen led Watt to conclusions based upon wholly fallacious data, whilst strengthening his own belief in the invalidity of Cavendish's interpretation of the experiments which established the compound nature of water and the nature and proportion of its constituents.

That common air was a mixture of two dissimilar "airs," one only of which was concerned with the phenomena of respiration and combustion, had been recognised, more or less clearly, since the time of Mayow, but by the followers of Stahl it was surmised that the essential distinction between the two airs consisted in the presence or absence in them of phlogiston—the respirable and fire-air being devoid of phlogiston whereas the other air—the mephitic air in the atmosphere (nitrogen)—was saturated with it. Combustion, decay and putrefaction, the breathing of animals, and other processes, resulting as we now recognise in the abstraction of oxygen, were considered as phlogisticating the air, and the relative goodness or badness of common air was for a time considered to depend upon the degree of its phlogistication. Accordingly the efforts of chemists were directed to finding a method whereby the extent to which common air was phlogisticated by the various vitiating processes to which it was subjected could be, if possible, quantitatively ascertained. It was imagined that in this manner a definite numerical value could be associated with the relative salubrity of air in different localities and in different climates. Priestley's discovery that *nitrous air* (nitric oxide) when brought in contact with common air in a vessel standing over water, combined with that portion devoid of phlogiston, the product dissolving in water, seemed to afford the basis of a method for determining the amount of air not already phlogisticated—in other words, a method for ascertaining the relative goodness of air.

A considerable number of arrangements based upon this principle were accordingly devised by chemists, notably by Priestley, Fontana, Magellan, Dobson, and Landriani, the last-named coining the word *Eudiometer* by which they were commonly designated. The chemical theory of the method was, of course, unknown, as was the fact that nitric oxide and oxygen are capable of uniting in different proportions by volume, according to circumstances. It was quickly recognised however that concordant and comparable results could only be obtained by operating in a uniform manner, and the differences in the various methods suggested from time to time consisted for the most part in the mode in which uniformity of procedure was thought to be secured.

Among these experimenters was Cavendish, who, in 1781, entered upon a critical inquiry at his laboratory in Great Marlborough St., on the working of the nitric oxide eudiometer, and made considerable modifications in the apparatus with a view of obviating certain sources of irregularity which he noticed in the course of his trials. These he sets out at length in his paper. His eudiometer suffered in portability, ease and rapidity of working when compared with other eudiometers, such as Priestley's, but it was far more capable of giving constant results, and in his hands effected what was nothing less than a revolution in current doctrine concerning the constitution and functions of the atmosphere. He conclusively established that common air so far from being widely variable in character, as might reasonably be supposed from the numerous vitiating influences to which it was exposed, was, in reality, remarkably uniform.

"During the last half of the year 1781," he says, "I tried the air of near 60 different days, in order to find whether it was sensibly more phlogisticated at one time than another; but found no difference that I could be sure of, though the wind and weather on those days were very various; some of them being very fair and clear, others very wet, and others very foggy....I made some experiments also to try whether the air was sensibly more dephlogisticated at one time of the day than another, but could not find any difference. I also made several trials with a view to examine whether there was any difference between the air of London and the country, by filling bottles with air on the same day, and nearly at the same hour, at Marlborough-Street, and at Kensington. The result was, that sometimes the air of London appeared rather the purest, and sometimes that of Kensington; but the difference was never more than might proceed from the error of the experiment; and by taking a mean of all, there did not appear to be any difference between them. The number of days compared was 20, and a great part of them taken in winter, when there are a greater number of fires, and on days when there was very little wind to blow away the smoke."

Having thus proved that there existed "very little difference in the purity of common air at different times and places," Cavendish proceeded to point out that

it is very much to be wished, that those gentlemen who make experiments on factitious airs, and have occasion to ascertain their purity [that is their degree of phlogistication] by the nitrous test would reduce their observations to one common scale, as the different instruments employed for that purpose differ so much, that at present it is almost impossible to compare the observations of one person with those of another.

Accordingly he suggested that a common scale applicable to all nitric oxide eudiometers might be made "by assuming common air and perfectly phlogisticated air as fixed points." By perfectly phlogisticated air Cavendish implicitly means nitrogen, as shown by the methods he subsequently describes for obtaining it. On this scale perfectly phlogisticated air was zero and pure dephlogisticated air (oxygen) was 4·8.

If the test of any air [factitious air] be found to be the same as that of a mixture of equal parts of common and phlogisticated air, I would say, that it was half as good as common air; or, for shortness, I would say, that its standard was $\frac{1}{2}$: and, in general, if its test was the same as that of a mixture of one part of common air and x of phlogisticated air, I would say, that its standard was $\frac{1}{1+x}$. In like manner, if one part of this air would bear being mixed with x of phlogisticated air in order to make its test the same as that of common air, I would say, that it was $1 + x$ times as good as common air, or that its standard was $1 + x$; consequently, if common air, as Mr Scheele and La Voisier suppose, consist of a mixture of dephlogisticated and phlogisticated air, the standard of any air is in proportion to the quantity of pure dephlogisticated air in it. In order to find what test on the Eudiometer answers to different standards below that of common air, all which is wanted is, to mix common and perfectly phlogisticated air in different proportions, and to take the test of those mixtures; but in standards above that of common air, it is necessary to procure some good dephlogisticated air, and to find its standard by trying what proportion of phlogisticated air it must be mixed with, in order to have the same test as common air, and then to mix this dephlogisticated air with different proportions of phlogisticated air, and find the test of those mixtures[1].

These extracts are of importance for several reasons. In the first place they show that the eudiometer was an instrument designed to determine the degree of phlogistication of "air," whether common or factitious. In the next they serve to indicate what was in the mind of Cavendish con-

[1] The rule for computing the standard of any mixture of dephlogisticated and phlogisticated air is as follows. Suppose the test of a mixture of D parts of dephlogisticated air and P of phlogisticated air, is the same as that of common air, then is the standard of the dephlogisticated air $\frac{D+P}{D}$. Let now δ parts of this dephlogisticated air be mixed with ϕ parts of phlogisticated air, the standard of the mixture will be $\frac{D+P}{D} \times \frac{\delta}{\delta+\phi}$.

cerning the essential nature and constitution of the various factitious airs which had been prepared prior to 1783. They were fairly numerous and very different in characteristics. Priestley had discovered nitric oxide in 1772; ammonia, hydrogen chloride and oxygen in 1774; sulphur dioxide and silicon tetrafluoride in 1775 and nitrous oxide in 1776. The existence of some of them might indeed have been inferred from the work of some of Priestley's predecessors, particularly Stephen Hales. But to Cavendish as also to Priestley, the shadow of phlogiston lay over them all. With the exception of oxygen which seemed to be wholly devoid of that obscure and protean principle, they were all more or less "infected" with phlogiston and their properties were imagined to depend upon the degree of phlogistication, although such attempts as were made to explain them in terms of this inconceptible and elusive entity were occasionally felt to be more ingenious than satisfactory. In seeking to get at Cavendish's views concerning gaseous phenomena it must not be forgotten that he was never able to shake himself free from the trammels of phlogistonism. Its conceptions dominated and coloured all his attempts to arrive at the true meaning of the facts he observed. As we shall see, they largely obscured his recognition of the full significance of his great discovery of the compound nature of water. His guarded reference to the views of Scheele and Lavoisier concerning the composition of common air does not imply that he shared them. Nowhere in the paper on the new eudiometer does he commit himself to a definite statement of belief that common air is a mixture of essentially different elastic fluids. His statements as regards common air invariably refer to its "degree of phlogistication," that is, extent of vitiation; or conversely, "degree of purity," which only implicitly and incidentally meant amount of oxygen.

Cavendish concludes his paper with some remarks as to the limited value of the nitric oxide eudiometer in affording information concerning the salubrity or extent of vitiation of common air.

Where the impurities mixed with the air have any considerable smell, our sense of smelling may be able to discover them, though the quantity is vastly too small to phlogisticate the air in such a degree as to be perceived by the nitrous test, even though those impurities impart their phlogiston to the air very freely. For instance, the great and instantaneous power of nitrous air in phlogisticating common air is well known; and yet ten ounce measures of nitrous air, mixed with the air of a room upwards of twelve feet each way, is sufficient to communicate a strong smell to it, though its effect in phlogisticating the air must be utterly insensible to the nicest Eudiometer; for that quantity of nitrous air is not more than 140,000th part of the air of the room, and therefore can hardly alter its test by more than $\frac{3}{140000}$ or $\frac{1}{47000}$ part.... In like manner it is certain, that putrefying animal and vegetable substances, paint mixed with oil, and flowers, have a great tendency to phlogisticate the air; and yet it has been found, that the air of an house of office, of a fresh painted room, and of a room in which such a number of flowers were kept as to be very disagreeable to many persons, was not sensibly more phlogisticated than common air. There is no reason to suppose from these instances, either that these substances have not much tendency to phlogisticate the air, or that nitrous air is not a true test of its phlogistication, as both these points have been sufficiently proved by experiment; it only shews, that our sense of smelling can, in many cases, perceive infinitely smaller alterations in the purity of the air than can be perceived by the nitrous test, and that in most rooms the air is so frequently changed, that a considerable quantity of phlogisticating materials may be kept in them without sensibly impairing the air. But it must be observed, that the nitrous test shews the degree of phlogistication of air, and that only; whereas our sense of smelling cannot be considered as any test of its phlogistication, as there are many ways of phlogisticating air without imparting much smell to it; and I believe there are many strong smelling substances which do not sensibly phlogisticate it.

In spite of its limitations, and its imperfect theory, the paper on the New Eudiometer is a very notable contribution to the history of our knowledge of the atmosphere. It clearly established, for the first time, that common air was sensibly uniform in its character. Cavendish would have expressed this by saying that the extent to which it was phlogisticated was practically constant, and independent of locality or meteorological conditions: Scheele and Lavoisier would regard it as proving that the relative proportions of the dephlogisticated and phlogisticated air were invariable and this view gradually gained acceptance. It served to sweep away all the attempts which had been made by eudiometrical tourists to establish, in the words of Landriani, that the air of all those "places which from the long experience of the inhabitants had been reputed unwholesome," could be shown to be so by the instrument. Etymologically the name had no longer any significance as there were no degrees of goodness to be measured. Curiously enough it survives in our literature as the only remnant of the terminology of phlogistonism.

As might be expected from his halting views concerning the real nature of atmospheric air, Cavendish made no attempts to determine the relative volume of that portion of the air which he found to require a constant degree of phlogistication, which seems to show what little importance he attached to the views of Scheele and Lavoisier. Nevertheless the data he furnishes, as demonstrated by Wilson, would have enabled him by the formula he gives in connection with his method of graduation to obtain the relative proportion of the dephlogisticated air (oxygen) and phlogisticated air (nitrogen).

Taking 100 volumes of air, $D + P = 100$. By Cavendish's formula $\frac{D+P}{D} = 4{\cdot}8$; substituting the value of $D + P$;

$$\frac{100}{D} = 4{\cdot}8 \text{ and } D = \frac{100}{4{\cdot}8} = 20{\cdot}83.$$

Therefore 100 volumes of common air consist, on Cavendish's showing, of

Dephlogisticated air (oxygen)	20·83
Phlogisticated air (nitrogen)	79·17
	100·00

which is remarkably close to the truth, and a striking proof of his care and manipulative skill.

In a small packet of papers among the Chatsworth MSS. which had been previously examined by Dr George Wilson and labelled, "This parcel contains various interesting tables of the analysis of air, in 1780–1781, connected with the eudiometrical researches of that period," forming the analytical material for Cavendish's memoir on "The New Eudiometer," is a quarto sheet, the significance of which would appear to have escaped his biographer's notice. On it is written in Cavendish's handwriting:

Air taken by Dr Jeffries: tried Dec. 3, 1784

		1st trials		
No. 2	....	1·042	2nd tr.	1·037
No. 1	....	1·05	2d tr.	1·049
No. 5	3rd trial 1·05	1·041	2nd trial	1·05
No. 6	....	1·052	2d tr.	1·052
No. 3	....	1·05	2d tr.	1·052

Air taken at Hampstead at the time of the trial Dec. 3, 1784

	....	1·067	2 tr.	1·068

On November 30th, 1784, Jean Pierre Blanchard, a native of Le Petit Andelys, on the Seine, and one of the most successful of the earlier aeronauts, made a balloon ascent in the neighbourhood of London, accompanied by Dr J. Jeffries, an American physician, who subsequently, also with Blanchard, made the first Channel crossing by balloon from Dover to Calais[1]. There can be little doubt that Cavendish had made arrangements with Jeffries to collect samples of air at various heights during the ascent of November 30th, 1784. The usual method of procuring air for analysis at that period was to empty stoppered bottles filled with water at the spot at which the air was to be collected, and the Nos. 1–6 evidently refer to the samples so taken. The "trials" (analyses) were made by "the new Eudiometer" three days after the ascent, and the results compared with the air "taken out at Mr Cavendish's S. window at the same time." Nothing is said concerning the heights at which the several collections were made, but it will be observed that the "trials" proved that the samples were fairly uniform in composition and showed little variation from that at the ground-level.

These are the first analyses of the upper air of which there is any record. Cavendish thus anticipated Gay Lussac by about twenty years.

[1] *Encycl. Brit.*, 9th edit., art. Aeronautics, p. 191.

Incidentally, it may be remarked, Cavendish showed great interest in aeronautics. Among his papers are observations on the track of Blanchard's balloon in an ascent prior to that on which the samples of air were taken, with letters from a correspondent in Paris who supplied him with information concerning Montgolfier's aerostatic machines.

In the year following the publication of his memoir on "A New Eudiometer," Cavendish communicated to the Royal Society a paper entitled "Experiments on Air," which is printed in the *Phil. Trans.* for 1784. It is concerned with an inquiry which occupied him during 1781, of which the paper on the eudiometer is to be looked upon as a side issue. When regard is had to its contents the title of the new paper is not without significance as a further exemplification of Cavendish's views as to the essential nature of "air." Although its chief importance consists in the fact that it afforded the first clear and incontestable proof of the compound nature of water, and of the nature and relative proportion of its constituents, this discovery was an unlooked-for incident in the inquiry and not its primary object. This object was, in Cavendish's words, "principally with a view to find out the cause of the diminution which common air is well known to suffer by all the various ways in which it is phlogisticated, and to discover what becomes of the air thus lost or condensed." Its relation to the paper immediately preceding it thus becomes apparent. The lack of accurate means of measuring the diminution by phlogistication of common air led him to make a critical examination of the nitric oxide eudiometer, with a view to remedying its acknowledged defects, and as a necessary preliminary to the general inquiry.

Cavendish begins his paper by recalling all the various modes of diminution of common air by phlogistication known to him. It had been surmised that fixed air was either generated or separated from atmospheric air by phlogistication and his first experiments were made in order to ascertain if such were the case. He rejects all experiments with animal and vegetable substances as affording no certain proof of the origin of the fixed air. "The only methods I know," he says, "which are not liable to objection are by the calcination of metals, the burning of sulphur or phosphorus, the mixture of nitrous air, and the explosion of inflammable air." Experiments by others had seemed to show that thé passage of electric sparks through common air produced fixed air, but he thinks the evidence inconclusive, owing to the conditions under which the trials were made.

With regard to the four unobjectionable methods he finds no reason to think that the calcination of metals, although phlogisticating common air, produces any fixed air; nor is it produced by the burning of sulphur or phosphorus. The allegation that it is formed by mixing nitrous air (nitric oxide) with common air he disproves by taking care to free the common air from any fixed air originally present, and the nitrous air from any fixed air in or derived from the calcareous earth (calcium carbonate) dissolved in the water of the trough over which it was collected, by passing the gases before admixture through lime-water. Nor was any fixed air produced by the explosion of the inflammable air obtained from metals, with either common or dephlogisticated air when all the "airs" had been previously washed with lime-water.

On the whole, though it is not improbable that fixed air may be generated in some chymical processes, yet it seems certain that it is not the general effect of phlogisticating air, and that the diminution of common air is by no means owing to the generation or separation of fixed air from it.

Although the relevance of this conclusion may not be very apparent to-day, nevertheless, in view of chemical opinion at the time, and especially of the speculations of Kirwan, whose theoretical opinions exercised a certain amount of influence at that period, altogether disproportionate to their intrinsic merit, it was a distinct step in advance and, although not without controversy, eventually settled an important point concerning the origin of fixed air and its relations to air in general. Kirwan, who was a specious but fallacious reasoner, made an attempt to substantiate his position; this, contrary to his practice of avoiding polemics, provoked a reply from Cavendish and, of course, a rejoinder from Kirwan, but time has sided with Cavendish as to the merits of the controversy which is now forgotten. Some experiments by Priestley appeared to show that dephlogisticated air (oxygen) could be obtained from nitrous (nitric) acid and from vitriolic (sulphuric) acid. Cavendish therefore "tried whether the dephlogisticated part of common air might not, by phlogistication, be changed into nitrous or vitriolic acid" by burning sulphur over lime-water, and by the action of liver of sulphur on phlogisticated common air. In neither case was any nitrous salt (nitre) obtained, but in addition to ordinary selenite there was obtained what Cavendish regarded as a form of selenite which was "very soluble, and even crystallised readily, and was intensely bitter," but which by repeated solution and evaporation was gradually changed into ordinary selenite. Concerning this phenomenon Cavendish makes the following observation:

The nature of the neutral salts made with the phlogisticated vitriolic [sulphites] and phlogisticated nitrous acid [nitrites] has not been much examined by the chemists, though it seems well worth their attention;...Nitre formed with the phlogisticated nitrous acid [potassium nitrite] has been found to differ considerably from common nitre, as well as sal polychrest [normal potassium sulphate] from vitriolated tartar [acid potassium sulphate].

Cavendish in fact had prepared calcium sulphite and thiosulphate, the properties of which correspond with the description he gives. The term phlogisticated vitriolic acid, as applied to sulphurous acid, was of course in conformity with the terminology of the phlogiston school, inasmuch as this substance could be prepared by heating oil of vitriol with charcoal or sulphur, both of which conceivably imparted their phlogiston, in which they were rich, being so combustible, to the oil of vitriol.

The same results were obtained when pure dephlogisticated air was substituted for common air.

No vitriolic acid could be detected in the solution when common air was phlogisticated by nitrous air over water. Only nitre, with no traces of vitriolated tartar, could be obtained when the solution was "exactly saturated with salt of tartar [potassium carbonate] and evaporated."

"Having now mentioned the unsuccessful attempts I made to find out what becomes of the air lost by phlogistication, I proceed to some experiments, which serve really to explain the matter." In Dr Priestley's last volume of experiments[1] is related an

experiment of Mr Warltire's, in which it is said that, on firing a mixture of common and inflammable air by electricity, in a close copper vessel holding about three pints, a loss of weight was always perceived, on an average about two grains, though the vessel was stopped in such a manner that no air could escape by the explosion. It is also related, that on repeating the experiment in glass vessels, the inside of the glass, though clean and dry before, immediately became dewy; which confirmed an opinion he [Priestley] had long entertained, that common air deposits its moisture by phlogistication. As the latter experiment seemed likely to throw great light on the subject I had in view, I thought it well worth examining more closely. The first experiment also, if there was no mistake in it, would be very extraordinary and curious; but it did not succeed with me; for though the vessel I used held more than Mr Warltire's, namely, 24,000 grains of water, and though the experiment was repeated several times with different proportions of common and inflammable air, I could never perceive a loss of weight of more than one-fifth of a grain, and commonly none at all. It must be observed however, that though there were some of the experiments in which it seemed to diminish a little in weight, there were none in which it increased. In all the experiments, the inside of the glass globe became dewy, as observed by Mr Warltire; but not the least sooty matter could be perceived. Care was taken in all of them to find how much the air was diminished by the explosion, and to observe its test. The result is as follows: the bulk of the inflammable air being expressed in decimals of the common air.

Common air	Inflammable air	Diminution	Air remaining after the explosion	Test of this air in first method	Standard
1	1·241	·686	1·555	·055	·0
	1·055	·642	1·413	·063	·0
	·706	·647	1·059	·066	·0
	·423	·612	·811	·097	·03
	·331	·476	·855	·339	·27
	·206	·294	·912	·648	·58

[1] *Experiments and Observations on Different Kinds of Air.* By Joseph Priestley, LL.D., F.R.S. London: Printed for J. Johnson, No. 72, in St Paul's Church-Yard.

In these experiments the inflammable air was procured from zinc, as it was in all my experiments, except where otherwise expressed; but I made two more experiments, to try whether there was any difference between the air from zinc and that from iron, the quantity of inflammable air being the same in both, namely, 0·331 of the common; but I could not find any difference to be depended on between the two kinds of air, either in the diminution which they suffered by the explosion, or the test of the burnt air [i.e. that remaining after the explosion].

In explanation of these numbers, the third column shows the diminution after the explosion in the aggregate relative volumes of the common air and inflammable air as given in the first and second columns; the fourth column gives the relative volume of the remaining "air," and columns five and six its test and standard, that is the amount of oxygen (if any) it still contained. These last numbers were doubtless obtained by the nitric oxide eudiometer described in the previous paper: thus the fifth column indicates the contraction which took place when the "air" remaining after the explosion was mixed with a little more than its own volume of nitric oxide; and the sixth the volume of oxygen in that air according to the scale suggested by Cavendish in that paper, phlogisticated air being zero, common air 1 and dephlogisticated air 4·8.

Cavendish concentrates attention on the fourth experiment. Having regard to the primary object of his inquiry, its significance was unmistakable. He says respecting it:

From the fourth experiment it appears, that 423 measures of inflammable air are nearly sufficient to completely phlogisticate 1000 of common air; and that the bulk of the air remaining after the explosion is then very little more than four-fifths of the common air employed; so that, as common air cannot be reduced to a much less bulk than that, by any method of phlogistication, we may safely conclude, that when they are mixed in this proportion, and exploded, almost all the inflammable air, and about one-fifth part of the common air, lose their elasticity, and are condensed into the dew which lines the glass.

There can, of course, be no question that this statement implicitly contains an announcement of the synthetical production of water by the union of hydrogen and oxygen, but it may be doubted whether Cavendish intended it to convey that meaning in the sense we now interpret it. This will be still more evident from subsequent expressions in his paper. It must never be forgotten that the object of his inquiry, which he kept steadily in view, was to follow the transference of phlogiston in the various reactions which he set out to study. At the same time he seized upon the formation of the dew as a fact of cardinal importance, and proceeded to collect larger quantities of it with a view to ascertain its real nature. He continues:

The better to examine the nature of this dew, 500000 grain measures of inflammable air were burnt with about 2½ times that quantity of common air, and the burnt air [i.e. the products of the combustion as well as the remaining "air"—mainly nitrogen] made to pass through a glass cylinder eight feet long and three-quarters of an inch in diameter, in order to deposit the dew. The two airs were conveyed slowly into this cylinder by separate copper pipes, passing through a brass plate which stopped up the end of the cylinder; and as neither inflammable nor common air can burn by themselves, there was no danger of the flame spreading into the magazines from which they were conveyed. Each of these magazines consisted of a large tin vessel, inverted into another vessel just big enough to receive it. The inner vessel communicated with the copper pipe, and the air was forced out of it by pouring water into the outer vessel; and in order that the quantity of common air expelled should be 2½ times that of the inflammable, the water was let into the outer vessel by two holes in the bottom of the same tin pan, the hole which conveyed the water into that vessel in which the common air was confined being 2½ times as big as the other.

In trying the experiment, the magazines being first filled with their respective airs, the glass cylinder was taken off, and water let, by the two holes, into the outer vessels, till the airs began to issue from the ends of the copper pipes; they were then set on fire by a candle, and the cylinder put on again in its place. By this means upwards of 135 grains of water were condensed in the cylinder, which had no taste nor smell, and which left no sensible sediment when evaporated to dryness; neither did it yield any pungent smell during the evaporation; in short, it seemed pure water.

...By the experiments with the globe it appeared, that when inflammable and common air are exploded in a proper proportion, almost all the inflammable air, and near one-fifth of the common air, lose their elasticity, and are condensed into dew. And by this experiment it appears, that this dew is plain water, and consequently that almost all the inflammable air, and about one-fifth of the common air, are turned into pure water.

That Cavendish surmised that it was the dephlogisticated portion of common air that was "turned into pure water" by uniting with the inflammable air, is evident from his next experiment, which he thus describes:

In order to examine the nature of the matter condensed on firing a mixture of dephlogisticated and inflammable air, I took a glass globe, holding 8800 grain measures, furnished with a brass cock, and an apparatus for firing air by electricity. This globe was well exhausted by an air-pump and then filled with a mixture of inflammable and dephlogisticated air, by shutting the cock, fastening a bent glass tube to its mouth and letting up the end of it into a glass jar inverted into water, and containing a mixture of 19500 grain measures of dephlogisticated air, and 37000 of inflammable; so that, on opening the cock, some of this mixed air rushed through the bent tube, and filled the globe[1]. The cock was then shut, and the included air fired by electricity, by which means almost

[1] "In order to prevent any water from getting into this tube, while dipped under water to let it up into the glass jar, a bit of wax was stuck on the end of it, which was rubbed off when raised above the surface of the water." [*Note*. This tube was filled with common air to begin with, the nitrogen of which would find its way into the globe.]

all of it lost its elasticity. The cock was then again opened, so as to let in more of the same air, to supply the place of that destroyed by the explosion, which was again fired, and the operation continued till almost the whole of the mixture was let into the globe and exploded. By this means, though the globe held not more than the sixth part of the mixture, almost the whole of it was exploded therein, without any fresh exhaustion of the globe.

Although it is not so stated, it may be presumed that the bent tube remained in the jar containing the mixed gases throughout the course of the experiment, otherwise it would have been refilled with more or less common air between successive explosions, and so have introduced nitrogen into the globe at each operation.

As Cavendish wished to ascertain the volume and character of the "air" remaining in the globe after the series of explosions, as well as the weight and nature of the liquid produced, he proceeded as follows:

As I was desirous to try the quantity and test of this burnt [residual] air, without letting any water into the globe, which would have prevented my examining the nature of the condensed matter [i.e. the water produced], I took a larger globe, furnished also with a stop-cock, exhausted it by an air-pump, and screwed it on upon the cock of the former globe; upon which, by opening both cocks, the air rushed out of the smaller globe into the larger, till it became of equal density in both; then, by shutting the cock of the larger globe, unscrewing it again from the former, and opening it under water, I was enabled to find the quantity of the burnt air in it; and consequently, as the proportion which the contents of the two globes bore to each other was known, could tell the quantity of burnt air in the small globe before the communication was made between them. By this means the whole quantity of the burnt air was found to be 2950 grain measures: its standard was 1·85.

The liquor condensed in the globe, in weight about 30 grains, was sensibly acid to the taste, and by saturation with fixed alkali, and evaporation, yielded near two grains of nitre; so that it consisted of water united to a small quantity of nitrous acid. No sooty matter was deposited in the globe. The dephlogisticated air used in this experiment was procured from red precipitate [not *mercurius calcinatus per se*], that is, from a solution of quicksilver in spirit of nitre distilled till it acquires a red colour.

Cavendish's procedure was ingenious but it was open to several sources of error which must have affected the quantitative measurements. To begin with, it depended upon the efficiency of the air-pump which at that period was not very high. Hence the volume of the residual "air" was almost certainly overestimated. The standard of the residual "air" was 1·85 which means that it contained more oxygen than common air in the ratio of 1·85 to 1. That is, in round numbers it contained about 39 per cent of oxygen, so that there must have been a notable amount of nitrogen present—not less than about 1800 grain measures. No standard is given for the oxygen employed: it was presumably regarded as pure: the greater part of the nitrogen must have arisen from the imperfect evacuation of the globes, leakage of stop-cocks, and difficulty of preparing and preserving hydrogen without admixture with common air.

Cavendish then proceeded to search for the origin of the nitrous (nitric) acid in the water.

As it was suspected, that the acid contained in the condensed liquor was no essential part of the dephlogisticated air, but was owing to some acid vapour which came over in making it, and had not been absorbed by the water, the experiment was repeated in the same manner. with some more of the same air, which had been previously washed with water, by keeping it a day or two in a bottle with some water and shaking it frequently....The condensed liquor was still acid.

Dephlogisticated air prepared by heating red lead with oil of vitriol also gave an acid liquor, as did that from the leaves of plants prepared "in the manner of Doctors Ingenhousz and Priestley."

Cavendish noted that one circumstance common to all the experiments was that "the proportion of inflammable was such, that the burnt [residual] air was not much dephlogisticated; and it was observed that the less phlogisticated it was, the more acid was the condensed liquor." In other words, the acid appeared when the residual air was mainly oxygen, and, within the limits he observed, it seemed that the amount of acid increased with the quantity of oxygen. He therefore increased the proportion of inflammable air,

so that the burnt air was almost completely phlogisticated, its standard being $\frac{1}{10}$ [that is, the residual gas contained not more than about 2 per cent. by volume of oxygen]. The condensed liquor was then not at all acid, but seemed pure water: so that it appears, that with this kind of dephlogisticated air [from plants], the condensed liquor is not at all acid, when the airs are mixed in such a proportion that the burnt air is almost completely phlogisticated, but is considerably so when it is not much phlogisticated.

The experiment was repeated with oxygen from red precipitate and with variable proportions of inflammable air.

In the first, in which the burnt air was almost completely phlogisticated [that is in which there was no substantial excess of oxygen, if any, in the residual gas] the condensed liquor was not at all acid. In the second, in which its standard was 1·86 [i.e. containing about 39 per cent. of oxygen] that is, not much phlogisticated, it was considerably acid; so that with this air, as well as with that from plants, the condensed liquor contains, or is entirely free from, acid, according as the burnt air is less or more phlogisticated; and there can be little doubt but that the same rule obtains with any other kind of dephlogisticated air.

In order to see whether the acid, formed by the explosion of dephlogisticated air obtained by means of the vitriolic acid, would also be of the nitrous kind, I procured some air [oxygen] from turbith mineral [basic mercuric sulphate $HgSO_4 . 2HgO$—the *turpethum minerale* of the iatro-chemists], and exploded it with inflammable air, the proportion being such that the burnt air was not much phlogisticated. The condensed liquor manifested an acidity, which appeared by saturation with a solution of salt of tartar [potassium carbonate], to be of the nitrous kind; and it was found, by the addition of some terra ponderosa salita [barium chloride] to contain little or no vitriolic acid.

On repeating the experiment with common air

in such proportion that the standard of the burnt air was about $\frac{4}{10}$, the condensed liquor was not in the least acid.

He next mixed dephlogisticated air from red precipitate with perfectly phlogisticated air [nitrogen]

in such a proportion as to reduce it to the standard of common air...and then exploded with the same proportion of inflammable air as the common air was in the foregoing experiment, the condensed liquor was not in the least acid.

The conclusions to be drawn from this laborious and protracted inquiry, stated in Cavendish's own words, are as follows:

From the foregoing experiments it appears, that when a mixture of inflammable and dephlogisticated air is exploded in such proportion that the burnt air is not much phlogisticated, the condensed liquor contains a little acid, which is always of the nitrous kind, whatever substance the dephlogisticated air is procured from; but if the proportion be such that the burnt air is almost entirely phlogisticated, the condensed liquor is not at all acid, but seems pure water, without any addition whatever; and as, when they are mixed in that proportion, very little air remains after the explosion, almost the whole being condensed, it follows, that almost the whole of the inflammable and dephlogisticated air is converted into pure water.

Cavendish clearly recognised to what sources the residual gas left after the detonation might be due. He goes on to say:

It is not easy, indeed, to determine from these experiments what proportion the burnt air, remaining after the explosions, bore to the dephlogisticated air employed, as neither the small nor the large globe could be perfectly exhausted of air, and there was no saying with exactness what quantity was left in them; but in most of them, after allowing for this uncertainty, the true quantity of burnt air seemed not more than $\frac{1}{17}$th of the dephlogisticated air employed, or $\frac{1}{50}$th of the mixture. It seems, however, unnecessary to determine this point exactly, as the quantity is so small, that there can be little doubt but that it proceeds only from the impurities mixed with the dephlogisticated and inflammable air, and consequently that, if those airs could be obtained perfectly pure, the whole would be condensed.

He then comments upon the difference observed, as regards the production of nitric acid, between the experiments in which pure oxygen was used, and those in which common air, or a mixture of oxygen and nitrogen in the proportion in which they are present in common air, was employed.

With respect to common air, and dephlogisticated air reduced by the addition of phlogisticated air to the standard of common air, the case is different; as the liquor condensed in exploding them with inflammable air, I believe I may say in any proportion, is not at all acid; perhaps, because if they are mixed in such a proportion as that the burnt air is not much phlogisticated [that is, still contains much oxygen] the explosion is too weak, and not accompanied with sufficient heat.

A surmise which has been abundantly verified by subsequent experience.

The foregoing lengthy extracts from a paper which is classical have been purposely made in order that Cavendish's great discovery and the conclusions he drew from his experiments may be described in his own words and without any attempt to read into them any interpretation based upon or biased by subsequent knowledge. The facts of his experiments are stated with a clearness and precision which leave nothing to be desired, and it is impossible not to admire the care, patience, skill and sagacity with which he traced and accounted for what was without doubt a disturbing factor. He seems, indeed, to have perceived at an early stage of the inquiry that the formation of the nitric acid was purely fortuitous, and no necessary concomitant of the union by explosion of oxygen and hydrogen; nevertheless he recognised that his proof of the true nature of water and the mode of its synthetical formation could not be considered complete until the cause of the accidental contamination had been satisfactorily explained. This section of the inquiry seems to have occupied the greater portion of the time spent upon it, and presumably delayed the announcement of the cardinal discovery of the non-elementary nature of water for possibly a couple of years, with the unfortunate consequence of raising a controversy concerning Cavendish's claim to priority.

This circumstance is alluded to in the following paragraph which was not in the paper as originally received by the Royal Society but was interpolated before it was printed, no doubt with Cavendish's cognisance, by his friend Blagden who became one of the secretaries of the Society, and who had personal knowledge of the facts.

All the foregoing experiments, on the explosion of inflammable air with common and dephlogisticated airs, except those which relate to the cause of the acid found in the water, were made in the summer of the year 1781, and were mentioned by me to Dr Priestley, who in consequence of it made some experiments of the same kind, as he relates in a paper printed in the preceding volume of the Transactions. During the last summer also, a friend of mine gave some account of them to M. Lavoisier, as well as of the conclusion drawn from them, that dephlogisticated air is only water deprived of phlogiston; but at that time so far was M. Lavoisier from thinking any such opinion warranted, that, till he was prevailed on to repeat the experiment himself, he found some difficulty in believing that nearly the whole of the two airs could be converted into water. It is remarkable that neither of these gentlemen found any acid in the water produced by the combustion; which might proceed from the latter [Lavoisier] having burnt the two airs in a different manner from what I did; and from the former [Priestley] having used a different kind of inflammable air, namely, that from charcoal, and perhaps having used a greater proportion of it.

Among the Cavendish papers preserved at Chatsworth is a detached quarto sheet apparently in Blagden's handwriting, but unsigned by him, which seems to throw light upon the history of this interpolation. From a number of similar memoranda relating to foreign chemical literature, in other handwriting than that of Cavendish, to be found among his papers, it would appear he was not familiar with German, and was accustomed to rely upon others for information from contemporary German literature on matters of interest to him. Blagden's memorandum runs as follows:

> In a number of Crell's Annals which I happened not to have looked over before (May, 1784) I found the following passage: Mr Cavendish in London has {repeated / imitated} the experiments of M. Lavoisier to produce water from dephlogisticated and inflammable air by combustion. He has laid before the Royal Society the result of his experiments which confirm that change of the airs, or the new generation of water. His memoir has met with great approbation and won the assent of such a well-informed chymist as Mr Kirwan.
>
> Nothing appears by which it is possible to judge from whom Mr Crell received this information.
>
> Thursday morning Mar. 10.
>
> It is right to mention, that in the next number of the Annals (for June) there is a letter from Mr Kirwan mentioning your paper in proper terms, without any notice of Mr Lavoisier's name or pretensions.

There can be little doubt that this memorandum was the immediate occasion of Blagden's interpolation, and that its introduction was sanctioned by Cavendish to protect himself from the insinuation implied in the communication from which Blagden had quoted.

It will be convenient to defer further comment on this paragraph, as well as on other passages relating to the connection of Priestley, Watt and Lavoisier with the history of the discovery of the composition of water, until the account of the contents of the paper is completed.

Before discussing the meaning of the phenomena he had observed as regards the formation of the nitric acid, Cavendish ventures upon an opinion as to the real nature of phlogisticated air [nitrogen] and its relation to nitrous [nitric] acid. Phlogisticated air, he conceives, must be

> nothing else than the nitrous acid united to phlogiston; for when nitre is deflagrated with charcoal, the acid is almost entirely converted into this kind of air....As far as I can perceive too, at present, the air into which much the greatest part of the acid is converted, differs in no respect from common air phlogisticated. A small part of the acid, however, is turned into nitrous air [nitric

oxide], and the whole is mixed with a good deal of fixed, and perhaps a little inflammable air [carbonic oxide] both proceeding from the charcoal.

This, of course, is the case and, although only a partial interpretation, so far as it goes, it is consistent with the teaching of the phlogiston school. Here the phlogiston is derived from the charcoal. He further points to the fact "that the nitrous [nitric] acid is also convertible by phlogistication into nitrous air [nitric oxide]" and he sees an analogy in this example of the effect of partial phlogistication to the behaviour of vitriolic acid which when united to a smaller proportion of phlogiston forms the volatile sulphureous acid [sulphur dioxide] but when united to a larger proportion of phlogiston forms sulphur, which shows

no signs of acidity.... In like manner, the nitrous acid, united to a certain quantity of phlogiston, forms nitrous fumes and nitrous air...but when united to a different, in all probability a larger quantity, it forms phlogisticated air, which shows no signs of acidity, and is still less disposed to part with its phlogiston than sulphur.

"This being premised," as Cavendish says, let us see how he applies these conceptions to the explanation of the cause of the acidity of the water.

There seem two ways by which the phænomena of the acid found in the condensed liquor may be explained: first by supposing that dephlogisticated air contains a little nitrous acid which enters into it as one of its component parts, and that this acid, when the inflammable air is in a sufficient proportion, unites to the phlogiston, and is turned into phlogisticated air, but does not when the inflammable air is in too small a proportion; and, secondly, by supposing that there is no nitrous acid mixed with, or entering into the composition of, dephlogisticated air, but that, when this air is in a sufficient proportion, part of the phlogisticated air with which it is debased is, by the strong affinity of phlogiston to dephlogisticated air, deprived of its phlogiston, and turned into nitrous acid; whereas when the dephlogisticated air is not more than sufficient to consume the inflammable air, none then remains to deprive the phlogisticated air of its phlogiston, and turn it into acid.

Although, as we have seen, Cavendish probably regarded the nitric acid as an accidental contamination, its formation in his experiments was evidently considered by him as significant in throwing light upon the true nature of water and of dephlogisticated air. Indeed from the way in which he labours this part of the inquiry he would appear to consider the questions of the formation of the water and acid as inseparable and of almost equal importance. It is obvious from the above extracts that he regarded the presence of the acid as affording a possible clue to the constitution of dephlogisticated air and that he considered it was necessary to know this in order to form a just conception of the true nature of water and the mode of its synthesis.

He pointed out that if the nitrous acid is not to be considered as an essential constituent of dephlogisticated air, then, he says,

I think we must allow that dephlogisticated air is in reality nothing but dephlogisticated water, or water deprived of its phlogiston; or, in other words, that water consists of dephlogisticated air united to phlogiston; and that inflammable air is either pure phlogiston, as Dr. Priestley and Mr. Kirwan suppose [and as Cavendish formerly supposed], or else water united to phlogiston; since, according to this supposition, these two substances united together form pure water. On the other hand, if the first explanation be true, we must suppose that dephlogisticated air consists of water united to a little nitrous acid and deprived of its phlogiston; but still the nitrous acid in it must make only a very small part of the whole, as it is found, that the phlogisticated air, which it is converted into, is very small in comparison of the dephlogisticated air.

It will be observed from the wording of this passage that Cavendish had changed the opinion which he held in 1766, and which he expressed in his paper on inflammable air from metals, that this air was in reality phlogiston. He now ascribes this view to Priestley and Kirwan, and inclines to the belief that inflammable air is a compound of water and phlogiston—a sort of phlogiston hydrate. In a long footnote he explains what had led him to alter his opinion. Its substance is this: whereas common or dephlogisticated air will combine more or less readily at ordinary temperatures with phlogiston already united to substances, these gases refuse to unite with inflammable air unless at a red heat,

and it seems inexplicable, that they should refuse to unite to pure phlogiston, when they are able to extract it from substances to which it has an affinity: that is, that they should overcome the affinity of phlogiston to other substances, and extract it from them, when they will not even unite to it when presented to them.

Another reason would seem to be that in all the operations known to him in which inflammable air is generated water is more or less concerned.

As regards the two views of the nature of dephlogisticated air, Cavendish inclines to the belief that the nitrous acid is not an essential constituent of it, and that in fact it was not directly derived from it, inasmuch as this acid was formed not only in the case of oxygen from red precipitate but also in that derived from plants and from turbith mineral;

and it seems not likely that air procured from plants, and still less likely that air procured from a solution of mercury in oil of vitriol should contain any nitrous acid. Another strong argument in favour of this opinion is, that dephlogisticated air yields no nitrous acid when phlogisticated by liver of sulphur; for if this air contains nitrous acid and yields it when phlogisticated by explosion with inflammable air, it is very extraordinary that it should not do so when phlogisticated by other means.

What Cavendish regarded as a "strong argument" as a matter of fact only shows how little the real nature of the change experienced when air or oxygen is brought into contact with liver of sulphur was known to him.

> But what forms a stronger, and I think almost decisive argument, in favour of this explanation is, that when the dephlogisticated air is very pure, the condensed liquor is made much more strongly acid by mixing the air to be exploded with a little phlogisticated air [nitrogen].

The details of two experiments are then given in which the same quantities of a mixture of hydrogen and oxygen are exploded, to one of which, however, a quantity of nitrogen was added; "the condensed liquor in both cases was acid, but that in the latter evidently more so," as appeared from the amount of lime required to neutralise it. In the case where the nitrogen was added the burnt air would be more phlogisticated than in the other, and in that case "from what has been before said" from the line of argument and the suppositious presence of nitric acid as an essential constituent of oxygen, the condensed liquor should be less acid; "and yet it was found to be much more so; which shows strongly that it was the phlogisticated air which furnished the acid."

Further comparative experiments of a similar kind were made with the same general result and leading to the same inference, and the conclusion of the whole matter is thus stated:

> From what has been said there seems the utmost reason to think, that dephlogisticated air is only water deprived of its phlogiston, and that inflammable air, as was before said, is either phlogisticated water, or else pure phlogiston; but in all probability the former.

This, then, is Cavendish's formal statement of his views of the nature of dephlogisticated and inflammable air, and of their several relations to water. The conclusion is expressed in terms of phlogiston, and it is impossible to gather from the statement as it stands, whether Cavendish was convinced that water was actually a compound substance. He does not explicitly say so. The issue is confused by his view as to the nature of inflammable air and by our ignorance of his own opinion as to the real nature of phlogiston. Did he regard it as a material entity, or an imponderable principle—simply an *affection* or quality which by transference to and fro, determined the characters of substances? As we read it his statement might imply that he considered that water was formed by the phlogiston of his hypothetical hydrate phlogisticating the dephlogisticated air, whereby the water of the hydrate was liberated. Evidently he did not regard hydrogen as a simple elementary substance, in the modern sense, and to that extent, at least, his interpretation of his results is incomplete and erroneous.

Had he still continued to regard phlogiston as identical with hydrogen, of which he had determined the relative weight, and of whose material existence he was therefore assured, our inference as to his real view of the nature of water would have had a surer basis. It would almost seem as if he still halted between two opinions, and that his judgement was biased by a lingering old-time belief in the essential and fundamental unity of all "airs" and of water as a primordial element.

The statement of Cavendish's views of the nature of dephlogisticated and inflammable airs and of their relation to water, is followed by a paragraph which occasioned much discussion in the course of the controversy to which his paper gave rise. It may be desirable therefore to reproduce it in full:

> As Mr. Watt, in a paper lately read before this Society, supposes water to consist of dephlogisticated air and phlogiston deprived of part of their latent heat, whereas I take no notice of the latter circumstance, it may be proper to mention in a few words the reason of this apparent difference between us. If there be any such thing as elementary heat, it must be allowed that what Mr. Watt says is true; but by the same rule we ought to say, that the diluted mineral acids consist of the concentrated acids united to water and deprived of part of their latent heat; that solutions of sal-ammoniac, and most other neutral salts, consist of the salt united to water and elementary heat; and a similar language ought to be used in speaking of almost all chemical combinations, as there are very few which are not attended with some increase or diminution of heat. Now I have chosen to avoid this form of speaking, both because I think it more likely that there is no such thing as elementary heat, and because saying so in this instance, without using similar expressions in speaking of other chemical unions, would be improper, and would lead to false ideas; and it may even admit of doubt, whether the doing it in general would not cause more trouble and perplexity than it is worth.

This passage was interpolated by Cavendish after his paper was received by the Royal Society and before it was published in the *Phil. Trans.* and this circumstance gave rise to the assertion that if Cavendish did not actually owe to Watt his conception of the inference to be drawn from his experiments, to Watt at least belongs the credit of having been the first to point out that water is a compound substance and that its components are dephlogisticated air and phlogiston.

At first sight it may be thought the point of difference between the two philosophers as regards heat in relation to the composition of water is irrelevant. In reality this is not so, as something turns on their respective views of the nature of heat, Cavendish, at all events, being convinced, as will be subsequently evident, that it is not a material entity.

How it happened that Watt came into conflict with Cavendish on the question of priority will be stated later.

The greater part of the rest of the paper is concerned with speculations on the nature of common air and of dephlogisticated air which are of interest as throwing some light upon the extent to which his opinions had been modified by experience and reflection since the publication of his paper on inflammable air, seventeen years previously. He is now more inclined to the opinion of Lavoisier and Scheele that dephlogisticated and phlogisticated airs are "quite distinct substances, and not differing only in their degree of phlogistication; and that common air is a mixture of the two." He finds support for this view in his discovery that pure dephlogisticated air by complete phlogistication is converted into water instead of into phlogisticated air. It is interesting to note how closely he approaches to the view-point of the French school in the following passage: "From what has been said, it follows, that instead of saying air is phlogisticated or dephlogisticated by any means, it would be more strictly just to say, it is deprived of, or receives, an addition of dephlogisticated air." But however "strictly just" it may be, Cavendish, confirmed phlogistian as he was, cannot bring himself to say it, for he immediately adds "but as the other expression is convenient, and can scarcely be considered as improper, I shall still frequently make use of it in the remainder of this paper."

Considerable space is occupied by a discussion of the theory of the various methods employed in the production of dephlogisticated air, and he traverses Priestley's opinion that nitrous and vitriolic acids are convertible into dephlogisticated air: "Their use in preparing it is owing only to the great power they possess of depriving bodies of their phlogiston." He shows that the production of oxygen from red precipitate is not owing to any nitrous acid in that substance "and consequently that, in procuring dephlogisticated air from it, no acid is converted into air; and it is reasonable to conclude, therefore, that no such change is produced in procuring it from any other substance"—a sweeping generalisation on too limited a basis.

The way in which the nitrous [nitric] acid acts, in the production of it [dephlogisticated air] from red precipitate, seems to be as follows. On distilling the mixture of quicksilver and spirit of nitre, the acid comes over, loaded with phlogiston, in the form of nitrous vapour [oxides of nitrogen], and continues to do so till the remaining matter [mercuric nitrate] acquires its full red colour, by which time all the nitrous acid is driven over, but some of the watery part still remains behind, and adheres strongly to the quicksilver; so that the red precipitate may be considered, either as quicksilver deprived of part of its phlogiston, and united to a certain portion of water, or as quicksilver united to dephlogisticated air[1], after which, on further increasing the heat, the water in it rises deprived of its phlogiston, that is, in the form of dephlogisticated air, and at the

[1] In a footnote to this passage he says "It would be ridiculous to say, that it is the quicksilver in the red precipitate which is deprived of its phlogiston, and not the water, or that it is the water and not the quicksilver; all that we can say is that red precipitate consists of quicksilver and water, one or both of which are deprived of part of their phlogiston. In like manner, during the preparation of the red precipitate, it is certain that the acid absorbs phlogiston, either from the quicksilver or the water; but we are by no means authorised to say from which."

same time the quicksilver distils over in its metallic form.... Mercurius calcinatus appears to be only quicksilver which has absorbed dephlogisticated air from the atmosphere during its preparation; accordingly, by giving it a sufficient heat, the dephlogisticated air is driven off, and the quicksilver acquires its original form. It seems therefore that mercurius calcinatus and red precipitate, though prepared in a different manner, are very nearly the same thing.

It seems to be pretty obvious from this somewhat laboured account that Cavendish was disposed to think *mercurius calcinatus* and red precipitate, although "very nearly," were not in reality quite "the same thing," but that red precipitate differed from *mercurius calcinatus* in containing "a certain portion [amount] of water" and that when heated the water is deprived of its phlogiston, thereby liberating the dephlogisticated air—the phlogiston presumably attaching itself to the quicksilver which consequently resumes its metallic form.

Apart from the light it throws—however dimly—on Cavendish's views as to the essential nature of the relation between dephlogisticated air and water, this passage, as already pointed out by Dr Wilson, is instructive

as showing what many other passages in the papers of Cavendish, and his contemporaries also show, that the discovery of the composition of water would not, in the hands of the disciples of the phlogiston school, have materially altered the aspect of chemistry. The difference it introduced was little more than this, that where formerly transferences of phlogiston, from one body to another, were assumed to take place, now water instead of phlogiston was shifted backwards and forwards, and decomposed and recomposed as the exigencies of theory required[1].

An exemplification of the truth, as elsewhere remarked by Dr Wilson, of the necessity under which a false theory lies of multiplying falsities.

The production of oxygen from nitre, in the course of which Cavendish observed the formation of potassium nitrite, is accounted for in a similar way. Now that water was known to contain or to be capable of yielding dephlogisticated air, that "air" is to be regarded as coming from the water which the nitre is assumed to contain as an essential constituent. The same line of reasoning led him to conclude that "the rationale of the production of dephlogisticated air from turbith mineral, and from red precipitate, are nearly similar."

Cavendish then comments on the action of light on substances, such as the bleaching of an alcoholic solution of chlorophyll, the colouring of nitric acid, and of silver chloride, etc. which he attributes to the absorption of phlogiston from the water, and the liberation of dephlogisticated air. He conceives a similar action to take place in plants under the influence of sunshine: "it seems likely," he says, "that the use of light, in promoting the growth of plants and the production of dephlogisticated air from them,

[1] Wilson, *Life of Cavendish*, p. 248.

is that it enables them to absorb phlogiston from the water." He attempts to answer certain objections which may be urged against this view of the origin of the oxygen, and he seeks to explain the fact observed by Senebier "that plants yield much more dephlogisticated air in distilled water impregnated with fixed air, than in plain distilled water," by suggesting that "as fixed air is a principal constituent part of vegetable substances, it is reasonable to suppose that the work of vegetation will go on better in water containing this substance, than in other water."

This sentence concludes the paper as it was originally received by the Royal Society. Before it was actually printed off Cavendish made a very significant addition from which it is desirable to quote pretty fully, as it shows that he was quite alive to the interpretation which the new school might put upon his results. It would almost seem as if he had anticipated the use which the opponents of phlogistonism would make of his discovery, and that it fell to him as a leading upholder of Stahl's doctrine to combat their arguments. He says:

There are several memoirs of Mr. Lavoisier published by the Academy of Sciences, in which he intirely discards phlogiston, and explains those phænomena which have been usually attributed to the loss or attraction of that substance, by the absorption or expulsion of dephlogisticated air; and as not only the foregoing experiments, but most other phænomena of nature, seem explicable as well, or nearly as well, on this as upon the commonly believed principle of phlogiston, it may be proper briefly to mention in what manner I would explain them on this principle, and why I have adhered to the other. In doing this, I shall not conform strictly to his theory, but shall make such additions and alterations as seem to suit it best to the phænomena; the more so, as the foregoing experiments may, perhaps, induce the author himself to think some such additions proper.

To seek to modify an opponent's theory by making such additions and alterations as in one's judgement may best suit the phenomena would seem at first sight an unwarrantable procedure on the part of a controversialist who strives to combat that theory; but, as a matter of fact, no reasonable objection could be urged against the manner in which Lavoisier's views are stated by Cavendish in explanation of the various phenomena related in his paper. Cavendish says that water according to Lavoisier, "consists of inflammable air united to dephlogisticated air"; and in effect that nitric oxide, sulphurous acid, and phosphoric acid are combinations respectively of nitrogen, sulphur and phosphorus with oxygen; and that by further oxidation nitric oxide and sulphurous acid may be converted respectively into nitric acid and oil of vitriol; that the metallic calces are oxides of the metals. The rationale of the production of red precipitate from mercury and nitric acid and its decomposition by heat into oxygen and mercury is correctly stated by him, as is the production of oxygen from nitre. The only ambiguity is in the production of oxygen from plants which is partially, at least, explicable by the circumstance that the composition of fixed air was unknown to Cavendish and the real significance of Senebier's observation was therefore not recognised by him.

But in spite of this perspicuous and impartial statement of Lavoisier's views, which of course had the crowning merit of explaining "the phænomena of nature" without the help of a purely hypothetical principle, concerning which no two of those who believed in it could agree as to its real character or attributes, Cavendish could not shake himself free from his orthodoxy.

It seems, therefore, from what has been said, as if the phænomena of nature might be explained very well on this principle, without the help of phlogiston; and indeed, as adding dephlogisticated air to a body comes to the same thing as depriving it of its phlogiston and adding water to it, and as there are, perhaps, no bodies entirely destitute of water, and as I know no way by which phlogiston can be transferred from one body to another, without leaving it uncertain whether water is not at the same time transferred, it will be very difficult to determine by experiment which of these opinions is the truest; but as the commonly received principle of phlogiston explains all phænomena, at least as well as Mr. Lavoisier's, I have adhered to that.

It is unnecessary to make any lengthened comment upon these statements. It might appear that in reality there is no essential difference in the views of Lavoisier and Cavendish as to the chemical nature of water. But that is not so. Lavoisier had a clear conception of the individuality of hydrogen, even before he or his associates coined that word; on the other hand Cavendish thought inflammable air was a common principle of bodies rich in phlogiston, and was capable of assuming an elastic form, either alone or in combination with water. He may have drawn a clearer distinction than Priestley between the various forms of "inflammable air," but he was no less convinced than Priestley that all of them were compounds of phlogiston. As regards his reference to Lavoisier's explanations of the phenomena of nature without reference to phlogiston, he makes no real attempt to combat them, and is apparently unaware of the *petitio principii* involved in his statement: he simply contents himself with reiterating his belief in the existence of phlogiston. "The human mind," wrote Davy, "is always governed not by what it knows, but by what it believes."

It can hardly be doubted, as his biographer admits, that Cavendish's defence of his views was an afterthought, added after his opinions had been made public, as a justification of what could not then be withdrawn.

We now turn to the allusions in Cavendish's paper to Lavoisier and Watt, in reference to their share in the experimental proof of the compound nature of water.

Some time before the results of Cavendish's work were made known to him, Lavoisier had enunciated the hypothesis that dephlogisticated air was the principle of acidity, since its combinations, more especially with substances supposed by his contemporaries to be rich in phlogiston, such as carbon, sulphur, phosphorus, etc., were, as he thought, invariably acids. Hence his scepticism as to the validity of Cavendish's results. It seemed to him incredible that the inflammable air from metals—which some indeed had regarded as phlogiston itself, should by its union with oxygen furnish a perfectly neutral substance. Lavoisier, in fact, had publicly stated only a year or two previously, "que l'air inflammable en brûlant devoit donner de l'acide vitriolique, ou de l'acide sulfureux."

In Lavoisier's own account of the experiment he was prevailed upon to repeat, he admits his prior acquaintance with Cavendish's results, and states that Blagden was his informant:

Ce fut le 24 Juin, 1783, que fîmes cette expérience, M. de la Place et moi, en présence de Mm. le Roi, de Vandermonde, de plusieurs autres Académiciens, et de M. Blagden, aujourd'hui Secrétaire de la Société Royale de Londres; ce dernier nous apprit que M. Cavendish avoit déjà essayé, à Londres, de brûler de l'air inflammable dans des vaisseaux fermés, et qu'il avoit obtenu une quantité d'eau très sensible (*Mém. de l'Acad.* 1781, p. 472).

Blagden felt constrained to point out in a letter addressed to Crell, the editor of the *Chemische Annalen*, and published in that journal for 1786, that the above was a partial and somewhat disingenuous account of what had actually transpired. The letter which is not dated (according to Muirhead's translation) runs as follows:

I can certainly give you the best account of the little dispute about the first discoverer of the artificial generation of water, as I was the principal instrument through which the first news of the discovery that had been already made was communicated to Mr. Lavoisier. The following is a short statement of the history:

In the spring of 1783, Mr. Cavendish communicated to me and other members of the Royal Society, his particular friends, the result of some experiments with which he had for a long time been occupied. He showed us, that out of them he must draw the conclusion, that dephlogisticated air was nothing else than water deprived of its phlogiston; and *vice versâ* that water was dephlogisticated air united with phlogiston. About the same time the news was brought to London, that Mr. Watt of Birmingham had been induced by some observations [by Priestley] to form a similar opinion. Soon after this I went to Paris, and in the company of Mr. Lavoisier, and of some other members of the Royal Academy of Sciences, I gave some account of these new experiments and of the opinions founded upon them. They replied, that they had already heard something of these experiments, and particularly, that Dr. Priestley had repeated them. They did not doubt, that in such manner a considerable quantity of water might be obtained; but they felt convinced that it did not come near to the weight of the two species of air employed; on which account it was not to be regarded as water formed or produced out of the two kinds of air, but was already contained in and united with the airs, and deposited in their combustion. This opinion was held by Mr. Lavoisier, as well as by the rest of the gentlemen who conferred on the subject; but as the experiment itself appeared to them very remarkable in all points of view, they unanimously requested Mr. Lavoisier, who possessed all the necessary preparations, to repeat the experiment on a somewhat larger scale, as early as possible. This desire he complied with on the 24th June, 1783 (as he relates in the latest volume of the *Paris Memoirs*). From Mr. Lavoisier's own account of his experiment, it sufficiently appears, that at that period he had not yet formed the opinion that water was composed of dephlogisticated and inflammable airs; for he expected that a sort of acid would be produced by their union. In general, Mr. Lavoisier cannot be convicted of having advanced anything contrary to truth; but it can less be denied, that he concealed a part of the truth. For he should have acknowledged that I had, some days before, apprised him of Mr. Cavendish's experiments, instead of which, the expression "il nous apprit" gives rise to the idea that I had not informed him earlier than that very day. In like manner, Mr. Lavoisier has passed over a very remarkable circumstance, namely, that the experiment was made in consequence of what I had informed him of. He should likewise have stated in his publication, not only that Mr. Cavendish had obtained "une quantité d'eau très sensible" but that the water was equal to the weight of the two airs added together. Moreover, he should have added, that I had made him acquainted with Messrs Cavendish and Watt's conclusions; namely, that water, and not an acid, or any other substance, arose from the combustion of the inflammable and dephlogisticated airs. But *those* conclusions opened the way to Mr. Lavoisier's present theory, which perfectly agrees with that of Mr. Cavendish; only that Mr. Lavoisier accommodates it to his old theory, which banishes phlogiston. Mr. Monge's experiments (of which Mr. Lavoisier speaks as if made about the same time) were really not made until pretty long, I believe at least two months, later than Mr. Lavoisier's own, and were undertaken on receiving information of them.

The course of all this history will clearly convince you, that Mr. Lavoisier (instead of being led to the discovery by following up the experiments which he and Mr. Bucquet had commenced in 1777) was induced to institute again such experiments solely by the account he received from me, and of our English experiments; and that he really discovered nothing but what had before been pointed out to him to have been previously made out and demonstrated in England.

That at least one of Lavoisier's colleagues was aware of the real facts and merits of the case is evident from a letter which La Place addressed to Deluc on June 28th, 1783, which contains the following passage:

M. Lavoisier and I have repeated recently before Mr. Blagden and several other persons, the experiment of Mr. Cavendish upon the conversion into water of dephlogisticated and inflammable airs, by their combustion; with this difference, that we have burned them without the assistance of the electric spark, by bringing together two currents, the one of pure air, the other of inflammable air. We have obtained in this way more than 2½ drachms of pure water, or which, at least, had no character of acidity, and was insipid to the taste; but we do not yet know if this quantity of water represents the weight of the airs consumed. It is an experiment to be recommenced with all possible attention, and which appears to me of the greatest importance.

These statements, which have never been contradicted, would seem to leave no doubt on the question of priority as between Cavendish and Lavoisier concerning the experimental facts, or, indeed, as to the inference which each drew from them as to the non-elementary nature of water.

How Dr Crell regarded the matter is evident from a footnote which he appended to a translation of Cavendish's memoir which appeared in the *Chemische Annalen* for 1785, Part IV, p. 324.

This communication contains the substance of a paper presented to the Royal Society in London, by Henry Cavendish, Esq. and which has not only been inserted in the *Philosophical Transactions*; but has also been published separate under the title of "Experiments on Air" (London, J. Nichols, 1784, 4°, p. 37). Soon after, Sir Joseph Banks, Bart. (President of the Royal Society) was so obliging as to send me a copy, for the purpose of mentioning it in these *Chemical Annals*. This becomes a twofold duty upon me, because I have committed the same error as most of my compatriots and other men of letters, by ascribing to Mr. Lavoisier the discovery of the water resulting from the different kinds of inflamed air (see *Chem. Annals*, 1785, Part I, p. 48). Justice alone, therefore, demands of me, to return to Mr. Cavendish (whom I take this opportunity to assure of my most sincere esteem) the well earned honour of *The First Discovery* of this so very important and remarkable Phenomenon (which appears clearly from this paper) and at the same time to correct some other circumstances in mine above-mentioned publication[1].

We have now to consider the significance of the allusion to Watt. This has no reference to any experimental proof afforded by Watt himself but to certain inferences he deduced from the observations of his friend Priestley. The essential facts, stated shortly, are as follows. Watt, as we gather from his correspondence with Black, had long been of the opinion that "air" was a modification of water: he thought that as steam parts with its latent heat as it acquires sensible heat, or is more compressed, when it arrives at a certain point, it will have no latent heat, and may, under proper compression, be an elastic fluid nearly as specifically heavy as water; at which point it would again change its state and become air. He finds some confirmation of this belief in experiments on which Priestley was engaged at the period when Cavendish was occupied with the work which has just been described. Priestley's results on the seeming conversion of water into air were wholly fallacious, as he subsequently found.

Although Priestley's discovery of the source of his error may have shaken, and indeed did shake, Watt's belief in the experimental proof of the conversion of water into "air," it apparently had no influence on his

[1] Extract from a translation to be found among the Cavendish MSS.

conviction of the essential unity of all forms of "air." This is abundantly evident from the few chemical papers he published, and from the tenor of his correspondence with Black, Priestley, Kirwan and others of his contemporaries. Watt was no doubt familiar with the "mere random experiment" which Priestley made in conjunction with Warltire, but he seems to have attached no more importance than they did to the formation of the dew on detonation, but like them, to have regarded it as evidence that common air deposits its moisture when phlogisticated. Cavendish, as we have seen, communicated to Priestley the facts arising from his repetition of Warltire's experiments, as Priestley relates in a paper published in the *Phil. Trans.* for 1783. Priestley interpreted Cavendish's experiment as proving the conversion of air into water, thus strengthening his belief in the intimate connection between water and air of which hitherto he had been unable to acquire satisfactory proof. "Still hearing," he says, "of many objections to the conversion of water into air, I now gave particular attention to an experiment of Mr. Cavendish's concerning the *reconversion* of air into water by *decomposing* it in conjunction with inflammable air." He therefore repeated, as he thought, Cavendish's experiments but with certain modifications which he imagined, although quite erroneously, would remove objections which might be urged against them. He says:

In order to be sure that the water I might *find in the air was really a constituent part of it,* and not what it might have imbibed after its formation, I made a quantity of both dephlogisticated and inflammable air, in such a manner as that neither of them should ever come into contact with water, receiving them as they were produced in mercury; the former from nitre, and in the middle of the process (long after the water of crystallisation was come over[1]), and the latter from perfectly made charcoal. The two kinds of air thus produced I decomposed by firing them together by the electric explosion and found a manifest deposition of water, and to appearance in the same quantity as if both the kinds of air had been previously confined by water...the result was such as to afford a strong presumption that the air was reconverted into water, and *therefore that the origin of it had been air.*

It is unnecessary to examine this passage very minutely. The better, or rather the seeming better, is here the enemy of the good. In the attempt to prepare pure dry gases Priestley only succeeded in making them more impure: it was physically impossible that he could have obtained, as he surmised, "the weight of the decomposed air in the moisture." But it was upon this wholly fallacious experiment that Watt theorised: it clearly proved to him that water and air are mutually convertible and are therefore essentially the same. Under date April 21st, 1783, he writes to Black: "Dr. Priestley has made many more experiments on the conversion of water into air, and I believe I have found out the cause of it; which I have

[1] Nitre contains no water of crystallisation.

put in the form of a letter to him which will be read at the Royal Society with his paper on the subject." He then gives Black a summary of the facts, or supposed facts, on which he bases his deductions:

In the deflagration of the inflammable and dephlogisticated airs, the airs unite with violence, become red hot, and, on cooling totally disappear. The only fixed matter which remains is *water*; and water, light and heat are all the products. Are we not then authorised to conclude that water is composed of dephlogisticated and inflammable air, or phlogiston, deprived of part of their latent heat, and that dephlogisticated, or pure air [oxygen] is composed of water deprived of its phlogiston, and united to heat and light; and if light be only a modification of heat, or a component part of phlogiston, then pure air [oxygen] consists of water deprived of its phlogiston and of latent heat.

On learning of the fallacy of Priestley's experimental proof of the conversion of water into air Watt desired that his letter to Priestley should not be publicly read and it was temporarily withdrawn on account of what Watt styled in a letter to Black, Priestley's "ugly experiment." In the meantime knowledge of Watt's letter, or of his views, seems to have been conveyed to Paris. In a letter from Watt to Deluc dated November 30th, 1783, we read:

I was at Dr. Priestley's last night. He thinks, as I do, that Mr. Lavoisier, having heard some imperfect account of the paper I wrote in the spring, has run away with the idea and made up a memoir hastily, without any satisfactory proof.... I, therefore, put the query to you of the propriety of sending my letter to pass through their hands to be printed; for even if this theory is Mr. Lavoisier's own, I am vain enough to think that he may get some hints from my letter, which may enable him to make experiments, and to improve his theory, and produce a memoir to the Academy before my letter can be printed, which may be so much superior as to eclipse my poor performance, and sink it into utter oblivion; nay, worse, I may be condemned as a plagiary, for I certainly cannot be heard in opposition to an Academician and a Financier....But, after all, I may be doing Mr. Lavoisier injustice.

Cavendish's paper was read to the Royal Society on January 15th, 1784, and some of Watt's friends, Deluc in particular, hastened to imply that its conclusions were framed in the light of knowledge derived from Watt's letter. In reply to Deluc Watt writes:

On the slight glance I have been able to give your extract of the paper, I think his theory very different from mine; which of the two is the right I cannot say; his is more likely to be so, as he has made many more experiments, and consequently has more facts to argue upon.

Watt's letter to Priestley, as well as one he wrote subsequently to Deluc, were by their author's direction subsequently merged into a single communication and published in the *Phil. Trans.* under the title "Thoughts on the Constituent Parts of Water, and of Dephlogisticated Air; with an account of some Experiments on that Subject. In a letter from Mr. James Watt, Engineer, to Mr. Deluc, F.R.S."

The controversy as regards priority died down during the lifetime of the parties principally concerned, and it seems to have made no difference in their friendly relations. It was however revived by the action of Arago, who as Perpetual Secretary of the French Academy read an *Eloge* on Watt who, like Cavendish, had been elected a member of the Institute. This provoked a reply from the Rev. W. Vernon Harcourt, who in the course of his Presidential Address to the British Association at the Birmingham Meeting in 1839, set out in detail all the facts in support of Cavendish's claims, including a lithographed reprint of the original laboratory notes, giving the dates and details of the experiments, thus occasioning what is known in the history of science as the Water Controversy, in which Brougham, Peacock, Muirhead, Whewell, Brewster and Jeffrey took part. The considerable body of literature to which this gave rise was critically examined by George Wilson and constitutes a section of his *Life of Cavendish.*

Time has now set its seal upon the matter and there is practical agreement as to its merits. As regards Lavoisier, it cannot be claimed that he was the first to obtain the facts. To Cavendish belongs the merit of having first supplied the true experimental basis upon which accurate knowledge could alone be founded. Watt, on the other hand, although reasoning from imperfect and indeed altogether erroneous data, was the first, so far as we can prove from documentary evidence, to state distinctly that water is not an element, but is composed, weight for weight, of two other substances, one of which he regarded as phlogiston and the other as dephlogisticated air. It would be a mistake, however, to suppose that Watt taught precisely the same doctrine of the true nature of water that we hold to-day. Nor did Cavendish utter a more certain sound. What we regard to-day as the expression of the truth, we owe to Lavoisier, who stated it with a directness and a precision that ultimately swept all doubt and hesitation aside—except in the mind of Priestley, whose "random experiment" gave the first glimmer of the truth[1].

As already stated, Cavendish's paper, or rather one section of it, was subjected to criticism immediately after it was communicated, and before it was printed, by Kirwan, Mr Cavendish, as he says, having had the politeness to permit him to read it. The criticism was directed not to the question of the formation of water but to that of fixed air in the various processes which Kirwan alleged it to be produced, but which Cavendish was unable to verify. As regards water Kirwan says "when inflammable air from metals and dephlogisticated air are fired, as a great diminution takes place, and yet no fixed air is found, I am nearly convinced, by Mr. Cavendish's experiments that water is really produced"; and he goes

[1] *Essays in Historical Chemistry*, James Watt, p. 120.

on to say that he is not surprised at this fact as he should have expected on *a priori* grounds that this particular method of phlogistication would have produced "a compound very different from that which it forms in other instances of phlogistication"; and he proceeds to develop his reasons with the dialectical skill and specious reasoning characteristic of his nimble intellect. Kirwan's paper is, in fact, a tissue of misstatements, false analogies, and loose reasoning, and is remembered only from the circumstance that it provoked a reply from Cavendish, and incidentally led to the production of his second paper on "Experiments on Air."

Cavendish's reply is characteristic of him. He begins by saying:

> In a paper lately read before this Society containing many experiments on air, I gave my reasons for supposing that the diminution which respirable air suffers by phlogistication, is not owing either to the generation or separation of fixed air from it; but without any arguments of a personal nature, or which related to any one person who espouses the contrary doctrine more than to another. This being contrary to the opinion maintained by Mr. Kirwan, he has written a paper in answer to it which was read on the fifth of February. As I do not like troubling the Society with controversy, I shall take no notice of the arguments used by him, but shall leave them for the reader to form his own judgement of; much less will I endeavour to point out any inconsistencies or false reasonings, should any such have crept into it; but as there are two or three experiments mentioned there, which may perhaps be considered as disagreeing with my opinion, I beg leave to say a few words concerning them.

The two or three experiments are then discussed and Kirwan's inferences from them refuted. The Irish chemist, who was certainly a well-read and accomplished man, a keen critic and remarkably familiar with the chemical literature of his time, both at home and abroad, returned to the attack, but without adducing any fresh facts, and Cavendish therefore took no notice of the rejoinder.

It is not improbable, however, that Kirwan's criticism of his remarks on the action of the electric spark on air, which it must be admitted are not very conclusive, induced him to undertake further inquiry on this matter. His results, which led to the discovery of the true nature of nitric acid, were communicated to the Royal Society in 1785, and are published in the *Phil. Trans.* for that year under the same title as his preceding paper, of which it is professedly a continuation.

He begins by a reference to his previous paper in which he gave reasons for his belief that the diminution produced in atmospheric air by phlogistication was not due to the formation of fixed air. As regards the action of the electric spark on air he admits that, as he had made no experiments himself on that subject, his opinion had been formed on the experiments of others. As the result of further inquiry he now finds that although he was right in supposing the diminution in volume of the air was not due to its phlogistication by the spark, and that no fixed air was formed by its action, the real cause was very different from what he expected, and that it depends upon the conversion of phlogisticated air [nitrogen] into nitrous [nitric] acid.

He then proceeds to give an account of the apparatus he employed to demonstrate this fact. A small quantity of the "air" to be experimented upon was introduced into a Λ-shaped tube of about $\frac{1}{10}$th of an inch in bore filled with mercury, the limbs of which were placed in separate vessels of mercury, the length of the column of the "air" being in general from $\frac{3}{4}$ to $1\frac{1}{2}$ inches. Various solutions, e.g. litmus, lime-water, soap-lees, etc. could be introduced into the two limbs.

When the electric spark was made to pass through common air, included between short columns of a solution of litmus, the solution acquired a red colour, and the air was diminished, conformably to what was observed by Dr. Priestley. When lime-water was used, instead of the solution of litmus, and the spark was continued till the air could be no further diminished, not the least cloud could be perceived in the lime-water; but the air was reduced to two-thirds of its original bulk; which is a greater diminution than it could have suffered by mere phlogistication, as that is very little more than one-fifth of the whole.

By continued passage of the spark Cavendish found that the whole of the lime could be neutralised, after which the free acid in the liquid began to attack the mercury. "When the air is confined by soap-lees [solution of caustic potash] the diminution proceeds rather faster than when it is confined by lime-water." Accordingly in the rest of the trials this solution was employed to absorb the acid produced. In the case of pure oxygen "the diminution was but small": in the case of nitrogen,

no sensible diminution took place; but when five parts of pure dephlogisticated air were mixed with three parts of common air, almost the whole of the air was made to disappear.

It must be considered, that common air consists of one part of dephlogisticated air, mixed with four of phlogisticated; so that a mixture of five parts of pure dephlogisticated air, and three of common air is the same thing as a mixture of seven parts of dephlogisticated air with three of phlogisticated....

As fast as the air was diminished by the electric spark, I continued adding more of the same kind, till no further diminution took place....The soap-lees being then poured out of the tube, and separated from the quicksilver, seemed to be perfectly neutralized, as they did not at all discolour paper tinged with the juice of blue flowers. Being evaporated to dryness, they left a small quantity of salt, which was evidently nitre, as appeared by the manner in which paper, impregnated with a solution of it, burned.

For more satisfaction, I tried this experiment over again on a larger scale.... the spark was continued till no more air could be made to disappear. The liquor when poured out of the tube, smelled evidently of phlogisticated nitrous acid [nitrous acid or nitric oxide], and being evaporated to dryness, yielded $1\frac{4}{10}$ grains of salt, which is pretty exactly equal in weight to the nitre which that amount of soap-lees would have afforded if saturated with nitrous [nitric] acid. This salt was found, by the manner in which paper dipped into a solution of it burned, to be true nitre. It appeared, by the test of *terra ponderosa salita* [barium chloride], to contain not more vitriolic acid than the soap-lees themselves contained, which was excessively little; and there is no reason to think that any other acid entered into it, except the nitrous [nitric].

In testing for the possible formation of hydrochloric acid Cavendish incidentally obtained a precipitate of silver nitrite: it is characteristic of his acute power of observation that he was not misled by it. He says:

A circumstance, however, occurred, which at first seemed to shew, that this salt [the nitre from the soap-lees] contained some marine acid; namely, an evident precipitation took place when a solution of silver was added to some of it dissolved in water; though the soap-lees used in its formation were perfectly free from marine acid, and though, to prevent all danger of any precipitate being formed by an excess of alkali in it, some purified nitrous [nitric] acid had been added to it, previous to the addition of the solution of silver. On consideration, however, I suspected that this precipitation might arise from the nitrous acid in it being phlogisticated; and therefore I tried whether nitre, much phlogisticated, would precipitate silver from its solution. For this purpose I exposed some nitre to the fire, in an earthen retort, till it had yielded a good deal of dephlogisticated air; and then, having dissolved it in water, and added to it some well purified spirit of nitre till it was sensibly acid, in order to be certain that the alkali did not predominate, I dropped into it some solution of silver, which immediately made a very copious precipitate. This solution however being deprived of some of its phlogiston by evaporation to dryness, and exposure for a few weeks to the air, lost the property of precipitating silver from its solution; a proof that this property depended only on its phlogistication, and not on its having absorbed sea-salt from the retort, or by any other means.

Hence it is certain that nitre, when much phlogisticated, is capable of making a precipitate with a solution of silver; and therefore there is no reason to think that the precipitate, which our salt [i.e. the salt formed in the soap-lees] occasioned with a solution of silver, proceeded from any other cause than that of its being phlogisticated; especially as it appeared by the smell, both on first taking it out of the tube, and on the addition of the spirit of nitre, previous to dropping in the solution of silver, that the acid in it was much phlogisticated. This property of phlogisticated nitre is worth the attention of chemists; as otherwise they may sometimes be led into mistakes, in investigating the presence of marine acid by a solution of silver.

No apology is needed for this somewhat lengthy extract. It is significant of the care, patience and skill with which Cavendish followed up and unravelled phenomena which to less cautious operators would have been so many pitfalls. All that is necessary to make the account consistent with modern terminology is to read "abstraction of oxygen" for addition of phlogiston: otherwise it is a perfectly accurate statement of the relation of nitrous acid to nitric acid; of the modes of their synthetical formation;

of the behaviour of nitre when heated; of the instability of the alkaline nitrites; and of the method of producing silver nitrite.

Having thus satisfactorily demonstrated that oxygen and nitrogen under the influence of heat and in presence of an alkali will unite to form nitrites and nitrates, Cavendish proceeds to explain what he considers to be the rationale of their production. As might be anticipated, his theory is obscured by the mists of phlogistonism. He had stated in his previous paper that when nitre is deflagrated with charcoal the acid [nitric acid] is converted into phlogisticated air identical with that contained in our atmosphere;

from which I concluded, that phlogisticated air is nothing else than nitrous acid united to phlogiston. According to this conclusion, phlogisticated air ought to be reduced to nitrous acid by being deprived of its phlogiston. But as dephlogisticated air is only water deprived of phlogiston, it is plain, that adding dephlogisticated air to a body, is equivalent to depriving it of phlogiston, and adding water to it; and therefore phlogisticated air ought also to be reduced to nitrous [nitric] acid, by being made to unite to, or form a chemical combination with, dephlogisticated air; only the acid formed this way will be more dilute, than if the phlogisticated air was simply deprived of phlogiston.

This inverted method of representing the facts is characteristic of the logic of phlogistonism. Cavendish implies that nitrogen is a compound of nitric acid and phlogiston, and as, according to him, phlogiston is a hydrate of inflammable air, it follows that in modern terminology nitrogen should be regarded as a *hydrated hydrogen nitrate*. When therefore oxygen acts upon this combination, it combines with the phlogiston forming water and liberating and then diluting the nitrous [nitric] acid. He then recapitulates the facts of his experiments already stated and interprets them in the light of his theory, and in like manner explains the formation of the nitric acid observed in the course of his observations on the synthesis of water.

He next reviews the properties of phlogisticated air as it exists in the atmosphere, noting their negative character, and then raises the question "whether there are not in reality many different substances confounded together by us under the name of phlogisticated air." This latter point has already been noted by the late Lord Rayleigh; it is of such special interest in the light of subsequent work that the passage merits quotation. Cavendish thus describes how he proceeds to find an answer to his query:

I therefore made an experiment to determine, whether the whole of a given portion of the phlogisticated air of the atmosphere could be reduced to nitrous acid, or whether there was not a part of a different nature from the rest, which would refuse to undergo that change. The foregoing experiments indeed in some measure decided this point, as much the greatest part of the air let up into the tube lost its elasticity; yet, as some remained unabsorbed, it did not appear for certain whether that was of the same nature as the rest or not. For this purpose

I diminished a similar mixture of dephlogisticated and common air, in the same manner as before, till it was reduced to a small part of its original bulk. I then, in order to decompound as much as I could of the phlogisticated air which remained in the tube, added some dephlogisticated air to it, and continued the spark till no further diminution took place. Having by these means condensed as much as I could of the phlogisticated air, I let up some solution of liver of sulphur to absorb the dephlogisticated air; after which only a small bubble of air remained unabsorbed, which certainly was not more than $\frac{1}{120}$ of the bulk of the phlogisticated air let up into the tube; so that if there is any part of the phlogisticated air of our atmosphere which differs from the rest, and cannot be reduced to nitrous acid, we may safely conclude, that it is not more than $\frac{1}{120}$ part of the whole.

No doubt, according to Newton's second rule that "to natural effects of the same kind the same causes are to be assigned, as far as it may be done," Cavendish might be warranted in concluding that the phlogisticated air of the atmosphere is of uniform character with the exception of at least the $\frac{1}{120}$th part. At the same time the conclusion implicitly assumes the absence of forms of phlogisticated air which might equally have the power to unite with oxygen under the conditions of the experiment.

The late Lord Rayleigh, as is well known, repeated the Cavendish experiment on a large scale and showed that the "small bubble" must have consisted substantially of argon, doubtless mixed with the other inert gases of the atmosphere subsequently discovered by Sir William Ramsay.

Cavendish then reverts to his original supposition that in the presence of "inflammable" [combustible or organic] matter "some of this matter might be burned by the spark, and thereby diminish the air." Oxygen confined over distilled water, soap-lees and litmus solution was therefore "sparked," but only a very slight diminution was observed, due probably to air in the solutions or to the formation of ozone. In the case of the litmus "the solution soon acquired a red colour, which became paler and paler as the spark was continued, till at last it was quite colourless and transparent." Fixed air was formed, as shown by lime-water becoming cloudy.

In this experiment therefore the litmus was, if not burnt, at least decompounded, so as to lose entirely its purple colour, and to yield fixed air...and so very likely might the solutions of many other combustible [organic] substances. But there is nothing, in any of these experiments, which favours the opinion of the air being at all diminished by means of phlogiston communicated to it by the electric spark.

There is also nothing in Cavendish's electrical researches to show that he ever associated phlogiston with electricity, and although he attempted to explain some of the principal phenomena of electricity by the assumption of an elastic fluid, no mention of phlogiston occurs in his memoirs, nor can we gather how he supposed that phlogiston could be communicated to oxygen by means of the electric spark. We are again met with the ever-recurrent difficulty: what exactly was Cavendish's conception of phlogiston?

This paper, as throwing light upon the true nature of nitric acid, and its relations to oxygen and nitrogen, naturally attracted considerable attention, and became the subject of further inquiry. Other observers "of distinguished ability" in attempting to repeat Cavendish's experiment were not equally successful, and accordingly he "thought it right to take some measures to authenticate the truth of it." He therefore requested Mr Gilpin, Clerk to the Royal Society, to repeat the experiment "in presence of some of the Gentlemen most conversant with these subjects." It appeared that the chemists who had endeavoured to repeat the synthesis of the nitric acid were Van Marum and Paets Van Trootswyk in Holland; and Lavoisier, Hassenfratz and Monge in France. "I am not acquainted," says Cavendish, "with the method which the three latter Gentlemen employed, and am at a loss to conceive what could prevent such able philosophers from succeeding, except want of patience."

The details of the repetition are set forth in a paper read to the Royal Society on April 17th, 1788, and published in the *Phil. Trans.* for that year under the title "On the Conversion of a Mixture of dephlogisticated and phlogisticated Air into nitrous Acid, by the Electric Spark." We learn from this paper that the electrical machine employed by Cavendish

was one of Mr. Nairne's patent machines, the cylinder of which is 12½ inches long, and 7 in diameter. A conductor of 5 feet long, and 6 inches in diameter, was adapted to it, and the ball which received the spark was placed at two or three inches from another ball, fixed to the end of the conductor. Now when the machine worked well, Mr. Gilpin supposes he got about two or three hundred sparks a minute, and the diminution of the air during the half hour which he continued working at a time, varied in general from 40 to 120 measures, but was usually greatest when there was most air in the tube, provided the quantity was not so great as to prevent the spark from passing readily.

The experiment was repeated twice and the formation of nitric acid was fully confirmed. As Van Marum had described the details of his method Cavendish was able to point out the cause of his want of success. A diminution in volume of the included gas was actually observed by him, but the alkali was only imperfectly neutralised and the touch-paper prepared from it was not sufficiently quick-burning. The experiment has been frequently repeated and with simpler apparatus. Faraday showed that if paper moistened with caustic potash solution be suspended between two brass balls from which a spark discharge is passing, nitre is rapidly formed and the paper becomes touch-paper.

It is not necessary to dwell upon the importance of Cavendish's discovery simply as a scientific fact, or to point out the manifold results which have flowed from it in connection with the natural occurrence of nitrates, and the nutrition of plants. His method of synthesis is to-day the basis of an industry which bids fair to revolutionise large and important departments of chemical and agricultural procedure.

This paper constitutes the last of Cavendish's published chemical researches. In point of time they range from 1766 to 1788. The Chatsworth manuscripts contain results of other chemical inquiries which will be referred to later.

We now proceed to give some account of his published labours in other departments of science with the exception of his electrical researches which have already been fully dealt with by Professor Clerk Maxwell.

As already stated, Cavendish, as regards science, was a remarkably many-sided man. Practically every field of scientific inquiry opened up in his time attracted him, and he carried on more or less simultaneously investigations in very different branches. As his published work in departments other than electricity and chemistry does not allow itself to be conveniently classified, it may be desirable to treat of these papers in the order of their appearance.

Cavendish's memorable paper on electricity, published in the *Phil. Trans.* for 1771, established his position as an authority on that branch of physical science, and no doubt led to his inclusion on a committee appointed by the Royal Society at the request of the Board of Ordnance, to consider the best method of protecting the powder magazine at Purfleet from lightning. This matter has already been dealt with by Professor Clerk Maxwell in the earlier volume of this work and need not therefore be further referred to.

During the latter half of the eighteenth century meteorological observations began to attract an increasing amount of attention and the early volumes of the *Phil. Trans.* contain numerous communications on the subject. In 1773 the Royal Society instituted under Cavendish's superintendence and direction systematic and regular observations on atmospheric temperature, pressure, humidity, rain-fall, and wind, as well as on magnetic variation and inclination, at their house in Crane Court and subsequently at Somerset Place. These records were tabulated and discussed in successive volumes of the *Transactions* down to 1843 when on the recommendation of the Council of the Society, the Government established a meteorological and magnetic observatory in association with the Royal Observatory at Greenwich.

A couple of years after their installation in Crane Court the Council of the Society requested Cavendish to examine the condition and mode of working of the instruments, and his report was published in the *Phil. Trans.* for 1776 under the title of "An Account of the Meteorological Instruments used at the Royal Society's House." It contains a description of the thermometers, barometer, rain-gauge, hygrometer, variation-compass and dipping-needle.

Cavendish, like his father, had paid considerable attention to thermometry and he took advantage of the opportunity afforded by the report to indicate certain sources of error in the mode of graduation of the thermometer and in the manner of its use. He was the first to draw attention to the necessity of correcting for the emergent column; that is, for the portion of mercury in the stem not heated to the temperature it is desired to ascertain. He points out that a thermometer,

dipped into a liquor of the heat of boiling-water, will stand at least 2° higher, if it is immersed to such a depth that the quicksilver in the tube is heated to the same degree as that in the ball, than if it is immersed no lower than the freezing-point, and the rest of the tube is not much warmer than the air. The only accurate method is, to take care that all parts of the quicksilver should be heated equally. For this reason, in trying the heat of liquor much hotter or colder than the air, the thermometer ought, if possible, to be immersed almost as far as to the top of the column of quicksilver in the tube.

But as this procedure is not always practicable, Cavendish gives a table showing the amount to be added or subtracted for the number of degrees not immersed "owing to the supposed difference of heat of the quicksilver in that part of the tube and in the ball," based on the assumption "that quicksilver expands one 11500th part of its bulk by each degree [F.] of heat."

This table has long since been superseded by others based upon similar principles, but even subsequent tables, although more accurate, are affected, like that of Cavendish, by uncertainty as to the true temperature of the mercury in the emergent column.

In spirit thermometers, as he points out, the error of the emergent column may be much greater owing to the greater expansibility by heat of spirits of wine.

He indicates the necessity of immersing the whole of the mercury in the steam from boiling water when determining the upper fixed point of a thermometer and describes a simple apparatus for this purpose.

"At present," he says, "there is so little uniformity observed in the manner of adjusting thermometers, that the boiling-point, in instruments made by our best artists, differ from each other by not less than 2½°; owing partly to a difference in the height of the barometer at which they were adjusted, and partly to the quicksilver in the tube being more heated in the method used by some persons, than in that used by others. It is very much to be wished therefore, that some means were used to establish a uniform method of proceeding; and there are none which seem more proper, or more likely to be effectual, than that the Royal Society should take it into consideration, and recommend that method of proceeding which shall appear to them to be most expedient."

The barometer in use by the Royal Society was of the "cistern kind," and Cavendish discusses the best method of reading it, and he states his reasons for preferring it to the syphon barometer. He gives a table of corrections for capillary depression depending upon the diameter of the tube, based upon observations made by his father, Lord Charles Cavendish, which as Professor Clerk Maxwell has pointed out "have furnished the basis not only for the correction of the reading of barometers, etc. but for the verification of the theory of capillary action by Young, Laplace, Poisson and Ivory." From notes and memoranda to be found among his papers it would appear that the actual measurements made to determine the degree of depression were made by Cavendish himself.

The results of the barometric observations were published by the Society as monthly means. It would seem to have been the practice of the observer to reduce each observation, taken twice a day, to the standard temperature, and to take the mean of the whole. Cavendish points out that

> it is sufficient to take the mean height of the barometer, and correct that according to the mean heat of the thermometer; the result will be exactly the same as if each observation had been corrected separately, and a mean of the corrected observations taken.

Cavendish seems to have been satisfied with the character and position of the "rain-gage," and notes that the strength of the wind is registered as "gentle, brisk, and violent or stormy, which are distinguished by the figures 1, 2, and 3. When there is no sensible wind, it is distinguished by a cypher." Observations of humidity were made by Smeaton's hygrometer (*Phil. Trans.* 61. 198), depending upon the variations in length of a stretched string.

The construction and mode of use of the variation-compass and dipping-needle are discussed at considerable length, and the sources of error and methods of eliminating them are set out in detail. Cavendish would appear to have been familiar with magnetic observations of this kind, and to have assisted his father in making and recording them. The variation-compass was constructed by Nairne after the pattern designed by Knight with certain modifications introduced by Lord Charles Cavendish and Sisson. An attempt was made to determine the error due to local attraction in the Society's house by comparison with an instrument at Great Marlborough Street, presumably belonging to Lord Charles Cavendish.

The dipping-needle was made by Nairne "after a plan of the Rev. Mr. Michell, F.R.S., Rector of Thornhill[1]" and is described in *Phil. Trans.* 1772, 62. 476. The method of observation was precisely the same as that in use to-day, viz. reading the needle east and west and reversing the poles. Cavendish discusses the theory of this procedure and points out its practical limitations due to mechanical difficulties in construction, more particularly to the ends of the axis not being truly cylindrical—an imperfection not infrequently still met with in modern dipping-needles. In his report he gives the result of comparisons between needles of different construction, made

[1] Cf. the short memoir of Michell by Sir A. Geikie, Camb. Univ. Press, 1918

in the garden of his house in Great Marlborough Street, "partly with a view to determine the true dip at this time [1775] in a place out of reach of the influence of any iron work, and partly to see how nearly different needles would agree."

This report to the Society displays Cavendish at his best. It reveals the range of his knowledge, his painstaking care, his sense of accuracy, his perspicacity and the thoroughness with which he studied any problem he attacked. Instruments designed for measurement had always a special attraction for him, and he seems to have taken a peculiar pleasure in working with them and in studying their behaviour with a view to getting the best results out of them. He would appear to have had no great interest in the technical side of invention; at all events his name is not now associated with any instrument of precision, or any formal piece of apparatus now in use[1]. His main concern seemed to be to make such instruments as he could construct out of rough material, or as the "artists" of his day provided, serve his purpose by skilful and intelligent use. With him it was a case of "the man behind the gun."

His suggestion that the Royal Society should take steps to standardise the method of determining the fixed-points of mercurial thermometers led the Council to appoint a committee consisting of himself, Dr Heberden, Mr Alex. Aubert, Dr Deluc, Rev. Nevil Maskelyne, Dr Horsley and Mr Planta to consider further and advise them on the matter. The report, which was presented in 1777 was published in the *Phil. Trans.* 67. 816. The committee adopted substantially Cavendish's methods of ascertaining the upper fixed point, after a careful experimental inquiry, the details of which are fully described and the notes of which are still preserved among his papers. In regard to the correction for the emergent column, it is suggested that its mean temperature may be ascertained with sufficient accuracy by attaching a second thermometer to the stem, in the manner still practised. The standard atmospheric pressure adopted is 29·8 inches, and full instructions are given as to the corrections to be applied to the observed boiling-point when the barometer differs from this height. The report, which seems to have been largely drawn up by Cavendish,—considerable sections of it in his own handwriting are to be found among his MSS. papers,—is a notable contribution to thermometry; it exercised an immediate influence on the construction and use of the mercurial thermometer and incidentally on the development of the science of heat.

Cavendish's interest in mercury as a thermometric agent was doubtless the reason that induced him to occupy himself with the question of its solidification. That mercury could be frozen was first clearly demonstrated

[1] As might be supposed, Cavendish paid considerable attention to the improvement of the chemical balance and he possessed good instruments of Ramsden's construction. We owe to him the first suggestion to use agate planes for the bearing of the central and terminal knife-edges.

in 1759 by Braun of St Petersburg who effected its solidification by means of a freezing-mixture of snow and nitric acid, when he "obtained a solid, shining metallic mass, which extended under the strokes of a pestle; in hardness rather inferior to lead, and yielding a dull dead sound like that metal." This observation excited great interest, mercury being regarded from the time of the alchemists as a substance of quite peculiar properties, and possessed in a preeminent degree of the "essential principle of fluidity." At the instance of the Royal Society, Braun's experiments were repeated and confirmed in 1775 by Mr Thomas Hutchins, the Governor of Albany Fort, in Hudson's Bay, but the most exaggerated estimates of the degree of cold necessary for the solidification of the metal were current. The proper method of ascertaining the freezing-point was pointed out, independently, by Black and Cavendish, the latter of whom furnished Hutchins with thermometers and a simple apparatus in which the experiment might be repeated, and the temperature of solidification accurately determined. The second series of experiments were made in 1781–1782 and the results were communicated to the Royal Society in 1783 and are published in the *Phil. Trans.* 73. 303. They were commented upon by Cavendish in a paper published in the same volume, entitled "Observations on Mr. Hutchins's Experiments for determining the Degree of Cold at which Quicksilver freezes." He begins by explaining the apparatus he suggested, and which was employed by Hutchins.

It consisted of a small mercurial thermometer, the bulb of which reached about 2½ inches below the scale, and was inclosed in a glass cylinder swelled at bottom into a ball, which, when used, was filled with quicksilver, so that the bulb of the thermometer was intirely surrounded with it. If this cylinder is immersed in a freezing mixture till great part of the quicksilver in it is frozen, it is evident, that the degree shewn at that time by the inclosed thermometer, is the precise point at which mercury freezes.

Cavendish then points out the fallacy in the preceding attempts in which the degree of cold was estimated by noting the degree of contraction in the thermometer itself containing the frozen mercury, as no account was taken of the diminution in volume which the mercury suffers in passing from the liquid to the solid state. He compares the phenomena of freezing mercury with those observed on the solidification of water and of melted tin or lead. He draws attention to the super-cooling of water, explains the rise of temperature at the moment of solidification and points out its natural effect. His explanation agrees with that of Black;

only instead of using the expression, heat is generated or produced, he [Black] says, latent heat is evolved or set free; but as this expression relates to an hypothesis depending on the supposition, that the heat of bodies is owing to their containing more or less of a substance called the matter of heat; and as I think Sir Isaac Newton's opinion, that heat consists in the internal motion of the particles of bodies, much the most probable, I chose to use the expression, heat is generated.

Cavendish then discusses *seriatim* the accounts furnished by Hutchins of his several experiments; and after correcting the results so as to accord with a thermometer

adjusted in the manner recommended by the Committee of the Royal Society, it follows, that all the experiments agree in shewing that the true point at which quicksilver freezes is $38\frac{2}{3}°$, or in whole numbers 39° [F.] below 0[1].

With regard to the contraction of mercury in freezing, Cavendish, after discussing certain observations by Hutchins and Braun, concludes that it is "almost $\frac{1}{23}$ of its whole bulk," which agrees with the value deduced from Mallet's determination of its specific gravity at its melting point (*Proc. Roy. Soc.* 1877, 26. 71).

The remainder of the paper consists of a discussion on the cold produced by the freezing mixture of snow and nitric acid employed. This he says is owing to the melting and solution of the snow in the acid.

Now, in all probability, there is a certain degree of cold in which the spirit of nitre, so far from dissolving snow, will yield out part of its own water, and suffer that to freeze, as is the case with solutions of common salt; so that if the cold of the materials before mixing be equal to this, no additional cold can be produced. If the cold of the materials is less, some increase of cold will be produced; but the total cold will be less than in the former case, since the additional cold cannot be generated without some of the snow being dissolved, and thereby weakening the acid, and making it less able to dissolve more snow; but yet the less the cold of the materials is, the greater will be the additional cold produced....

However extraordinary it may at first appear, there is the utmost reason to think, that a rather greater degree of cold would have been obtained if the spirit of nitre had been weaker.

[1] The following are recorded determinations of the freezing point of mercury (*Science Abstracts*, 20, II, 1917, 58):

Authority	Date	Thermometer	Value °C.
Hutchins	1776		−39·44
Cavendish	1783		−39·26
Regnault	1862		−38·50
B. Stewart	1863	Gas thermometer	−38·85
Vicentini and Omodei	1888	Mercury thermometer previously compared with the gas thermometer	−38·80
Chappuis	1896	Mercury thermometer previously compared with the gas thermometer	−38·85
Chree	1898	Mercury thermometer previously compared with the gas thermometer	−38·86
Henning	1914	Platinum thermometer	−38·89
Wilhelm	1916	Resistance thermometer	−38·87

As Blagden states in his "History of the Congelation of Quicksilver" (*Phil. Trans.* 73. 1783, 329) Cavendish was the first to effect its solidification in England, at his house at Hampstead, on February 26th, 1783 by a freezing mixture of pounded ice or snow and dilute nitric acid.

He points out that strong nitric acid generates heat by combining with water and that it is only when a certain amount of water has been added that this generation of heat ceases, when the addition of snow produces cold. The amount of water which needs to be added to the strong acid before the generation of heat ceases was found by Cavendish to be about ¼ the weight of the acid, which points to the formation of a monohydrate: this when mixed with snow appears to form the most effective of the freezing mixtures afforded by mixtures of this acid and snow.

This paper was followed by a lengthy communication from Blagden on a "History of the Congelation of Quicksilver" (*Phil. Trans.* 73. 329) which gives an account of all the observations on the subject previously published, with an examination of the statements concerning extraordinary low temperatures which had been recorded by travellers and others in Siberia, Lapland and elsewhere, and which he shows to be fallacious on grounds stated by Cavendish. These papers taken together are of importance as refuting the exaggerated conceptions of the intensity of cold in the neighbourhood of the polar regions and as putting an end to erroneous speculations concerning the influence of low temperatures on animal and vegetable life.

Cavendish's interest in the subject of freezing mixtures, and in the theory of their action, induced him to institute a further series of experiments at Hudson's Bay with the assistance of Mr John McNab, to whom he sent solutions of nitric and sulphuric acids of various strengths, as well as of ordinary alcohol, with "accurately adjusted" thermometers, with instructions for their use. The results of the observations were communicated by Cavendish to the Royal Society in 1786 and are published in the *Phil. Trans.* for that year (Vol. 76, p. 241) under the title "An Account of Experiments made by Mr John McNab, at Henley House, Hudson's Bay, relating to freezing Mixtures."

In connection with these experiments Cavendish furnishes a table showing the specific gravities and corresponding strengths of the acids which were to be employed, the specific gravities being taken at 60° F., compared with water at the same temperature, and the strengths ascertained by determining the weight of marble which 1 part by weight of the acid would dissolve. The numbers he gives enable us to gain an idea of the accuracy with which he worked. Spirit of nitre of the specific gravity he states, viz. 1·4371 at 60° F., would contain 73·5 per cent. of real nitric acid and 1 part of it by weight would dissolve 0·583 parts of pure marble: Cavendish found 0·582. Similarly, spirit of nitre of specific gravity 1·4043 at 60°/60° F. would contain 66 per cent. of nitric acid, and 1 part by weight would dissolve 0·524 parts of marble: Cavendish found 0·525. Strong oil of vitriol of specific gravity 1·8437 at 60°/60° F. would contain 97·35 per cent. H_2SO_4, 1 part by weight of which would be theoretically capable of dissolving 0·990 parts of marble: Cavendish found indirectly 0·98. These numbers are a striking exemplification of the care, patience and manipulative skill which he spent upon all quantitative determinations.

Cavendish had a two-fold object in instituting these experiments. It appeared from some observations of Fahrenheit, who did a considerable amount of work on freezing mixtures, that nitric acid could be frozen, and that the frozen acid when mixed with ice produced cold, a result confirmed by Braun. It seemed doubtful, however, "whether it was the whole acid, or only the watery part, which froze," and to clear up this point Cavendish "desired Mr. McNab to expose it to the cold, and if it froze, to ascertain the temperature, and decant the fluid part into another bottle, and send both home to be examined." Similar experiments were to be made with the solutions of sulphuric acid and spirits of wine. His second object was to ascertain whether by proceeding as he directed, "a greater degree of cold might be produced than had been done hitherto."

In the experiments with the spirit of nitre it was found that this acid was "capable of a kind of congelation, in which the whole, and not merely the watery part, freezes." The freezing point "also differs greatly according to the strength and varies according to a very unexpected law." The acid, like water, may be supercooled without solidification: white crystals are formed on solidification *which are heavier than the still liquid portion.*

The difference indeed is so great, that in one case where it froze into solid crystals on the surface, these crystals, when detached by agitation, fell with force enough to make a tinkling noise against the bottom of the glass....It is this contraction of the acid in freezing which makes the frozen part subside in the fluid part; as it was found, in the undiluted acid, that the latter [the fluid part] consisted of a stronger and consequently heavier acid than the former [the frozen part]. But still the subsidence of the frozen part shows that the ice [the frozen part] is not mere water, or even a very dilute acid; which indeed was proved by the examination of the liquors sent home.

Neglecting the observations with the "dephlogisticated spirit of nitre," owing to the uncertainty as to its real composition, and confining our attention to the "common spirit of nitre" the results are thus summarised by Cavendish:

	Strength	Freezing point
Common spirit of nitre	0·54	$-31\frac{1}{2}$° F. (− 35·3° C.)
" " "	0·411	$-1\frac{1}{2}$° F. (− 18·6° C.)

The first corresponds fairly closely as regards strength and freezing point with the acid $N(OH)_5$ or $HNO_3 \cdot 2H_2O$, which according to Erdmann (*Zeitsch. Anorg. Chem.* 1902, 32. 431) crystallises in needles melting at − 35° C. The strength of the acid does not correspond with that of the monohydrate which is said to freeze at − 38° C. The second would seem to be identical with the trihydrate $HNO_3 \cdot 3H_2O$, which according to Pickering (*Chem. Soc. Trans.* 1893, 63. 436) melts at −18·2° C. and according to Küster and Kremann (*Zeitsch. anorg. Chem.* 1904, 41. 1) at − 18·5° C.

The trihydrate may be formed by adding snow little by little to the cooled acid so long as a rise of temperature is noted. The maximum point observed was – $1\frac{1}{4}$° F. [– 18·5° C.].

The snow did not appear to dissolve, but formed thin white cakes, which however did not float on the surface, but fell to the bottom, and when broke by the spatula formed a gritty sediment; so that it appears, that these cakes are not simply undissolved snow, but that the adjoining acid absorbed so much of the snow in contact with it as to become diluted sufficiently to freeze with that degree of cold and then congealed into these cakes. The quantity of congealed matter seems to have kept increasing till the end of the experiment.

From the minute description he gives "of the phenomena observed on mixing snow with the acid" there can be no doubt that he also obtained the cryohydrate Ice + $HNO_3 . 3H_2O$, or $HNO_3 . 3H_2O + HNO_3 . H_2O$, the melting points of which (– 43° C. and – 42° C.) agree closely with that noted by him, viz. – $45\frac{1}{4}$° F. (– 42·9° C.).

From these experiments it appears that spirit of nitre is subject to two kinds of congelation, which we may call the aqueous and spirituous; as in the first it is chiefly, if not intirely, the watery part which freezes, and in the latter the spirit itself. Accordingly, when the spirit is cooled to the point of aqueous congelation, it has no tendency to dissolve snow and produce cold thereby, but on the contrary is disposed to part with its own water; whereas its tendency to dissolve snow and produce cold, is by no means destroyed by being cooled to the point of spirituous congelation, or even by being actually congealed. When the acid is excessively dilute, the point of aqueous congelation must necessarily be very little below that of freezing water; when the strength is ·21 it is at – 17°, and at the strength of ·243 it seems from Art. 16 to be at – $44°\frac{1}{4}$. Spirit of nitre, of the foregoing degrees of strength, is liable only to the aqueous congelation, and it is only in greater strengths that the spirituous congelation can take place. This seems to be performed with the least degree of cold, when the strength is ·411 in which case the freezing point is at – $1°\frac{1}{2}$. When the acid is either stronger or weaker, it requires a greater degree of cold; and in both cases the frozen part seems to approach nearer to the strength of ·411 than the unfrozen part; it certainly does so when the strength is greater than ·411, and there is little doubt but what it does so in the other case. At the strength of ·54, the point of spirituous congelation is – $31°\frac{1}{2}$ and at ·33 probably – $45°\frac{1}{4}$; at least one kind of congelation takes place at that point, and there is little doubt but that it is of the spirituous kind. In order to present this matter more at one view, I have added the following table of the freezing point of common spirit of nitre answering to different strengths:

Strength	Freezing point	
·54	– $31\frac{1}{2}$° F.	spirituous congelation
·411	– $1\frac{1}{2}$[1]	spirituous congelation
·38	– $45\frac{1}{4}$	spirituous congelation
·243	– $44\frac{1}{4}$	aqueous congelation
·21	– 17	aqueous congelation

[1] The point of easiest freezing.

From the conditions under which these experiments were made they are necessarily not of a very high degree of accuracy; thus Cavendish had to calculate the degree of dilution of the acid in one or two cases from the weight of snow Mr McNab added, and this obviously could only be approximately known. It seemed however worth while to compare the results with those obtained independently by Pickering and Küster and Kremann, representing them in the form of curves so far as the observations are comparable. The general character of the curves is strikingly similar. An examination of the *Phil. Trans.* paper of 1786 and of the subsequent one in 1788 leaves little room for doubt that Cavendish was actually the first to indicate the existence of these particular hydrates of nitric acid.

Observations were then made upon the vitriolic acid. Strong oil of vitriol (sp. gr. 1·8437 at 60° F. = 97 per cent. strength) froze "to the colour and consistence of hog's-lard," contracting on solidification. It was not completely melted until the temperature rose to 20° F. (– 6·6° C.). According to Pictet and Knietsch pure (100 per cent.) H_2SO_4 melts at 10° C. The difference is due to the slight quantity of water in Cavendish's acid. He points out, as already observed by the Duc d'Ayen and De Morveau, that sulphuric acid "freezes with a less degree of cold when strong than when much diluted." Nevertheless

it is not certain whether it has any point of easiest freezing, like spirit of nitre, or whether the cold required to freeze it does not continually diminish as the strength increases, without limitation; but the latter opinion is the most probable.

Cavendish's "points of easiest freezing" correspond to the points of inflection in the curves shown on plotting his results. As will be seen subsequently he found reason to modify this opinion: further experiments showed that

oil of vitriol has not only a strength of easiest freezing, but that at a strength superior to this, has another point of contrary flexure [the expression is Cavendish's], beyond which, if the strength be increased, the cold necessary to freeze it again begins to diminish.

He seems to have suspected that the observations in the present paper might possibly be affected by the formation of what was known as glacial oil of vitriol (Nordhausen acid).

It appears also, both from Art. 21 and from M. De Morveau's experiment, that during the congelation of the oil of vitriol, some separation of its parts takes place, so that the congealed part differs in some respect from the rest, in consequence of which it freezes with a less degree of cold; and as there is reason to think from Art. 21 that these two parts do not differ much in strength, it seems as if the difference between them depended on some less obvious quality, and probably on that, whatever it is, which forms the difference between glacial and common oil of vitriol. The oil of vitriol prepared from green vitriol, has sometimes been obtained in such a state as to remain constantly congealed, except when exposed to a heat considerably greater than that of the atmosphere, whence it acquired its name of *glacial*. It is not known indeed upon what this property depends, but it is certainly something else than its strength; for oil of vitriol of this kind is always smoking, and the fumes it emits are particularly oppressive and suffocating, though very different from those of the volatile sulphureous acid [sulphur dioxide]. On rectification likewise it yields, with the gentlest heat, a peculiar concrete substance, in the form of saline crystals [sulphur trioxide]; and after this volatile part has been driven off, the remainder is no longer smoking, and has lost its glacial character.

The mixture of oil of vitriol and spirit of nitre when mixed with snow was found to offer no advantages over oil of vitriol alone and no phenomena of importance were noticed concerning it. Cavendish contents himself with the remark: "as the Society will most likely have less curiosity about the disposition to freeze of this mixture than of the simple acids, I shall spare the particulars."

Nor did the experiments with spirits of wine afford any very definite information. The snow seemed to be dissolved much less readily by spirits of wine than by nitric and sulphuric acids and no great degree of cold could be produced by its addition.

Cavendish was well aware that several questions might be raised concerning which his experiments afford no adequate answer. "But," he says, "as this would lead me into disquisitions of considerable length, without my being able to say anything very satisfactory on the subject, I shall forbear entering into it."

Nevertheless he was not content to remain satisfied with his work and Mr McNab was commissioned to institute further experiments.

As some of these properties were deduced from reasoning not sufficiently easy to strike the generality of readers with much conviction, Mr. McNab was desired to try some more experiments to ascertain the truth of it.

Fresh samples of acids of different strengths were sent out to him with a new set of instructions.

He was desired to expose each of these liquors to the cold till they froze; then to try their temperature by a thermometer; afterwards to keep them in a warm room till the ice [the solid portion] was almost melted, and then again expose them to the cold, and when a considerable part of the acid had frozen, to try the temperature a second time; then to decant the unfrozen part into another bottle, and send both parts back to England, that their strength might be examined.... The intent of decanting the fluid part, and sending both parts back, that their strength might be determined, was partly to examine the truth of the supposition laid down in my former Paper, that the strength of the frozen part approaches nearer to ·411 than that of the unfrozen; but it is also a necessary step towards determining the freezing point answering to a given strength of the acid; for as the frozen part is commonly of a different strength from the unfrozen, the strength of the fluid part, and the cold necessary to make it freeze, is continually altering during the progress of the congelation. In consequence of this, the temperature of the liquor is not that with which the frozen part congealed; but it is that necessary to make the remainder, or the fluid part, begin to freeze, or, in other words, it is the freezing point of the fluid part. This is the reason that a thermometer, placed in spirit of nitre, continually sinks during the progress of congelation; which is contrary to what is observed in pure water, and other fluids in which no separation of parts is produced by freezing.

The results of this second series of observations were communicated by Cavendish to the Royal Society in 1788 and are printed in *Phil. Trans.* 78. 166 under the title of "An Account of Experiments made by Mr. John McNab, at Albany Fort, Hudson's Bay, relative to the Freezing of Nitrous and Vitriolic Acids." From the results of these observations Cavendish deduces the following table showing the freezing point of aqueous solutions of nitric acid of various strengths:

Strength	Freezing point
·561	– 41·6° F.
·445	– 3·8
·390	– 4
·353	– 11
·343	– 13·8
·310	– 23
·276	– 40·3

By interpolation from these data, according to Newton's method (*Princip. Math.* Lib. 3, prop. 40, lem. 5) it appears that the strength at which the acid freezes with the least cold is ·418, and that the freezing point answering to that strength is – $2\frac{4}{10}$°....These experiments confirm the truth of the conclusions I drew from Mr. McNab's former experiments; for, first, there is a certain degree of strength at which spirit of nitre freezes with a less degree of cold than when it is either stronger or weaker; and when spirit of nitre, of a different strength from that, is made to congeal, the frozen part approaches nearer to the foregoing degree of strength than the unfrozen. Likewise this strength, as well as the freezing point corresponding thereto, and the freezing point answering to the strength of ·54 come out very nearly the same as I concluded from those experiments; for by the present experiments they come out ·418, – $2\frac{4}{10}$° and – 31°, and by the former ·411, – $1\frac{1}{2}$° and – 31°.

If these observations are plotted, the strengths of the acid being expressed in terms of molecular percentages, it will be found that the character of the curve is identical with that, over the same range, representing the values obtained by Pickering (*Chem. Soc. Trans.* 1893, 63. 436) and by Küster and Kremann (*Zeitsch. anorg. Chem.* 1904, 41. 1). The points of "contrary flexure," as Cavendish calls them, occur at substantially the same temperatures and the corresponding strengths are not very dissimilar. Cavendish found the substance crystallising at −18·8° C. to contain, as the mean of the two determinations of strength, 52·3 per cent. of nitric acid. The trihydrate which crystallises at − 18·5° C. contains 53·9 per cent. The coupled hydrate $HNO_3 . 3H_2O + HNO_3 . H_2O$, said to melt at − 42° C., has the same composition as the acid $N(OH)_5$ which according to Erdmann (*Zeitsch. anorg. Chem.* 1902, 32. 431) crystallises in needles melting at − 35° C. and contains 63·6 per cent. HNO_3.

In the interval between the publication of Cavendish's first and second papers on freezing mixtures, the subject of the freezing points of aqueous solutions of sulphuric acid had been attacked by Keir, a chemical manufacturer living near Birmingham, a friend of Priestley, and a Fellow of the Royal Society, who contributed a paper "On the Congelation of the Vitriolic Acid" to the *Phil. Trans.* 77. 267. Keir, from experiments made during the severe frost of 1784–5,

was led to believe that there must be some certain strength at which the vitriolic acid was more disposed to freeze than at any other, greater or less....I have found that the point of strength most favourable to congelation is very determinate, and that a very small variation above or below that point renders the acid incapable of freezing without a considerable augmentation of cold.

The sulphuric acid of "easiest freezing" was found by Keir to have a density of 1·78, and the freezing and melting points of the acid were identical, viz. 46° F. (7·8° C.).

These numbers agree with those subsequently found by Lunge (*Ber.* 14. 1881, 2649) and by Knietsch (*Ber.* 34. 1901, 4069). Mr McNab's experiments confirmed Keir's observations. From his experiments "it would seem," says Cavendish, "that the freezing point of oil of vitriol, answering to different strengths, is nearly as annexed":

Strength	Freezing point
·977	+ 1° F.
·918	− 26°
·846	+ 42°
·758	− 45°

From hence we may conclude, that oil of vitriol has not only a strength of easiest freezing, as Mr Keir has shown; but that, at a strength superior to this, it has another point of contrary flexure, beyond which, if the strength be increased, the cold necessary to freeze it again begins to diminish.

The strength answering to this latter point of contrary flexure must in all probability be rather more than ·918, as the decanted or unfrozen part of No. 2 seemed rather stronger than the undecanted part; and for a like reason the strength of easiest freezing is rather more than ·846.

As already stated, the "strengths" of the acid solutions express the weight of marble which 1 part by weight of the liquid is theoretically capable of dissolving. As a matter of fact, Cavendish, in the case of sulphuric acid solutions,

did not find their strength by actually trying how much marble they would dissolve; as that method is too uncertain, on account of the selenite [calcium sulphate] formed in the operation, and which in good measure defends the marble from the action of the acid. The method I used was, to find the weight of the plumbum vitriolatum [lead sulphate] formed by the addition of sugar of lead, and from thence to compute the strength, on the supposition that a quantity of oil of vitriol, sufficient to produce 100 parts of plumbum vitriolatum, will dissolve 33 of marble; as I found by experiment that so much oil of vitriol would saturate as much fixed alkali as a quantity of nitrous acid [nitric acid] sufficient to dissolve 33 of marble.

This estimation by Cavendish that 100 parts of lead sulphate may be produced from as much oil of vitriol as would be equivalent to 33 parts of marble, that is, that 100 of lead sulphate are equivalent to 33 of marble, is perfectly accurate.

We are now in a position to see how far subsequent work confirms Cavendish's observations. Of modern observations the most accurate are probably those of Knietsch (*Ber.* 34. 1901, 4069). The following table contains the results of the comparison, so far as it is applicable:

	Cavendish		Knietsch	
Strength	SO_3 p. c.	M. P.	SO_3 p. c.	M. P.
·977	78·1	− 17°·2 C.	78	− 16°·5 C.
·918	74·5	− 32°·3	74	− 25°
·846	67	+ 6°	67	+ 8°
·758	61	− 42°·8	61	below − 40°

Considering the circumstances, it is nothing short of marvellous that Cavendish should have succeeded in getting results so closely approximating to the truth. He not only determined the points of "easiest freezing" and of "contrary flexure" with precision, but his estimations of the corresponding strengths and temperatures are a remarkable testimony to his skill and accuracy of work, in spite of his limited means and the imperfections of his appliances.

It would appear from the absence of all reference to it on the part of later observers that Cavendish's work on the freezing of aqueous solutions of nitric and sulphuric acid was either unknown to them or that its significance was not recognised.

As regards the reasoning by which Cavendish deduced the "strengths" of his acid solutions, Dr Wilson has already pointed out that he was not only cognisant of what we now know as the "law of constant proportion," and acted upon it, but in the special case cited, he was practically applying also the "law of reciprocal proportion," thus showing that although the principles underlying these laws were not actually formulated, they were clearly recognised by him as at the basis of all quantitative analytical work. If he had only pursued the path of inquiry which this recognition opened up we might not have had to wait twenty years for the promulgation of the new departure we associate with the name and fame of Dalton.

Cavendish was now in his fifty-seventh year. Chemical inquiry continued to interest him, as may be proved by notes among his MSS., but he published no further contributions towards it. It has been surmised that the revolution effected by Lavoisier and his followers repelled him from its further prosecution. There would seem to be no adequate ground for this supposition. Biased as he might be towards Stahl's doctrine, he was not so prejudiced as to neglect the study of chemical phenomena because of its seeming insufficiency to explain them. Such an assumption would be wholly opposed to what we know to be his character as a natural philosopher. At no period of his activity as an experimentalist was his energy entirely absorbed by chemical pursuits. His published work, and still more his unpublished manuscripts, show that meteorology, astronomy, electricity, heat, geology, geodesy, mathematics—in fact nearly every branch of natural science known in his time—in addition to chemistry—attracted him, and he seemed to turn from one to the other with equal zeal as their several problems interested him.

In 1790 he communicated a paper "On the Height of the Luminous Arch that was seen Feb. 23, 1784," to the Royal Society which is printed in *Phil. Trans.* 80. 101. The measurements were based upon observations of an aurora made almost simultaneously by the Rev. F. J. H. Wollaston at Cambridge, the Rev. B. Hutchinson at Kimbolton, and Mr J. Franklin at Blockley in Worcestershire (*Phil. Trans.* 90. 43–46) and communicated to the Royal Society in 1786. It would appear from Professor Loomis's paper in *Silliman's Journ.* 1873 that manifestations of the aurora were particularly frequent at about this period. After discussing and correcting the data, Cavendish calculates from the geographical position of the places of observation that the height of the aurora "could hardly be less than 52 miles, and is not likely to have much exceeded 71." Many like calculations made since Cavendish's time have given similar results, and the estimates of Dalton, Backhouse, H. A. Newton, Alex. S. Herschel and others point to values of the same order as that found by Cavendish. With reference to this paper Dr Chree, F.R.S., of the Kew Observatory, writes as follows:

There is a very full historical discussion of ideas on Aurora by the late Prof. Cleveland Abbe in *Terrestrial Magnetism*, Vol. 3 (1898) in three parts, p. 5, p. 53 and p. 149. The first name he gives is Halley (*Phil. Trans.* 1716) but does not say he calculated heights. The first names he gives as calculators are: Mairan, *Traité de l'Aurore Boreale*, Paris 1731 to 1733; Maier, St Petersburg read 1728, published 1735. He also (p. 12) refers to the work by Cavendish (*Phil. Trans.*

1790) as quoted by Dr Thomas Young. Abbe again refers on p. 54 to Cavendish in connection with the fact that his views appeared unknown to Dalton when calculating auroral heights in 1791 and 1793.

Cavendish's height seems more in accordance with present ideas than those given by Mairan and Maier. Cavendish's method seems practically that used by Prof. Størmer of Christiania, now the leading authority on the subject. Størmer was the first who managed to photograph auroras. He takes simultaneous photographs from the ends of a base (say 25 kilometres long). The photograph includes stars as well as aurora, and the relative positions of the aurora among the stars from the two ends of the base give the parallax. From end B of the base the aurora A is seen in direction of star β, from the other end C it is seen in direction of star α. The angle $\alpha B\beta$ (or $\alpha C\beta$) is known from astronomical data, and hence the parallax angle BAC. This I fancy is how Cavendish actually worked it out. The only difference is that the apparent star positions of the aurora were observed (and not at absolutely identical times, the *assumption* being made that the arch remained stationary which may not, of course, have been true). Cavendish seems to have had more critical views upon questions of perspective and other uncertainties than his predecessors, and though I should hardly regard the paper as an epoch-making one it seems deserving of attention in any compendious treatment of Cavendish's writings.

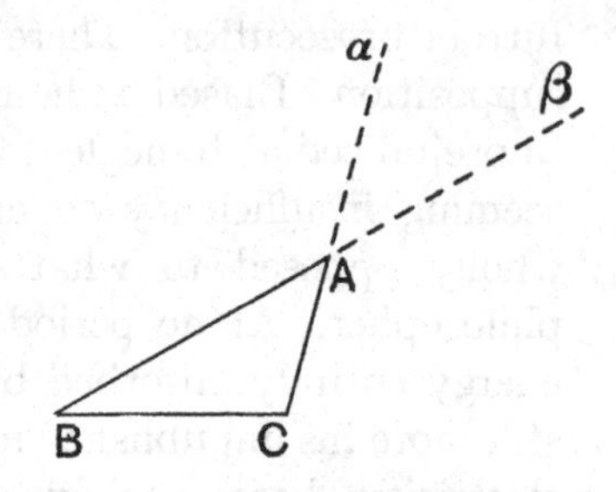

Størmer in his earlier work got aurora as low, I think, as 40 kms. In his more recent work he gets a great number of heights of 90 to under 100 kms. and less numerous instances of heights up to several hundred kms. Thus the height Cavendish got is of the same order as the heights Størmer gets as usual for the lower borders of Aurora.

With regard to Professor Abbe's reference to Dalton's supposed ignorance of Cavendish's prior work, we read in the First Edition (1793) of the *Meteorological Essays*, p. 155:

A very moderate skill in optics was sufficient to convince the author [Dalton], that as the luminous beams at all places appear to tend towards one point about the zenith, they must in reality be straight beams, parallel to each other, and nearly perpendicular to the horizon; and from the appearance of their breadth, they must be cylindrical.

To this sentence there is a footnote. "The author did not see before May 1793, the *Philosophical Transactions* for 1790 in which he finds this idea is suggested by H. Cavendish, Esq., F.R.S. and A.S."

The following letter from his younger brother Frederick, found among Cavendish's papers, may be worth reproduction, as evidence of their joint interest in its subject-matter, and as throwing a little light upon their personal relations, referred to in Wilson's *Life*.

Market Street
Wed. Mar. 1st 1780.

Dear Brother,

As I know you observe the Aurora Borealis with much attention, I send you an account of one which appeared last Night, and which in some respects was the most remarkable I have known. It had the most perfect Corona I ever beheld, with Radii streaming down on all sides, and overspreading *the whole* Hemisphere. The Corona was situated almost close to the hinder foot of Ursa Major, very near to the two stars ♏ and ♈; but rather on that side which is nearest to the Stars ψ and β and in Line with them. The Aurora was of a pale colour, tho' I am inform'd that before I observ'd it, the Sky was very Red in the Eastern quarter (as describ'd to me) sometimes in a Flush, and sometimes darting up the Heavens, and I myself occasionally observ'd a Flush of Red, in the West, and in other directions; 'tho it was not the general tenour of its' appearance. It was a little tremulous in its' motion, by no means darting quick, as I have sometimes observ'd it; but varying its Figure sometimes, and sometimes disappearing. The Situation of the Corona was always the same, its' Radii always concentering in the same Point. Sometimes the Space within the Corona was pretty clear; at other times fill'd nearly, with irregular Streams of luminous matter, hurried confusedly together, darting quick, and again instantly disappearing.

It was near 10 o'clock when I first perceiv'd it (it had been observ'd by others, an hour or two sooner) but it disappearing soon after, I did not attend to it, 'till looking at the Thermometer a little before 12 o'clock, I found the Aurora exceeding bright. I accordingly took my Plan of the Stars, in order to determine the precise situation of the Corona. I attended to it for near an hour, and am certain as to the Situation I have describ'd.

Give my Duty to my Father. I hope ye are both in good Health.

I am your affectionate
Brother
FREDERICK CAVENDISH.

Honble Henry Cavendish
at Lord Charles Cavendish's
Great Marlborough Street
London

At the foot of this letter Henry Cavendish has written

Altitude } ♈ about 72° } at 12 o'clock.
Azim. } 11° East

Alt. ψ about 85°. All 4 Stars nearly in the same vertical circle

In 1792 Cavendish contributed a short paper to the Royal Society "On the Civil Year of the Hindoos, and its Divisions," which is printed in

Vol. 82, 1792, of the *Philosophical Transactions*. This paper is now, doubtless, only of historical value, but to the biographer of Cavendish it is not without significance as evidence of its author's catholicity and the wide range of his studies and interests. Dr E. Denison Ross, of the School of Oriental Studies, was good enough to consult Dr Barnett of the British Museum concerning its merits as a contribution to our knowledge of Indian chronology. Having regard to the date of its appearance, Dr Barnett writes that on the whole "it marks a distinct advance." There are, he adds, "naturally errors in it,...but he is on the right way, and pursues it intelligently."

In the *Philosophical Transactions* for 1797 appears a paper by Mendoza y Rios (Vol. 87, 1797, p. 43) entitled "Researches on the Chief Problems of Nautical Astronomy. From the French" in which there is given an extract of a letter from Cavendish to the author, dated January, 1795. Concerning this paper, and Cavendish's contribution to it, Sir Frank Dyson the Astronomer Royal reports:

> It seems that in 1805 very elaborate tables were published by Mendoza, the expenses being partly defrayed by Sir Joseph Banks and other Englishmen interested in navigation. In this work no reference is made to Cavendish. As far as I can see he did not use the suggestions put forward by Cavendish, but adopted a wholly different method. Mendoza's work had several editions; we have at the Observatory the one of 1805 and one published at Madrid in 1850. I do not think that Cavendish's method was put to practical use. He advocates what looks at first sight the better method, finding the apparent distance between the sun and a star by a series of corrections to the so-called true distance. But the calculation, owing to refraction and especially parallax, is very complex and the corrections are not very small. Consequently instead of this differential method, Mendoza preferred to calculate the apparent distance directly and his procedure has been generally followed.

In 1798 appeared the famous memoir on "Experiments to determine the Density of the Earth" (*Phil. Trans.* 1798, 88. 469) which, by its appeal to wider interests, greatly enlarged the scientific fame of its author. There is little doubt the subject had been in Cavendish's mind for many years previously. He had taken much interest in Maskelyne's work on Schehallien[1], and, as his unpublished papers show, had furnished memoranda and calculations relating to it, as a member of the "Committee of Attraction" of the Royal Society appointed "to consider of a proper hill on which to try the experiment," to supervise generally the conduct of the inquiry, as well as the expenditure of the royal bounty which defrayed its cost.

[1] "An Account of Observations made on the Mountain Schehallien for finding its Attraction," by the Rev. Nevil Maskelyne, B.D., F.R.S., and Astronomer Royal, *Phil. Trans.* 1775, 65. p. 500.

The principle of the method employed by Cavendish was suggested by his friend the Rev. John Michell, F.R.S., Rector of Thornhill, near Dewsbury, who also contrived an apparatus for carrying it into effect. Michell[1], who had been a Fellow of Queens' College, Cambridge, was fourth in the Tripos List for 1748–49 and was made Woodwardian Professor of Geology in 1762, holding the office for two years. He was elected into the Royal Society in 1760, the same year as Cavendish, and his signature appears in the Charter-book under that of Sir Joshua Reynolds who was made a Fellow in 1761. He was interested in astronomy and geodesy, and made important communications on these subjects to the *Philosophical Transactions*. He was a good geometrician and skilled in the use of instruments, and was the first to suggest the use of Hadley's quadrant in surveying and pilotage. He was also the author of a noteworthy paper on the cause and phenomena of earthquakes (*Phil. Trans.* 1760) and was an acute and skilful geological observer. "In his generalizations, derived in great part from his own observations on the geological structure of Yorkshire, he anticipated many of the views more fully developed by later naturalists." (Lyell, *Principles of Geology*, 7th ed., p. 43.)

The apparatus devised by Michell for determining the mean density of the earth by observing the attraction of small masses of matter was based upon a principle which he had suggested and used as far back as 1768. It was subsequently employed for measuring small attractions and repulsions by Coulomb with whose name it is usually associated. Owing to Michell's other engagements its adaptation for measuring the density of the earth was not completed until a short time before his death and no actual observations were made by him with it.

The arrangement came into the possession of the Rev. Francis John Hyde Wollaston, Jacksonian Professor at Cambridge—one of the observers who furnished Cavendish with data for his computation of the height of the aurora. As Wollaston was unable to make use of it, he transferred it to Cavendish who proceeded to erect it, after very considerable modifications, in an outhouse in his garden at Clapham. The principle of the method consisted in measuring the angle of torsional deflection of a horizontal beam, suspended at the centre by a long thin wire, and provided at each extremity with a small leaden ball, when a much larger ball of the same metal is brought near so as to attract it. The main difficulty of the experiment consists in eliminating the various disturbing factors which interfere with trustworthy measurements of the torsion due to the force of the attracting masses. The sources of disturbance are due, partly to faults inherent in the arrangement of the apparatus, but more especially to the difficulty of securing absolute uniformity in temperature and freedom from air currents. From Cavendish's account of his experiments he would appear to have thought out pretty completely the theory of the method, and to have

[1] See short memoir by Sir A. Geikie, 1918.

striven, so far as his means permitted, to obviate or to correct for such sources of error as he perceived. In all seventeen series of observations were made. From the table giving the results it "appears" says Cavendish

that though the experiments agree pretty well together, yet the difference between them, both in quantity of motion of the arm and in the time of vibration, is greater than can proceed merely from the error of observation. As to the difference in the motion of the arm, it may very well be accounted for, from the current of air produced by the difference of temperature; but whether this can account for the difference in the time of vibration is doubtful. If the current of air was regular, and of the same swiftness in all parts of the vibration of the ball, I think it could not; but as there will most likely be much irregularity in the current, it may very likely be sufficient to account for the difference.

Two different wires were used to suspend the beam.

By a mean of the experiments made with the wire first used, the density of the earth comes out 5·48 times greater than that of water; and by a mean of those made with the latter wire, it comes out the same, and the extreme difference of the results of the 23 observations made with this wire, is only ·75; so that the extreme results do not differ from the mean by more than ·38 or $\frac{1}{14}$ of the whole, and therefore the density should seem to be determined hereby, to great exactness.

The "Cavendish experiment," as it is called, has been frequently repeated, viz. by Reich in 1837 and again in 1852, by Baily in 1841, by Cornu and Baille in 1872, by Boys in 1894, by Braun in 1894, by Eötvös in 1896, and by Burgess in 1901.

It is, of course, possible that Cavendish's method, as, indeed, he points out, may be affected by some error, which, as he says, "may perhaps act always, or commonly, in the same direction," thus making it desirable to use different methods and other instruments. Accordingly attempts have been made to obtain an independent value by means of the chemical balance. This instrument was employed for this purpose by von Jolly of Munich in 1878–80, by Poynting in 1890, by Richarz and Krigar-Menzel in 1898. Wilsing, at Potsdam, has also made a series of observations by causing a pendulum 1 metre in length armed at each end with balls of 540 grams in weight to oscillate between moveable cylinders weighing 325,000 grams. The values obtained by these several experimenters are as follows:

By the torsion-balance:

1798	Cavendish, recalculated by Baily	5·45
1837	Reich	5·49
1841	Baily	5·674
1852	Reich	5·583
1872	Cornu and Baille	5·56
1894	Boys	5·527

1894	Braun	5·527
1896	Eötvös	5·55
1901	Burgess	5·44–5·74

By the chemical balance:

1878–80	von Jolly	5·692
1890	Poynting	5·49
1898	Richarz and Krigar-Menzel ...	5·505

By the pendulum:

1886–88	Wilsing	5·579

Burgess, in a paper published in the *Phys. Rev.* 14. 1902, pp. 257–264, has given a summary of the most trustworthy determinations of the mean density of the earth and calculates that the most probable value is 5·5247, which is remarkably close to the concordant and independent values of Boys and Braun.

Hutton's calculations from Maskelyne's observations of the deviation of the plumb-line at Schehallien had shown that the ratio of the density of the earth to that of the mountain is as 9 to 5. From a lithological examination of the mountain, Playfair concluded that its mean density was between 2·7 and 2·8, which gives about 5 for the mean density of the earth, a result confirmed by the observations, in 1856, of James and Clarke on Arthur's Seat (5·3–5·4), and of Mendenhall on Fusiyama in 1880 (5·77).

Although these observations are of very unequal value, as regards accuracy those obtained by the Cavendish method being the more trustworthy, they are all in remarkable accordance with Newton's "guess" that

> the super-stratum of the Earth being about twice as dense as the water, and the sub-strata becoming, in proportion to their depth, three, four, and even five times more dense, it is probable that the whole mass of the Earth is five or six times more dense than if it were formed of water[1].

The last paper published by Cavendish was "On an improvement in the manner of dividing Astronomical Instruments," printed in the *Phil. Trans.* for 1809, pp. 221–231—the year before his death.

Edward Troughton, the well-known mathematical instrument maker, had communicated to the Royal Society an account of a method of dividing astronomical and other instruments (*Phil. Trans.*, 1809, 105) which attracted considerable attention at the time of its appearance on account of its ingenuity and the excellence of its performance, and as a distinct advance on the methods of Sisson, Bird and Ramsden—all eminent craftsmen in their day as instrument makers. Cavendish, for

[1] For an account of the observations on the determination of the mean density of the earth made prior to 1894, see Poynting's essay on that subject (Adams Prize), London, Griffin & Co. Also C. V. Boys, "La Constante de la gravitation," *Congrès Internationale de Physique*, Paris, 1900.

whom all instruments of precision had a sort of fascination, had evidently studied Troughton's method with care, and in the paper referred to he suggests a method of obviating certain disadvantages in the mode of dividing as at that time practised. Sir Horace Darwin reports concerning this paper:

Cavendish suggests a method which he has not actually tried, but which would seem to be an improvement on the method that appeared to have been used at that time for dividing large circles of astronomical instruments. In his method the circle was first divided into 6 parts by setting a beam compass with the points apart at a distance equal to the radius. These spaces were divided again by the beam compass, sometimes into two equal parts, and sometimes into three and five equal parts, and so on till quite small spaces were left. Errors have to be calculated and allowed for, and the process is most laborious and slow.

I do not think Cavendish's paper is of any practical value to anyone now dividing circles, as they are always made on a dividing engine, which is a method of copying an existing divided circle. But if anyone wished to make an original divided circle he should certainly read Cavendish's paper.

PAPERS COMMUNICATED BY CAVENDISH TO THE ROYAL SOCIETY AND PUBLISHED IN THE *PHILOSOPHICAL TRANSACTIONS*

ARRANGED IN CHRONOLOGICAL ORDER

[NOTE. Cavendish's published scientific communications are here reproduced exactly as originally printed, except that certain numerical tables have had to be placed somewhat differently in the text owing to the difference in the size of the page of this volume and that of the *Philosophical Transactions*. Obvious errors and necessary typographical corrections have also been indicated.

The responsibility for editing the *Philosophical Transactions* rests with the Secretaries of the Society and their practice, or that of the Printers, has varied from time to time. Thus Cavendish's earlier papers are printed without head-lines to the pages. Accordingly, for the sake of uniformity, in the present reproduction, page head-lines have been introduced in the case of the papers on *Factitious Air* and on the *Rathbone Place Water*. These are of the same general character as the head-lines appearing on the papers subsequently published.]

PHILOSOPHICAL TRANSACTIONS. VOL. 56, 1766, p. 141

Received May 12, 1766

XIX. *Three Papers, containing Experiments on factitious Air, by the Hon.* Henry Cavendish, *F.R.S.*

Read May 29, Nov. 6 and Nov. 13, 1766

By factitious air, I mean in general any kind of air which is contained in other bodies in an unelastic state, and is produced from thence by art.

By fixed air, I mean that particular species of factitious air, which is separated from alcaline substances by solution in acids or by calcination; and to which Dr. Black has given that name in his treatise on quicklime.

As fixed air makes a considerable part of the subject of the following papers; and as the name might incline one to think, that it signified any sort of air which is contained in other bodies in an unelastic form; I thought it best to give this explanation before I went any farther.

Before I proceed to the experiments themselves, it will be proper to mention the principal methods used in making them.

In order to fill a bottle with the air discharged from metals or alcaline substances by solution in acids, or from animal or vegetable substances by fermentation, I make use of the contrivance represented in TAB. VII. Fig. 1. where A represents the bottle, in which the materials for producing air are placed; having a bent glass tube C ground into it, in the manner of a stopper. E represents a vessel of water. D the bottle to receive the air, which is first filled with water, and then inverted into the vessel of water, over the end of the bent tube. F*f* represents the string, by which the bottle is suspended. When I would measure the quantity of air, which is produced by any of these substances, I commonly do it by receiving the air in a bottle, which has divisions marked on its sides with a diamond, shewing the weight of water, which it requires to fill the bottle up to those divisions: but sometimes I do it by making a mark on the side of the bottle in which I have received the air, answering to the surface of the water therein; and then, setting the [bottle] upright, find how much water it requires to fill it up to that mark.

In order to transfer the air out of one bottle into another, the simplest way, and that which I have oftenest made use of, is that represented in

Fig. 2. where A is the bottle, into which the air is to be transferred: it is supposed to be filled with water and inverted into the vessel of water DEFG, and suspended there by a string: the line DG is the surface of the water: B represents a tin funnel held under the mouth of the bottle: C represents the inverted bottle, out of which the air is to be transferred; the mouth of which is lifted up till the air runs out of it into the funnel, and from thence into the bottle A.

In order to transfer air out of a bottle into a bladder, the contrivance Fig. 3. is made use of. A is the bottle out of which the air is to be transferred, inverted into the vessel of water FGHK: B is a bladder whose neck is tied fast over the hollow piece of wood C*c*, so as to be air-tight. Into the piece of wood is run a bent pewter pipe D, and secured with lute[1]. The air is then pressed out of the bladder as well as possible, and a bit of wax E stuck upon the other end of the pipe, so as to stop up the orifice. The pipe, with the wax upon it, is then run up into the inverted bottle, and the wax torn off by rubbing it against the sides. By this means, the end of the pipe is introduced within the bottle, without suffering any water to get within it. Then, by letting the bottle descend, so as to be totally immersed in the water, the air is forced into the bladder.

The weights used in the following experiments, are troy weights, 1 ounce containing 480 grains. By an ounce or grain measure, I mean such a measure as contains one ounce or grain Troy of water.

EXPERIMENTS ON FACTITIOUS AIR

PART I

Containing Experiments on Inflammable Air.

I know of only three metallic substances, namely, zinc, iron and tin, that generate inflammable air by solution in acids; and those only by solution in the diluted vitriolic acid, or spirit of salt.

Zinc dissolves with great rapidity in both these acids; and, unless they are very much diluted, generates a considerable heat. One ounce of zinc produces about 356 ounce measures of air: the quantity seems just the same whichsoever of these acids it is dissolved in. Iron dissolves readily

[1] The lute used for this purpose, as well as in all the following experiments, is composed of almond powder, made into a paste with glue, and beat a good deal with a heavy hammer. This is the strongest and most convenient lute I know of. A tube may be cemented with it to the mouth of a bottle, so as not to suffer any air to escape at the joint; though the air within is compressed by the weight of several inches of water.

in the diluted vitriolic acid, but not near so readily as zinc. One ounce of iron wire produces about 412 ounce measures of air: the quantity was just the same, whether the oil of vitriol was diluted with 1½, or 7 times its weight of water: so that the quantity of air produced seems not at all to depend on the strength of the acid.

Iron dissolves but slowly in spirit of salt while cold: with the assistance of heat it dissolves moderately fast. The air produced thereby is inflammable; but I have not tried how much it produces.

Tin was found to dissolve scarce at all in oil of vitriol diluted with an equal weight of water, while cold: with the assistance of a moderate heat it dissolved slowly, and generated air, which was inflammable: the quantity was not ascertained.

Tin dissolves slowly in strong spirit of salt while cold: with the assistance of heat it dissolves moderately fast. One ounce of tinfoil yields 202 ounce measures of inflammable air.

These experiments were made, when the thermometer was at 50° and the barometer at 30 inches.

All these three metallic substances dissolve readily in the nitrous acid, and generate air; but the air is not at all inflammable. They also unite readily, with the assistance of heat, to the undiluted acid of vitriol; but very little of the salt, formed by their union with the acid, dissolves in the fluid. They all unite to the acid with a considerable effervescence, and discharge plenty of vapours, which smell strongly of the volatile sulphureous acid, and which are not at all inflammable. Iron is not sensibly acted on by this acid, without the assistance of heat; but zinc and tin are in some measure acted on by it, while cold.

It seems likely from hence, that, when either of the above-mentioned metallic substances are dissolved in spirit of salt, or the diluted vitriolic acid, their phlogiston flies off, without having its nature changed by the acid, and forms the inflammable air; but that, when they are dissolved in the nitrous acid, or united by heat to the vitriolic acid, their phlogiston unites to part of the acid used for their solution, and flies off with it in fumes, the phlogiston losing its inflammable property by the union. The volatile sulphureous fumes, produced by uniting these metallic substances by heat to the undiluted vitriolic acid, shew plainly, that in this case their phlogiston unites to the acid; for it is well known, that the vitriolic sulphureous acid consists of the plain vitriolic acid united to phlogiston[1]. It is highly probable too, that the same thing happens in dissolving these metallic substances in the nitrous acid; as the fumes produced during the

[1] Sulphur is allowed by chymists, to consist of the plain vitriolic acid united to phlogiston. The volatile sulphureous acid appears to consist of the same acid united to a less proportion of phlogiston than what is required to form sulphur. A circumstance which I think shews the truth of this, is that if oil of vitriol be distilled, from sulphur, the liquor, which comes over, will be the volatile sulphureous acid.

solution appear plainly to consist in great measure of the nitrous acid, and yet it appears, from their more penetrating smell and other reasons, that the acid must have undergone some change in its nature, which can hardly be attributed to anything else than its union with the phlogiston. As to the inflammable air, produced by dissolving these substances in spirit of salt or the diluted vitriolic acid, there is great reason to think, that it does not contain any of the acid in its composition; not only because it seems to be just the same whichsoever of these acids it is produced by; but also because there is an inflammable air, seemingly much of the same kind as this, produced from animal substances in putrefaction, and from vegetable substances in distillation, as will be shewen hereafter; though there can be no reason to suppose, that this kind of inflammable air owes its production to any acid. I now proceed to the experiments made on inflammable air.

I cannot find that this air has any tendency to lose its elasticity by keeping, or that it is at all absorbed, either by water, or by fixed or volatile alcalies; as I have kept some by me for several weeks in a bottle inverted into a vessel of water, without any sensible decrease of bulk; and as I have also kept some for a few days, in bottles inverted into vessels of sope leys and spirit of sal ammoniac, without perceiving their bulk to be at all diminished.

It has been observed by others, that, when a piece of lighted paper is applied to the mouth of a bottle, containing a mixture of inflammable and common air, the air takes fire, and goes off with an explosion. In order to observe in what manner the effect varies according to the different proportions in which they are mixed, the following experiment was made.

Some of the inflammable air, produced by dissolving zinc in diluted oil of vitriol, was mixed with common air in several different proportions, and the inflammability of these mixtures tried one after the other in this manner. A quart bottle was filled with one of these mixtures, in the manner represented in Fig. 2. The bottle was then taken out of the water, set upright on a table, and the flame of a lamp or piece of lighted paper applied to its mouth. But, in order to prevent the included air from mixing with the outward air, before the flame could be applied, the mouth of the bottle was covered, while under water, with a cap made of a piece of wood covered with a few folds of linnen; which cap was not removed till the instant that the flame was applied. The mixtures were all tried in the same bottle; and, as they were all ready prepared, before the inflammability of any of them was tried, the time elapsed between each trial was but small: by which means I was better able to compare the loudness of the sound in each trial. The result of the experiment is as follows.

With one part of inflammable air to 9 of common air, the mixture would not take fire, on applying the lighted paper to the mouth of the bottle; but, on putting it down into the belly of the bottle, the air took fire, but made very little sound.

With 2 parts of inflammable to 8 of common air, it took fire immediately, on applying the flame to the mouth of the bottle, and went off with a moderately loud noise.

With 3 parts of inflammable air to 7 of common air, there was a very loud noise.

With 4 parts of inflammable to 6 of common air, the sound seemed very little louder.

With equal quantities of inflammable and common air, the sound seemed much the same. In the first of these trials, namely, that with one part of inflammable to 9 of common air, the mixture did not take fire all at once, on putting the lighted paper into the bottle; but one might perceive the flame to spread gradually through the bottle. In the three next trials, though they made an explosion, yet I could not perceive any light within the bottle. In all probability, the flame spread so instantly through the bottle, and was so soon over, that it had not time to make any impression on my eye. In the last mentioned trial, namely, that with equal quantities of inflammable and common air, a light was seen in the bottle, but which quickly ceased.

With 6 parts of inflammable to 4 of common air, the sound was not very loud: the mixture continued burning a short time in the bottle, after the sound was over.

With 7 parts of inflammable to 3 of common air, there was a very gentle bounce or rather puff: it continued burning for some seconds in the belly of the bottle.

A mixture of 8 parts of inflammable to 2 of common air caught fire on applying the flame, but without any noise: it continued burning for some time in the neck of the bottle, and then went out, without the flame ever extending into the belly of the bottle.

It appears from these experiments, that this air, like other inflammable substances, cannot burn without the assistance of common air. It seems too, that, unless the mixture contains more common than inflammable air, the common air therein is not sufficient to consume the whole of the inflammable air; whereby part of the inflammable air remains, and burns by means of the common air, which rushes into the bottle after the explosion.

In order to find whether there was any difference in point of inflammability between the air produced from different metals by different acids, five different sorts of air, namely, 1. Some produced from zinc by diluted oil of vitriol, and which had been kept about a fortnight; 2. Some of the same kind of air fresh made; 3. Air produced from zinc by spirit of salt; 4. Air from iron by the vitriolic acid; 5. Air from tin by spirit of salt; were each mixed separately with common air in the proportion of 2 parts of inflammable air to $7\frac{7}{10}$ of common air, and their inflammability tried in the same bottle, that was used for the former experiment, and with the same precautions. They each went off with a pretty loud noise, and without any difference in the sound that I could be sure of. Some more of each of the above parcels of air were then mixed with common air, in the proportion of 7 parts of inflammable air to $3\frac{1}{5}$ of common air, and tried in the same way as before. They each of them went off with a gentle bounce, and burnt some time in the bottle, without my being able to perceive any difference between them.

In order to avoid being hurt, in case the bottle should burst by the explosion, I have commonly, in making these sort of experiments, made use of an apparatus contrived in such manner, that, by pulling a string, I drew the flame of a lamp over the mouth of the bottle, and at the same time pulled off the cap, while I stood out of the reach of danger. I believe, however, that this precaution is not very necessary; as I have never known a bottle to burst in any of the trials I have made.

The specific gravity of each of the above-mentioned sorts of inflammable air, except the first, was tried in the following manner. A bladder holding about 100 ounce measures was filled with inflammable air, in the manner represented in Fig. 3. and the air pressed out again as perfectly as possible. By this means the small quantity of air remaining in the bladder was almost intirely of the inflammable kind. 80 ounce measures of the inflammable air, produced from zinc by the vitriolic acid, were then forced into the bladder in the same manner: after which, the pewter pipe was taken out of the wooden cap of the bladder, the orifice of the cap stopt up with a bit of lute, and the bladder weighed. A hole was then made in the lute, the air pressed out as perfectly as possible, and the bladder weighed again. It was found to have increased in weight $40\frac{3}{4}$ grains. Therefore the air pressed out of the bladder weighs $40\frac{3}{4}$ grains less than an equal quantity of common air: but the quantity of air pressed out of the bladder must be nearly the same as that which was forced into it, i.e. 80 ounce measures: consequently 80 ounce measures of this sort of inflammable air weigh $40\frac{3}{4}$ grains less than an equal bulk of common air. The three other sorts of inflammable air were then tried in the same way, in the same bladder, immediately one after the other. In the trial with the air from zinc by spirit of salt, the bladder increased $40\frac{1}{2}$ grains on forcing out the air. In the trial with the air from iron, it increased $41\frac{1}{2}$ grains, and in that with the air from tin, it increased 41 grains. The heat of the air, when this experiment was made, was 50°; the barometer stood at $29\frac{3}{4}$ inches.

There seems no reason to imagine, from these experiments, that there is any difference in point of specific gravity between these four sorts of inflammable air; as the small difference observed in these trials is in all probability less than what may arise from the unavoidable errors of the experiment. Taking a medium therefore of the different trials, 80 ounce measures of inflammable air weigh 41 grains less than an equal bulk of common air. Therefore, if the density of common air, at the time when this experiment was tried, was 800 times less than that of water, which, I imagine, must be near the truth[1], inflammable air must be 5490 times lighter than water, or near 7 times lighter than common air. But if the density of common air was 850 times less than that of water, then would inflammable air be 9200 times lighter than water, or $10\frac{8}{10}$ lighter than common air.

This method of finding the density of factitious air is very convenient and sufficiently accurate, where the density of the air to be tried is not much less than that of common air, but cannot be much depended on in the present case, both on account of the uncertainty in the density of common air, and because we cannot be certain but what some common air might be mixed with the inflammable air in the bladder, notwithstanding the precautions used to prevent it; both which causes may produce a considerable error, where the density of the air to be tried is many times less than that of common air. For this reason, I made the following experiments.

I endeavoured to find the weight of the air discharged from a given

[1] Mr. Hawksbee, whose determination is usually followed as the most exact, makes air to be more than 850 times lighter than water; vid. Hawksbee's experiments, p. 94, or Cotes's Hydrostatics, p. 159. But his method of trying the experiment must in all probability make it appear lighter than it really is. For having weighed his bottle under water, both when full of air and when exhausted, he supposes the difference of weight to be equal to the weight of the air exhausted; whereas in reality it is not so much: for the bottle, when exhausted, must necessarily be compressed, and on that account weigh heavier in water than it would otherwise do. Suppose, for example, that air is really 800 times lighter than water, and that the bottle is compressed $\frac{1}{12000}$ part of its bulk; which seems no improbable supposition: the weight of the bottle in water will thereby be increased by $\frac{1}{12000}$ of the weight of a quantity of water of the same bulk, or more than $\frac{1}{15}$ of the weight of the air exhausted: whence the difference of weight will not be so much as $\frac{14}{15}$ of the weight of the air exhausted: and therefore the air will appear lighter than it really is in the proportion of more than 15 to 14, i.e. more than 857 times lighter than water: whereas, if the ball had been weighed in air in both circumstances, the error arising from the compression would have been very trifling.

It appears, from some experiments that have been made by weighing a ball in air, while exhausted, and also after the air was let in, that air, when the thermometer is at 50°, and the barometer at $29\frac{3}{4}$, is about 800 times lighter than water. Though the weight of the air exhausted was little more than 50 grains, no error could well arise near sufficient to make it agree with Hawksbee's experiment. Air seems to expand about $\frac{1}{500}$ part by 1° of heat, whence its density in any other state of the atmosphere is easily determined. The density here assumed agrees very well with the rule given by the gentlemen, who measured the length of a degree in Peru, for finding the height of mountains barometrically, and which is given in the Connoissance des mouvemens celestes, année 1762. To make that rule agree accurately with observation, the density of air, whose heat is the same as that of the places where these observations were made, and which I imagine we may estimate at about 45°, should be 798 times less than that of water, when the barometer stands at $29\frac{3}{4}$.

quantity of zinc by solution in the vitriolic acid, in the manner represented in Fig. 4. A is a bottle filled near full with oil of vitriol diluted with about six times its weight of water: B is a glass tube fitted into its mouth, and secured with lute: C is a glass cylinder fastened on the end of the tube, and secured also with lute. The cylinder has a small hole at its upper end to let the inflammable air escape, and is filled with dry pearl-ashes in coarse powder. The whole apparatus, together with the zinc, which was intended to be put in, and the lute which was to be used in securing the tube to the neck of the bottle, were first weighed carefully; its weight was 11930 grains. The zinc was then put in, and the tube put in its place. By this means, the inflammable air was made to pass through the dry pearl-ashes; whereby it must have been pretty effectually deprived of any acid or watery vapours that could have ascended along with it. The use of the glass tube B was to collect the minute jets of liquor, that were thrown up by the effervescence, and to prevent their touching the pearl-ashes; for which reason, a small space was left between the glass-tube and the pearl-ashes in the cylinder. When the zinc was dissolved, the whole apparatus was weighed again, and was found to have lost $11\frac{3}{4}$ grains in weight[1]; which loss is principally owing to the weight of the inflammable air discharged. But it must be observed, that, before the effervescence, that part of the bottle and cylinder, which was not occupied by other more solid matter, was filled with common air; whereas, after the effervescence, it was filled with inflammable air; so that, upon that account alone, supposing no more inflammable air to be discharged than what was sufficient to fill that space, the weight of the apparatus would have been diminished by the difference of the weight of that quantity of common air and inflammable air. The whole empty space in the bottle and cylinder was about 980 grain measures, there is no need of exactness; and the difference of the weight of that quantity of common and inflammable air is about one grain: therefore the true weight of the inflammable air discharged, is $10\frac{3}{4}$ grains. The quantity of zinc used was 254 grains, and consequently the weight of the air discharged is $\frac{1}{23}$ or $\frac{1}{24}$ of the weight of the zinc.

It was before said, that one grain of zinc yielded 356 grain measures of air: therefore 254 grains of zinc yield 90427 grain measures of air; which we have just found to weigh $10\frac{3}{4}$ grains; therefore inflammable air is about 8410 times lighter than water, or $10\frac{1}{2}$ times lighter than common air.

The quantity of moisture condensed in the pearl-ashes was found to be about $1\frac{1}{4}$ grains.

By another experiment, tried exactly in the same way, the density of inflammable air came out 8300 times less than that of water.

[1] As the quantity of lute used was but small, and as this kind of lute does not lose a great deal of its weight by being kept in a moderately dry room, no sensible error could arise from the drying of the lute during the experiment.

The specific gravity of the air, produced by dissolving zinc in spirit of salt, was tried exactly in the same manner. 244 grains of zinc being dissolved in spirit of salt diluted with about four times its weight of water, the loss in effervescence was $10\frac{3}{4}$ grains; the empty space in the bottle and cylinder was 914 grain measures; whence the weight of the inflammable air was $9\frac{3}{4}$ grains, and consequently its density was 8910 times less than that of water.

By another experiment, its specific gravity came out 9030 times lighter than water.

A like experiment was tried with iron. $250\frac{1}{2}$ grains of iron being dissolved in oil of vitriol diluted with four times its weight of water, the loss in effervescence was 13 grains, the empty space 1420 grain measures. Therefore the weight of the inflammable air was $11\frac{3}{8}$ grains, i.e. about $\frac{1}{22}$ of the weight of the iron, and its density was 8973 times less than that of water. The moisture condensed was $1\frac{1}{4}$ grains.

A like experiment was tried with tin. 607 grains of tinfoil being dissolved in strong spirit of salt, the loss in effervescence was $14\frac{3}{4}$ grains, the empty space 873 grain measures: therefore the weight of the inflammable air was $13\frac{3}{4}$ grains, i.e. $\frac{1}{44}$ of the weight of the tin, and its density 8918 times less than that of water. The quantity of moisture condensed was about three grains.

It is evident, that the truth of these determinations depend[s] on a supposition, that none of the inflammable air is absorbed by the pearl-ashes. In order to see whether this was the case or no, I dissolved 86 grains of zinc in diluted acid of vitriol, and received the air in a measuring bottle in the common way. Immediately after, I dissolved the same quantity of zinc in the same kind of acid, and made the air to pass into the same measuring bottle, through a cylinder filled with dry pearl-ashes, in the manner represented in Fig. 5. I could not perceive any difference in their bulks.

It appears from these experiments, that there is but little, if any, difference in point of density between the different sorts of inflammable air. Whether the difference of density observed between the air procured from zinc, by the vitriolic and that by the marine acid is real, or whether it is only owing to the error of the experiment, I cannot pretend to say. By a medium of the experiments, inflammable air comes out 8760 times lighter than water, or eleven times lighter than common air.

In order to see whether inflammable air, in the state in which it is, when contained in the inverted bottles, where it is in contact with water, contains any considerable quantity of moisture dissolved in it, I forced 192 ounce measures of inflammable air, through a cylinder filled with dry pearl-ashes, by means of the same apparatus, which I used for filling the bladders with inflammable air, and which is represented in Fig. 3. The cylinder was weighed carefully before and after the air was forced through;

whereby it was found to have increased 1 grain in weight. The empty space in the cylinder was 248 grains, the difference of weight of which quantity of common and inflammable air is $\frac{1}{4}$ of a grain. Therefore the real quantity of moisture condensed in the pearl-ashes is $1\frac{1}{4}$ grain. The weight of 192 ounce measures of inflammable air deprived of its moisture appears from the former experiments to be $10\frac{1}{2}$ grains; therefore its weight when saturated with moisture would be $11\frac{3}{4}$ grains. Therefore inflammable air, in that state in which it is in, when kept under the inverted bottles, contains near $\frac{1}{9}$ its weight of moisture; and its specific gravity in that state is 7840 times less than that of water.

I made an experiment with design to see, whether copper produced any inflammable air by solution in spirit of salt. I could not procure any inflammable air thereby: but the phenomena attending it seem remarkable enough to deserve mentioning. The apparatus used for this experiment was of the same kind as that represented in Fig. 1. The bottle A was filled almost full of strong spirit of salt, with some fine copper wire in it. The wire seemed not at all acted on by the acid, while cold; but, with the assistance of a heat almost sufficient to make the acid boil, it made a considerable effervescence, and the air passed through the bent tube, into the bottle D, pretty fast, till the air forced into it by this means seemed almost equal to the empty space in the bent tube and the bottle A: when, on a sudden, without any sensible alteration of the heat, the water rushed violently through the bent tube into the bottle A, and filled it almost intirely full.

The experiment was repeated again in the same manner, except that I took away the bottle D, and let out some of the water of the cistern: so that the end of the bent tube was out of water. As soon as the effervescence began, the vapours issued visibly out of the bent tube; but they were not at all inflammable, as appeared by applying a piece of lighted paper to the end of the tube. A small empty phial was then inverted over the end of the bent tube, so that the mouth of the phial was immersed in the water, the end of the tube being within the body of the phial and out of water. The common air was by degrees expelled out of the phial, and its room occupied by the vapours; after which, having chanced to shake the inverted phial a little, the water suddenly rushed in, and filled it almost full; from thence it passed through the bent tube into the bottle A, and filled it quite full. It appears likely from hence that copper, by solution in the marine acid, produces an elastic fluid, which retains its elasticity as long as there is a barrier of common air between it and the water, but which immediately loses its elasticity, as soon as it comes in contact with the water. In the first experiment, as long as any considerable quantity of common air was left in the bottle containing the copper and acid, the vapours, which passed through the bent tube, must have contained a good deal of common air. As soon therefore as any part of these vapours came to the farther end of the bent tube, where they were in contact with the water, that part of them, which consisted of the air from copper, would be immediately condensed, leaving the common air unchanged; whereby the end of the tube would be filled with common air only; by which means the vapours, contained in the rest of the tube and bottle A, seem to have been defended from the action of the water. But when almost all the common air was driven out of the bottle, then the proportion of common air contained in the vapours, which passed through the tube, seems to have been too small to defend them from the action of the water. In the second experiment, the narrow space left between the neck of the inverted phial and the tube would answer much the same end, in defending the vapours within the inverted phial from the action of the water, as the bent tube in the first experiment did in defending the vapours within the bottle from the action of the water.

EXPERIMENTS ON FACTITIOUS AIR

Part II

Containing Experiments on Fixed Air, or that Species of Factitious Air, which is produced from Alcaline Substances, by Solution in Acids or by Calcination.

Experiment I

The air produced, by dissolving marble in spirit of salt, was caught in an inverted bottle of water, in the usual manner. In less than a day's time, much the greatest part of the air was found to be absorbed. The water contained in the inverted bottle was found to precipitate the earth from lime-water; a sure sign that it had absorbed fixed air[1].

Experiment II

I filled a Florence flask in the same way with the same kind of fixed air. When full, I stopt up the mouth of the flask with my finger, while under water, and removed it into a vessel of quicksilver, so that the mouth of the flask was intirely immersed therein. It was kept in this situation upwards of a week. The quicksilver rose and fell in the neck of the flask, according to the alterations of heat and cold, and of the height

[1] Lime, as Dr. Black has shewn, is no more than a calcareous earth rendered soluble in water by being deprived of its fixed air. Lime water is a solution of lime in water: therefore, on mixing lime water with any liquor containing fixed air, the lime absorbs the air, becomes insoluble in water, and is precipitated. This property of water, of absorbing fixed air, and then making a precipitate with lime water, has been taken notice of by Mr. M'Bride.

of the barometer; as it would have done if it had been filled with common air. But it appeared, by comparing together the heights of the quicksilver at the same temper of the atmosphere, that no part of the fixed air had been absorbed or lost its elasticity. The flask was then removed, in the same manner as before, into a vessel of sope leys. The fixed air, by this means, coming in contact with the sope leys, was quickly absorbed.

I also filled another Florence flask with fixed air, and kept it with its mouth immersed in a vessel of quicksilver in the same manner as the other, for upwards of a year, without being able to perceive any air to be absorbed. On removing it into a vessel of sope leys, the air was quickly absorbed like the former.

It appears from this experiment, that fixed air has no disposition to lose its elasticity, unless it meets with water or some other substance proper to absorb it, and that its nature is not altered by keeping.

Experiment III

In order to find how much fixed air water would absorb, the following experiment was made. A cylindrical glass, with divisions marked on its sides with a diamond, shewing the quantity of water which it required to fill it up to those marks, was filled with quicksilver, and inverted into a glass filled with the same fluid. Some fixed air was then forced into this cylindrical glass, in the same manner that it was into the inverted bottles of water, in the former experiments; except that, to prevent any common air from being forced into the glass along with the fixed, I took care not to introduce the end of the bent tube within the cylindrical glass, till I was well assured that no common air to signify could remain within the bottle. This was done by first introducing the end of the bent tube within an inverted bottle of water, and letting it remain there, till the air driven into this bottle was at least 10 times as much as would fill the empty space in the bent tube, and the bottle containing the marble and acid. By this means one might be well assured, that the quantity of common air remaining within the bent tube and bottle must be very trifling. The end of the bent tube was then introduced within the cylindrical glass, and kept there till a sufficient quantity of fixed air was let up. After letting it stand for a few hours, the division answering to the surface of the quicksilver in the cylinder was observed and wrote down, by which it was known how much fixed air had been let up. A little rain water was then introduced into the cylindrical glass, by pouring some rain water into the vessel of quicksilver, and then lifting up the cylindrical glass so as to raise the bottom of it a little way out of the quicksilver. After having suffered it to stand a day or two, in which time the water seemed to have absorbed as much fixed air as it was able to do, the division answering to the upper surface of the water, and also that answering to the surface of the quicksilver, were observed: by which it was known how much air remained not absorbed, and also how much water had been introduced: the division answering to the surface of the water telling how much air remained not absorbed, and the difference of the two divisions telling how much water had been let up. More water was then let up in the same manner, at different times, till almost the whole of the fixed air was absorbed. As all water contains a little air, the water used in this experiment was first well purged of it by boiling, and then introduced into the cylinder while hot. The result of the experiment is given in the following table; in which the first column shews the bulk of the water let up each time; the second shews the bulk of air absorbed each time; the third the whole bulk of water let up; the fourth the whole bulk of air absorbed; and the fifth column shews the bulk of air remaining not absorbed. In order to set the result in a clearer light, the whole bulk of air introduced into the cylinder is called 1, and the other quantities set down in decimals thereof.

Bulk of air let up = 1.

Bulk of water let up each time	Bulk of air absorbed each time	Whole bulk of water let up	Whole bulk of air absorbed	Whole bulk of air remaining
·322	·374	·322	·374	·626
·481	·485	·803	·859	·141
·082	·048	·885	·907	·093
·145	·079	1·030	·986	·014

I imagine that the quantities of water let up and of the air absorbed could be estimated to about three or four 1000th parts of the whole bulk of air introduced. The height of the thermometer, during the trial of this experiment, was at a medium 55°.

This experiment was tried once before. The result agreed pretty nearly with this; but, as it was not tried so carefully, the result is not set down.

It appears from hence, that the fixed air contained in marble consists of substances of different natures, part of it being more soluble in water than the rest: it appears too, that water, when the thermometer is about 55°, will absorb rather more than an equal bulk of the more soluble part of this air.

It appears, from an experiment which will be mentioned hereafter, that water absorbs more fixed air in cold weather than warm; and, from the following experiment, it appears, that water heated to the boiling point is so far from absorbing air, that it parts with what it has already absorbed.

Experiment IV

Some water, which had absorbed a good deal of fixed air, and which made a considerable precipitate with lime water, was put into a phial, and kept about ¼ of an hour in boiling water. It was found when cold not to make any precipitate, or to become in the least cloudy on mixing it with lime water.

Experiment V

Water also parts with the fixed air, which it has absorbed by being exposed to the open air. Some of the same parcel of water, that was used for the last experiment, being exposed to the air in a saucer for a few days, was found at the end of that time to make no clouds with lime water.

Experiment VI

In like manner it was tried how much of the same sort of fixed air was absorbed by spirits of wine. The result is as follows.

Bulk of air introduced = 1.

Spirit let up each time	Air absorbed each time	Whole bulk of spirit let up	Whole bulk of air absorbed	Bulk of air remaining
·207	·453	·207	·453	·547
·146	·274	·353	·727	·273
·074	·103	·427	·830	·170
·046	·030	·473	·860	·140

The mean height of the thermometer, during the trial of the experiment, was 46°. Therefore spirits of wine, at the heat of 46°, absorbs near $2\frac{1}{4}$ times its bulk of the more soluble part of this air.

Experiment VII

After the same manner it was tried how much fixed air is absorbed by oil. Some olive oil, equal in bulk to $\frac{1}{8}$ part of the fixed air in the cylindrical glass, was let up. It absorbed rather more than an equal bulk of air; the thermometer being between 40 and 50°. The experiment was not carried any farther. The oil was found to absorb the air very slowly.

Experiment VIII

The specific gravity of fixed air was tried by means of a bladder, in the same manner which was made use of for finding the specific gravity of inflammable air; except that the air, instead of being caught in an inverted bottle of water, and thence transferred into the bladder, was thrown into the bladder immediately from the bottle which contained the marble and spirit of salt, by fastening a glass tube to the wooden cap of the bladder, and luting that to the mouth of the bottle containing the effervescing mixture, in such manner as to be air-tight. The bladder was kept on till it was quite full of fixed air: being then taken off and weighed, it was found to lose 34 grains, by forcing out the air. The bladder was previously found to hold 100 ounce measures. Whence if the outward air, at the time when this experiment was tried, is supposed to have been 800 times lighter than water, fixed air is 511 times lighter than water, or $1\frac{57}{100}$ times heavier than common air. The heat of the air during the trial of this experiment was 45°.

By another experiment of the same kind, made when the thermometer was at 65°, fixed air seemed to be about 563 times lighter than water.

Experiment IX

Fixed air has no power of keeping fire alive, as common air has; but, on the contrary, that property of common air is very much diminished by the mixture of a small quantity of fixed air; as appears from hence.

A small wax candle burnt 80″ in a receiver, which held 190 ounce measures, when filled with common air only.

The same candle burnt 51″ in the same receiver, when filled with a mixture of one part of fixed air to 19 of common air, i.e. when the fixed air was $\frac{1}{20}$ of the whole mixture.

When the fixed air was $\frac{3}{40}$ of the whole mixture, the candle burnt 23″.

When the fixed air was $\frac{1}{10}$ of the whole, it burnt 11″.

When the fixed air was $\frac{6}{55}$ or $\frac{1}{9\frac{1}{6}}$ of the whole mixture, the candle went out immediately.

Hence it should seem, that, when the air contains near $\frac{1}{9}$ its bulk of fixed air, it is unfit for small candles to burn in. Perhaps indeed, if I had used a larger candle and a larger receiver, it might have burnt in a mixture containing a larger proportion of fixed air than this; as I believe that large flaming bodies will burn in a fouler air than small ones. But this is sufficient to shew, that the power, which common air has of keeping fire alive, is very much diminished by a small mixture of fixed air.

This experiment was tried, by setting the candle in a large cistern of water, in such manner that the flame was raised but a little way above the surface; the receiver being inverted full of water into the same cistern. The proper quantity of fixed air was then let up, and the remaining space filled with common air, by raising the receiver gradually out of water; after which, it was immediately whelmed gently over the burning candle.

Experiments on the Quantity of Fixed Air, contained in Alcaline Substances.

Experiment X

The quantity of fixed air contained in marble was found by dissolving some marble in spirit of salt, and finding the loss of weight, which it suffered in effervescence, in the same manner as I found the weight of the inflammable air discharged from metals by solution in acids, except that the cylinder was filled with shreds of filtering paper instead of dry pearl ashes; for pearl ashes would have absorbed the fixed air that passed through them. The weight of the marble dissolved was $311\frac{1}{2}$ grains. The loss of weight in effervescence was $125\frac{1}{2}$ grains. The whole empty space in the bottle and cylinder was about 2700 grain measures: the excess of weight of that quantity of fixed, above an equal quantity of common, air is $1\frac{3}{4}$ grains. Therefore the weight of the fixed air discharged is $127\frac{1}{4}$ grains. The cylinder with the filtering paper was found to have increased $1\frac{3}{4}$ grains in weight during the effervescence. The empty space in the cylinder was about 1160 grain measures: the excess of weight of which quantity of fixed air above an equal bulk of common air is $\frac{3}{4}$ grains. Therefore the quantity of moisture condensed in the filtering paper is one grain, or about $\frac{1}{125}$ part of the weight of the air discharged.

As water has been already shewn to absorb fixed air, it seemed not improbable, but what there might be some fixed air contained in the solution of marble in spirit of salt; in which case the air discharged, during the effervescence, would not be the whole of the fixed air in the marble. In order to see whether this was the case, I poured some of the solution into lime water. It made scarce any precipitate; which, as the acid was intirely saturated with marble, it would certainly have done if the solution had contained any fixed air. It appears therefore from this experiment, first, that marble contains $\frac{127\frac{1}{4}}{311\frac{1}{2}}=\frac{407}{1000}$ of its weight of fixed air; and secondly, that the quantity of moisture, which flies off along with the fixed air in effervescence, is but trifling; as I imagine that the greatest part of what did fly off must have been condensed in the filtering paper.

By another experiment tried much in the same way, marble was found to contain $\frac{408}{1000}$ of its weight of fixed air.

Experiment XI

Volatile sal ammoniac dissolves with too great rapidity in acids, and makes too violent an effervescence, to allow one to try what quantity of fixed air it contains in the foregoing manner: I therefore made use of the following method.

Three small phials were weighed together in the same scale. The first contained some weak spirit of salt, the second contained some volatile sal ammoniac in moderate sized lumps without powder, corked up to prevent evaporation, and the third, intended for mixing the acid and alcali in, contained only a little water, and was covered with a paper cap, to prevent the small jets of liquor, which are thrown up during the effervescence, from escaping out of the bottle. In order to prevent too violent an effervescence, the acid and alcali were both added by a little at a time, care being taken that the acid should always predominate in the mixture. Care was also taken always to cover the bottle with the paper cap, as soon as any of the acid or alcali were added. As soon as the mixture was finished, the three phials were weighed again; whereby the loss in effervescence was found to be 134 grains. The weight of the volatile salt made use of was 254 grains, and was pretty exactly sufficient to saturate the acid. The solution appeared, by pouring some of it into lime water, to contain scarce any fixed air. Therefore 254 grains of the volatile sal ammoniac contain 134 grains of fixed air, i.e. $\frac{528}{1000}$ of their weight. It appeared from the same experiment, that 1680 grains of the volatile salt saturate as much acid as 1000 grains of marble.

By another experiment, tried with some of the same parcel of volatile salt, it was found to contain $\frac{538}{1000}$ of its weight of fixed air, and 1643 grains of it saturated as much acid as 1000 grains of marble. By a medium, the salt contained $\frac{533}{1000}$ of its weight of fixed air; and 1661 grains of it saturated as much acid as 1000 grains of marble.

One thousand grains of marble were found to contain $407\frac{1}{2}$ grains of air, and 1661 grains of volatile sal ammoniac contain 885 grains. Therefore this parcel of volatile sal ammoniac contains more fixed air, in proportion to the quantity of acid that it can saturate, than marble does, in the proportion of 885 to $407\frac{1}{2}$, or of 217 to 100.

N.B. It is not unlikely, that the quantity of fixed air may be found to differ considerably in different parcels of volatile sal ammoniac; so that any one, who was to repeat these experiments, ought not to be surprized if he was to find the result to differ considerably from that here laid down. The same thing may be said of pearl ashes.

Experiment XII

This serves to account for a remarkable phenomenon, which I formerly met with, on putting a solution of volatile sal ammoniac in water into a solution of chalk in spirit of salt. The earth was precipitated hereby, as might naturally be expected: but what surprized me, was, that it was attended with a considerable effervescence; though I was well assured, that the acid in the solution of chalk was perfectly neutralized. This is very easily accounted for, from the above-mentioned circumstance of volatile sal ammoniac containing more fixed air in proportion to the quantity of acid that it can saturate, than calcareous earths do. For the volatile alcali, by uniting to the acid, was necessarily deprived of its fixed air. Part of this air united to the calcareous earth, which was at the same time separated from the acid; but, as the earth was not able to absorb the whole of the fixed air, the remainder flew off in an elastic form, and thereby produced an effervescence.

Experiment XIII

The same solution of volatile sal ammoniac made no precipitate, when mixed with a solution of Epsom salt; though a mixture thereof with a little spirit of sal ammoniac, made with lime, immediately precipitated the magnesia from the same solution of Epsom salt; as it ought to do according to Dr. Black's account of the affinity of magnesia and volatile alcalies to acids. This experiment is not so easily accounted for as the last; but I imagine, that the magnesia is really separated from the acid by the volatile alcali; but that it is soluble in water, when united to so great a proportion of fixed air, as is contained in a portion of volatile sal ammoniac, sufficient to saturate the same quantity of acid. The reason, why the mixture of the solution of volatile sal ammoniac, with the spirits of sal ammoniac made with lime, precipitates the magnesia from the Epsom salt, is that, as the spirits made with lime contain no fixed air, the mixture of these spirits with the solution of volatile sal ammoniac contains less air in proportion to the quantity of acid which it can saturate, than the solution of volatile sal ammoniac by itself does.

Volatile sal ammoniac requires a great deal of water to dissolve it, and the solution has not near so strong a smell as the spirits of sal ammoniac made with fixed alcali; the reason of which is, that the latter contain much less fixed air. But volatile sal ammoniac dissolves in considerable quantity in weak spirits of sal ammoniac made with lime, and the solution differs in no respect from the spirits made with fixed alcali. This is a convenient way of procuring the mild spirits of sal ammoniac, as those made with fixed alcali are seldom to be met with in the shops.

Experiment XIV

The quantity of fixed air contained in pearl ashes was tried, by mixing a solution of pearl ashes with diluted oil of vitriol, in the same manner as was used for volatile sal ammoniac. As much of the solution was used as contained $328\frac{1}{4}$ grains of dry pearl ashes. The loss of effervescence was 90 grains. The mixture, which was perfectly neutralized, being then added to a sufficient quantity of lime water, in order to see whether it contained any fixed air, a precipitate was made, which being dried weighed $8\frac{1}{2}$ grains. Therefore, if we suppose this precipitate to contain as much fixed air as an equal weight of marble, which I am well assured cannot differ very considerably from the truth, the fixed air therein is $3\frac{1}{2}$ grains, and consequently the air in $328\frac{1}{4}$ grains of the pearl ashes, is $93\frac{1}{2}$ grains, i.e. $\frac{284}{1000}$ of their weight.

By another experiment tried in the same way, they appeared to contain $\frac{287}{1000}$ of their weight of fixed air.

1558 grains of the pearl ashes were found to saturate as much acid as 1000 grains of marble. Therefore this parcel of pearl ashes contains more air in proportion to the quantity of acid that it can saturate, than marble does, in the proportion of 109 to 100.

EXPERIMENT XV

Dr. Black says, that, by exposing a solution of salt of tartar for a long time to the open air, some crystals were formed in it, which seemed to be nothing else than the vegetable alcali united to more than its usual proportion of fixed air. This induced me to try, whether I could not perform the same thing more expeditiously, by furnishing the alcali with fixed air artificially; which I did in the manner represented in Fig. 6: where A represents a wide-mouthed bottle, containing a solution of pearl ashes; B*b* represents a round wooden ring fastened over the mouth of the bottle, and secured with luting; C is a bladder bound tight over the wooden ring. This bladder, being first pressed close together, so as to drive out as much of the included air as possible, was filled with fixed air, by means of the bent tube D; one end of which is fixed into the wooden ring, and the other fastened into the mouth of the bottle E, containing marble and spirit of salt. By this means the fixed air thrown into the bladder mixed with the air in the bottle, and came in contact with the fixed alcali. The fixed air was by degrees absorbed, and crystals were formed on the surface of the fixed alcali, which were thrown to the bottom by shaking the bottle. When the alcali had absorbed as much fixed air as it would readily do, the crystals were taken out and dryed on filtered paper, and the remaining solution evaporated; by which means some more crystals were procured.

N.B. It seemed, as, if not all the air discharged from the marble was of a nature proper to be absorbed by the alcali, but only part of it; for when the alcali had absorbed somewhat more than $\frac{1}{2}$ of the air first thrown into the bladder, it would not absorb any more: but, on pressing the remaining air out of the bladder, and supplying it's place with fresh fixed air, a good deal of this new air was absorbed. I cannot, however, speak positively as to this point; as I am not certain whether the apparatus was perfectly airtight[1].

These crystals do not in the least attract the moisture of the air; as I have kept some, during a whole winter, exposed to the air in a room without a fire, without their growing at all moist or increasing in weight.

Being held over the fire in a glass vessel, they did not melt as many salts do, but rather grew white and calcined.

They dissolve in about four times their weight of water when the weather is temperate, and dissolve in greater quantity in hot water than cold.

[1] Pearl ashes deprived of their fixed air, i.e. sope leys, will absorb the whole of the air discharged from marble; as I know by experience. But yet it is not improbable, but that the same alcali, when near saturated with fixed air, may be able to absorb only some particular part of it. For as it has been already shewn, that part of the air discharged from marble is more soluble in water than the rest; so it is not unlikely, but that part of it may have a greater affinity to fixed alcali, and be absorbed by it in greater quantity than the rest.

It was found, by the same method, that was made use of for the volatile sal ammoniac, that these crystals contain $\frac{423}{1000}$ of their weight of fixed air, and that 2035 grains of them saturate as much acid as 1000 grains of marble. Therefore these crystals contain more air in proportion to the quantity of acid they saturate, than marble does, in the ratio of 211 to 100.

EXPERIMENT XVI

As these crystals contain about as much fixed air in proportion to the quantity of acid, that they can saturate, as volatile sal ammoniac does, it was natural to expect, that they should produce the same effects with a solution of Epsom salt, or a solution of chalk in spirit of salt; as those effects seemed owing only to the great quantity of fixed air contained in volatile sal ammoniac. This was found to be the real case: for a solution of these crystals in five times their weight of water, being dropt into a solution of chalk in spirit of salt, the earth was precipitated, and an effervescence was produced. No precipitate was made on dropping some of the same solution into a solution of Epsom salt, though the mixture was kept upwards of twelve hours. But, upon heating this mixture over the fire, a great deal of air was discharged, and the magnesia was precipitated.

EXPERIMENTS ON FACTITIOUS AIR

PART III

Containing Experiments on the Air, produced by Fermentation and Putrefaction.

Mr. M'Bride has already shewn, that vegetable and animal substances yield fixed air by fermentation and putrefaction. The following experiments were made chiefly with a view of seeing, whether they yield any other sort of air besides that.

EXPERIMENT I

The air produced from brown sugar and water, by fermentation, was caught in an inverted bottle of sope leys in the usual manner, and which is represented in Fig. 1. As the weather was too cold to suffer the sugar and water to ferment freely, the bottle containing it was immersed in water, which, by means of a lamp, was kept constantly at about 80° of heat. The quantity of sugar put into the bottle was 931 grains: it was dissolved in about 6½ times its weight of water, and mixed with 100 grains of yeast, by way of ferment. The empty space left in the fermenting bottle and tube together measured 1920 grains. The mixture fermented freely, and generated a great deal of air, which was forced up in bubbles into the inverted bottle, but was absorbed by the sope leys, as fast as it rose up. It frothed greatly; but none of the froth or liquor ran over. In about ten days, the fermentation seeming almost over, the vessels were separated. The bottle with the fermented liquor was found to weigh 412 grains less than it did, before the fermentation began. As none of the liquor ran over, and as little or no moisture condensed within the bent tube, I think one may be well assured, that the loss of weight was owing intirely to the air forced into the inverted bottle; for the matter discharged, during the fermentation, must have consisted either of air, or of some other substance, changed into vapour: if this last was the case, I think it could hardly have failed, but that great part of those vapours must have condensed in the tube. The air remaining unabsorbed in the inverted bottle of sope leys was measured, and was found to be exactly equal to the empty space left in the bent tube and fermenting bottle. It appears therefore, that there is not the least air of any kind discharged from the sugar and water by fermentation, but what is absorbed by the sope leys, and which may therefore be reasonably supposed to be fixed air. It seems also, that no part of the common air left in the fermenting bottle was absorbed by the fermenting mixture, or suffered any change in its nature from thence: for a small phial being filled with one part of this air, and two of inflammable air; the mixture went off with a bounce, on applying a piece of lighted paper to the mouth, with exactly the same appearances, as far as I could perceive, as when the phial was filled with the same quantities of common and inflammable air.

The sugar used in this experiment was moist, and was found to lose $\frac{228}{1000}$ parts of its weight by drying gently before a fire. Therefore the quantity of dry sugar used was 715 grains; and the weight of the air discharged by fermentation appears to be near 412 grains, i.e. near $\frac{57}{100}$ parts of the weight of the dry sugar in the mixture.

The fermented liquor was found to have intirely lost its sweetness; so that the vinous fermentation seemed to be compleated; but it was not grown at all sour.

Experiment II

The air, discharged from apple-juice by fermentation, was tried exactly in the same manner. The quantity set to ferment was 7060 grains, and was mixed with 100 grains of yeast. Some of the same parcel of apple-juice, being evaporated gently to the consistence of a moderately hard extract, was reduced to $\frac{1}{7}$ of its weight; so that the quantity of extract, in the 7060 grains of juice employed, was 1009 grains. The liquor fermented much faster than the sugar and water. The loss of weight during the fermentation was 384 grains. The air remaining unabsorbed in the inverted bottle of sope leys was lost by accident, so that it could not be measured; but, from the space it took up in the inverted bottle, I think I may be certain that it could not much exceed the empty space in the bent tube and fermenting bottle, if it did at all. Therefore there is no reason to think that the apple-juice, any more than the sugar and water, produced any kind of air during the fermentation, except fixed air. It appears too, that the fixed air was near $\frac{381}{1000}$ of the weight of the extract contained in the apple-juice. The fermented liquor was very sour; so that it had gone beyond the vinous fermentation, and made some progress in the acetous fermentation.

In order to compare more exactly the nature of the air produced from sugar by fermentation, with that produced from marble by solution in acids, I made the three following experiments.

Experiment III

I first tried in what quantity the air from sugar was absorbed by water, and at the same time made a like experiment on the air discharged from marble, by solution in spirit of salt. This was done exactly in the same way as the former experiments of this kind. The result is as follows, beginning with the air from sugar and water.

Air from sugar and water let up = 1000.

Bulk of water let up each time	Bulk of air absorbed each time	Whole bulk of water let up	Whole bulk of air absorbed	Bulk of air remaining	Height of thermometer when observ. was made
375	517	375	517	483	40°
143	164	518	681	319	45
153	164	673	845	154	45
82	103	755	948	52	46

Air from marble let up = 1000.

Bulk of water let up each time	Bulk of air absorbed each time	Whole bulk of water let up	Whole bulk of air absorbed	Bulk of air remaining	Height of thermometer when observ. was made
391	473	391	473	527	40
143	133	534	606	394	45
284	115	818	811	189	45
194	80	1012	891	109	46

The apparatus used in this experiment was suffered to remain in the same situation till summer, when the thermometer stood at 65°. The bulk of the air from sugar, not absorbed by the water, was then found to be 287; so that the matter had remitted 235 parts of air. The bulk of the air from marble not absorbed, was 194; so that 85 parts were remitted; which is therefore a proof, that water absorbs less fixed air in warm weather than cold.

It appears from this experiment, that the air produced from sugar by fermentation, as well as that discharged from marble by solution in acids, consists of substances of different nature: part being absorbed by water in greater quantity than the rest. But, in general, the air from sugar is absorbed in greater quantity than that from marble.

In forcing the air from sugar into the cylindrical glass, no sensible quantity of moisture was found to condense on the surface of the quick-silver, or sides of the glass; which is a proof that no considerable quantity of any thing except air could fly off from the sugar and water in fermentation.

Experiment IV

The specific gravity of the air produced from sugar was found in the same way as that produced from marble. A bladder holding 102 ounce measures, being filled with this kind of air, lost $29\frac{1}{8}$ grains on forcing out the air, the thermometer standing at 62°, and the barometer at $29\frac{1}{2}$ inches. Whence, supposing the outward air during the trial of this experiment to be 826 times lighter than water, as it should be, according to the supposition made use of in the former parts of this paper, the air from sugar should be 554 times lighter than water. Its density therefore appears to be much the same as that of the air contained in marble; as that air appeared to be 511 times lighter than water, by a trial made when the thermometer was at 45°; and 563 times lighter, by another trial when the thermometer was at 65°.

This air seems also to possess the property of extinguishing flame, in much the same degree as that produced from marble; as appears from the following experiment.

Experiment V

A small wax candle burnt 15″ in a receiver filled with $\frac{1}{10}$ of air from sugar, the rest common air.

In a mixture containing $\frac{6}{55}$ or $\frac{1}{9\frac{1}{6}}$, of air from sugar, the rest common air, the candle went out immediately. When the receiver was filled with common air only, the same candle burnt 72″.

The receiver was the same as that used in the former experiment of this kind, and the experiment tried in the same way, except that the air from sugar was first received in an empty bladder, and thence transferred into the inverted bottles of water, in which it was measured: for the air is produced from the sugar so slowly, that, if it had been received in the inverted bottles immediately, it would have been absorbed almost as fast as it was generated.

It appears from these experiments, that the air produced from sugar by fermentation, and in all probability that from all the other sweet juices of vegetables, is of the same kind as that produced from marble by solution in acids, or at least does not differ more from it than the different parts of that air do from each other, and may therefore justly be called fixed air. I now proceed to the air generated by putrefying animal substances.

EXPERIMENT VI

The air produced from gravy broth by putrefaction, was forced into an inverted bottle of sope leys, in the same way as in the former experiment. The quantity of broth used, was 7640 grains, and was found, by evaporating some of the same to the consistence of a dry extract, to contain 163 grains of solid matter. The fermenting bottle was immersed in water kept constantly to the heat of about 96°. In about two days the fermentation seemed intirely over. The liquor smelt very putrid, and was found to have lost $11\frac{1}{2}$ grains of its weight. The sope leys had acquired a brownish colour from the putrid vapours, and a musty smell. The air forced into the inverted bottle, and not absorbed by the sope leys, measured 6280 grains; the air left in the bent tube and fermenting bottle was 1100 grains; almost all of which must have been forced into the inverted bottles: so that this unabsorbed air is a mixture of about one part of common air and $4\frac{7}{10}$ of factitious air.

This air was found to be inflammable; for a small phial being filled with 109 grain measures of it, and 301 of common air, which comes to the same thing as 90 grains of pure factitious air, and 320 of common air, it took fire on applying a piece of lighted paper, and went off with a gentle bounce, of much the same degree of loudness as when the phial was filled with the last mentioned quantities of inflammable air from zinc and common air. When the phial was filled with 297 grains of this air, and 113 of common air, i.e. with 245 of pure factitious air, and 165 of common air, it went off with a gentle bounce on applying the lighted paper; but I think not so loud as when the phial was filled with the last-mentioned quantities of air from zinc and common air.

5500 grain measures of this air, i.e. 4540 of pure factitious air, and 960 of common air, were forced into a piece of ox-gut furnished with a small brass cock, which I find more convenient for trying the specific gravity of small quantities of air, than a bladder: the gut increased $4\frac{1}{2}$ grains in weight on forcing out the air. A mixture of 4540 grains of air from zinc and 960 of common air being then forced into the same gut, it increased $4\frac{3}{4}$ grains on forcing out the air. So that this factitious air should seem to be rather heavier than air from zinc; but the quantity tried was too small to afford any great degree of certainty.

N.B. The weight of 4540 grain measures of inflammable air, is $\frac{58}{100}$ grains, and the weight of the same quantity of common air is $5\frac{7}{10}$ grains.

On the whole it seems that this sort of inflammable air is nearly of the same kind as that produced from metals. It should seem, however, either to be not exactly the same, or else to be mixed with some air heavier than it, and which has in some degree the property of extinguishing flame, like fixed air.

The weight of the inflammable air discharged from the gravy appears to be about one grain, which is but a small part of the loss of weight which it suffered in putrefaction. Part of the remainder, according to Mr. M'Bride's experiments, must have been fixed air. But the colour and smell, communicated to the sope leys, shew, that it must have discharged some other substance besides fixed and inflammable air.

Raw meat also yields inflammable air by putrefaction, but not in near so great a quantity, in proportion to the loss of weight which it suffers, as gravy does. Four ounces of raw meat mixed with water, and treated in the same manner as the gravy, lost about 100 grains in putrefaction; but it yielded hardly more inflammable air than the gravy. This air seemed of the same kind as the former; but, as the experiments were not tried so exactly, they are not set down.

I endeavoured to collect in the same manner the air discharged from bread and water by fermentation, but I could not get it to ferment, or yield any sensible quantity of air; though I added a little putrid gravy by way of ferment.

PHILOSOPHICAL TRANSACTIONS. VOL. 57, 1767, p. 92

Received Dec. 11, 1766

XI. *Experiments on* Rathbone-Place *Water: By the Hon.* Henry Cavendish, *F.R.S.*

Read Feb. 19, 1767

DR. LUCAS has given a short examination of this water in the first part of his treatise of waters. It is the produce of a large spring at the end of Rathbone-Place, and used a few years ago to be raised by an engine for supplying part of the town. The engine is now destroyed; but there is a pump, nearly in the same situation, which yields the same kind of water. It is the water of this pump, which was used in these experiments.

Most waters, though ever so transparent, contain some calcareous earth, which is separated from them by boiling, and which seems to be dissolved in them without being neutralized by any acid, and may therefore not improperly be called their unneutralized earth. The following experiments were made chiefly with a view of enquiring into the cause of the suspension of this earth, for which purpose this water seemed well adapted; as it contains more unneutralized earth than most others.

These experiments were made towards the latter end of September 1765, after a very dry summer; whereby the water was most likely more impregnated with saline and other matters than it usually is.

The water, at the time I used it, looked rather foul to the eye. On exposing some of it for a few days to the open air, a scurf was formed on its surface, which was nothing else but some of the unneutralized earth separated from the water. On dropping into it a solution of corrosive sublimate, it grew cloudy in a few seconds; it quickly became opake, and let fall a sediment. This is a property, which I believe does not take place, in any considerable degree, in most of the London waters.

EXPERIMENT I

494 ounces of this water were distilled in a copper still, till about 150 oz. were drawn off. A good deal of earth was precipitated during the distillation, which being collected and dried, weighed 271 grains. It proved to be entirely a calcareous earth, except a small part, which was

magnesia. This I found in the following manner. A little of this earth, being mixed with spirit of salt, dissolved entirely; which shews it to consist solely of an absorbent earth, but does not shew whether it is a calcareous earth or magnesia. The remainder was saturated with oil of vitriol: a great deal of matter remained undissolved, which, as the earth was shewn to be entirely of the absorbent kind, must have been selenite, or a calcareous earth saturated with the oil of vitriol. The clear liquor strained from off the selenite yielded on evaporation only eighteen grains of solid matter, which proved to be Epsom salt; so that all the earth, except that contained in the eighteen grains of Epsom salt, must have been of the calcareous kind. That contained in the Epsom salt is well known to be magnesia.

The water remaining after distillation, and from which the earth was separated, was evaporated, first in a silver pan, and afterwards in a glass cup, till it was reduced to about three ounces. Not the least earth was precipitated during the evaporation, till it was reduced to a small quantity; there then fell 39 grains, which were entirely selenite: so that all the unneutralized earth in the water was separated during the distillation. The liquor thus evaporated was of a reddish colour, like an infusion of soot.

Many waters contain a good deal of neutral salt composed of the nitrous acid united to a calcareous earth; the most convenient way of ascertaining the quantity of which, is to drop a solution of fixed alcali into the evaporated water, till all the earth is precipitated; whereby this salt is changed into true nitre, and is capable of being crystallized. For this reason, some fixed alcali was dropt into the evaporated water till it made no farther precipitation. The earth precipitated thereby weighed thirty-six grains, and was entirely magnesia. The liquor was then farther evaporated, but no nitre could be made to shoot: being then evaporated to dryness, it weighed 256 grains. It gave not the least signs of containing any nitrous salt, either by putting some of it upon lighted charcoal, or by making a match with a solution of it, but appeared to be a mixture of sea salt and vitriolated tartar, or some other salt composed of the vitriolic acid. As I have heard of no other London water, that has been examined with this view, but what has been found to contain a considerable proportion of nitrous salt, it seems very remarkable that this should be intirely destitute of it. I now proceed to the experiments made on the distilled water.

The distilled water, especially that part of it which came over first, became opake, and let fall a precipitate, on drop[p]ing into it a solution of sugar of lead. It also became opake by the addition of corrosive sublimate, much in the same manner that the plain water did before distillation.

It was found, by dropping into it a little acid of vitriol and committing it to evaporation, to contain a small quantity of volatile alcali; as it left four grains of a brownish salt, which being re-dissolved in water, yielded a smell of volatile alcali on the addition of lime. It is doubtless this volatile alcali, which is the cause of the precipitate, which the distilled water makes with sugar of lead and corrosive sublimate.

What first suggested to me that the distilled water contained a volatile alcali, was the distilling some of it over again in a retort; whereby the first runnings were so much impregnated with volatile alcali, as to turn paper died with the juice of blue flowers, to a green colour, and in some measure to yield a smell of volatile alcali.

In the foregoing experiment, the salt procured from the distilled water was perfectly neutral; so that the quantity of acid employed was certainly not more than sufficient to saturate the alcali, but it may very likely have been less; as in that case the superfluous volatile alcali would have flown off in the evaporation. The following experiment shews pretty nearly the quantity of volatile alcali in the distilled water.

Experiment II

1128 ounces of Rathbone-place water were distilled in the same manner as the former. The distilled water was divided into two parcels, that parcel which came over first weighing 121 ounces, the other 146. A preparatory experiment was first made, in order to form a judgement of the comparative strength of each parcel, and also of the quantity of acid which it would require to saturate them. This was done by dropping sugar of lead into each parcel till it ceased to make a precipitate. It was judged from hence that the first parcel contained about 2½ times as much volatile alcali as an equal quantity of the second. Into 30 ounces of the first parcel, mixed with as much of the second, was then put 43 grains of oil of vitriol, which was supposed to be about ½ more than sufficient to saturate the alcali therein. The mixture was then evaporated. When reduced to a small quantity, it was found to be rather acid: sixteen grains of volatile sal ammoniac were therefore added, which seemed nearly sufficient to neutralize it. Being then evaporated to dryness, it left sixty-six grains of a brownish salt, which dissolved readily in water, leaving only a trifling quantity of brown sediment. A little of this salt was found to make no precipitate on the addition of fixed alcali, and the remainder, being boiled with lime, was converted into selenite; a sure sign that the salt was merely vitriolic ammoniacal salt. The volatile alcaline salt contained in sixty-six grains of vitriolic ammoniacal salt is 58½ grains; from whence deducting sixteen grains, the weight of the volatile sal ammoniac added, it appears that the distilled water used in this experiment contains 42½ grains of volatile salt; and therefore the whole quantity of volatile salt driven over by distillation seems to be about sixty-eight grains, which, as the second parcel was so much weaker than the first, is probably nearly the whole volatile alkali contained in the water.

EXPERIMENT III

Dr. Brownrigg, in a paper printed in the Philosophical Transactions, for the year 1765, shews that a great deal of fixed air is contained in Spa water. This induced me to try whether I could not find any in that of Rathbone-place; which I did by means of the contrivance represented in the drawing.

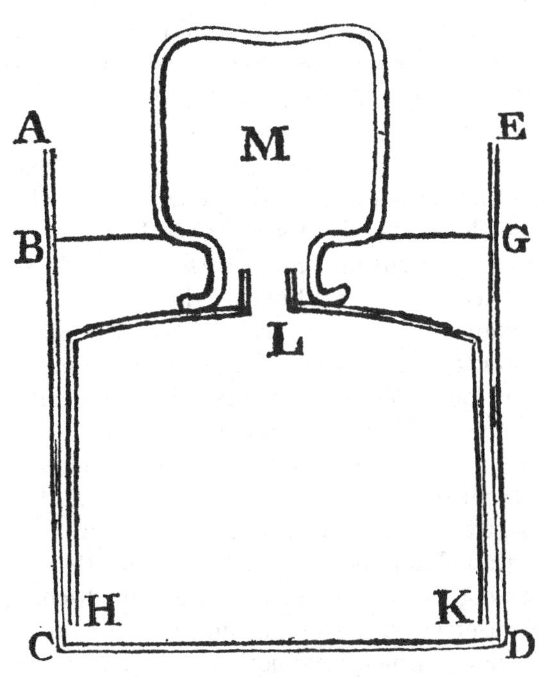

ACDE represents a tin pan, filled with Rathbone-place water as high as BG. HKL is another tin pan, within the first, in the manner of an inverted funnel, and made in such a manner as to leave as little room as possible between that and the sides of the outward vessel. M represents a bottle, full of the same water, inverted over the mouth of the funnel. By this means, as fast as the air is disengaged by heat from the water within the funnel, it must necessarily rise up into the bottle. The Rathbone-place water, put into the vessel, weighed 411 ounces, the funnel held 353 ounces. A bottle full of water being inverted over the mouth of the funnel, as in the figure, the water was heated, and kept boiling about $\frac{1}{4}$ of an hour. As soon as one bottle was filled with air, it was removed by putting a small ladle under its mouth, while under water, and set with its mouth immersed in the same manner in another vessel of water, taking care not to suffer any communication between the included air and the outward air during the removal. At the same time, another bottle full of water was inverted over the mouth of the funnel, in the same manner as the former. It was not easy telling how much air was discharged from the water; as the air in the bottles, when first removed, was hot and expanded; and, before I could be sure it was cold, there was some of it absorbed by the water: but there seemed to be above 75 ounce measures discharged, scarce twenty of which arose before the water began to boil. The water continued discharging air after the experiment was discontinued. In about a day's time, much the greatest part of the air was absorbed, scarce sixteen ounce measures remaining. That which was absorbed appeared to be fixed air, as the water which had absorbed it made a precipitate with lime-water. But, in order to absorb all the fixed air more perfectly, the air which remained not absorbed was transferred into another bottle of water, in the manner described in my first paper on factitious air, page 142 of the preceding volume[1]. This bottle was then set with its mouth immersed in a bottle of sope-leys; after which, by shaking the bottle, the sope-leys was mixed with the included water; whereby the air in the bottle was brought in contact with the sope-leys, which is well-known to absorb fixed air very readily. By this means the air was reduced to 8¾ ounce measures. A small vial being filled with equal quantities of this and inflammable air, and a piece of lighted paper applied to its mouth, it went off with as loud a bounce, as when the same vial was filled with equal quantities of common air and inflammable air. The specific gravity of the remainder was tried by a bladder, in the manner described in the above-mentioned paper; as well as could be judged from so small a quantity, it was just the same as that of common air. From these two circumstances, I think we may fairly conclude that this unabsorbed part was intirely common air; consequently the air discharged from the Rathbone-place water consisted of 8¾ ounces of common air and about 66 of fixed air. The air which was discharged before the water began to boil contained much more common air, than that which was discharged afterwards; that which was discharged towards the latter end seeming to containing scarce any but fixed air.

As so much fixed air is discharged from this water by boiling, it seemed reasonable to suppose, that the distilled water should contain fixed air. I accordingly found it to make a precipitate with lime-water.

Experiment IV

The following experiment shews that the fixed air was not generated during the boiling, but was contained in the water before. Into 30 ounces of Rathbone-place water was poured some lime-water, which immediately made a precipitate. More lime-water was added, till it ceased to make any farther precipitate. It required 20¼ ounces. The precipitated earth being dried weighed 39 grains.

The unneutralized earth contained in 30 ounces of Rathbone-place water is 16½ grains, and the earth contained in 20¼ ounces of lime-water

[1] See p. 78 of this volume.

(as was found by precipitating the earth by volatile sal ammoniac) is 21 grains. Therefore the earth precipitated from the mixture of Rathbone-place water, and lime-water, is about equal to the sum of the weights, of the earth contained in the lime-water, and of the unneutralized earth in the Rathbone-place water; and consequently all the unneutralized earth seems to be precipitated from Rathbone-place water by the addition of a proper quantity of lime-water. But a more convincing proof that this is the case, is that the clear liquor, after the precipitate had subsided, did not deposit any earth on boiling, or become in the least cloudy on the addition of fixed alkali; whereas Rathbone-place water in its natural state becomes opake thereby. It might perhaps be expected, that the clear liquor should still make a precipitate on the addition of fixed alcali, though the unneutralized earth is precipitated; as in all probability there is still a good deal of earth remaining in it in a neutralized state. The reason why it does not, seems to be, that the remaining earth is most likely intirely magnesia; and Epsom salt, when dissolved in a great quantity of water, does not make any precipitate on the addition of fixed alcali.

There is great reason to suppose that the earth precipitated on mixing the Rathbone-place water and lime-water, was very nearly saturated with fixed air, i.e. that it contained very near as much fixed air, as is naturally contained in the same quantity of calcareous earth. If so, 30 ounces of Rathbone-place water contain as much fixed air as 39 grains of calcareous earth; whereas the unneutralized earth, in that quantity of water, is only 16½ grains; so that Rathbone-place water contains near 2⅓ times as much fixed air as is sufficient to saturate the unneutralized earth in it.

It seems likely from hence, that the suspension of the earth in the Rathbone-place water, is owing merely to its being united to more than its natural proportion of fixed air; as we have shewn that this earth is actually united to more than double its natural proportion of fixed air, and also that it is immediately precipitated, either by driving off the superfluous fixed air by heat, or absorbing it by the addition of a proper quantity of lime water.

Calcareous earths, in their natural state, i.e. saturated with fixed air, are totally insoluble in water; but the same earths, entirely deprived of their fixed air, i.e. converted into lime, are in some measure soluble in it; for lime-water is nothing more than a solution of a small quantity of lime in water. It is very remarkable, therefore, that calcareous earths should also be rendered soluble in water, by furnishing them with more than their natural proportion of fixed air, i.e. that they should be rendered soluble, both by depriving them of their fixed air, and by furnishing them with more than their natural quantity of it. Yet, strange as this may appear, the following experiments, I think, shew plainly that it is the real case.

Experiment V

In order to see whether I could suspend a calcareous earth in water, by furnishing it with more than its natural proportion of fixed air, I took 30 ounces of rain water, and divided it into two parts: into one part I put as much spirit of salt, as would dissolve $30\frac{3}{10}$ grains of calcareous earth, and as much of a saturated solution of chalk, in spirit of salt, as contained 20 grains of calcareous earth: into the other part I put as much fixed alcali, as was equivalent to $46\frac{8}{10}$ grains of calcareous earth, i.e. which would saturate as much acid. This alcali was known to contain as much fixed air as 39 grains of calcareous earth. The whole was then mixed together and the bottle immediately stopped. The alcali was before said to be equivalent to $46\frac{8}{10}$ grains of calcareous earth, and was, therefore, sufficient to saturate all the spirit of salt, and also to decompound as much of the solution of chalk as contains $16\frac{1}{2}$ grains of earth. This mixture, therefore, supposing I made no mistake in my calculation, contained $16\frac{1}{2}$ grains of unneutralized earth, with as much fixed air as is contained in 39 grains of calcareous earth; which is the quantity which was found to be in the same quantity of Rathbone-place water. The mixture became turbid on first mixing, but the earth was quickly re-dissolved on shaking, so that the liquor became almost transparent. After standing some time, a slight sediment fell to the bottom, leaving the liquor perfectly transparent. The mixture was kept three or four days stopped up, during which time it remained perfectly clear, without depositing any more sediment. The clear liquor was then poured off from the sediment, and boiled for a few minutes, in a Florence flask; it grew turbid before it began to boil, and discharged a good deal of air; some earth was precipitated during boiling, which being dried weighed 13 grains.

This shews that there was really, at least 13 grains of earth suspended in this mixture, without being neutralized by any acid; the suspension of which could be owing only to its being united to more than its natural proportion of fixed air. But, as a further proof of this, I made the following experiment.

Experiment VI

I took the same quantities of rain water, solution of chalk, spirit of salt, and fixed alcali, as in the last experiment, but mixed them in a different order. The fixed alcali was first dropped into the spirit of salt, and when the effervescence was over, was diluted with $\frac{1}{2}$ the rain water. The solution of chalk was then diluted with the remainder of the rain water, the whole mixed together, and the bottle immediately stoppered, and shook vehemently. A precipitate was immediately formed on mixing, which could not be re-dissolved on shaking.

It must be observed, that, in the first of the two foregoing experiments, all the fixed air contained in the alcali was retained in the mixture, none being lost by effervescence; whereas, in the last experiment, the greatest part of the fixed air was dissipated in the effervescence; no more being retained than what was contained in that portion of the fixed alcali, which was not neutralized by the acid; and consequently the unneutralized earth, in the mixture, contained not much more fixed air than what was sufficient to saturate it. As the latter of these mixtures differed no otherwise from the former, than that it contained less fixed air; the suspension of the earth in the former must necessarily be owing to the fixed air.

In the two foregoing experiments the water contained, besides the unneutralized earth, and fixed air, some sal sylvii, and a little solution of chalk in the marine acid; which, it may be supposed, contributed to the suspension of the earth: but the following experiment shews that a calcareous earth may be suspended in water, without the addition of any other substance than fixed air.

Experiment VII

A bottle full of rain water was inverted into a vessel of rain water, and some fixed air forced up into the bottle, at different times, till the water had absorbed as much fixed air as it would readily do; 11 ounces of this water were mixed with 6½ of lime water. The mixture became turbid on first mixing, but quickly recovered its transparency, on shaking, and has remained so for upwards for a year.

This mixture contains 7 grains of calcareous earth; and, from a subsequent experiment, I guess it to contain as much fixed air, as there is in 14 grains of calcareous earth.

Experiment VIII

Least it should be supposed, that the reason why the earth was not precipitated in the foregoing experiment, was, that it was not furnished with a sufficient quantity of fixed air, the following mixture was made, which contains the same proportion of earth as the former, but a less proportion of fixed air: 4¾ ounces of the above-mentioned water, containing fixed air, were diluted with 6¼ of rain water, and then mixed with 6½ ounces of limewater. A precipitate was immediately made on mixing, which could not be re-dissolved on shaking.

Experiment IX

I made some experiments to find whether the unneutralized earth could be precipitated from other London waters, by the addition of lime water, as well as from Rathbone-place water. It is necessary for this purpose, that the quantity of lime water should be adjusted very exactly; for, if it is too little, it does not precipitate all the unneutralized earth; if it is too great, some of the earth in the lime water remains suspended. For this reason, as I found it almost impossible to adjust the quantity with sufficient exactness, I added such a quantity of lime water, as I was well assured, was more than sufficient to precipitate the whole of the unneutralized earth; and when the precipitate was subsided, decanted off the clear liquor, and exposed it to the open air, till all the lime remaining in the water was precipitated, by attracting fixed air from the atmosphere. The clear liquor was then decanted and evaporated, which is much the most exact way I know of seeing whether any unneutralized earth remains suspended in the water. The result of the experiments was as follows:

200 ounces of water, from a pump in Marlborough-street, were mixed with 38 ounces of lime water. The earth precipitated thereby weighed 38 grains. The clear liquor, exposed to the air, and evaporated in a silver pan till it was reduced to 6 or 7 ounces, deposited no more than 2 or 3 grains of unneutralized earth.

A like quantity of the same pump water, evaporated by itself without the addition of lime water, deposited about 19 grains of unneutralized earth.

200 ounces of water, from a pump in Hanover-square, being mixed with 67 ounces of lime water, the precipitate weighed 93 grains. The clear liquor, treated in the same way as the former, deposited about 2 grains of earth. 200 ounces of the same water, evaporated by itself, deposited 28 grains of earth.

The same quantity of water from a pump in St. Martin's church-yard, being mixed with 82 ounces of lime water, the precipitate weighed 108 grains. The clear liquor deposited scarce any unneutralized earth on evaporation.

The same quantity of water, evaporated by itself, yielded 45 grains of unneutralized earth.

The way, by which I found the quantity of unneutralized earth deposited on evaporation, was, after having decanted the clear liquor, and washed the residuum with rain water, to pour a little spirit of salt into the silver pan, which dissolves all the calcareous earth, but does not corrode the silver. Then, having separated the solution from the insoluble matter, the earth was precipitated by fixed alcali.

In this way of finding the quantity of unneutralized earth, care must be taken to add very little more acid than is necessary to dissolve the unneutralized earth, and to use as little water in washing out the solution as possible; for otherwise a good deal of the selenite, which is deposited in the evaporation of most water[s], will be dissolved; the earth of which will be precipitated by the fixed alcali, and by that means make the quantity of unneutralized earth appear greater than it really is.

It appears from these experiments, that the unneutralized earth is intirely precipitated from these three waters, by the addition of a proper quantity of lime water; as the trifling quantity found to be deposited, on the evaporation of two of them, most likely proceeded only from not exposing the water to the air, long enough for all the lime to be precipitated. So that I think it seems reasonable to conclude, that the unneutralized earth, in all waters, is suspended merely by being united to more than its natural proportion of fixed air.

To return to Rathbone-place water; it appears from the foregoing experiments, that one pint of it, or 7315 grains, contains, first, as much volatile alcali as is equivalent to about $\frac{9}{10}$ grains of volatile sal ammoniac: secondly, $8\frac{4}{10}$ grains of unneutralized earth, a very small part of which is magnesia, the rest a calcareous earth: thirdly, as much fixed air, including that in the unneutralized earth, as is contained in $19\frac{8}{10}$ grains of calcareous earth: fourthly, $1\frac{2}{10}$ of selenite: fifthly, $7\frac{9}{10}$ of a mixture of sea salt, and Epsom salt; and the whole solid contents of 1 pint of the water is $17\frac{1}{2}$ grains.

One pint of water, from the pump in Marlborough-street, contains $1\frac{4}{10}$ grains of unneutralized earth, and as much fixed air as is contained in $2\frac{9}{10}$ grains of calcareous earth.

The same quantity of water, from the pump in Hanover-square, contains $2\frac{1}{10}$ grains of unneutralized earth, with as much fixed air as is contained in $7\frac{2}{10}$ of earth.

The same quantity of water, from St. Martin's Church-yard, contains $3\frac{4}{10}$ grains of unneutralized earth, with as much fixed air as is contained in $8\frac{2}{10}$ of earth.

PHILOSOPHICAL TRANSACTIONS. VOL. 66, 1776, p. 375

XXI. *An Account of the Meteorological Instruments used at the* Royal Society's *House. By the Hon.* Henry Cavendish, *F.R.S.*

Read March 14th, 1776

Of the thermometers, with reflections concerning some precautions necessary to be used in making experiments with those instruments, and in adjusting their fixed points.

THE thermometers are both adjusted to Fahrenheit's scale: that without doors is placed out of a two-pair-of-stairs window, looking to the North, and stands about two or three inches from the wall, that it may be the more exposed to the air, and the less affected by the heat and cold of the house. The situation is tolerably airy, as neither the buildings opposite to it, nor those on each side, are elevated above it in an angle of more than 12°; but as the opposite building is only twenty-five feet distant, perhaps the heat may be a little increased at the time of the afternoon observation by the reflection from thence. In the middle of summer the Sun shines on the wall of the house, against which the thermometer is fixed, for an hour or two before the morning observation, but never shines on the thermometer itself, or that part of the wall close to it, except in the afternoon, long after the time of observing. On the whole, the situation is not altogether such as could be wished, but is the best the house afforded.

The thermometer within doors is intended chiefly for correcting the heights of the barometer, and is therefore placed close to it. The room in which it is kept looks to the North, and has sometimes a fire in it, but not often.

It has been too common a custom, both in making experiments with thermometers and in adjusting their fixed points, to pay no regard to the heat of that part of the quicksilver which is contained in the tube, though this is a circumstance which ought by no means to be disregarded; for a thermometer, dipped into a liquor of the heat of boiling water, will stand at least 2° higher, if it is immersed to such a depth that the quicksilver in the tube is heated to the same degree as that in the ball, than if it is immersed no lower than the freezing point, and the rest of the tube is not much warmer than the air. The only accurate method is, to take care that all parts of the quicksilver should be heated equally. For this reason, in trying the heat of liquors much hotter or colder than the air, the thermometer ought, if possible, to be immersed almost as far as to the top of the column of quicksilver in the tube. As this, however, would frequently be attended with great inconvenience, the observer will often be obliged to content himself with immersing it to a much less depth; but then, as the quicksilver in a great part of the tube will be of a different heat from that in the ball, it will be necessary to apply a correction on that account to the heat shewn by the thermometer; to facilitate which the following table is given, in which the upper horizontal line is the

Diff. of Heat	Degrees not immersed in the liquors							
	50	100	150	200	250	300	350	400
50	,2	,4	,7	,9	1,1	1,3	1,5	1,7
100	,4	,9	1,3	1,8	2,2	2,6	3,0	3,5
150	,7	1,3	2,0	2,6	3,3	3,8	4,6	5,2
200	,9	1,8	2,6	3,5	4,4	5,1	6,1	7,0
250	1,1	2,2	3,3	4,4	5,5	6,4	7,6	8,7
300	1,3	2,6	3,8	5,1	6,4	7,7	9,1	10
350	1,5	3,0	4,6	6,1	7,6	9,1	11	12
400	1,7	3,5	5,2	7,0	8,7	10	12	14
450	2	3,9	5,9	7,8	9,8	12	14	16
500	2,2	4,4	6,5	8,7	11	13	15	17
550	2,4	4,8	7,2	9,6	12	14	17	19
	450	500	550	600	650	700	750	
50	2	2,2	2,4	2,6	2,8	3,1	3,3	
100	3,9	4,4	4,8	5,2	5,7	6,1	6,6	
150	5,9	6,5	7,2	7,9	8,4	9,2	9,8	
200	7,8	8,7	9,6	10	11	12	13	
250	9,8	11	12	13	14	15	16	
300	12	13	14	16	17	18	20	
350	14	15	17	18	20	21	23	
400	16	17	19	21	23	24	26	
450	18	20	22	24	25	27	29	
500	20	22	24	26	28	31	33	
550	22	24	26	29	31	34	36	

length of the column of quicksilver contained in that part of the tube which is not immersed in the liquor expressed in degrees; the first perpendicular column is the supposed difference of heat of the quicksilver in that part of the tube and in the ball; and the corresponding numbers in the table shew how much higher or lower the thermometer stands than it ought to do. The foundation on which the table is computed is, that quicksilver expands one 11500th part of its bulk by each degree of heat.

But as the generality of observers will be apt to neglect this correction, it would be proper to form two sets of divisions on such thermometers as are intended for trying the heat of liquors; one of which should be used when the tube is immersed almost to the top of the column of quicksilver; and the other, when not much more than the ball is immersed; in which last case the observer should be careful, that the tube should be as little heated by the steam of the liquor as possible. It must be observed, however, that the heat of the liquor may be estimated with much more accuracy by the first set of divisions, with the help of the correction, than it can by the second set, as the latter method is just only in one particular heat of the atmosphere, namely, that to which the divisions are adapted; but, if they are adapted to the mean heat of the climate for which the thermometer is intended, the error can never be very great, and, when the liquor is much hotter or colder than the air of that climate ever is, will be much less than if the first set of divisions were used without any correction; but, when the liquor is within the limits of the heat of the atmosphere, greater accuracy will sometimes be obtained by using the first set of divisions than the second, for which reason the latter set should not be continued within those limits. I would willingly have given rules for the construction of this second set of divisions, but am obliged to omit it, as it cannot be done properly without first determining, by experiment, how much the quicksilver in the tube is heated by immersing the ball in hot liquors.

In a spirit thermometer, the error proceeding from the fluid in the tube being not of the same heat as that in the ball, is much greater; as spirits of wine expand much more by heat than quicksilver: for which reason spirit thermometers are not so proper for trying the heat of liquors as those of quicksilver.

Another circumstance which ought to be attended to in adjusting the boiling point of a thermometer is, that the ball should not be immersed deep in the water; for, if it is, the fluid which surrounds it will be compressed by considerably more than the weight of the atmosphere, and will therefore acquire a sensibly greater heat than it would otherwise do. The most convenient vessel I know for adjusting the boiling point is represented in Fig. 1. *ABCD* is the vessel; *AB* the cover, made to take on and off readily; *E* a chimney to carry off the steam; *FG* the thermometer, passed through a hole *Mm* in the cover, and resting in a little bag fastened to the wire *HK*, intended to prevent the ball from being broken by accidentally falling to the bottom. This wire is made so as to be raised higher or lower at pleasure, and must be placed at such a height that the boiling point shall rise very little above the cover. The hole *Mm* is stopped with bits of cork or tow. By this means, as the tube is inclosed in a vessel intirely filled with the steam of boiling water, the quicksilver in it is heated to the same degree as that in the ball; and besides, that part of the tube, on which the boiling point is to be placed, is defended from the vapour, so that it is easy making a mark on the glass with ink. If such a vessel as this is used, the thermometer will be found to stand not sensibly higher when the water boils vehemently than when it boils gently; and if the mouth of the chimney is covered by any light body, in such manner as to leave no more passage for the steam than what is necessary to prevent the body from being blown off by the pressure of the included vapour, the thermometer will stand only half or three quarters of a degree higher, if the ball is immersed a little way in the water, than if it is exposed only to the steam. But if the covering of the chimney is removed, the thermometer will immediately sink several degrees, when the ball is exposed only to the steam, at least if the cover does not fit close; whereas when the ball is immersed in the water, the removal of the covering has scarce any effect upon it. Whence it appears, that the steam of water boiling in a vessel, from which the air is perfectly excluded, is a little but not much cooler than the water itself, but is considerably so if the air has the least admission to the vessel. Perhaps a still more convenient method of adjusting the boiling point would be not to immerse the ball in the water at all, but to expose it only to the steam, as thereby the trouble of keeping the water in the vessel to the right depth would be avoided; and besides, several thermometers might be adjusted at the same time, which cannot be done with proper accuracy when they are immersed in the water, unless the distance of the boiling point from the ball is nearly the same in all of them. At present there is so little uniformity observed in the manner of adjusting thermometers, that the boiling points, in instruments made by our best artists, differ from one another by not less than $2\frac{1}{2}°$; owing partly to a difference in the height of the barometer at which they were adjusted, and partly to the quicksilver in the tube being more heated in the method used by some persons than in that used by others. It is very much to be wished, therefore, that some means were used to establish an uniform method of proceeding; and there are none which seem more proper, or more likely to be effectual, than that the Royal Society should take it into consideration, and recommend that method of proceeding which shall appear to them to be most expedient.

Of the barometer, rain-gage, wind, and hygrometer.

The barometer is of the cistern kind, and the height of the quicksilver is estimated by the top of its convex surface, and not by the edge where it touches the glass, the index being properly adapted for that purpose. This manner of observing appears to me more accurate than the other; because if the quicksilver should adhere less to the tube, or be less convex at one time than another, the edge will, in all probability, be more affected by this inequality than the surface. I prefer the cistern to the syphon barometer, because both the trouble of observing and error of observation are less; as in the latter we are liable to an error in observing both legs. Moreover, the quicksilver can hardly fail of settling truer in the former than in the latter; for the error in the settling of the quicksilver can proceed only from the adhesion of its edge to the sides of the tube; now the latter is affected by the adhesion in two legs, and the former by that in only one: and, besides, as the air has necessarily access to the lower leg of the syphon barometer, the adhesion of the quicksilver in it to the tube will most likely be different, according to the degree of dryness or cleanness of the glass. It is true, as Mr. De Luc observes, that the cistern barometer does not give the true pressure of the atmosphere; the quicksilver in it being a little depressed on the same principle as in capillary tubes. But this does not appear to me a sufficient reason for rejecting the use of them. It is better, I think, where so much nicety is required, to determine, by experiment, how much the quicksilver is depressed in tubes of a given bore, and to allow accordingly.

By some experiments which have been made on this subject by my father Lord Charles Cavendish, the depression appears to be as in the following table:

Inside diameter of tube	Grains of quicksilver in one inch of tube	Depress. of surface of quicksilver	Inside diameter	Grains of quicksilver	Depress. of surface	Inside diameter	Grains of quicksilver	Depress. of surface
,6	972	,005	,35	331	,025	,20	108	,067
,5	675	,007	,30	243	,036	,15	61	,092
,4	432	,015	,25	169	,050	,10	27	1,40

The first column is the inside diameter of the tube, expressed in decimals of an inch; the second is the weight of a quantity of quicksilver sufficient to fill one inch in length of it; and the third is the corresponding depression of the convex surface of the quicksilver in a cistern barometer, whose tube is of that size. The reason of giving the second column is, because the easiest way of ascertaining the inside diameter of the tube is, by finding the quantity of quicksilver sufficient to fill a given length of it. It is needless saying, that the part of the tube, whose diameter is to be measured, is that answering to the upper part of the column of quicksilver; and that the table can be of no use but to those only who observe by the convex surface.

In this barometer, the inside diameter of the tube is about ,25 of an inch, and consequently the depression is ,05; the area of the cistern is near 120 times as great as that of the bore of the tube; so that as the quantity of quicksilver was adjusted when the barometer stood at $29\frac{3}{4}$, the error arising from the alteration of the height of the quicksilver in the cistern can scarce ever amount to so much as $\frac{1}{100}$th of an inch. As the tube appeared to be well filled, it was thought unnecessary to have the quicksilver boiled in it; but that is certainly the surest way of filling a barometer well.

The principal reason of setting down the mean heat of the thermometer within doors, during each month, in the journal of the weather, is this: suppose that any one desires to find the mean height of the barometer in any month, corrected on account of the heat of the quicksilver in the tube; that is, to find what would have been the mean height, if the quicksilver in the tube had been constantly of a certain given heat. To do this it is sufficient to take the mean height of the barometer, and correct that according to the mean heat of the thermometer; the result will be exactly the same as if each observation had been corrected separately, and a mean of the corrected observations taken. For example, suppose it is desired to find what would have been the mean height of the barometer in the month of August 1775, if the quicksilver during that time had been always at 50 degrees of heat: the mean of the observed heights is 29,86 inches, and the mean heat of the thermometer is 65° or 50 + 15. The alteration of the height of the barometer by 15° of heat, according to M. De Luc's rule, is ,047 inch; consequently, the corrected mean height is 29,813.

The vessel which receives the rain is a conical funnel, strengthened at the top by a brass ring, twelve inches in diameter. The sides of the funnel and inner lip of the brass ring are inclined to the horizon, in an angle of above 65°; and the outer lip in an angle of above 50°[1]; which are such degrees of steepness, that there seems no probability either that any rain which falls within the funnel, or on the inner lip of the ring, should dash out, or that any which falls on the outer lip should dash into the funnel. This vessel is placed on some flat leads on the top of the Society's House. It can hardly be screened from any rain by the chimnies, as none of them are elevated above it in an angle of more than 25°; and as it is raised 3½ feet above the roof, there seems no danger of any rain dashing into it by rebounding from the lead.

The strength of the wind is divided in the journal into three degrees; namely, gentle, brisk, and violent or stormy, which are distinguished by the figures 1, 2, and 3. When there is no sensible wind it is distinguished by a cypher.

In the future journals of the weather will be given observations of the hygrometer. The instrument intended to be used is of Mr. Smeaton's construction, and is described in *Phil. Trans.* vol. LXI. p. 198. It is kept in a wooden case, made so as to exclude the rain, but to leave a free passage for the wind, and placed in the open air, where the Sun scarce ever shines on it. The instrument and case are both a present to the

[1] To make what is here said the more intelligible, there is, in Fig. 2, given a vertical section of the funnel, *ABC* and *abc* being the brass ring, *BA* and *ba* the inner lip, and *BC* and *bc* the outer.

Society from Mr. Smeaton. The hygrometer was last adjusted in Dec. 1775, and as the string has now been in use upwards of five years, it is not likely to want re-adjusting soon.

Of the Variation Compass.

In this instrument, the box which holds the needle is not fixed, but turns horizontally on a center, and has an index fastened to it, pointing to a divided arch on the brass frame on which it turns; and the method of observing is to move the box, till a line drawn on it points exactly to the end of the needle; which being done, the angle that the needle makes with the side of the frame is shewn by the index. Fig. 3. is the plan of the instrument; *ABba* is the brass frame, the sides *AB* and *ab* being parallel; *Ee* is a circular plate fastened thereto, on which *CDdc*, the box which holds the needle, turns as on a center; *Nn* is the needle, the pin on which it vibrates, being fixed in the center of the plate *Ee*; *Bb* is the division on the brass frame; and *G* the index fastened to the box *CDdc*, furnished with a vernier division; the division and vernier being constructed so as to shew the angle which the line *Ff* makes with *AB* or *ab*. The instrument is placed in the meridian by the telescope *Mm*, the line of collimation of which is parallel to *AB*, and is pointed to a mark fixed due North of it.

Fig. 4. is a vertical section of the instrument passing along the line *Ff*; *AB* is the brass frame; *CDdc* the box which holds the needle; *Ee* the circular plate on which it turns; *Nn* is the needle; *P* and *p* are small plates of brass fixed to the ends of it, on each of which is drawn a line serving by way of index. These pieces of brass are raised to such a height that their tops are on a level with the point of the pin on which the needle turns. The use of them is, that it is much easier observing this way than when the lines, serving by way of index, are drawn on the needle itself, as by this means the inconvenience proceeding from one kind of vibration in the needle is avoided. *S* and *s* are two brass plates, on each of which is drawn a line to which the index at the end of the needle is to point; there is also a line parallel to these drawn on the bottom of the box; these three lines form the line *Ff* in Fig. 3. *R* is a double microscope intended to assist us in judging when the index *P* points exactly to the line *F*, that is, to the line drawn on the plate *S*. It is placed so, that a wire *Ww* in its *focus* appears to coincide with this line; and in observing, the box is moved till the wire appears also to coincide with the index *P*.

The cap in the center of the needle is made to take on and off readily, and to fit on upon either face; so that we may on occasion observe with the under face of the needle uppermost, as is represented in Fig. 5. But the regular observations are always made with the needle in its upright position, and by the help of the index *P* only; the intention of the other index and of inverting the needle is, to shew whether the line joining the indices P and p, or the line Pp as I shall call it, is parallel to the direction of magnetism in the needle, and thereby to find whether, in the usual method of observing, the index G shews the true angle which the direction of magnetism makes with the side AB. The way of doing this is as follows; having suffered the needle to settle, the observer moves the box by means of the adjusting screw T, till the index P coincides with the line F, and reads off the angle shewn by the vernier. He then moves the box till the other index p coincides with the line f, which, as the pin on which the needle is suspended is fixed to the brass frame, may be done without any danger of altering the position of the needle or making it vibrate, and reads off the angle as before. The mean of these two is the true angle which the line Pp makes with the side AB, supposing the division and vernier to be rightly constructed, even though neither the lines Pp nor Ff should pass through the center of the pin. Having done this, he takes off the cap and inverts the needle, and observes by both indices as before. It is plain, that if the line Pp is parallel to the direction of magnetism in the needle, this mean will agree with the former, supposing that the magnetic variation has not altered between the observations. On the other hand, if it is not parallel to the direction of magnetism, but makes the variation appear greater than it ought to do when the needle is upright, it will make it appear as much less when the needle is inverted; so that the mean of the two abovementioned means is the true angle which the direction of magnetism in the needle makes with the side AB; that is, the true variation of the needle at that time and place, supposing AB to be placed accurately in the meridian. Having thus found the true angle which the direction of magnetism makes with AB, he subtracts that shewn by the index P in the upright position of the needle; the difference is the error of the instrument in the usual manner of observing.

It was by this method that the error of the instrument, at the time of the observations in 1774, was found to be 10′. For example, by a mean of the observations made on Sept. 5. the variation with the needle, in its upright position, was 21·36 by the South end, and 21·27 by the North; with the needle inverted it was 21·19 by the South end, and 21·29 by the North. The mean of all four is 21·28, which is the true variation at that time and place[1], and is 8′ less than that shewn in the upright position of the needle by the South end, which is the end always used in observing; so that by this day's experiment the error of the instrument appeared to be 8′; but by a mean of the observations of this and two other days it came out 10′. Since that time the needle has been altered; and, at the time of the observations in 1775, the error was so small as to be scarcely sensible.

[1] The quantity found by taking a mean of all the four numbers is evidently the same as that got by taking a mean of the two first and of the two last, and taking a mean of those two means.

Great care was taken that the metal, of which this variation compass is composed, should be perfectly free from magnetism. There is a contrivance in it for lifting the needle from off the point, and letting it down gently, to prevent injury in carrying from one room to another. The instrument is constructed nearly on the same plan as some made by the late Dr. Knight. The principal difference is, that in his the pin which carried the needle was not fixed to the lower frame as in this, but to the box; the consequence of which was, that when the needle had settled, and the box was moved to make the index on the needle point to the proper mark, it was again put into vibration, which caused great trouble to the observer. This inconvenience is intirely removed by the present construction. There is no other material difference, except that of the needle being made to invert, and the addition of the telescope. The contrivance of fixing the pin which carries the needle to the lower frame, is taken from an instrument of Lord Charles Cavendish; that of making the needle invert I have seen in some compasses made by Sisson.

There is a very common fault in the agate-caps usually made for needles, which is, that they are not hollowed to a regular concave, but have a little projecting part in the center of the hollow; the consequence of which is, that the point of the pin will not always bear against the same part of the agate, and consequently the needle will not always stand horizontal; but sometimes one end will stand highest, and sometimes the other, which causes a difficulty in observing. There is also another inconvenience attends it when the indices of the needle are on a level with the point of the pin, which is of more consequence; namely, that it causes the two indices not to agree, and consequently makes a sensible error, when only one index is made use of, at least in nice observations: but when the lines, serving by way of index, are drawn on the needle itself, and therefore are nearly on a level with its center of gravity, it can cause very little error. The agate cap, which was first made for this instrument, was of this kind; and was so faulty, that, if no better could have been procured, it would have been necessary either to have drawn the lines serving by way of index on the needle itself, or to have observed by both ends, either of which would have been attended with a considerable increase of trouble to the observer; but Mr. Nairne, the artist who made the instrument, has since ground some himself, which are perfectly free from this fault, the concave surface being of an extremely regular shape and well polished, and also of a very small radius of curvature; which is a matter of considerable consequence, as otherwise the point of the pin will not easily slip sufficiently near to the bottom of the hollow.

Care was taken to place the variation compass in a part of the house where it is as little likely to be affected by the attraction of the iron work as in any that could be found. As it seemed, however, to be not intirely out of the reach of the influence of that metal, I took the following method to examine how much it was influenced thereby. The instrument was removed into a large garden belonging to a house in Marlborough Street, distant from the Society's House about one mile and a quarter towards the West, where there seemed no danger of its being affected by any iron-work. Here it was placed exactly in the meridian, and compared for a few days with a very exact compass, placed in an adjoining room, and kept fixed constantly in the same situation. It was then removed back to the Society's House, and compared again with the same compass. The observations were as follow:

Observations made with the Society's instrument in the garden.

Time		Variation by Society's instrument	Variation by Compass in room	Difference
1775	h ′	° ′	° ′	′
July 21	4 48 V	21 31	21 33	− 2
	5 0	32	35	− 3
	5 26	30	28	+ 2
	5 43	31	32	− 1
	5 48	30	30	0
22	10 45 M	33	33	0
	11 2	29	30	− 1
	11 18	31	29	+ 2
	11 37	31	31	0
	11 55	31	33	− 2
	4 36 V	31	32	− 1
	4 53	27	30	− 3
	5 22	24	26	− 2
	5 54	26	26	0
July 31	11 4 M	21 28	21 32	− 4
	11 20	28	30	− 2
	11 38	30	30	0
	11 57	29	32	− 3
	0 13 V	29	33	− 4
	0 32	30	31	− 1
	2 24	32	35	− 3
	2 54	32	31	+ 1
Aug. 1	10 34 M	26	28	− 2
	3 13 V	32	33	− 1
	4 33	29	29	0
	4 46	29	31	− 2
	5 12	27	29	− 2
	5 35	27	28	− 1
	5 57	28	30	− 2

The instrument being removed back to the Society's house.

	Time	Variation by Society's instrument	Variation by Compass in room	Difference
1775	h ′	° ′	° ′	′
Aug. 2	1 8 V	21 45	21 32	+ 13
	1 10	44	30	+ 14
	1 20	46	29	+ 17
	1 30	47	29	+ 18
	1 40	47	32	+ 15
	1 50	47	31	+ 16
	2 0	47	31	+ 16
Aug. 4	10 50 M	21 47	21 33	+ 14
	11 0	47	34	+ 13
	11 10	47	35	+ 12
	11 20	47	35	+ 12
	11 30	46	35	+ 11
	11 40	47	34	+ 13

By a mean of the observations, the variation shewn by the compass in the room is 1′,3 greater than by the Society's instrument in the garden, and 14′,1 less than by the same instrument placed in its proper situation; so that the variation appears to be 15′,4 greater in that part of the Society's House where the compass is placed, than in the abovementioned garden; and therefore, as there is no likelihood of its being affected by any iron in the latter place, the needle seems to be drawn aside 15′½ towards the N.W. by the iron work of the house and adjacent buildings.

On comparing the observations of the two last years together, the variation appears, after allowing for the error of the instrument, to have been 27′ greater in 1775 than in 1774; though I have been informed by Dr. Heberden, who has made observations of this kind for several years past, that the annual alteration of the variation has, in general, been not more than 10′; and in particular, that the alteration in the last year appears to be only 11½′; so that the great difference observed at the Society's House seems to be owing, not solely to the real alteration in the variation, but partly to some other cause; though what that should be I cannot conceive, unless some change was made in the iron work either of this or the adjoining houses between the two periods; but I do not find that any such change has been made. During the last year, indeed, there have been two large magnets in the house, each consisting of several great bars joined together, being what the late Dr. Knight used for making artificial magnets, and at the time of the observations in 1774 there was only one; but their distance from the compass is above fifty feet: and I am well assured, that in the situation in which they are actually placed, they cannot draw the needle aside more than 3′, and not more than 15′, when the line joining their poles is placed in such a direction as to act with most force[1]. The single magnet in the year 1774 was placed nearly in the same situation and direction that the two were in 1775, so that the difference of their effect in these two years can hardly have been so much as 3′; and therefore, the great apparent alteration of the variation between the two periods cannot have been owing to them. Neither can it have been owing to the fault of the agate cap used in the year 1774, as the error proceeding from thence could hardly be more than 2 or 3′. It is intended that, for the future, the abovementioned magnets shall be kept always in the same situation and direction that they are in at present, and in which they were in 1775.

Of the Dipping-needle.

In this instrument the ends of the axis of the needle roll on horizontal agate planes, a contrivance being applied, by which the needle is at pleasure lifted off from the planes, and let down on them again, in such manner as to be supported always by the same points of the axis resting on the same parts of the agate planes; and the motion with which it is let down is very gradual and without shake. The general form of the instrument, the size and shape of the needle, and the cross used for balancing it, are the same as in the dipping-needle described in *Phil. Trans.* vol. LXII. p. 476. It is also made by the same artist Mr. Nairne.

It may be seen in the Meteorological Journal, that the dip was observed first with the front of the instrument to the West, and then to the East; after which the poles of the needle were reversed, and the dip observed both ways as before. The reason of this is, that the mean of the observed dips, in these four situations, differs very little from the truth, though the needle is not well balanced, and even though a great many other errors are committed in the construction of the instrument; provided the needle is made equally magnetical after the poles are reversed as before[2]; and that the difference of the observed dip, in these four situations, is not very great, as will appear from the following considerations.

First, let Fig. 7 [6] be a front view of the needle; AB a line parallel to the direction of magnetism therein; and CD a perpendicular thereto, meeting

[1] The principle by which this was determined is, that if a magnet is placed near a variation compass, with its poles equi-distant from it, and situated so that each shall act equally oblique to the length of the needle, it can have no tendency to alter the variation; and that the situation in which it alters it most, except when placed nearly North or South of the compass, is when the line joining its poles points almost directly towards the needle. This experiment I tried purposely on the occasion, and found it answer; but, I believe, any one skilled in magnetism would have granted the truth of the position without that precaution.

[2] It is easy to see whether the needle is made equally magnetical after the poles are reversed as before, by counting the number of vibrations which it makes in a minute.

it in the line joining the centers of the cylindrical ends of the axis, or in the axis of motion as we may call it. If the needle was truly balanced, its center of gravity would be in *d*, the intersection of *AB* and *CD*. Suppose now, that the needle is not truly balanced, but that its center of gravity is in *g*; draw *gn* perpendicular to *AB*, cutting it in *m*; and let the parts *mn* and *mg* be equal. When the instrument is turned half-way round, so that the contrary face of the needle is presented towards us, the edge *ADB*, which is now lowest, will become uppermost, and the center of gravity will be in that situation in which the point *n* now is; therefore, the mean between the forces with which the needle is drawn out of its true position in these two situations, in consequence of its not being truly balanced, is accurately the same; and the mean between the two observed dips is very nearly the same, as if the center of gravity was at *m*. But if the center of gravity is at *m*, the dip will be very nearly as much too great in the present state of the needle, as it will be too little when the poles are reversed. Therefore, the mean of the observed dips in these four situations will be very nearly the same as if the needle was truly balanced.

Secondly, if the planes on which the axis rolls are not horizontal, the dip will be very nearly as much greater than it would otherwise be, when one face is turned to the West, as it is less when the other is; for if these planes dip towards the South in one case, they will dip as much towards the North in the other, supposing the levels by which the instrument is set to remain unaltered. Consequently, the mean of the two observations will be very nearly the same as if they were placed truly horizontal.

Thirdly, by the same method of reasoning it appears, that the mean of the two abovementioned observations will be not at all altered, though the line, joining the mark on that end of the needle by which we observe, with the axis of motion, is not parallel to the direction of magnetism in the needle; that is, though the mark does not coincide with the point *A* or *B*, or though the line joining the two divisions of 90° is not perpendicular to the horizon, or though the axis of motion does not pass through the center of the divided circle, provided it is in the same horizontal plane with it. If, indeed, the axis of motion is not in the same horizontal plane with the center of the divided circle, the error proceeding from thence will not be compensated by this method of observing, unless both ends of the needle are made use of. This, however, is of no consequence as it is easy to examine whether they are in the same horizontal plane or not.

But the error which is most difficult to be avoided is, that which proceeds from the ends of the axis being not truly cylindrical. I before said, that the parts of them which rest on the agate planes are always exactly the same. The instrument is so contrived, however, that we may on occasion, by giving the axis a little liberty in the notches by which it is lifted up and down, make those planes bear against a part of the axis distant about $\frac{1}{100}$ or $\frac{1}{50}$th of an inch from their usual point of bearing.

Now, I find, that when the axis is confined so as to have none of this liberty, and when care is taken, by previously making the needle stand at nearly the right dip, that it shall vibrate in very small arches when let down on the planes; that then, if the needle is lifted up and down any number of times, it will commonly settle exactly at the same point each time, at least the difference is so small as to be scarcely sensible; but if it is not so confined, there will often be a difference of 20′ in the dip, according as different parts of the axis rest on the planes, and that though care is taken to free the axis and planes from dust as perfectly as possible, which can be owing only to some irregularity in the axis. Moreover, if the needle vibrates in arches of five or more degrees, when let down on the planes, there will frequently be as great an error in the dip. It is true, that the part of the agate planes, which the axis rests on when the vibrations are stopped, will be a little different according to the point which the needle stood at before it was let down; which will make a small difference in the dip as shewn by the divided circles, when only one end of the needle is observed, though the real dip or inclination of the needle to the horizon is not altered: but this difference is by much too small to be perceived; so that the abovementioned error cannot be owing to this cause. Neither does it seem owing to any irregularity in the surface of the agate planes, for they were ground and polished with great accuracy; but it most likely proceeds from the axis slipping in the large vibrations, so as to make the agate planes bear against a different part of it from what they would otherwise do. I have great reason to think, that this irregularity is not owing either to want of care or skill in the execution, but to the unavoidable imperfection of this kind of work. I imagine too, that this instrument is at least as exact, if not more so, than any which has been yet made.

The following table contains the result of some observations which I made, partly with a view to determine the true dip at this time in a place out of reach of the influence of any iron work, and partly to see how nearly different needles would agree. The instruments were all tried in the same garden in which the variation compass was observed, and all on the 10th, 11th, 13th and 14th days of October, 1775, except that marked *, which was tried on the 15th of the preceding April.

Each of the numbers set down in the above table [see p. 126] is the mean of two observations, the instruments being observed first with the front to the East, then to the West; then a second time to the East, and then again to the West; and in all the observations, except those with the two last instruments, which are of a different construction, care was taken that the needle should vibrate in very small arches when let down on the agate planes. By a mean of all, the true dip at London, at this time, comes out 72° 30′, the different needles all agreeing within 14′, which is a difference considerably less than I should have expected. It appears also, that the

	East	West	Poles reversed East	Poles reversed West	Poles restored to their first situation East	Poles restored to their first situation West	True dip
	° ′	° ′	° ′	° ′	° ′	° ′	° ′
The Society's needle ...	72 32	72 8	72 9	72 40	72 59	71 50	72 23
Another of the same construction, belonging to Mr. Nairne ...	72 56	72 29	71 45	73 21	72 51	72 27	72 37
One of mine on nearly the same construction	72 33	72 22	71 41	73 23	72 34	72 18	72 30
Another needle in the same frame	72 22	72 7	71 40	73 53	72 16	72 30	72 33
A needle of mine, made by Sisson, partly on the same construction as Mr. Lorimer's[1]	*73 1	71 49	71 57	73 0			72 27
Another of Mr. Nairne's on the same construction	73 8	72 0	73 15	71 57			72 35

dipping-needle, in the situation in which it is placed at the Society's House, is not much affected by any iron work, as the dip shewn by it in the garden differs only 7′ from that set down in the journal of the weather.

According to Norman, the inventor of the dipping-needle, the dip at London in the year 1576 was 71° 50′[2]; in 1676 it was 73° 47′, according to Mr. Bond[3]; Mr. Whiston in 1720 made it 75° 10′[4]; Mr. Graham in 1723 made it between 73½ or 75°[5], his different trials varying so much; and at present it appears to be 72° 30′. I do not know how much Mr. Bond's determination is to be depended on, as he does not say by what means he arrived at it; but, I believe, Mr. Whiston's is pretty accurate, for he observed the dip in many parts of the kingdom, and the observations agree well together; so that it is reasonable to suppose, that his instrument was a good one, and that he observed in places where the needle was not much influenced by iron work. The dip, therefore, seems to have been considerably greater about the year 1720, than it was in Norman's time, or is at present: it appears, however, to alter very slowly in comparison of the variation.

[1] See *Phil. Trans.* vol. LXV. p. 79.
[2] New Attractive, c. 4.
[3] Longitude found, p. 65.
[4] Longitude and Latitude found by Dipping-needle, pp. 7, 49, and 94.
[5] *Phil. Trans.* No. 389, p. 332.

Philosophical Transactions. Vol. 73, 1783, p. 106

VIII. *An Account of a new Eudiometer.* *By Mr* Cavendish, *F.R.S.*

Read January 16, 1783

Dr. Priestley's discovery of the method of determining the degree of phlogistication of air by means of nitrous air, has occasioned many instruments to be contrived for the more certain and commodious performance of this experiment; but that invented by the Abbé Fontana is by much the most accurate of any hitherto published. There are many ingenious contrivances in his apparatus for obviating the smaller errors which this experiment is liable to; but the great improvement consists in this, that as the tube is long and narrow, and the orifice of the funnel not much less than the bore of the tube, and the measure is made so as to deliver its contents very quick, the air rises slowly up the tube in one continued column; so that there is time to take the tube off the funnel, and to shake it before the airs come quite in contact, by which means the diminution is much greater and much more certain than it would otherwise be. For instance, if equal measures of nitrous and common air are mixed in this manner, the bulk of the mixture will, in general, be about one measure, whereas, if the airs are suffered to remain in contact about one-fourth of a minute before they are shaken, the bulk of the mixture will be hardly less than one measure and two-tenths, and will be very different according as it is suffered to remain a little more or a little less time before it is shaken. In like manner, if through any fault in the apparatus, the air rises in bubbles, as in that case it is almost impossible to shake the tube soon enough, the diminution is less than it ought to be.

Another great advantage in this manner of mixing is, that thereby the mixture receives its full diminution in the short time during which it is shaken, and is not sensibly altered in bulk after that; whereas, if the airs are suffered to remain some time in contact before they are shaken, they will continue diminishing for many hours.

The reason of the abovementioned differences seems to be, that in the Abbé Fontana's method the water is shaken briskly up and down in the tube while the airs are mixing, whereby each small portion of the nitrous air must be in contact with water, either at the instant it mixes with the common air, or at least immediately after; and it should seem, that when the airs are in contact with water during the mixing, the diminution is much greater and more certain than when there is no water ready to absorb the nitrous acid produced by the mixture. This induced me to try whether the diminution would not be still more certain and regular if one of the two kinds of air was added slowly to the other in small bubbles, while the vessel containing the latter was kept continually shaking. I was not disappointed in my expectations, as, I think, this method is really more accurate than the Abbé Fontana's; and, moreover, in the course of my experiments I had occasion to observe a circumstance which is necessary to be attended to by those who would examine the purity of air with exactness by any kind of eudiometer, besides some others which tend very much to explain many of the phenomena attending the mixture of common and nitrous air.

The apparatus I use is as follows. *A* (Fig. 1) is a cylindrical glass vessel, with brass caps at top and bottom; to the upper cap is fitted a brass cock *B*; the bottom cap is open, but is made to fit close into the brass socket *Dd*, and is fixed in it in the same manner as a bayonet is on a musquet. The socket *Dd* has a small hole *E* in its bottom, and is fastened to the board of my tub by the bent brass *F f G*, in such manner that *b*, the top of the cock, is about half an inch under water; consequently if the vessel *A* is placed in its socket, with any quantity of air in it, and the cock is then opened, the air will run out by the cock, but will do so very slowly, as it can escape no faster than the water can enter by the small hole *E* to supply its place.

Besides this vessel, I have three glass bottles like *M* (Fig. 2) each with a flat brass cap at bottom to make it stand steady, and a ring at top to suspend it by, and also some measures of different sizes such as *B* (Fig. 3); these are of glass with a flat brass cap at bottom and a wooden handle. In using them they are filled with the air wanted to be measured, and then set upon the brass knob *C* fitted upon the board of my tub below the surface of the water, which drives out some of the air, and leaves only the proper quantity. This measure is easier made, and more expeditious in using, than the Abbé Fontana's, and, I believe, is equally accurate; but if it was not it would not signify, as I determine the exact quantity of air used by weight.

There are two different methods of proceeding which I have used; the first is to add the respirable air slowly to the nitrous; and the other, to add the nitrous air in the same manner to the respirable. The first is what I have commonly used, and which I shall first describe. In this method a proper quantity of nitrous air is put into one of the bottles *M*, by means of one of the measures above described, and a proper quantity of respirable is let into the vessel *A*, by first filling it with this air, and then setting it on the knob *C*, as was done by the measure. The vessel *A* is then fixed in the socket, and the bottle *M* placed with its mouth over the cock. Then on opening the cock, the air in the vessel *A* runs slowly in small bubbles into the bottle *M*, which is kept shaking all the time by moving it backwards and forwards horizontally while the mouth still remains over the cock.

Notwithstanding the precautions used by the Abbé Fontana in measuring the quantity of air used, I have sometimes found that method liable to very considerable errors, owing to more water sticking to the sides of the measure and tube at one time than at another: for this reason I determine the quantities of air used and the diminution, by weighing the vessels containing it under water in this manner. From one end of a balance, placed so as to hang over the tub of water, is suspended a forked wire, to each end of which fork is fixed a fine copper wire; and in trying the experiment the vessel *A*, with the respirable air in it, is first weighed, by suspending it from one of these copper wires, in such manner as to remain intirely under water. The bottle *M*, with the proper quantity of nitrous air in it, is then hung on in the same manner to the other wire, and the weight of both together found. The air is then let out of the vessel *A* into the bottle *M*, and the weight of both vessels together found again, by which the diminution of bulk which they suffer on mixing is known. Lastly, the bottle *M* is taken off, and the vessel *A* weighed again by itself, which gives the quantity of respirable air used. It is needless to determine the quantity of nitrous air by weight; because, as the quantity used is always sufficient to produce the full diminution, a small difference in the quantity makes no sensible difference in the diminution[1].

[1] Mr. De Saussure also determines the quantity of air which he uses by weight; but does it by weighing the vessels containing it in air. This method is liable to some inaccuracy, as the air in the vessel is apt to be compressed by putting in the stopper; though, I believe, that, if care is taken to push in the stopper slowly, the error arising from thence is but small. It is also less expeditious than weighing them under water, as some time is necessarily lost in wiping the wet off the vessels; but, on the other hand, it requires less apparatus, which makes it fitter for a portable apparatus as Mr. De Saussure's was. If any gentleman is desirous of adapting this method of determining the quantities to the above described manner of mixing the airs, nothing more is required than to have glass stoppers fitted to the vessel *A* and to the bottle *M*.

It is needless to mention, that in both these methods no sensible error can arise from any difference in the specific gravity of the air; for the thing found by weighing the vessel is the difference of weight of the included air and of an equal bulk of water, which, as no air is less than 500 times lighter than water, is very nearly equal to the weight of a quantity of water, equal in bulk to the included air.

It must be observed, that a common balance is not convenient for weighing the vessels of air under water, without some addition to it; for the lower the vessel of air sinks under the water, the more the air is compressed, which makes the vessel heavier, and thereby causes that end of the beam to preponderate. This makes it necessary either to have the index placed below the beam, as in many assay balances; or by some other means to remove the center of gravity of the beam so much below the center of suspension as to make the balance vibrate, notwithstanding the tendency which the compressibility of the air in the vessels has to prevent it.

In this manner of determining the quantities by weight, care should be taken to proportion the lengths of the copper wires in such manner that the surface of the water in *A* and *M* shall be on the same level when both have the usual quantity of air in them, as otherwise some errors will arise from the air being more compressed in one than in the other. This precaution indeed does not intirely take away the error, as the level of the water in *M* is not the same after the airs are mixed as it was before; but in vessels of the same size as mine, the error arising from thence can never amount to the 500th part of the whole, which is not worth regarding; and indeed if it were much greater, it would be of very little consequence, as it would be always the same in trying the same kind of air.

There are several contrivances which I use, in order to diminish the trouble of weighing the vessels; but I omit them, as the description would take up too much room.

The vessel *A* holds 282 grains of water, and is the quantity which I shall distinguish by the name of one measure. I have three bottles for mixing the airs in, with a measure *B* for the nitrous air adapted to each. The first bottle holds three measures, and the corresponding measure $1\frac{1}{4}$; the second bottle holds six, and the corresponding measure $2\frac{1}{2}$; and the third bottle holds 12, and the corresponding measure 5. The first bottle and measure are used in trying common air, or air not better than that; the two others in trying dephlogisticated air. The quantity of respirable air used, as was said before, is always the same, namely, one measure; consequently, in trying common air I used $1\frac{1}{4}$ measures of nitrous air to one of common; and in trying very pure dephlogisticated air I used five measures of nitrous air to one of the dephlogisticated. I believe there is no air so much dephlogisticated as to require a greater proportion of nitrous than that. The way by which I judge whether the quantity of nitrous air used is sufficient, is by the bulk of the two airs when mixed; for if that is not less than one measure, that is, than the respirable air alone, it is a sign that the quantity of nitrous air is sufficient, or that it is sufficient to produce the full diminution, unless it is very impure.

Though the quantity of respirable air used will be always nearly the same, as being put in by measure; yet it will commonly be not exactly so, for which reason the observed diminution will commonly require some correction: for example, suppose that the observed diminution was 2·353 measures, and that the quantity of respirable air was found to be ·985 of a measure; then the observed diminution must be increased by $\frac{15}{1000}$ of the whole or ·035, in order to have the true diminution, or that which would have been produced if the respirable air used had been exactly one measure; consequently, the true diminution is 2·388.

The method of weighing, described in p. [129], is that which I use in trying air much different in purity from common air; but in trying common air, I use a shorter method, namely, I do not weigh the vessel *A* at all, but only weigh the bottle *M* with the nitrous air in it; then mix the airs, and again weigh the same bottle with the mixture in it, and find the increase of weight. This, added to one measure, is very nearly the true diminution, whether the quantity of common air used was a little more or a little less than one measure. The reason of this is, that as the diminution produced on mixing common and nitrous air is only a little greater than the bulk of the common air, the bulk of the mixture will be very nearly the same, whether the bulk of the common air is a little greater or a little less than one measure: for example, let us first suppose, that the quantity of common air used is exactly one measure, and that the diminution of bulk on mixing is 1·08 of a measure, then must the increase of weight of the bottle *M*, on adding the common air, be ·08 of a measure. Let us now suppose, that the quantity of common air used is 1·02 of a measure, then will the diminution, on adding the common air, be $1{\cdot}08 \times \frac{1{\cdot}02}{1{\cdot}00}$ or 1·1016 of a measure, and consequently the increase of weight of the bottle *M* will be $1{\cdot}1016 - 1{\cdot}02$ or ·0816 of a measure, which is very nearly the same as if the common air used had been exactly one measure.

In the second method of proceeding, or that in which the nitrous air is added to the respirable, I use always the same bottle, namely, that which holds three measures, and use always one measure of respirable air; and in trying common air I use the same vessel *A* as in the first method; but for dephlogisticated air I use one that holds $3\frac{3}{4}$ measures.

In trying the experiment I first weigh the bottle *M* without any air in it, and then weigh it again with the respirable air in it, which gives the quantity of respirable air used. I next put the nitrous air into the vessel *A*, and weigh that and the bottle *M* together, and then having mixed the airs, weigh them again, which gives the diminution.

From what has been just said, it appears, that in this method of proceeding I use a less quantity of nitrous air in trying the same kind of respirable air than in the former; the reason of which is, that the same quantity of nitrous air goes further in phlogisticating a given quantity of respirable air in this than in the former method, as will be shewn further on.

In both these methods I express the test of the air by the diminution which they suffer in mixing; for example, if the diminution on mixing them is two measures and $\frac{353}{1000}$, I call its test 2·353, and so on.

In the first method of proceeding I found, that the diminution was scarce sensibly less when I used one measure of nitrous air than when I used a much greater quantity; so that one measure is sufficient to produce the full diminution. I chuse, however, to use $1\frac{1}{4}$, for fear the nitrous air may be impure; $\frac{7}{8}$ths of a measure of nitrous air produced about $\frac{15}{16}$, and $\frac{3}{4}$ths of a measure about $\frac{7}{9}$ths of the full diminution.

I found also, that there was no sensible difference in the diminution whether the orifice by which the air passed out of the vessel *A* into the bottle *M* was only $\frac{1}{25}$th of an inch in diameter, or whether it was $\frac{1}{8}$th of an inch; that is, whether the air escaped in smaller or larger bubbles. The diminution was rather less when the bottle was shook gently than when briskly; but the difference between shaking it very gently and as briskly as I could was not more than $\frac{1}{100}$dth of a measure. But if it was not shaken at all the diminution was remarkably less, being at first only ·9; in about 3′, indeed, it increased to ·93, and after being shaken for about a minute it increased to ·99; whereas, when the bottle was shaken gently, the diminution was 1·08 at first mixing, and did not increase sensibly after that time. The difference proceeding from the difference of time which the air took up in passing into the bottle was rather greater; namely, in some trials, when it took up 80″ in passing, the diminution was $\frac{5}{100}$dths greater than when it took up only 22″, and about $\frac{2}{100}$dths greater than when it took up 45′; in some other trials, however, the difference was less. It appears, therefore, that the difference arising from the difference of time which the air takes up in passing into the bottle is considerable; but, as with the same hole in the plate *Dd* it will take up always nearly the same time, and as it is easy adjusting the size of the hole, so as to make it take up nearly the time we desire, the error proceeding from thence is but small. The time which it took up in passing in my experiments was usually about 50″

The difference proceeding from the difference of size of the bottle, and the nature of the water made use of is greater; for when I use the small bottle which holds three measures, and fill it with distilled water, the usual diminution in trying common air is 1·08; whereas, if I fill the bottle with water from my tub, the diminution is usually about ·05 less. If I use the bottle which holds twelve measures, filled with distilled water, the diminution is about 1·15; and if I use the same bottle, filled with water from my tub, about 1·08.

The reason of this difference is, that water has a power of absorbing a small quantity of nitrous air; and the more dephlogisticated the water is, the more of this air it can absorb. If the water is of such a nature also as to froth or form bubbles on letting in the common air, the diminution is remarkably less than in other waters.

The following table [p. 133] contains the diminution produced in trying common air in the bottle containing three measures, with several different kinds of water, and also the diminution which the same quantity of nitrous air suffered by being only shook in the same bottle, without the addition of any common air, tried by stopping the mouth of the bottle with my finger, and shaking it briskly for one minute, and afterwards for one minute more.

In general, the diminution was nearly as great with rain water as distilled water; but sometimes I have found rain water froth a good deal, and then the diminution was not much greater than by the water fouled with oak shavings.

Diminution in trying common air	Diminution on shaking nitrous air for: One minute	Two minutes	
1·099	·118	·122	Distilled water
1·049	·083	·088	Water from tub
1·036	·090	·098	Pump water
1·062	·090	·099	Distilled water, in which a few drops of liver of sulphur were kept for a few days
1·045	·052	·056	Distilled water impregnated with nitrous air, by keeping it with about $\frac{1}{4}$ of its bulk of nitrous air for two days, and frequently shaking it
·897	·082	·085	Water fouled by oak shavings. N.B. it frothed very much

This difference in the diminution, according to the nature of the water, is a very great inconvenience, and seems to be the chief cause of uncertainty in trying the purity of air; but it is by no means peculiar to this method, as I have found as great a difference in Fontana's method, according as I have filled the tube with different waters[1]. But it shews plainly, how little all the experiments which have hitherto been made for determining the variations in the purity of the atmosphere can be relied on, as I do not know that any one before has been attentive to the nature of the water he has used, and the difference proceeding from the difference of waters is much greater than any I have yet found in the purity of air.

The best way I know of obviating this inconvenience is to be careful always to use the same kind of water: that which I always use is distilled, as being most certain to be always alike. I should have used rain water, as being easier procured, if it had not been that this water is sometimes apt to froth, which I have never known distilled water do.

As I found that the power with which the distilled water I used absorbed nitrous air was greater at some times than others, which must necessarily make an error in the observation, I was in hopes that, by observing the quantity of nitrous air which the water absorbed in the same manner as in the preceding experiment, together with the heat of the water, as that also seems to affect the experiment, one might be able to correct the observed test, and thereby obviate the error which would otherwise arise from any little difference in the nature of the water employed. With this view I made the following experiment.

I purged some distilled water of its air by boiling, and kept one part

[1] I do not find that it makes much difference in Fontana's method whether the water is disposed to froth or not; but the advantage which it has in that respect over this method is not of much consequence, as it is easy finding water which will not froth.

of it for a week in a bottle along with some dephlogisticated air, and shook it frequently; the other part was treated in the same manner with phlogisticated air. At the end of this time I found, by a mean of three different trials, that the test of common air tried with the first of these waters was 1·139, the diminution which nitrous air suffered by being shook 2′ in it in the usual manner was ·285. The test of the same air tried with the last of these waters was only 1·054, and the diminution of nitrous air only ·090, the heat of the water in the tub and of the distilled waters being 45°. I then raised both the water of the tub and the distilled waters to the heat of 67°, and found that the test of the same air, tried by the first water, was then 1·100, and by the latter 1·044; and that the diminution of nitrous air was ·235 by the first water, and ·089 by the latter.

It should seem from hence, as if the observed test ought to be corrected by subtracting $\frac{4}{10}$ths of the diminution which nitrous air suffers by being shaken in the water, and adding ·002 for every 3° of heat above 0, as the foregoing trials will agree very well together, if they are corrected by this rule, and better than if corrected by any different rule, as will appear by the following table.

	Heat	Diminution of nitrous air	Observed test	Correction for Diminution	Correction for Heat	Corrected test
Former water	45	·285	1·139	·114	·030	1·055
	67	·235	1·100	·094	·045	1·051
Latter water	45	·090	1·054	·036	·030	1·048
	67	·089	1·044	·036	·045,	1·053

Though in all probability this correction will diminish the error proceeding from a difference in the nature of the distilled water employed, yet I have reason to think, that it will by no means entirely take it away; for which reason I do not in general make use of it. In almost all the trials, indeed, in which I have applied the correction, it has come out very nearly the same; which seems to shew, that there was no other difference in the absorbing power of the distilled water I employed, than what proceeded from its difference of heat. The above experiment, however, shews plainly, that distilled water is capable of a very great difference in this respect independent of its heat.

In the second method of proceeding, or that in which the nitrous air is added to the respirable, I found nearly the same difference in the diminution, according as the bottle was shaken briskly or gently, as in the former method: I found also nearly the same difference, or perhaps rather less, according to the nature of the water employed, only it seemed to be of not much consequence whether the water frothed or not; but there seemed to be much less difference in the diminution, according to the time which the air took up in passing into the bottle. The usual diminution on trying common air with different quantities of nitrous air, when distilled water was employed, was as follows:

Common air	Nitrous air	Diminution
1·	·6	·74
	·8	·88
	1·	·89
	1·5	·90

It appears, therefore, that $\frac{8}{10}$ths of a measure of nitrous air is sufficient to produce very nearly the full diminution. I chuse, however, always to use one measure. It appears also, that the diminution is always much less in this method than when the common air is added to the nitrous; as in that method it was before said, that the usual diminution was 1·08. The reason of this is, that when nitrous and common air are mixed together, the nitrous air is robbed of part of its phlogiston, and is thereby turned into phlogisticated nitrous acid, and is absorbed by the water in that state, and besides that, the common air is phlogisticated, and thereby diminished: so that the whole diminution on mixing is equal to the bulk of nitrous air, which is turned into acid, added to the diminution which the common air suffers by being phlogisticated. Now it appears, that when a small quantity of nitrous air comes in contact with a large quantity of common air, it is more completely deprived of its phlogiston, and is absorbed by the water in a more dephlogisticated state than when a small quantity of common air comes in contact with a large quantity of nitrous; consequently, in the second method, where small portions of nitrous air come in contact with a large quantity of common air, the nitrous air is more deprived of its phlogiston, and therefore a less quantity of it is required to phlogisticate the common air than in the first method, where small portions of common air come in contact with a large quantity of nitrous air; so that a less quantity of the nitrous air is absorbed in the second method than in the first. As to the common air, as it is completely phlogisticated in both methods, it most likely suffers an equal diminution in both.

A clear proof that a less quantity of nitrous is required to phlogisticate a given quantity of common air in the second method than in the first, is, that if common air is mixed with a quantity of nitrous air not sufficient to completely phlogisticate it, the mixture will be more phlogisticated if the nitrous air is added slowly to the common, as in the second method, than if the common air is added to the nitrous; and if the nitrous air is added slowly to the common, without being in contact with water, the mixture will be found to be still more phlogisticated than in the second method, where the two airs are in contact with water at the time of mixing.

The following table contains the result of the experiments I have made on this subject.

First method			Second method			Nitrous air added slowly to common without being in contact with water		
Nitrous air	Bulk of mixture	Test	Nitrous air	Bulk of mixture	Test	Nitrous air	Bulk of mixture	Test
·716	·856	·244	·635	·849	·137			
·474	·915	·513	·430	·867	·352	·294	·836	·337
			·280	·930	·599			

The two-first sets of experiments were not tried with the apparatus above described, as that held too small a quantity, but with another upon the same principle. The last set was tried by the apparatus represented in Fig. 4 where *A* is a bottle containing nitrous air, inverted into the tub of water *DE*; *B* is a bottle with a bent glass tube *C* fitted to its mouth. This bottle is filled with common air, without any water, and is first slightly warmed by the hand; the end of the glass tube is then put into the bottle of nitrous air, as in the figure; consequently, as the bottle *B* cools, a little nitrous air runs into it, which, by the common air in it, is deprived of its elasticity, so that more nitrous air runs in to supply its place. By this means the nitrous air is added slowly to the common without coming in contact with water, till the whole of the nitrous air has run out of the bottle *A* into *B*; then, indeed, the water runs through the glass tube into *B*, to supply the vacancy formed by the diminution of the common air.

It appears from the foregoing table, that a quantity of nitrous air, used in the first method, does not phlogisticate common air more than three-fourths of that quantity used in the second way does, and not so much as half that quantity used in the third way: so that we may safely conclude, that it is this circumstance of the nitrous air going further in phlogisticating common air in some circumstances than others, which is the cause that the diminution in trying the purity of air by the nitrous test is so much greater in some methods of mixing them than in others.

From what was said in p. [135] it should seem as if the second method was more exact than the first, as the error proceeding from the air employing more or less time in passing into the bottle was found to be less, and that proceeding from a difference in the water, and from the bottle being shaken more or less strongly was not greater. I, however, have found, that the trials of the same air on the same day have commonly differed more when made in this manner than in the first; for which reason, and because in trying common air the first method takes up the least time, I have commonly used that.

It should be observed, that in trying dephlogisticated air by the first method it is convenient to use different bottles, according to the different purity of the air; and the same air will appear purer, if tried by a larger bottle than by a smaller. For example, if its test, tried by the large bottle, comes out 2·54, it will appear not more than 2·44, if tried by the middle bottle; and, in like manner, if its test by the middle bottle comes out 1·11, it will appear to be about 1·08, if tried by the least bottle; for this reason it is right always to set down which bottle it is tried by.

I think I may confidently assert, that either of the above methods are considerably more accurate than Fontana's, supposing the experiment to be made exactly in his manner, that is, determining the quantities by measure. But, in order to judge which method of mixing the airs is most exact, it was necessary to determine the quantities in his method also by weight, as otherwise it would be uncertain whether my method of mixing the airs is really better than his, or whether the apparent greater exactness proceeds only from the superiority of weighing above measuring: for this reason I made some experiments in which common and nitrous air were mixed in his manner, except that I used only one measure of each, as Dr. Ingen-Housz did, and that the nitrous air was put up first, the true diminution being determined by weight, by first weighing the tube under water with the nitrous air in it, and then adding the common air, and weighing the tube again under water. It was unnecessary, for the reasons given in [pp. 130 and 131], to determine the quantity of either the nitrous or common air by weight. My reason for this variation was, that it afforded a much easier method of determining the quantities by weight, was less trouble, and, I believe, must be at least as exact: for I have always found, that the experiments made with the Abbé Fontana's apparatus, in which I used only one measure of each air, agreed better together than those in which I used two of common, and added the nitrous air by one at a time; and I imagine it can be of no signification whether the nitrous or common air is put in first, as I cannot perceive the diminution to be sensibly greater in one of those ways than the other[1].

From the result of these experiments I am persuaded, that my method of mixing the airs is really rather more accurate than Fontana's, as in trying the same bottle of air six or seven times in my method the different trials would not often differ more than $\frac{1}{200}$dth part, and very seldom more than $\frac{1}{100}$dth; whereas in his there would commonly be a difference of $\frac{1}{100}$dth, and frequently near twice that quantity, though I endeavoured to be as regular as I could in my manner of trying the experiment. My method also certainly requires less dexterity in the operator than his.

[1] It is not extraordinary, that in this method the diminution is just the same whether the common or nitrous air is put up first, notwithstanding that in mine it is very different; since in this method the two airs mix in the same manner whichever is put up first: whereas in mine, the manner in which they mix is very different in those two cases; as in one, small portions of common air come in contact with large portions of the nitrous; and in the other, small portions of nitrous air come in contact with large portions of common air.

It is of much importance towards forming a right judgement of the degree of accuracy to be expected in the nitrous test, to know how much it is affected by a difference in the nitrous air employed. Now it must be observed, that nitrous air may differ in two respects; first, it may vary in purity, that is, in being more or less mixed with phlogisticated or other air; and, secondly, it is possible, that out of two parcels equally pure one may contain more phlogiston than the other. If it differs in the second respect, it will evidently cause an error in the test, in whatever proportion it is mixed with the respirable air; but if it differs only in the first respect, it will hardly cause any sensible error, unless it is more than usually impure, provided care is taken to use such a quantity as is sufficient to produce the full diminution. This has been observed by the Abbé Fontana, and agrees with my own experiments; for the test of common air tried in my usual method, with some nitrous air which had been debased by the mixture of common air, came out only 18 thousandths less than when tried with air of the best quality, though this air was so much debased that the diminution, on mixing two parts of this with five of common, was one-sixth part less than when good nitrous air was employed; which shews, that the error proceeding from the difference of purity of the nitrous air is much less when it is used in the full quantity than in a smaller proportion; and also shews, that if it is used in the full quantity it can hardly cause any sensible error, unless it is more impure than usual. One does not easily see, indeed, why it should cause any error; for no reason appears why the mixture of phlogisticated or other air, not absorbable by water, and not affected by respirable air, should prevent the nitrous air from diminishing and being diminished by the respirable air in just the same manner that it would otherwise be. It must be observed, however, that if the nitrous air is mixed with fixed air, it will cause an error, as part of the fixed air will be absorbed by the water while the test is trying; for which reason care should be taken that the nitrous air should not be much mixed with this substance, which it will hardly be, unless either the metal it is procured from is covered with rust; or unless the water in which it is received contains much calcareous earth suspended by fixed air, as in that case, if any of the nitrous acid comes over with the air, it will dissolve the calcareous earth, and separate some fixed air.

In order to see whether it is possible for nitrous air to differ in the second respect, I procured some from quicksilver, copper, brass, and iron, and observed the test of the same parcel of common air with them, on the same day, making four trials with each, when the difference between the tests tried with the three first kinds of air was not greater than might proceed from the error of the experiment; but those tried with the air from iron were $\frac{15}{1000}$ths greater than the rest. I then took the test of some more common air with them in the same manner, only using four parts of common to one of nitrous air, when the tests tried with the air from iron came out smaller than the rest by not less than $\frac{130}{1000}$ths. It should seem, therefore, from these experiments, that the nitrous air procured from iron, besides being much more impure than the others, differs from them also in the second respect; that is, that the pure nitrous air in it contains rather less phlogiston than that in the others: whence it happens, that a greater quantity is necessary to phlogisticate a given portion of common air, and consequently that the diminution is greater when a sufficient quantity of it is used, though with a less proportion the diminution is much less than with other nitrous air, on account of its greater impurity. As for the air procured from the three other substances, I cannot be sure that there is any difference between them. The nitrous air I always use is made from copper, as it is procured with less trouble than from quicksilver, and I have no reason to think it more likely to vary in its quality.

During the last half of the year 1781, I tried the air of near 60 different days, in order to find whether it was sensibly more phlogisticated at one time than another; but found no difference that I could be sure of, though the wind and weather on those days were very various; some of them being very fair and clear, others very wet, and others very foggy.

My way was to fill bottles with glass stoppers every now and then with air from without doors, and preserve them stopped and inverted into water, till I had got seven or eight, and then take their test; and whenever I observed their test, I filled two bottles, one of which was tried that day, and the other was kept till the next time of trying, in order to see how nearly the test of the same air, tried on different days, would agree. The experiment was always made with distilled water, and care was always taken to observe the diminution which nitrous air suffered by being shaken in the water, as mentioned in p. [133]. The heat of the water in the tub also was commonly set down. Most of the bottles were tried only in the first method; but some of them were also tried by the second, and by the method just described in the manner of Fontana.

The result was, that the test of the different bottles tried on the same day never differed more than ·013, and in general not more than half that quantity. The test, indeed, of those tried on different days differed rather more; for taking a mean between the tests of the bottles tried on the same day, there were two of those means which differed ·025 from each other; but, except those two, there were none which differed more than ·013. Though this difference is but small, yet as each of these means is the mean of seven or eight trials, it is greater than can be expected to proceed from the usual errors of the experiment. This difference also is not much diminished by correcting the observations on account of the heat and absorbing power of the water, according to the rule in p. [134]. This might incline one to think, that the parcels of air examined on some of those days of trial were really more dephlogisticated than the rest; but yet, I believe, that they were not: for whenever there was any considerable difference between the means of two successive days of trial, there was nearly the same difference between the tests of the two bottles of the very same air tried on those two days. For example, the mean of the trials on July 7 was ·016 less than that of those on the 15th of the same month; but then the test of the air caught and tried on the 7th was equally less than that of the air of the same day tried on the 15th; which shews, that this difference between the means of those two days was not owing to the parcels of air tried on the former day being really more dephlogisticated than those tried on the latter, but only to some unperceived difference in the manner of trying the experiment; or else to some unknown difference in the nature of the water or nitrous air employed. A circumstance which seems to shew that it was owing to the first of these two causes is, that it frequently happened, that on those days in which the tests taken in the first method came out greater than usual, those taken in Fontana's manner, or in the second method, did not do so; the trials, however, made in these two methods were too few to determine any thing with certainty. On the whole there is great reason to think, that the air was in reality not sensibly more dephlogisticated on any one of the sixty days on which I tried it than the rest.

The highest test I ever observed was 1·100, the lowest 1·068, the mean 1·082.

I would by all means recommend it to those who desire to compare the air of different places and seasons, to fill bottles with the air of those places, and to try them at the same time and place, rather than to try them at the time they were filled, as all the errors to which this experiment is liable, as well those which proceed from small differences in the manner of trying the experiment, as those which proceed from a difference in the nature of the water and nitrous air, will commonly be much less when the different parcels of air are tried at the same time and place than at different ones; provided only, that air can be kept in this manner a sufficient time without being injured, which I believe it may, if the bottles are pretty large, and care is taken that they, as well as the water used in filling them with air, are perfectly clean. I have tried air kept in the abovementioned manner for upwards of three-quarters of a year in bottles holding about a pint, which I have no reason to think was at all injured; but then I have tried some kept not more than one-third part of that time which seemed to have been a little impaired, though I do not know what it could be owing to, unless it was that the bottles were smaller, namely, holding less than one-fourth of a pint, and that in all of them, except two, which were smaller than the rest, the stopper which, however, fitted in very tight, was tied down by a piece of bladder.

I made some experiments also to try whether the air was sensibly more dephlogisticated at one time of the day than another, but could not find any difference. I also made several trials with a view to examine whether there was any difference between the air of London and the country, by filling bottles with air on the same day, and nearly at the same hour, at Marlborough-street and at Kensington. The result was, that sometimes the air of London appeared rather the purest, and sometimes that of Kensington; but the difference was never more than might proceed from the error of the experiment; and by taking a mean of all, there did not appear to be any difference between them. The number of days compared was 20, and a great part of them taken in winter, when there are a greater number of fires, and on days when there was very little wind to blow away the smoke.

It is very much to be wished, that those gentlemen who make experiments on factitious airs, and have occasion to ascertain their purity by the nitrous test, would reduce their observations to one common scale, as the different instruments employed for that purpose differ so much, that at present it is almost impossible to compare the observations of one person with those of another. This may be done, as there seems to be so very little difference in the purity of common air at different times and places, by assuming common air and perfectly phlogisticated air as fixed points. Thus, if the test of any air is found to be the same as that of a mixture of equal parts of common and phlogisticated air, I would say, that it was half as good as common air; or, for shortness, I would say, that its standard was $\frac{1}{2}$: and, in general, if its test was the same as that of a mixture of one part of common air and x of phlogisticated air, I would say, that its standard was $\frac{1}{1+x}$. In like manner, if one part of this air would bear being mixed with x of phlogisticated air, in order to make its test the same as that of common air, I would say, that it was $1 + x$ times as good as common air, or that its standard was $1 + x$; consequently, if common air, as Mr. Scheele and La Voisier suppose, consists of a mixture of dephlogisticated and phlogisticated air, the standard of any air is in proportion to the quantity of pure dephlogisticated air in it. In order to find what test on the Eudiometer answers to different standards below that of common air, all which is wanted is to mix common and perfectly phlogisticated air in different proportions, and to take the test of those mixtures; but in standards above that of common air, it is necessary to procure some good dephlogisticated air, and to find its standard by trying what proportion of phlogisticated air it must be mixed with, in order to have the same test as common air, and then to mix this dephlogisticated air with different proportions of phlogisticated air, and find the test of those mixtures[1].

[1] The rule for computing the standard of any mixture of dephlogisticated and phlogisticated air is as follows. Suppose the test of a mixture of D parts of de-

On this principle I found the standard answering to different tests on both my Eudiometers, and also on Fontana's, to be as follows:

Standard	Test by first method	Test by second method	Test by Fontana abridged 1	2	3	4	Total diminution
4·8	5·02	3·62	·73	·44	·13	1·02	3·98
3·61	3·72	2·70	·75	·49	1·00	—	3·
2·39	2·55 by large bottle 2·45 by middle bottle	1·87	·76	·96	1·92	—	2·08
1·00	1·11 by middle bottle 1·08 by least bottle	·89	1·00	—	—	—	1·00
·75	·81	·69	1·23	—	—	—	·77
·5	·57	·51	1·45	—	—	—	·55
·25	·32	·31	1·66	—	—	—	·34
·0	·07	·08	1·94	—	—	—	·06

Standard	Test by Fontana's method 1	2	3	4	5	6	7	Total diminution
4·8	1·75	1·43	1·11	·78	·46	·21	1·18	7·82
3·61	1·75	1·46	1·17	·89	1·16	2·13	—	5·87
2·39	1·76	1·50	1·25	2·06	—	—	—	3·94
1·	1·81	2·12	3·12	—	—	—	—	1·88
·75	1·82	2·54	—	—	—	—	—	1·46
·5	1·98	2·94	—	—	—	—	—	1·06
·25	2·42	3·39	—	—	—	—	—	·61
·0	2·91	—	—	—	—	—	—	·09

The phlogisticated air used in these experiments was procured by means of liver of sulphur.

The trials, called Fontana abridged, were made in the Abbé Fontana's manner, except that only one measure of respirable air was used, the nitrous air being added by one measure at a time as usual. The column marked 1 at top is the bulk of the mixture after one measure of nitrous air was added; that marked 2, its bulk after two measures were added, and so on.

It must be observed, that in these experiments a considerable diminution took place in taking the test of the unmixed phlogisticated air, or that whose standard is marked 0 in the table; but, notwithstanding this, the air, as far as I could perceive, was perfectly phlogisticated, the diminution being caused merely by the absorbtion of the nitrous air by the water.

phlogisticated air with P of phlogisticated air is the same as that of common air, then is the standard of the dephlogisticated air $\frac{D+P}{D}$. Let now δ parts of this dephlogisticated air be mixed with ϕ parts of phlogisticated air, the standard of the mixture will be $\frac{D+P}{D} \times \frac{\delta}{\delta + \phi}$.

What shews this to be the case is, that if common and nitrous air are mixed in such proportions as that the nitrous should be predominant, so as to be considerably diminished by the mixture of common air, this mixture will produce as great a diminution with nitrous air as the phlogisticated air used in these experiments; and if plain nitrous air is added to nitrous air, the diminution is still greater. This shews, that a considerable diminution is produced by mixing perfectly phlogisticated air with nitrous air, and also that air may be perfectly phlogisticated by liver of sulphur.

These experiments also shew the necessity of using such a quantity of nitrous air as is sufficient to produce the full diminution, in order to form a proper estimate of the goodness of air; for if the quantity of nitrous air is much less than that, the air you try will appear very little better than air of a much inferior quality. For example, if in taking the test of very good dephlogisticated air, only an equal bulk of nitrous air is used, it will appear very little better than a mixture of equal parts of this and phlogisticated air; and if twice that quantity of nitrous air is used, it will appear very little better than a mixture of three parts of this air with one of phlogisticated. Another great advantage of using the full quantity of nitrous air is, that thereby the error arising from any difference in its purity is very much diminished.

Perfectly phlogisticated air may be conveniently procured by putting some solution of liver of sulphur into a bottle of air well stopped, and shaking it frequently till the air is no longer diminished, which, unless it is shaken very frequently, will take up some days. Care must be taken, however, to loosen the stopper now and then, so as to let in air to supply the place of the diminished air. In order to know when the air is as much diminished as it can be, the best way is, when the air is supposed to be nearly phlogisticated, to place the bottle with its mouth under water, still keeping it stopped, and to loosen the stopper now and then, while under water, so as to let in water to supply the place of the diminished air, by which means the alteration of weight of the bottle shews whether the air is diminished or not. If the solution of liver of sulphur is made by boiling together fixed alkali, lime, and flowers of sulphur, which is the most convenient way of procuring it, the air phlogisticated by it will be perfectly free from fixed air: whether it will be so if the liver of sulphur is made without lime, I am not sure.

A still more convenient way, however, of procuring phlogisticated air is by a mixture of iron filings and sulphur; and, as far as I can perceive, the air procured this way is as completely phlogisticated as that prepared by liver of sulphur.

Where the impurities mixed with the air have any considerable smell, our sense of smelling may be able to discover them, though the quantity is vastly too small to phlogisticate the air in such a degree as to be perceived by the nitrous test, even though those impurities impart their phlogiston to the air very freely. For instance, the great and instantaneous power of nitrous air in phlogisticating common air is well known; and yet ten ounce measures of nitrous air, mixed with the air of a room upwards of twelve feet each way, is sufficient to communicate a strong smell to it, though its effect in phlogisticating the air must be utterly insensible to the nicest Eudiometer; for that quantity of nitrous air is not more than the 140000th part of the air of the room, and therefore can hardly alter its test by more than $\frac{3}{140000}$ or $\frac{1}{47000}$th part. Liver of sulphur also phlogisticates the air very freely, and yet the air of a room will acquire a very strong smell from a quantity of it vastly too small to phlogisticate it in any sensible degree. In like manner it is certain, that putrifying animal and vegetable substances, paint mixed with oil, and flowers, have a great tendency to phlogisticate the air; and yet it has been found, that the air of an house of office, of a fresh painted room, and of a room in which such a number of flowers were kept as to be very disagreeable to many persons, was not sensibly more phlogisticated than common air. There is no reason to suppose from these instances, either that these substances have not much tendency to phlogisticate the air, or that nitrous air is not a true test of its phlogistication, as both these points have been sufficiently proved by experiment; it only shews, that our sense of smelling can, in many cases, perceive infinitely smaller alterations in the purity of the air than can be perceived by the nitrous test, and that in most rooms the air is so frequently changed, that a considerable quantity of phlogisticating materials may be kept in them without sensibly impairing the air. But it must be observed, that the nitrous test shews the degree of phlogistication of air, and that only; whereas our sense of smelling cannot be considered as any test of its phlogistication, as there are many ways of phlogisticating air without imparting much smell to it; and, I believe, there are many strong smelling substances which do not sensibly phlogisticate it.

PHILOSOPHICAL TRANSACTIONS. VOL. 73, 1783, p. 303

XX. *Observations on Mr.* Hutchins's *Experiments for determining the Degree of Cold at which Quicksilver freezes. By* Henry Cavendish, *Esq., F.R.S.*

Read May 1, 1783

THE design of the following paper is to explain some particulars in the apparatus sent by me to Mr. Hutchins, the intention of which does not readily appear; and also to endeavour to shew the cause of some phenomena which occurred in his experiments; and point out the consequences to be drawn from them.

This apparatus was intended to determine the precise degree of cold at which quicksilver freezes: it consisted of a small mercurial thermometer, the bulb of which reached about 2½ inches below the scale, and was inclosed in a glass cylinder swelled at bottom into a ball, which, when used, was filled with quicksilver, so that the bulb of the thermometer was intirely surrounded with it. If this cylinder is immersed in a freezing mixture till great part of the quicksilver in it is frozen, it is evident, that the degree shewn at that time by the inclosed thermometer is the precise point at which mercury freezes; for as in this case the ball of the thermometer must be surrounded for some time with quicksilver, part of which is actually frozen, it seems impossible, that the thermometer should be sensibly above that point; and while any of the quicksilver in the cylinder remains fluid, it is impossible that it should sink sensibly below it. The ball of the thermometer was kept constantly in the middle of the swelled part of the cylinder, without danger of ever touching the sides, by means of some worsted wound round the tube. This worsted also served to prevent the access of the air to the quicksilver in the cylinder, which, if not prevented, would have made it more difficult to have communicated a sufficient degree of cold. The diameter of the bulb of the thermometer was rather less than one-fourth of an inch, that of the swelled part of the cylinder was two-thirds, so that there was nowhere a much less thickness of quicksilver between the ball and cylinder than one-sixth of an inch. The bulb of the thermometer was purposely made as small as it conveniently could, in order to leave a sufficient space between it and the cylinder, without making the swelled part thereof larger than necessary, which would have caused more difficulty in freezing the quicksilver in it. Two of these instruments were sent for fear of accidents.

One of the most striking circumstances in the experiments which have been made for freezing mercury, is the excessively low degree to which the thermometers sunk, and which, if it had proceeded, as was commonly supposed from the freezing mixture having actually produced such a degree of cold, would have been really astonishing. The experiments, however, made at Petersburg afforded the utmost reason to suppose, and Mr. Hutchins's last experiments have put beyond a possibility of doubt, that quicksilver contracts in the act of freezing, or in other words, that it takes up less room in a solid than in a fluid state; and that the very low degree to which the thermometers sunk was owing to this contraction, and not to the intensity of the cold produced: for example, in one of Mr. Hutchins's experiments a mercurial thermometer, placed in the freezing mixture, sunk to 450° below nothing, though the cold of the mixture was never more than − 46°; so that the quicksilver was contracted not less than 404° by the action of freezing.

If a glass of water, with a thermometer in it, is exposed to the cold, the thermometer will remain perfectly stationary from the time the water begins to freeze till it is intirely congealed, and will then begin to sink again. In like manner, if a thermometer is dipped into melted tin or lead, it will remain perfectly stationary, as I know by experience, from the time the metal begins to harden round the edges of the pot till it is all become solid, when it will again begin to descend; and there was no reason to doubt that the same thing would obtain in quicksilver.

From what has been just said it was concluded, that if this apparatus was put into a freezing mixture of a sufficient coldness, the thermometer would immediately sink till the quicksilver in the cylinder began to freeze, and would then continue stationary, supposing the mixture still to keep cold enough, till it was intirely congealed. This stationary height of the thermometer is the point at which mercury freezes, though in order to make the experiment convincing, it was necessary to continue the process till so much of the quicksilver in the cylinder was frozen as to put the fact out of doubt.

If the experiment had been tried with no further precautions, I apprehended that considerable difficulties would have occurred, from want of knowing whether the cold of the mixture was sufficiently great, and when a sufficient quantity of the quicksilver was frozen; for, in the first place, there would be no judging when a sufficient quantity was frozen without taking out the apparatus now and then to examine it, which could not be done without a loss of cold; and what is still worse, if before the experiment was completed the cold of the mixture was so much abated as to become less than that of congealing mercury, the frozen quicksilver would begin to melt, and the operator would have no way of detecting it, but by finding that great part of his labour was undone. For this reason two other mercurial thermometers were sent called *A* and *B* by Mr. Hutchins, the scales of which were of wood, for which reason I shall call them, for shortness, the wooden thermometers, as I shall call the two others the ivory ones, their scales being of that material; they were graduated to about 600° below nothing, and their balls were nearly equal in diameter to the swelled part of the cylinders, in order that the quicksilver in both should cool equally fast; and it was recommended to Mr. Hutchins to put one of these into the freezing mixture along with the apparatus: for then, if the cold of the mixture was sufficient, both thermometers would sink fast till the quicksilver in the cylinder began to freeze, when the ivory thermometer would become stationary, but the wooden one would still continue to sink, on account of the contraction of the quicksilver in its ball by freezing; but if this last thermometer, after having continued to sink for some time after the ivory one had become stationary, ceased at last to descend, it would shew, that the mixture was no longer cold enough to freeze mercury; for as long as that was the case, the wooden thermometer would continue to descend by the freezing of fresh portions of quicksilver in its ball, but would cease to do so as soon as the cold was at all less than that. As I was afraid, however, that the quicksilver might possibly freeze and stick tight in the tube of this thermometer, and prevent its sinking, which would make the cold of the mixture appear too small when in reality it was not, one of these thermometers instead of having a vacuum above the quicksilver as usual, was made with a bulb at top filled with air, in order that the pressure might serve to force down the quicksilver.

If the degree of cold at which mercury freezes had been known, a spirit thermometer would have answered better; but that was the point to be determined.

Another advantage which I expected from the wooden thermometer was, that it would afford a guess when a sufficient quantity of the quicksilver in the cylinder was frozen; for if the cold was continued long enough to make that thermometer sink to near 400° below nothing, I supposed, a very visible portion of the quicksilver would be frozen.

It must be observed, however, that in Mr. Hutchins's experiments the natural cold approached so near to the point of mercurial congelation, and in consequence the freezing mixture retained its cold so long as to make these precautions of not so much use as they would otherwise have been.

As it appeared, from Mr. Hutchins's table of comparison, that these thermometers did not agree well together, they were all examined after they came back, except the ivory thermometer *F*, which was broke before it arrived. This loss, however, is of little consequence, as it appeared from the abovementioned table, that *F* and *G* agreed well together. The boiling and freezing points were first examined in the presence of Sir Joseph Banks, Dr. Blagden, Mr. Hutchins, Mr. Nairne, and myself, when the divisions on the scale answering thereto were found to be as follows:

	Boiling point	Freezing point
A	220·3	29·9
B	218·8	30·9
G	215·3	32

The boiling point was tried in the manner recommended in the report of the Committee of the Royal Society, printed in the *Philosophical Transactions* for the year 1777, and allowance made, as there directed, for the height of the barometer at that time. In fixing the freezing point also allowance was made for the temperature of the room in which it was tried.

The great difference in the position of the boiling point on these thermometers seems owing only to care not having been taken to keep the quicksilver in the tube of the same heat as that in the ball, which is a circumstance that was very little attended to when they were made; and I am afraid is not so much observed at present as it ought to be, and which in *A* and *B*, whose tubes contained upwards of 900° of quicksilver, caused an excessively great error, and much more than it did in *G*, which contained fewer degrees in its tube.

In order to see whether the inequalities of the bore of the tube were properly allowed for, a column of quicksilver, about 100° long, was separated from the rest; and it was examined, whether its length comprehended the same number of degrees on the scale in different parts of the tube; when no sensible error could be found in this respect in *G*, and none worth regarding in *B*. The thermometer *A*, by reason of its being constructed with a bulb filled with air at top, could not be examined in this manner; but there is no reason to think, that it was faulty in this respect.

From what has been said it appears, that 183°·3 on the scale of *G* are equal to only 180° on a thermometer adjusted as recommended by the Committee, and therefore 72° are equal to 70°$\frac{2}{3}$; so that the point of − 40° answers really to − 38°$\frac{2}{3}$; that is, the cold shewn by this thermometer at the temperature of about − 40° is 1°$\frac{1}{3}$ too great. In like manner it appears, that the cold shewn at that temperature by *B* is 4°$\frac{1}{3}$, and by *A* 6°$\frac{1}{3}$, too great.

On the whole, these thermometers seem to have been carefully made, their disagreement being owing only to a faulty manner of adjusting the boiling point, and to not allowing for the temper of the air in settling the degree of freezing; and as these points were examined after they came back, the experiments made with them are just as much to be depended on as if they had been truly adjusted at first.

These instruments were made in the year 1776, and were intended to have been sent to Mr. Hutchins that year, through the hands of the late

Dr. Maty, who promised to recommend the experiment to him; but, by not being got ready time enough to be sent that year, and a mistaken supposition that Mr. Hutchins was to come back the next summer, they were prevented from being sent till 1781; when Sir Joseph Banks was informed by Mr. Wegg, that there was a gentleman at Hudson's Bay who was willing to undertake any experiments of that kind; and that the Hudson's Bay Company would be at the expence of any instruments necessary for the purpose. Then, as Sir Joseph thought the abovementioned apparatus well adapted to the purpose, I gladly embraced the opportunity of sending it. It appears, however, from the letter inserted by Mr. Hutchins, that Dr. Black, without being acquainted with what I had done, recommended nearly the same method of determining the degree of cold at which mercury freezes.

Besides the abovementioned instruments, there were sent to Mr. Hutchins two spirit thermometers and a thermometer marked *C*, made at the expence of the Hudson's Bay Company. The two spirit thermometers were made at the recommendation, and under the inspection of Dr. Blagden, and were of great use, as they serve to ascertain several circumstances relating to the experiments, which could not otherwise have been determined. The intention of the thermometer *C* will be mentioned in the course of this paper.

Before I enter into the examination of Mr. Hutchins's experiments, it will be proper to take notice of a phenomenon which occurs in the freezing of water, and is now found to take place in that of quicksilver, and which occasioned many remarkable appearances in these experiments.

It is well known, that if a vessel of water, with a thermometer in it, is exposed to the cold, the thermometer will sink several degrees below the freezing point, especially if the water is covered up so as to be defended from the wind, and care is taken not to agitate it; and then, on dropping in a bit of ice, or on mere agitation, spiculæ of ice shoot suddenly through the water, and the inclosed thermometer rises quickly to the freezing point where it remains stationary[1].

This shews, that water is capable of being cooled considerably below the freezing point, without any congelation taking place; and that, as soon as by any means a small part of it is made to freeze, the ice spreads rapidly

[1] Though I here say conformably to the common opinion, that mere agitation may set the water a freezing, yet some experiments, lately made by Dr. Blagden, seem to shew, that it has not much, if any, effect of that kind, otherwise than by bringing the water in contact with some substance colder than itself. Though in general also the ice shoots rapidly, and the inclosed thermometer rises very quick; yet I once observed it to rise very slowly, as, to the best of my remembrance, it took up not less than half a minute before it rose to the freezing point; but in this experiment the water was cooled not more than one or two degrees below freezing; and it should seem, that the more the water is cooled below that point, the more rapidly the ice shoots, and the inclosed thermometer rises.

through the remainder of the water. The cause of the rise of the thermometer, when the water begins to freeze, is the circumstance now pretty well known to philosophers, that all, or almost all, bodies by changing from a fluid to a solid state, or from the state of an elastic to that of an unelastic fluid, generate heat; and that cold is produced by the contrary process. This explains all the circumstances of the phenomenon perfectly well; for as soon as any part of the water freezes, heat will be generated thereby in consequence of the abovementioned law, so that the new formed ice and remaining water will be warmed, and must continue to receive heat by the freezing of fresh portions of water, till it is heated exactly to the freezing point, unless the water could become quite solid before a sufficient quantity of heat was generated to raise it to that point, which is not the case; and it is evident, that it cannot be heated above the freezing point, for as soon as it comes thereto, no more water will freeze, and consequently no more heat will be generated.

The reason why the ice spreads all over the water, instead of forming a solid lump in one part, is, that as soon as any small portion of ice is formed, the water in contact with it will be so much warmed as to be prevented from freezing; but the water at a little distance from it will still be below the freezing point, and will consequently begin to freeze.

If it was not for this generation of heat by the act of freezing, whenever a vessel of water, exposed to the cold, was arrived at the freezing point, and began to freeze, the whole would instantly be turned into solid ice; for as the new formed ice is not sensibly colder than water beginning to freeze, it follows, that as soon as all the water in the vessel was cooled to that point, the least addition of cold would convert the whole into ice; whereas it is well known, that though the whole vessel of water is cooled to, or even below, the freezing point, there is a long interval of time between its beginning to freeze and being intirely frozen, during all which time it does not grow at all colder.

In like manner, it is the cold generated by the melting of ice which is the cause of the long time required to thaw ice or snow. It is this also which is the cause of the cold produced by freezing mixtures; for no cold is produced by mixing snow with any substance, unless part of the snow is dissolved.

I formerly found, by adding snow to warm water, and stirring it about till all was melted, that the water was as much cooled as it would have been by the addition of the same quantity of water, rather more than 150° colder than the snow; or, in other words, somewhat more than 150° of cold are generated by the thawing of snow; and there is great reason to think, that just as much heat is produced by the freezing of water. The cold generated was exactly the same whether I used ice or snow[1].

[1] I am informed, that Dr. Black explains the abovementioned phenomena in the same manner; only, instead of using the expression, heat is generated or produced,

I have formerly kept a thermometer in melted tin and lead till they became solid; the thermometer remained perfectly stationary from the time the metal began to harden round the sides of the pot till it was intirely solid; but I could not perceive it to sink at all below that point, and rise up to it when the metal began to harden. It is not unlikely, however, that the great difference of heat between the air and melted metal might prevent this effect from taking place; so that though I did not perceive it in those experiments, it is not unlikely that those metals, as well as water and quicksilver, may bear being cooled a little below the freezing or hardening point (for the hardening of melted metals and freezing of water seems exactly the same process) without beginning to lose their fluidity.

Mr. Hutchins's five first experiments were made with the apparatus, and in the manner above described. In the first experiment the ivory thermometer, inclosed in the cylinder, sunk to $-40°$, where it remained stationary for about half an hour, though the wooden thermometer, placed in the same mixture, kept sinking almost all the while. At the end of that time the apparatus was taken out of the mixture to be examined, and the quicksilver in the cylinder was found frozen. It seems evident, therefore, that the true point at which mercury freezes is 40° below nothing on the thermometer *F*, which was that made use of in the experiment. It cannot be lower than that, for if it was, the thermometer could not have remained so long stationary at that point, while surrounded with freezing quicksilver; and it cannot be higher, as the thermometer could not sink below the freezing point, while much of the quicksilver, with which it was surrounded, remained unfrozen.

To those who have attended to the former part of this paper it is needless saying, that the reason why the wooden thermometer continued sinking so long after the ivory thermometer became stationary is, that as the former was placed in the freezing mixture, the quicksilver in its ball froze, and therefore it continued descending during the greatest part of that half hour, by the continual freezing of fresh portions of quicksilver in its ball, and the contraction occasioned thereby; whereas the latter, which was placed only in freezing quicksilver, did not freeze.

There is a circumstance, however, in this experiment, the reason of which does not so readily appear; namely, on putting back the apparatus

he says, latent heat is evolved or set free; but as this expression relates to an hypothesis depending on the supposition, that the heat of bodies is owing to their containing more or less of a substance called the matter of heat; and as I think Sir Isaac Newton's opinion, that heat consists in the internal motion of the particles of bodies, much the most probable, I chose to use the expression, heat is generated. Mr. Wilke also, in the *Transactions of the Stockholm Academy of Sciences*, explains the phenomena in the same way, and makes use of an hypothesis nearly similar to that of Dr. Black. Dr. Black, as I have been informed, makes the cold produced by the thawing of snow 140°; Mr. Wilke, 130°.

into the freezing mixture, after it was taken out to be examined, the thermometer sunk to − 42°; but in about four or five minutes returned back to − 40°. The like happened on removing the apparatus into a fresh freezing mixture, and it then remained about ten minutes before it returned to − 40°. It seems probable from this, that the quicksilver in the cylinder became intirely frozen about the time that it was first taken out to be examined, and that it then grew 2° colder than the freezing point; and that this degree of cold was not sufficient to make the quicksilver in the inclosed thermometer freeze, since mercury, as was before said, will bear being cooled a little below its freezing point without freezing. What confirms this explanation is, that the spirit thermometers shew that the cold of the mixture was actually much the same as that shewn by the ivory thermometer.

In the second experiment, tried with the same apparatus, the ivory thermometer quickly sunk to − 43°; but, in about half a minute, rose to − 40°, where it remained stationary for upwards of 17′. It appears, therefore, that in this experiment the quicksilver was cooled 3° below the freezing point, without losing its fluidity; it then began to freeze, and the inclosed thermometer immediately rose to − 40°: so that this experiment, besides confirming the former, shews, that quicksilver is capable of being cooled a little below the freezing point without freezing; and that it suddenly rises up to it as soon as it begins to lose its fluidity.

In this experiment the cold was carried far enough to freeze the quicksilver in the ivory thermometer, which was not the case in the former: for after it had remained 17′ stationary at − 40°, it began to sink again, and in about a minute sunk to − 44°½; it then sunk instantaneously to − 92°, and soon after remained fixed for an hour and a quarter at 95° (*sic*) [−95°]; being then left without examination for three-quarters of an hour, the mercury was found to have sunk into the ball, the spirit thermometer shewing at that time that the mixture was rather above the point of freezing, whereas before it had been below it. It appears, therefore, that the quicksilver in the thermometer, after having descended to − 44°½, froze in the tube, and stuck there; but, being by some means loosened, sunk instantly to − 92°, and again stuck tight at − 95°, till at last the mixture rising above the freezing point, the quicksilver in the tube melted, and sunk into the ball, to supply the vacuum formed there by the frozen quicksilver. A similar accident of the quicksilver freezing in the tube of the thermometer, and sticking there, and then melting and sinking into the ball as the weather grew warmer, has been found by Dr. Blagden to have happened to several gentlemen whose thermometers froze by the natural cold of the atmosphere, and with reason caused much perplexity to some of them.

In this experiment the apparatus was not taken out to be examined till the ivory thermometer had sunk to − 95°; it was then found to be frozen solid.

The third experiment was tried while the former was carrying on, and was made by putting the other apparatus, namely, that with the thermometers *G* and *B*, into the first mixture made for the former experiment, and which may consequently be supposed to have lost a great part of its cold. The ivory thermometer quickly sunk to – 43°, where it remained stationary for near 12′. The apparatus being then taken out to be examined, the quicksilver in the cylinder was found fluid, but thick and in grains, like crumbs of bread. The apparatus was then put back into the mixture; and, on observing the thermometer, it was found to have risen to – 40°, where it remained stationary about 40′; being then examined, the quicksilver was found solid.

It appears, therefore, that the cold of the mixture was sufficient to cool the quicksilver in the cylinder about 3° below the point of freezing, but did not make it freeze till, on taking out the apparatus, the agitation suddenly set it a freezing, and produced the appearance described by Mr. Hutchins. This immediately made the inclosed thermometer rise; so that when it was re-placed in the mixture and observed, it stood exactly at the freezing point. It appeared, by the spirit thermometer, that the cold of the mixture, at the time the apparatus was first taken out to be examined, was only 2° below the point of freezing, which agrees very well with this explanation.

This experiment, therefore, affords a fresh confirmation that the point of mercurial congelation is – 40° on these thermometers; and that quicksilver will bear being cooled a little below that point without freezing.

As in these two experiments the quicksilver in the cylinder and ivory thermometer bore being cooled a few degrees below the freezing point without freezing, it is natural to conclude, that the same fluid in the wooden thermometer should do so too; and it may, perhaps, be supposed that, in consequence of it, this thermometer, after having sunk a little below the point of freezing, ought suddenly to have risen up to it, which was not observed. But there is great reason to think, that though the quicksilver in it did bear cooling in this manner, it would not have occasioned any such appearance: for suppose that it is cooled below the freezing point, and then suddenly freezes, its bulk will be increased, on account of the heat generated thereby; but then it will be diminished on account of the contraction in freezing; so that, unless the expansion by the heat generated exceeds the contraction by freezing it will cause no rise in the thermometer. I do not, indeed, know how much the heat generated by freezing in quicksilver is, but in water it is about 150°, and the contraction by freezing is at least as much as its expansion by 400°; so that, unless the heat generated by freezing is two or three times as great in quicksilver as in water, the thermometer ought not to rise on this account.

In the fourth, fifth, sixth, and seventh experiments a new phenomenon occurred, namely, the ivory thermometer sunk a great deal below the freezing point without ever becoming stationary at − 40°. In the fifth experiment, tried with the apparatus *G*, it quickly sunk to − 42°, and then, without remaining stationary at any point, sunk in half a minute to − 72°, and soon after remained fixed at − 79°. While it was at − 79°, the apparatus was twice examined, and the quicksilver found fluid; but being again examined after having been removed into a fresh mixture, it was found solid.

It seems likely from hence, that the quicksilver, in the cylinder was quickly cooled so much below the freezing point as to make that in the inclosed thermometer freeze, though it did not freeze itself. If so, it accounts for the appearances perfectly well; nor does there seem any thing improbable in the explanation, except that it is contrary to what happened in the three first experiments; but the degree to which fluids will bear being cooled below the freezing point without freezing seems to depend on such minute circumstances, that, I think, this forms no objection. It must be observed, that the cold of the mixture appeared by the spirit thermometer to be five or six degrees below the freezing point; so that if the quicksilver in the cylinder was as cold as the mixture, and I have no reason to think it was not, it is not at all extraordinary that the thermometer should have froze; the only thing extraordinary is, that the quicksilver in the cylinder should have borne that cold without freezing.

The same phenomenon occurred in the sixth and seventh experiments, on putting the same apparatus into the freezing mixture.

In the fourth experiment the ivory thermometer sunk quickly to − 42°; but soon after rose half a degree, probably from the cold of the mixture diminishing; it then, after having remained six or seven minutes at those two points, sunk very quick to − 77°. It does not appear, at what time the quicksilver in the cylinder began to freeze, as it was not examined till long after the thermometer had sunk to − 77°, when it was found solid; but from the resemblance of this to the three former experiments, I think it much most likely, that it did not begin to freeze till after the thermometer had sunk to − 77°.

In the fifth experiment the wooden thermometer was partly frozen before it was put into the freezing mixture, and the ivory one was at − 40°. On putting them into the mixture, they both rose; the latter, half a degree; the former, many degrees; which shews that the part of the mixture in which they were placed was rather warmer than the freezing point, though that in which the spirit thermometer was placed was colder; but as there seems nothing to be learnt from this, it is not worth while entering into a detail of the circumstances.

Though these experiments do not serve to shew what the freezing point of quicksilver is, yet they do not at all contradict the conclusion drawn from the three former.

If these experiments only had been made, I should have been inclined to suppose, that quicksilver froze with a less degree of cold in vacuo than in the open air, as the quicksilver in the ivory thermometer was in vacuo, and that in the cylinder was not; but, as in the three former experiments, the event was different, the quicksilver in the cylinder there freezing first, I have no reason to think that this is the case.

Though in the sixth experiment the thermometer in the apparatus *G* froze without the quicksilver with which it was surrounded freezing, yet in trying the apparatus *F* in the same mixture, this did not happen; but, on the contrary, it afforded as striking a proof that the point of freezing quicksilver answers to about – 40° on this thermometer as any of Mr. Hutchins's experiments; for, on taking out the apparatus after it had been two minutes in the mixture, the quicksilver in the cylinder was found frozen solid, the inclosed thermometer standing at 40° or 41° below nothing. After having been exposed for near an hour to the air, which was then very little above the point of freezing quicksilver, only a small quantity of the surface was become fluid; the rest formed a frozen globe round the ball of the thermometer, resembling polished silver, and in 17′ after this only a segment of a globe of frozen quicksilver, with a concavity on the inside, formed by the ball of the thermometer, was observed, the thermometer all this while continuing the same as before, namely, at 40° or 41° below nothing; so that in this experiment the ball of the thermometer was surrounded for more than an hour with quicksilver, which was visibly frozen and slowly melting, and during all which time it continued stationary at 40° or 41° below nothing.

It must be observed, however, that in the first and second experiments, which were both tried with this apparatus, the freezing point came out exactly – 40°, whereas in this it seemed about half a degree lower; the reason of which, in all probability, is, that the tube of this thermometer was not so well fitted to its scale but that it had a little play, which would make the freezing point appear near half a degree higher or lower, according as the tube was pushed up or down.

Though the foregoing experiments leave no reasonable room to doubt, that this is the true point at which quicksilver freezes, yet Mr. Hutchins has, if possible, made this still more evident by his two last experiments; as, in the first of them, he froze some quicksilver in a gally-pot immersed in a freezing mixture, so that the quicksilver was in contact with, and covered by, the snow and spirit of nitre; and in the latter in the open air, by the natural cold of the weather, and then dipping the ball of the thermometer into the unfrozen part, observed what degree it stood at. These experiments agree with the former in shewing the freezing point to be – 40° on the two mercurial thermometers; and also shew what degree on the spirit thermometers answers thereto, namely, 29°¾ or 28°½ on *D*, and 30° on *E*; for in these two experiments the spirit thermometers also were dipped into the frozen quicksilver.

In all the experiments, therefore, tried with the thermometer *G*, the freezing point came out – 40°. In those tried with *F*, it came out either – 40°, or about – 40°½; so that as it appears, from Mr. Hutchins's table of comparison, that *F* stood at a medium a quarter of a degree lower than *G*, the experiments made with that thermometer also shew the freezing point to be – 40° on *G*; and as it appeared from the examination of this thermometer after it came home, that – 40° thereon answers to – 38°⅔, on a thermometer adjusted in the manner recommended by the Committee of the Royal Society, it follows, that all the experiments agree in shewing that the true point at which quicksilver freezes is 38°⅔, or in whole numbers 39° below nothing.

From what has been said it appears, that the point at which quicksilver freezes has been determined by Mr. Hutchins in different ways, all perfectly satisfactory, and all agreeing in the same result. In the three first experiments the thermometer was surrounded by quicksilver, which continued freezing till it became solid. In the sixth experiment the quicksilver with which it was surrounded continued slowly melting till the whole was dissolved; and in both cases the thermometer remained stationary all the while at what we have just said to be the freezing point. In the ninth and tenth experiments, the ball of the thermometer was dipped into quicksilver, previously frozen and beginning to melt, as usually practised in settling the freezing point on thermometers, and agreed in the same result, the quicksilver in the last experiment being frozen by the natural cold of the atmosphere; and in the former, by being immersed in, and in contact with, a freezing mixture; so that this point appears to be determined in as satisfactory a manner as can be desired; and the more so, as it seems impossible that experiments should be made with more care and attention, or more faithfully and circumstantially related than these have been. The second and third experiments also shew, that quicksilver, as well as water, can bear being cooled a little below the freezing point without freezing, and is suddenly heated to that point as soon as it begins to congeal.

On the Contraction of quicksilver in freezing.

All these experiments prove, that quicksilver contracts or diminishes in bulk by freezing; and that the very low degrees to which the thermometers have been made to sink, is owing to this contraction, and not to the cold having been in any degree equal to that shewn by the thermometer. In the fourth experiment the thermometer *A* sunk to – 450°, though it appeared by the spirit thermometers that the cold of the mixture was not more than 5° or 6° below the point of freezing quicksilver. In the first experiment also, it sunk to – 448°, at a time when the cold of the mixture was only 2°½ below that point; so that it appears, that the contraction of quicksilver, by freezing, must be at least equal to its expansion by 404° of heat[1]. This, however, is not the whole contraction which it suffers; for it appears, by an extract which Mr. Hutchins was so good as to give me from a meteorological journal, kept by him at Albany Fort, that his thermometer once sunk to 490° below nothing, though it appeared, by a spirit thermometer, that the cold scarcely exceeded the point of freezing quicksilver. There are two experiments also of Professor Braun, in which the thermometer sunk to 544° and 556° below nothing, which is the greatest descent he ever observed without the ball being cracked. It is not indeed known how cold his mixtures were; but from Mr. Hutchins's, there is great reason to think that they could not be many degrees below − 40°. If so, the contraction which quicksilver suffers in freezing is sometimes not much less than its expansion by 500° or 510° of heat, that is almost $\frac{1}{23}$d of its whole bulk, and in all probability is never much more than that.

It is very likely, however, that the contraction which quicksilver suffers in freezing is no very determinate quantity; for a considerable difference may frequently be observed in the specific gravity of the same piece of metal, cast different times over, and almost all cast metals become heavier by hammering; and it is likely that the same thing may obtain in quicksilver, which is only a metal which melts with a much less degree of heat than the rest. I do not know, indeed, how much this variation can amount to; but, on casting the same piece of tin three times over, I found its density to vary from 7·252 to 7·294, though I have great reason to think that no hollows were left in it, and that only a small part of this difference could proceed from the error of the experiment. This variation of density is as much as is produced in quicksilver by an alteration of 66° of heat; and it is not unlikely, that the descent of a thermometer, on account of the contraction of the quicksilver in its ball by freezing, may vary as much in different trials, though the whole mass of quicksilver is frozen and without any vacuities.

The thermometer marked *C* was intended for trying how much the contraction of quicksilver is; but the experiments made with it were not attended with success, as in the first experiment it did not sink so low as *A* had done, owing, most likely, to the great cold of the weather which froze the quicksilver in the tube; and in the second experiment the ball broke.

On the cold of the freezing mixtures.

The cold produced by mixing spirit of nitre with snow is owing, as was before said, to the melting of the snow. Now, in all probability, there is a

[1] The numbers here given are those shewn by the thermometer without any correction; but if a proper allowance is made for the error of that instrument it will appear, that the true contraction was 25° less than here set down, and from the manner in which thermometers have been usually adjusted, it is likely, that in the following experiment of Mr Hutchins, as well as those of Professor Braun, the true contraction might equally fall short of that shewn by observation.

certain degree of cold in which the spirit of nitre, so far from dissolving snow, will yield out part of its own water, and suffer that to freeze, as is the case with solutions of common salt; so that if the cold of the materials before mixing is equal to this, no additional cold can be produced. If the cold of the materials is less, some increase of cold will be produced; but the total cold will be less than in the former case, since the additional cold cannot be generated without some of the snow being dissolved, and thereby weakening the acid, and making it less able to dissolve more snow; but yet the less the cold of the materials is, the greater will be the additional cold produced. This is conformable to Mr. Hutchins's experiments; for in the fifth experiment, in which the cold of the materials was – 40°, the additional cold produced was only 5°. In the first experiment, in which the cold of the materials was only – 23°, an addition of at least 19° of cold was obtained; and by mixing some of the same spirit of nitre with snow in this climate, when the heat of the materials was + 26°, I have sunk the thermometer to – 29°; so that an addition of 55° of cold was produced.

It is remarkable, that in none of Mr. Hutchins's experiments the cold of the mixture was more than 6° of the spirit thermometer below the point of freezing quicksilver, which is so little that it might incline one to think, that the spirit of nitre used by him was weak. This, however, was not the case, as its specific gravity at 58° of heat was 1·4923. It was able to dissolve $\frac{1}{1·42}$ its weight of marble, and contained very little mixture of the vitriolic or marine acid: as well as I could judge from what experience I have of spirit of nitre, it was as little phlogisticated as acid of that strength usually is.

But, however extraordinary it may at first appear, there is the utmost reason to think, that a rather greater degree of cold would have been obtained if the spirit of nitre had been weaker; for I found, by adding snow gradually to some of this acid, that the addition of a small quantity produced heat instead of cold; and it was not until so much was added as to increase the heat from 28° to 51°, that the addition of more snow began to produce cold; the quantity of snow required for this purpose being pretty exactly one-quarter of the weight of the spirit of nitre, and the heat of the snow and air of the room, as well as of the acid, being 28°. The reason of this is, that a great deal of heat is produced by mixing water with spirit of nitre, and the stronger the spirit is, the greater is the heat produced. Now it appears from this experiment, that before the acid was diluted, the heat produced by its union with the water formed from the melted snow was greater than the cold produced by the melting of the snow; and it was not till it was diluted by the addition of one-quarter of its weight of that substance, that the cold generated by the latter cause began to exceed the heat generated by the former. From what has been said, it is evident, that the cold of a freezing mixture, made with the undiluted acid, cannot be quite so great as that of one made with the same acid, diluted with a quarter of its weight of water, supposing the acid and snow to be both at 28° of heat, and there is no reason to think, that the event will be different if they are colder; for the undiluted acid will not begin to generate cold until so much snow is dissolved as to increase its heat from 28° to 51°, so that no greater cold will be produced than would be obtained by mixing the diluted acid heated to 51° with snow of the heat of 28°. This method of adding snow gradually to an acid is much the best way I know of finding what strength it ought to be of, in order to produce the greatest effect possible.

By means of this acid, diluted in the above-mentioned proportion, I froze the quicksilver in the thermometer called *G* by Mr. Hutchins, on the 26th of last February. I did not, indeed, break the thermometer to examine the state of the quicksilver therein; for as it sunk to − 110° it must certainly have been in part frozen; but immediately took it out, and put the spirit thermometer in its room, in order to find the cold of the mixture. It sunk only to − 30°; but, by making allowance for the spirit in the tube being not so cold as that in the ball, it appears, that if it had not been for this cause it would have sunk to − 35°[1], which is 5° below the point of freezing, and is as great a degree of cold, within 1°, as was produced in any of Mr. Hutchins's experiments.

In this experiment the thermometer *G* sunk very rapidly, and, as far as I could perceive, without stopping at any intermediate point, till it came to the above-mentioned degree of − 110°, where it stuck. The materials used in making the mixture were previously cooled, by means of salt and snow, to near nothing; the temper of the air was between 20° and 25°; the quantity of acid used was $4\frac{1}{4}$ oz.; and the glass in which the mixture was made was surrounded with wool, and placed in a wooden box, to prevent its losing its cold so fast as it would otherwise have done.

Some weeks before this, I made a freezing mixture with some spirit of nitre, much stronger than that used in the foregoing experiment, though not quite so strong as the undiluted acid, in which the cold was less intense by $4°\frac{1}{2}$, as the thermometer *G* sunk to − $40°\frac{1}{2}$. It is true, that the temper of the air was much less cold, namely, 35°; but the spirit of nitre was at least as cold, and the snow not much less so. The experiment was tried in the same vessel and with the same precautions as the former.

[1] As the surface of the freezing mixture answered to − 185° on the tube, there were 155° of spirit in the tube which could hardly be cooled much below the temper of the air, and which must, therefore, be warmer than that in the ball by about 55° of this thermometer, as the heat of the spirit in the ball was before said to be − 35°, and the temper of the air above + 20°. Therefore, the correction must be equal to the expansion of a column of spirits 155° long, by an alteration of heat equal to 55° on this thermometer, which, if 1° on the scale answers to $\frac{1}{1700}$th of the bulk of the spirit, is equal to $\frac{55 \times 155}{1700}$ or 5°.

The cold produced by mixing oil of vitriol, properly diluted with snow, is not so great as that procured by spirit of nitre, though it seems not to differ from it by so much as 8°; for a freezing mixture, prepared with diluted oil of vitriol, whose specific gravity, at 60° of heat, was 1·5642, sunk the thermometer *G* to − 37°, the experiment being tried at the same time, and with the same precautions, as the foregoing. It was previously found, by adding snow gradually to some of this acid, as was done by the spirit of nitre, that it was a little, but not much stronger than it ought to be, in order to produce the greatest effect.

PHILOSOPHICAL TRANSACTIONS. VOL. 74, 1784, p. 119

XIII. *Experiments on Air. By* Henry Cavendish, *Esq., F.R.S. & S.A.*

Read January 15, 1784

THE following experiments were made principally with a view to find out the cause of the diminution which common air is well known to suffer by all the various ways in which it is phlogisticated, and to discover what becomes of the air thus lost or condensed; and as they seem not only to determine this point, but also to throw great light on the constitution and manner of production of dephlogisticated air, I hope they may be not unworthy the acceptance of this society.

Many gentlemen have supposed that fixed air is either generated or separated from atmospheric air by phlogistication, and that the observed diminution is owing to this cause; my first experiments therefore were made in order to ascertain whether any fixed air is really produced thereby. Now, it must be observed, that as all animal and vegetable substances contain fixed air, and yield it by burning, distillation, or putrefaction, nothing can be concluded from experiments in which the air is phlogisticated by them. The only methods I know, which are not liable to objection, are by the calcination of metals, the burning of sulphur or phosphorus, the mixture of nitrous air, and the explosion of inflammable air. Perhaps it may be supposed, that I ought to add to these the electric spark; but I think it much most likely, that the phlogistication of the air, and production of fixed air, in this process, is owing to the burning of some inflammable matter in the apparatus. When the spark is taken from a solution of tournsol, the burning of the tournsol may produce this effect; when it is taken from lime-water, the burning of some foulness adhering to the tube, or perhaps of some inflammable matter contained in the lime, may have the same effect; and when quicksilver or metallic knobs are used, the calcination of them may contribute to the phlogistication of the air, though not to the production of fixed air.

There is no reason to think that any fixed air is produced by the first method of phlogistication. Dr. Priestley never found lime-water to become turbid by the calcination of metals over it[1]: Mr. Lavoisier also found only

[1] *Experiments on Air*, vol. I. p. 137.

a very slight and scarce perceptible turbid appearance, without any precipitation, to take place when lime-water was shaken in a glass vessel full of the air in which lead had been calcined; and even this small diminution of transparency in the lime-water might very likely arise, not from fixed air, but only from its being fouled by particles of the calcined metal, which we are told adhered in some places to the glass. This want of turbidity has been attributed to the fixed air uniting to the metallic calx, in preference to the lime; but there is no reason for supposing that the calx contained any fixed air; for I do not know that any one has extracted it from calces prepared in this manner; and though most metallic calces prepared over the fire, or by long exposure to the atmosphere, where they are in contact with fixed air, contain that substance, it by no means follows that they must do so when prepared by methods in which they are not in contact with it.

Dr. Priestley also observed, that quicksilver, fouled by the addition of lead or tin, deposits a powder by agitation and exposure to the air, which consists in great measure of the calx of the imperfect metal. He found too some powder of this kind to contain fixed air[1]; but it is by no means clear that this air was produced by the phlogistication of the air in which the quicksilver was shaken; as the powder was not prepared on purpose, but was procured from quicksilver fouled by having been used in various experiments, and may therefore have contained other impurities besides the metallic calces.

I never heard of any fixed air being produced by the burning of sulphur or phosphorus; but it has been asserted, and commonly believed, that lime water is rendered cloudy by a mixture of common and nitrous air; which, if true, would be a convincing proof that on mixing those two substances some fixed air is either generated or separated; I therefore examined this carefully. Now it must be observed, that as common air usually contains a little fixed air, which is no essential part of it, but is easily separated by lime water; and as nitrous air may also contain fixed air, either if the metal from which it is procured be rusty, or if the water of the vessel in which it is caught contain calcareous earth, suspended by fixed air, as most waters do, it is proper first to free both airs from it by previously washing them with lime water[2]. Now I found, by repeated experiments, that if the lime water was clean, and the two airs were previously washed with that substance, not the least cloud was produced, either immediately

[1] *Exper. in Nat. Phil.* vol. I. p. 144.

[2] Though fixed air is absorbed in considerable quantity by water, as I shewed in *Phil. Trans.* vol. LVI., yet it is not easy to deprive common air of all the fixed air contained in it by means of water. On shaking a mixture of ten parts of common air, and one of fixed air, with more than an equal bulk of distilled water, not more than half of the fixed air was absorbed, and on transferring the air into fresh distilled water only half the remainder was absorbed, as appeared by the diminution which it still suffered on adding lime water.

on mixing them, or on suffering them to stand upwards of an hour, though it appeared by the thick clouds which were produced in the lime water, by breathing through it after the experiment was finished, that it was more than sufficient to saturate the acid formed by the decomposition of the nitrous air, and consequently that if any fixed air had been produced, it must have become visible. Once indeed I found a small cloud to be formed on the surface, after the mixture had stood a few minutes. In this experiment the lime water was not quite clean; but whether the cloud was owing to this circumstance, or to the air's having not been properly washed, I cannot pretend to say.

Neither does any fixed air seem to be produced by the explosion of the inflammable air obtained from metals, with either common or dephlogisticated air. This I tried by putting a little lime-water into a glass globe fitted with a brass cock, so as to make it air tight, and an apparatus for firing air by electricity. This globe was exhausted by an air-pump, and the two airs, which had been previously washed with lime-water, let in, and suffered to remain some time, to shew whether they would affect the lime-water, and then fired by electricity. The event was, that not the least cloud was produced in the lime-water, when the inflammable air was mixed with common air, and only a very slight one, or rather diminution of transparency, when it was combined with dephlogisticated air. This, however, seemed not to be produced by fixed air; as it appeared instantly after the explosion, and did not increase on standing, and was spread uniformly through the liquor; whereas if it had been owing to fixed air, it would have taken up some short time before it appeared, and would have begun first at the surface, as was the case in the abovementioned experiment with nitrous air. What it was really owing to I cannot pretend to say; but if it did proceed from fixed air it would shew that only an excessively minute quantity was produced[1]. On the whole, though it is not improbable that fixed air may be generated in some chymical processes, yet it seems certain that it is not the general effect of phlogisticating air, and that the diminution of common air is by no means owing to the generation or separation of fixed air from it.

As there seemed great reason to think, from Dr. Priestley's experiments, that the nitrous and vitriolic acids were convertible into dephlogisticated air, I tried whether the dephlogisticated part of common air might not, by phlogistication, be changed into nitrous or vitriolic acid. For this purpose I impregnated some milk of lime with the fumes of burning sulphur, by putting a little of it into a large glass receiver, and burning sulphur therein, taking care to keep the mouth of the receiver stopt till the fumes were all absorbed; after which the air of the receiver was changed, and more sulphur burnt in it as before, and the process repeated

[1] Dr. Priestley also found no fixed air to be produced by the explosion of inflammable and common air. Vol. v. p. 124.

till 122 grains of sulphur were consumed. The milk of lime was then filtered and evaporated, but it yielded no nitrous salt, nor any other substance except selenite; so that no sensible quantity of the air was changed into nitrous acid. It must be observed, that as the vitriolic acid produced by the burning sulphur is changed by its union with the lime into selenite, which is very little soluble in water, a very small quantity of nitrous salt, or any other substance which is soluble in water, would have been perceived.

I also tried whether any nitrous acid was produced by phlogisticating common air with liver of sulphur; for this purpose I made a solution of flowers of sulphur by boiling it with lime, and put a little of it into a large receiver, and shook it frequently, changing now and then the air, till the yellow colour of the solution was quite gone; a sign that all the sulphur was, by the loss of its phlogiston, turned into vitriolic acid, and united to the lime, or precipitated; the liquor was then filtered and evaporated, but it yielded not the least nitrous salt.

The experiment was repeated in nearly the same manner with dephlogisticated air procured from red precipitate; but not the least nitrous acid was obtained.

It is well known that common selenite is very little soluble in water; whereas that procured in the two last experiments was very soluble, and even crystallized readily, and was intensely bitter; this however appeared to be owing merely to the acid with which it was formed being very much phlogisticated; for on evaporating it to dryness, and exposing it to the air for a few days, it became much less soluble, so that on adding water to it not much dissolved; and by repeating this process once or twice, it seemed to become not more soluble than selenite made in the common manner.

This solubility of the selenite caused some trouble in trying the experiment; for while it continued much soluble it would have been impossible to have distinguished a small mixture of nitrous salt; but by the abovementioned process I was able to distinguish as small a proportion as if the selenite had been originally no more soluble than usual.

The nature of the neutral salts made with the phlogisticated vitriolic and nitrous acids has not been much examined by the chymists, though it seems well worth their attention; and it is likely that many besides the foregoing may differ remarkably from those made with the same acids in their common state. Nitre formed with the phlogisticated nitrous acid has been found to differ considerably from common nitre, as well as Sal Polychrest from vitriolated tartar.

In order to try whether any vitriolic acid was produced by the phlogistication of air, I impregnated fifty ounces of distilled water with the fumes produced on mixing fifty-two ounce measures of nitrous air with a quantity of common air sufficient to decompound it. This was done by filling a bottle with some of this water, and inverting it into a bason of the same, and then, by a syphon, letting in as much nitrous air as filled it half-full; after which common air was added slowly by the same syphon, till all the nitrous air was decompounded. When this was done, the distilled water was further impregnated in the same manner till the whole of the above-mentioned quantity of nitrous air was employed. This impregnated water, which was very sensibly acid to the taste, was distilled in a glass retort. The first runnings were very acid, and smelt pungent, being nitrous acid much phlogisticated; what came next had no sensible taste or smell; but the last runnings were very acid, and consisted of nitrous acid not phlogisticated. Scarce any sediment was left behind. These different parcels of distilled liquor were then exactly saturated with salt of tartar, and evaporated; they yielded 87½ grains of nitre, which, as far as I could perceive, was unmixed with vitriolated tartar or any other substance, and consequently no sensible quantity of the common air with which the nitrous air was mixed was turned into vitriolic acid.

It appears, from this experiment, that nitrous air contains as much acid as 2¾ times its weight of saltpetre; for fifty-two ounce measures of nitrous air weigh 32 grains, and, as was before said, yield as much acid as is contained in 87½ grains of saltpetre; so that the acid in nitrous air is in a remarkably concentrated state, and I believe more than 1½ times as much so as the strongest spirit of nitre ever prepared.

Having now mentioned the unsuccessful attempts I made to find out what becomes of the air lost by phlogistication, I proceed to some experiments, which serve really to explain the matter.

In Dr. Priestley's last volume of experiments is related an experiment of Mr. Warltire's, in which it is said that, on firing a mixture of common and inflammable air by electricity in a close copper vessel holding about three pints, a loss of weight was always perceived, on an average about two grains, though the vessel was stopped in such a manner that no air could escape by the explosion. It is also related, that on repeating the experiment in glass vessels, the inside of the glass, though clean and dry before, immediately became dewy; which confirmed an opinion he had long entertained, that common air deposits its moisture by phlogistication. As the latter experiment seemed likely to throw great light on the subject I had in view, I thought it well worth examining more closely. The first experiment also, if there was no mistake in it, would be very extraordinary and curious; but it did not succeed with me; for though the vessel I used held more than Mr. Warltire's, namely, 24,000 grains of water, and though the experiment was repeated several times with different proportions of common and inflammable air, I could never perceive a loss of weight of more than one-fifth of a grain, and commonly none at all. It must be observed, however, that though there were some of the experiments in which it seemed to diminish a little in weight, there were none in which it increased[1].

[1] Dr. Priestley, I am informed, has since found the experiment not to succeed.

In all the experiments, the inside of the glass globe became dewy, as observed by Mr. Warltire; but not the least sooty matter could be perceived. Care was taken in all of them to find how much the air was diminished by the explosion, and to observe its test. The result is as follows: the bulk of the inflammable air being expressed in decimals of the common air,

Common air	Inflammable air	Diminution	Air remaining after the explosion	Test of this air in first method	Standard
	1,241	,686	1,555	,055	,0
1	1,055	,642	1,413	,063	,0
	,706	,647	1,059	,066	,0
	,423	,612	,811	,097	,03
	,331	,476	,855	,339	,27
	,206	,294	,912	,648	,58

In these experiments the inflammable air was procured from zinc, as it was in all my experiments, except where otherwise expressed: but I made two more experiments, to try whether there was any difference between the air from zinc and that from iron, the quantity of inflammable air being the same in both, namely, 0,331 of the common; but I could not find any difference to be depended on between the two kinds of air, either in the diminution which they suffered by the explosion, or the test of the burnt air.

From the fourth experiment it appears, that 423 measures of inflammable air are nearly sufficient to completely phlogisticate 1000 of common air; and that the bulk of the air remaining after the explosion is then very little more than four-fifths of the common air employed; so that as common air cannot be reduced to a much less bulk than that by any method of phlogistication, we may safely conclude, that when they are mixed in this proportion, and exploded, almost all the inflammable air, and about one-fifth part of the common air, lose their elasticity, and are condensed into the dew which lines the glass.

The better to examine the nature of this dew, 500,000 grain measures of inflammable air were burnt with about 2½ times that quantity of common air, and the burnt air made to pass through a glass cylinder eight feet long and three-quarters of an inch in diameter, in order to deposit the dew. The two airs were conveyed slowly into this cylinder by separate copper pipes, passing through a brass plate which stopped up the end of the cylinder; and as neither inflammable nor common air can burn by themselves, there was no danger of the flame spreading into the magazines from which they were conveyed. Each of these magazines consisted of a large tin vessel, inverted into another vessel just big enough to receive it. The inner vessel communicated with the copper pipe, and the air was forced out of it by pouring water into the outer vessel; and in order that the quantity of common air expelled should be $2\frac{1}{2}$ times that of the inflammable, the water was let into the outer vessels by two holes in the bottom of the same tin pan, the hole which conveyed the water into that vessel in which the common air was confined being $2\frac{1}{2}$ times as big as the other.

In trying the experiment, the magazines being first filled with their respective airs, the glass cylinder was taken off, and water let, by the two holes, into the outer vessels, till the airs began to issue from the ends of the copper pipes; they were then set on fire by a candle, and the cylinder put on again in its place. By this means upwards of 135 grains of water were condensed in the cylinder, which had no taste nor smell, and which left no sensible sediment when evaporated to dryness; neither did it yield any pungent smell during the evaporation; in short, it seemed pure water.

In my first experiment, the cylinder near that part where the air was fired was a little tinged with sooty matter, but very slightly so; and that little seemed to proceed from the putty with which the apparatus was luted, and which was heated by the flame; for in another experiment, in which it was contrived so that the luting should not be much heated, scarce any sooty tinge could be perceived.

By the experiments with the globe it appeared, that when inflammable and common air are exploded in a proper proportion, almost all the inflammable air, and near one-fifth of the common air, lose their elasticity, and are condensed into dew. And by this experiment it appears, that this dew is plain water, and consequently that almost all the inflammable air, and about one-fifth of the common air, are turned into pure water.

In order to examine the nature of the matter condensed on firing a mixture of dephlogisticated and inflammable air, I took a glass globe, holding 8800 grain measures, furnished with a brass cock and an apparatus for firing air by electricity. This globe was well exhausted by an air-pump, and then filled with a mixture of inflammable and dephlogisticated air, by shutting the cock, fastening a bent glass tube to its mouth, and letting up the end of it into a glass jar inverted into water, and containing a mixture of 19,500 grain measures of dephlogisticated air, and 37,000 of inflammable; so that, upon opening the cock, some of this mixed air rushed through the bent tube, and filled the globe[1]. The cock was then shut, and the included air fired by electricity, by which means almost all of it lost its elasticity. The cock was then again opened, so as to let in more of the same air, to supply the place of that destroyed by the explosion, which was again fired, and the operation continued till almost the whole of the mixture was let into the globe and exploded. By this means, though the globe held not more than the sixth part of the mixture, almost the

[1] In order to prevent any water from getting into this tube, while dipped under water to let it up into the glass jar, a bit of wax was stuck upon the end of it, which was rubbed off when raised above the surface of the water.

whole of it was exploded therein, without any fresh exhaustion of the globe.

As I was desirous to try the quantity and test of this burnt air, without letting any water into the globe, which would have prevented my examining the nature of the condensed matter, I took a larger globe, furnished also with a stop cock, exhausted it by an air-pump, and screwed it on upon the cock of the former globe; upon which, by opening both cocks, the air rushed out of the smaller globe into the larger, till it became of equal density in both; then, by shutting the cock of the larger globe, unscrewing it again from the former, and opening it under water, I was enabled to find the quantity of the burnt air in it; and consequently, as the proportion which the contents of the two globes bore to each other was known, could tell the quantity of burnt air in the small globe before the communication was made between them. By this means the whole quantity of the burnt air was found to be 2950 grain measures; its standard was 1,85.

The liquor condensed in the globe, in weight about 30 grains, was sensibly acid to the taste, and by saturation with fixed alkali, and evaporation, yielded near two grains of nitre; so that it consisted of water united to a small quantity of nitrous acid. No sooty matter was deposited in the globe. The dephlogisticated air used in this experiment was procured from red precipitate, that is, from a solution of quicksilver in spirit of nitre distilled till it acquires a red colour.

As it was suspected, that the acid contained in the condensed liquor was no essential part of the dephlogisticated air, but was owing to some acid vapour which came over in making it and had not been absorbed by the water, the experiment was repeated in the same manner, with some more of the same air, which had been previously washed with water, by keeping it a day or two in a bottle with some water, and shaking it frequently; whereas that used in the preceding experiment had never passed through water, except in preparing it. The condensed liquor was still acid.

The experiment was also repeated with dephlogisticated air, procured from red lead by means of oil of vitriol; the liquor condensed was acid, but by an accident I was prevented from determining the nature of the acid.

I also procured some dephlogisticated air from the leaves of plants, in the manner of Doctors Ingenhousz and Priestley, and exploded it with inflammable air as before; the condensed liquor still continued acid, and of the nitrous kind.

In all these experiments the proportion of inflammable air was such, that the burnt air was not much phlogisticated; and it was observed, that the less phlogisticated it was, the more acid was the condensed liquor. I therefore made another experiment, with some more of the same air from plants, in which the proportion of inflammable air was greater, so that the burnt air was almost completely phlogisticated, its standard being $\frac{1}{10}$. The condensed liquor was then not at all acid, but seemed pure water: so that it appears, that with this kind of dephlogisticated air, the condensed liquor is not at all acid, when the two airs are mixed in such a proportion that the burnt air is almost completely phlogisticated, but is considerably so when it is not much phlogisticated.

In order to see whether the same thing would obtain with air procured from red precipitate, I made two more experiments with that kind of air, the air in both being taken from the same bottle, and the experiment tried in the same manner, except that the proportions of inflammable air were different. In the first, in which the burnt air was almost completely phlogisticated, the condensed liquor was not at all acid. In the second, in which its standard was 1,86, that is, not much phlogisticated, it was considerably acid; so that with this air, as well as with that from plants, the condensed liquor contains, or is entirely free from, acid, according as the burnt air is less or more phlogisticated; and there can be little doubt but that the same rule obtains with any other kind of dephlogisticated air.

In order to see whether the acid, formed by the explosion of dephlogisticated air obtained by means of the vitriolic acid, would also be of the nitrous kind, I procured some air from turbith mineral, and exploded it with inflammable air, the proportion being such that the burnt air was not much phlogisticated. The condensed liquor manifested an acidity, which appeared, by saturation with a solution of salt of tartar, to be of the nitrous kind; and it was found, by the addition of some terra ponderosa salita, to contain little or no vitriolic acid.

When inflammable air was exploded with common air, in such a proportion that the standard of the burnt air was about $\frac{4}{10}$, the condensed liquor was not in the least acid. There is no difference, however, in this respect between common air, and dephlogisticated air mixed with phlogisticated in such a proportion as to reduce it to the standard of common air; for some dephlogisticated air from red precipitate, being reduced to this standard by the addition of perfectly phlogisticated air, and then exploded with the same proportion of inflammable air as the common air was in the foregoing experiment, the condensed liquor was not in the least acid.

From the foregoing experiments it appears, that when a mixture of inflammable and dephlogisticated air is exploded in such proportion that the burnt air is not much phlogisticated, the condensed liquor contains a little acid, which is always of the nitrous kind, whatever substance the dephlogisticated air is procured from; but if the proportion be such that the burnt air is almost entirely phlogisticated, the condensed liquor is not at all acid, but seems pure water, without any addition whatever; and as, when they are mixed in that proportion, very little air remains after the explosion, almost the whole being condensed, it follows, that almost the whole of the inflammable and dephlogisticated air is converted into pure water. It is not easy, indeed, to determine from these experiments what proportion the burnt air, remaining after the explosions, bore to the dephlogisticated air employed, as neither the small nor the large globe could be perfectly exhausted of air, and there was no saying with exactness what quantity was left in them; but in most of them, after allowing for this uncertainty, the true quantity of burnt air seemed not more than $\frac{1}{17}$th of the dephlogisticated air employed, or $\frac{1}{50}$th of the mixture. It seems, however, unnecessary to determine this point exactly, as the quantity is so small, that there can be little doubt but that it proceeds only from the impurities mixed with the dephlogisticated and inflammable air, and consequently that, if those airs could be obtained perfectly pure, the whole would be condensed.

With respect to common air, and dephlogisticated air reduced by the addition of phlogisticated air to the standard of common air, the case is different; as the liquor condensed in exploding them with inflammable air, I believe I may say in any proportion, is not at all acid; perhaps, because if they are mixed in such a proportion as that the burnt air is not much phlogisticated, the explosion is too weak, and not accompanied with sufficient heat.

All the foregoing experiments, on the explosion of inflammable air with common and dephlogisticated airs, except those which relate to the cause of the acid found in the water, were made in the summer of the year 1781, and were mentioned by me to Dr. Priestley, who in consequence of it made some experiments of the same kind, as he relates in a paper printed in the preceding volume of the *Transactions*. During the last summer also, a friend of mine gave some account of them to M. Lavoisier, as well as of the conclusion drawn from them, that dephlogisticated air is only water deprived of phlogiston; but at that time so far was M. Lavoisier from thinking any such opinion warranted, that, till he was prevailed upon to repeat the experiment himself, he found some difficulty in believing that nearly the whole of the two airs could be converted into water. It is remarkable, that neither of these gentlemen found any acid in the water produced by the combustion; which might proceed from the latter having burnt the two airs in a different manner from what I did; and from the former having used a different kind of inflammable air, namely, that from charcoal, and perhaps having used a greater proportion of it.

Before I enter into the cause of these phænomena, it will be proper to take notice, that phlogisticated air appears to be nothing else than the nitrous acid united to phlogiston; for when nitre is deflagrated with charcoal, the acid is almost entirely converted into this kind of air. That the acid is entirely converted into air, appears from the common process for making what is called clyssus of nitre; for if the nitre and charcoal are dry, scarce any thing is found in the vessels prepared for condensing the fumes; but if they are moist a little liquor is collected, which is nothing but the water contained in the materials, impregnated with a little volatile alkali, proceeding in all probability from the imperfectly burnt charcoal, and a little fixed alkali, consisting of some of the alkalized nitre carried over by the heat and watery vapours. As far as I can perceive too, at present, the air into which much the greatest part of the acid is converted, differs in no respect from common air phlogisticated. A small part of the acid, however, is turned into nitrous air, and the whole is mixed with a good deal of fixed, and perhaps a little inflammable air, both proceeding from the charcoal.

It is well known, that the nitrous acid is also converted by phlogistication into nitrous air, in which respect there seems a considerable analogy between that and the vitriolic acid; for the vitriolic acid, when united to a smaller proportion of phlogiston, forms the volatile sulphureous acid and vitriolic acid air, both of which, by exposure to the atmosphere, lose their phlogiston, though not very fast, and are turned back into vitriolic acid; but, when united to a greater proportion of phlogiston, it forms sulphur, which shews no signs of acidity, unless a small degree of affinity to alkalies can be called so, and in which the phlogiston is more strongly adherent, so that it does not fly off when exposed to the air, unless assisted by a heat sufficient to set it on fire. In like manner the nitrous acid, united to a certain quantity of phlogiston, forms nitrous fumes and nitrous air, which readily quit their phlogiston to common air; but when united to a different, in all probability a larger quantity, it forms phlogisticated air, which shews no signs of acidity, and is still less disposed to part with its phlogiston than sulphur.

This being premised, there seem two ways by which the phænomena of the acid found in the condensed liquor may be explained; first, by supposing that dephlogisticated air contains a little nitrous acid which enters into it as one of its component parts, and that this acid, when the inflammable air is in a sufficient proportion, unites to the phlogiston, and is turned into phlogisticated air, but does not when the inflammable air is in too small a proportion; and, secondly, by supposing that there is no nitrous acid mixed with, or entering into the composition of, dephlogisticated air, but that, when this air is in a sufficient proportion, part of the phlogisticated air with which it is debased is, by the strong affinity of phlogiston to dephlogisticated air, deprived of its phlogiston and turned into nitrous acid; whereas, when the dephlogisticated air is not more than sufficient to consume the inflammable air, none then remains to deprive the phlogisticated air of its phlogiston, and turn it into acid.

If the latter explanation be true, I think, we must allow that dephlogisticated air is in reality nothing but dephlogisticated water, or water deprived of its phlogiston; or, in other words, that water consists of dephlogisticated air united to phlogiston; and that inflammable air is either pure phlogiston, as Dr. Priestley and Mr. Kirwan suppose, or else water united to phlogiston[1]; since, according to this supposition, these two substances united together form pure water. On the other hand, if the first explanation be true, we must suppose that dephlogisticated air consists of water united to a little nitrous acid and deprived of its phlogiston; but still the nitrous acid in it must make only a very small part of the whole, as it is found, that the phlogisticated air, which it is converted into, is very small in comparison of the dephlogisticated air.

I think the second of these explanations seems much the most likely; as it was found, that the acid in the condensed liquor was of the nitrous kind, not only when the dephlogisticated air was prepared from red precipitate, but also when it was procured from plants or from turbith mineral: and it seems not likely, that air procured from plants, and still less likely that air procured from a solution of mercury in oil of vitriol, should contain any nitrous acid.

Another strong argument in favour of this opinion is, that dephlogisticated air yields no nitrous acid when phlogisticated by liver of sulphur; for if this air contains nitrous acid, and yields it when phlogisticated by explosion with inflammable air, it is very extraordinary that it should not do so when phlogisticated by other means.

But what forms a stronger and, I think, almost decisive argument in favour of this explanation is, that when the dephlogisticated air is very pure, the condensed liquor is made much more strongly acid by mixing the air to be exploded with a little phlogisticated air, as appears by the following experiments.

A mixture of 18,500 grain measures of inflammable air with 9750 of dephlogisticated air procured from red precipitate were exploded in the usual manner; after which, a mixture of the same quantities of the same dephlogisticated and inflammable air, with the addition of 2500 of air

[1] Either of these suppositions will agree equally well with the following experiments; but the latter seems to me much the most likely. What principally makes me think so is, that common or dephlogisticated air do not absorb phlogiston from inflammable air, unless assisted by a red heat, whereas they absorb the phlogiston of nitrous air, liver of sulphur, and many other substances, without that assistance; and it seems inexplicable, that they should refuse to unite to pure phlogiston, when they are able to extract it from substances to which it has an affinity; that is, that they should overcome the affinity of phlogiston to other substances, and extract it from them, when they will not even unite to it when presented to them. On the other hand, I know no experiment which shews inflammable air to be pure phlogiston rather than an union of it with water, unless it be Dr. Priestley's experiment of expelling inflammable air from iron by heat alone. I am not sufficiently acquainted with the circumstances of that experiment to argue with certainty about it; but I think it much more likely, that the inflammable air was formed by the union of the phlogiston of the iron filings with the water dispersed among them, or contained in the retort or other vessel in which it was heated; and in all probability this was the cause of the separation of the phlogiston, as iron seems not disposed to part with its phlogiston by heat alone, without being assisted by the air or some other substance.

phlogisticated by iron filings and sulphur, was treated in the same manner. The condensed liquor, in both experiments, was acid, but that in the latter evidently more so, as appeared also by saturating each of them separately with marble powder, and precipitating the earth by fixed alkali, the precipitate of the second experiment weighing one-fifth of a grain, and that of the first being several times less. The standard of the burnt air in the first experiment was 1,86, and in the second only 0,9.

It must be observed, that all circumstances were the same in these two experiments, except that in the latter the air to be exploded was mixed with some phlogisticated air, and that in consequence the burnt air was more phlogisticated than in the former; and from what has been before said, it appears, that this latter circumstance ought rather to have made the condensed liquor less acid; and yet it was found to be much more so, which shews strongly that it was the phlogisticated air which furnished the acid.

As a further confirmation of this point, these two comparative experiments were repeated with a little variation, namely, in the first experiment there was first let into the globe 1500 of dephlogisticated air, and then the mixture, consisting of 12,200 of dephlogisticated air and 25,900 of inflammable, was let in at different times as usual. In the second experiment, besides the 1500 of dephlogisticated air first let in, there was also admitted 2500 of phlogisticated air, after which the mixture, consisting of the same quantities of dephlogisticated and inflammable air as before, was let in as usual. The condensed liquor of the second experiment was about three times as acid as that of the first, as it required 119 grains of a diluted solution of salt of tartar to saturate it, and the other only 37. The standard of the burnt air was 0,78 in the second experiment, and 1,96 in the first.

The intention of previously letting in some dephlogisticated air in the two last experiments was, that the condensed liquor was expected to become more acid thereby, as proved actually to be the case.

In the first of these two experiments, in order that the air to be exploded should be as free as possible from common air, the globe was first filled with a mixture of dephlogisticated and inflammable air, it was then exhausted, and the air to be exploded let in; by which means, though the globe was not perfectly exhausted, very little common air could be left in it. In the first set of experiments this circumstance was not attended to, and the purity of the dephlogisticated air was forgot to be examined in both sets.

From what has been said there seems the utmost reason to think, that dephlogisticated air is only water deprived of its phlogiston, and that inflammable air, as was before said, is either phlogisticated water, or else pure phlogiston; but in all probability the former.

As Mr. Watt, in a paper lately read before this Society, supposes water to consist of dephlogisticated air and phlogiston deprived of part of their latent heat, whereas I take no notice of the latter circumstance, it may be proper to mention in a few words the reason of this apparent difference between us. If there be any such thing as elementary heat, it must be allowed that what Mr. Watt says is true; but by the same rule we ought to say, that the diluted mineral acids consist of the concentrated acids united to water and deprived of part of their latent heat; that solutions of sal ammoniac, and most other neutral salts, consist of the salt united to water and elementary heat; and a similar language ought to be used in speaking of almost all chemical combinations, as there are very few which are not attended with some increase or diminution of heat. Now I have chosen to avoid this form of speaking, both because I think it more likely that there is no such thing as elementary heat, and because saying so in this instance, without using similar expressions in speaking of other chemical unions, would be improper, and would lead to false ideas; and it may even admit of doubt, whether the doing it in general would not cause more trouble and perplexity than it is worth.

There is the utmost reason to think, that dephlogisticated and phlogisticated air, as M. Lavoisier and Scheele suppose, are quite distinct substances, and not differing only in their degree of phlogistication; and that common air is a mixture of the two; for if the dephlogisticated air is pretty pure, almost the whole of it loses its elasticity by phlogistication, and, as appears by the foregoing experiments, is turned into water, instead of being converted into phlogisticated air. In most of the foregoing experiments, at least $\frac{16}{17}$ths of the whole was turned into water; and by treating some dephlogisticated air with liver of sulphur, I have reduced it to less than $\frac{1}{30}$th of its original bulk, and other persons, I believe, have reduced it to a still less bulk; so that there seems the utmost reason to suppose, that the small residuum which remains after its phlogistication proceeds only from the impurities mixed with it.

It was just said, that some dephlogisticated air was reduced by liver of sulphur to $\frac{1}{30}$th of its original bulk; the standard of this air was 4,8, and consequently the standard of perfectly pure dephlogisticated air should be very nearly 5, which is a confirmation of the foregoing opinion; for if the standard of pure dephlogisticated air is 5, common air must, according to this opinion, contain one-fifth of it, and therefore ought to lose one-fifth of its bulk by phlogistication, which is what it is actually found to lose.

From what has been said, it follows, that instead of saying air is phlogisticated or dephlogisticated by any means, it would be more strictly just to say, it is deprived of, or receives, an addition of dephlogisticated air; but as the other expression is convenient, and can scarcely be considered as improper, I shall still frequently make use of it in the remainder of this paper.

There seemed great reason to think, from Dr. Priestley's experiments, that both the nitrous and vitriolic acids were convertible into dephlogisticated air, as that air is procured in the greatest quantity from substances containing those acids, especially the former. The foregoing experiments, however, seem to shew that no part of the acid is converted into dephlogisticated air, and that their use in preparing it is owing only to the great power which they possess of depriving bodies of their phlogiston. A strong confirmation of this is, that red precipitate, which is one of the substances yielding dephlogisticated air in the greatest quantity, and which is prepared by means of the nitrous acid, contains in reality no acid. This I found by grinding 400 grains of it with spirits of sal ammoniac, and keeping them together for some days in a bottle, taking care to shake them frequently. The red colour of the precipitate was rendered pale, but not entirely destroyed; being then washed with water and filtered, the clear liquor yielded on evaporation not the least ammoniacal salt.

It is natural to think, that if any nitrous acid had been contained in the red precipitate, it would have united to the volatile alkali and have formed ammoniacal nitre, and would have been perceived on evaporation; but in order to determine more certainly whether this would be the case, I dried some of the same solution of quicksilver from which the red precipitate was prepared with a less heat, so that it acquired only an orange colour, and treated the same quantity of it with volatile alkali in the same manner as before. It immediately caused an effervescence, changed the colour to grey, and yielded 52 grains of ammoniacal nitre. There is the utmost reason to think, therefore, that red precipitate contains no nitrous acid; and consequently that, in procuring dephlogisticated air from it, no acid is converted into air; and it is reasonable to conclude, therefore, that no such change is produced in procuring it from any other substance.

It remains to consider in what manner these acids act in producing dephlogisticated air. The way in which the nitrous acid acts, in the production of it from red precipitate, seems to be as follows. On distilling the mixture of quicksilver and spirit of nitre, the acid comes over, loaded with phlogiston, in the form of nitrous vapour, and continues to do so till the remaining matter acquires its full red colour, by which time all the nitrous acid is driven over, but some of the watery part still remains behind, and adheres strongly to the quicksilver; so that the red precipitate may be considered, either as quicksilver deprived of part of its phlogiston, and united to a certain portion of water, or as quicksilver united to dephlogisticated air[1]; after which, on further increasing the heat, the water in it

[1] Unless we were much better acquainted than we are with the manner in which different substances are united together in compound bodies, it would be ridiculous to say, that it is the quicksilver in the red precipitate which is deprived of its phlogiston, and not the water, or that it is the water and not the quicksilver; all that we can say is, that red precipitate consists of quicksilver and water, one or both of which are deprived of part of their phlogiston. In like manner, during the preparation of the red precipitate, it is certain that the acid absorbs phlogiston, either from the quicksilver or the water; but we are by no means authorised to say from which.

rises deprived of its phlogiston, that is, in the form of dephlogisticated air, and at the same time the quicksilver distils over in its metallic form. It is justly remarked by Dr. Priestley, that the solution of quicksilver does not begin to yield dephlogisticated air till it acquires its red colour.

Mercurius calcinatus appears to be only quicksilver which has absorbed dephlogisticated air from the atmosphere during its preparation; accordingly, by giving it a sufficient heat, the dephlogisticated air is driven off, and the quicksilver acquires its original form. It seems therefore that mercurius calcinatus and red precipitate, though prepared in a different manner, are very nearly the same thing.

From what has been said it follows, that red precipitate and mercurius calcinatus contain as much phlogiston as the quicksilver they are prepared from; but yet, as uniting dephlogisticated air to a metal comes to the same thing as depriving it of part of its phlogiston and adding water to it, the quicksilver may still be considered as deprived of its phlogiston; but the imperfect metals seem not only to absorb dephlogisticated air during their calcination, but also to be really deprived of part of their phlogiston, as they do not acquire their metallic form by driving off the dephlogisticated air.

In procuring dephlogisticated air from nitre, the acid acts in a different manner, as, upon heating the nitre red-hot, the dephlogisticated air rises mixed with a little nitrous acid, and at the same time the acid remaining in the nitre becomes very much phlogisticated; which shews that the acid absorbs phlogiston from the water in the nitre, and becomes phlogisticated, while the water is thereby turned into dephlogisticated air. On distilling 3155 grains of nitre in an unglazed earthen retort, it yielded 256,000 grain measures of dephlogisticated air[1], the standard of different parts of which varied from 3 to 3,65, but at a medium was 3,35. The matter remaining in the retort dissolved readily in water, and tasted alcaline and caustic. On adding diluted spirit of nitre to the solution, strong red fumes were produced; a sign that the acid in it was very much phlogisticated, as no fumes whatever would have been produced on adding the same acid to a solution of common nitre; that part of the solution also which was supersaturated with acid became blue; a colour which the diluted nitrous acid is known to assume when much phlogisticated. The solution, when saturated with this acid, lost its alcaline and caustic taste, but yet tasted very different from true nitre, seeming as if it had been mixed with sea-salt, and also required much less water to dissolve it; but on exposing it for some days to the air, and adding fresh acid as fast as by the flying off

[1] This is, about eighty-one grain measures from one grain of nitre; and the weight of the dephlogisticated air, supposing it 800 times lighter than water is one-tenth of that of the nitre. In all probability it would have yielded a much greater quantity of air, if a greater heat had been applied.

of the fumes the alcali predominated, it became true nitre, unmixed, as far as I could perceive, with any other salt[1].

It has been remarked, that the dephlogisticated air procured from nitre is less pure, than that from red precipitate and many other substances, which may perhaps proceed from unglazed earthen retorts having been commonly used for this purpose, and which, conformably to Dr. Priestley's discovery, may possibly absorb some common air from without, and emit it along with the dephlogisticated air; but if it should be found that the dephlogisticated air procured from nitre in glass or glazed earthen vessels is also impure, it would seem to shew that part of the acid in the nitre is turned into phlogisticated air, by absorbing phlogiston from the watery part.

From what has been said it appears, that there is a considerable difference in the manner in which the acid acts in the production of dephlogisticated air from red precipitate and from nitre; in the former case the acid comes over first, leaving the remaining substance deprived of part of its phlogiston; in the latter the dephlogisticated air comes first, leaving the acid loaded with the phlogiston of the water from which it was formed.

On distilling a mixture of quicksilver and oil of vitriol to dryness, part of the acid comes over, loaded with phlogiston, in the form of volatile sulphureous acid and vitriolic acid air; so that the remaining white mass may be considered as consisting of quicksilver deprived of its phlogiston, and united to a certain proportion of acid and water, or of plain quicksilver united to a certain proportion of acid and dephlogisticated air. Accordingly on urging this white mass with a more violent heat, the dephlogisticated air comes over, and at the same time part of the quicksilver rises in its metallic form, and also part of the white mass, united in all probability to a greater proportion of acid than before, sublimes; so that the rationale of the production of dephlogisticated air from turbith mineral, and from red precipitate, are nearly similar.

True turbith mineral consists of the abovementioned white mass, well washed with water, by which means it acquires a yellow colour, and contains much less acid than the unwashed mass. Accordingly it seems likely, that on exposing this to heat, less of it should sublime without being decompounded, and consequently that more dephlogisticated air should be procured from it than from the unwashed mass.

This is an instance, that the superabundant vitriolic acid may, in some cases, be better extracted from the base it is united to by water than by heat. Vitriolated tartar is another instance; for, if vitriolated tartar be mixed with oil of vitriol and exposed even to a pretty strong red heat, the mass will be very acid; but, if this mass is dissolved in water, and evaporated, the crystals will be not sensibly so.

[1] This phlogistication of the acid in nitre by heat has been observed by Mr. Scheele; see his experiments on air and fire, p. 45, English translation.

In all probability, the vitriolic acid acts in the same manner in the production of dephlogisticated air from alum, as the nitrous does in its production from nitre; that is, the watery part comes over first in the form of dephlogisticated air, leaving the acid charged with its phlogiston. Whether this is also the case with regard to green and blue vitriol, or whether in them the acid does not rather act in the same manner as in turbith mineral, I cannot pretend to say, but I think the latter more likely.

There is another way by which dephlogisticated air has been found to be produced in great quantities, namely, the growth of vegetables exposed to the sun or day-light; the rationale of which, in all probability, is, that plants, when assisted by the light, deprive part of the water sucked up by their roots of its phlogiston, and turn it into dephlogisticated air, while the phlogiston unites to, and forms part of, the substance of the plant.

There are many circumstances which shew, that light has a remarkable power in enabling one body to absorb phlogiston from another. Mr. Senebier has observed, that the green tincture procured from the leaves of vegetables by spirit of wine, quickly loses its colour when exposed to the sun in a bottle not more than one-third part full, but does not do so in the dark, or if the bottle is quite full of the tincture, or if the air in it is phlogisticated; whence it is natural to conclude, that the light enables the dephlogisticated part of the air to absorb phlogiston from the tincture; and this appears to be really the case, as I find that the air in the bottle is considerably phlogisticated thereby. Dephlogisticated spirit of nitre also acquires a yellow colour, and becomes phlogisticated, by exposure to the sun's rays[1]; and I find on trial that the air in the bottle in which it is contained becomes dephlogisticated, or, in other words, receives an increase of dephlogisticated air, which shews that the change in the acid is not owing to the sun's rays communicating phlogiston to it, but to their enabling it to absorb phlogiston from the water contained in it, and thereby to produce dephlogisticated air. Mr. Scheele also found, that the dark colour acquired by luna cornea on exposure to the light, is owing to part of the silver being revived; and that gold, dissolved in aqua regia and deprived by distillation of the nitrous and superfluous marine acid, is revived by the same means; and there is the utmost reason to think, that, in both cases, the revival of the metal is owing to its absorbing phlogiston from the water.

[1] If spirit of nitre is distilled with a very gentle heat, the part which comes over is high coloured and fuming, and that which remains behind is quite colourless, and fumes much less than other nitrous acid of the same strength, and the fumes are colourless. This is called dephlogisticated spirit of nitre, as it appears to be really deprived of phlogiston by the process. The manner of preparing it, as well as its property of regaining its yellow colour by exposure to the light, is mentioned by Mr. Scheele in the *Stockholm Memoirs*, 1774.

Vegetables seem to consist almost intirely of fixed and phlogisticated air, united to a large proportion of phlogiston and some water, since by burning in the open air, in which their phlogiston unites to the dephlogisticated part of the atmosphere and forms water, they seem to be reduced almost intirely to water and those two kinds of air. Now plants growing in water without earth, can receive nourishment only from the water and air, and must therefore in all probability absorb their phlogiston from the water. It is known also that plants growing in the dark do not thrive well, and grow in a very different manner from what they do when exposed to the light.

From what has been said it seems likely that the use of light, in promoting the growth of plants and the production of dephlogisticated air from them, is, that it enables them to absorb phlogiston from the water. To this it may perhaps be objected, that though plants do not thrive well in the dark, yet they do grow, and should therefore, according to this hypothesis, absorb water from the atmosphere, and yield dephlogisticated air, which they have not been found to do. But we have no proof that they grew at all in any of those cases in which they were found not to yield dephlogisticated air; for though they will grow in the dark, yet their vegetative powers may perhaps at first be intirely checked by it, especially considering the unnatural situation in which they must be placed in such experiments. Perhaps too plants growing in the dark may be able to absorb phlogiston from water not much impregnated with dephlogisticated air, but not from water strongly impregnated with it; and consequently, when kept under water in the dark, may perhaps at first yield some dephlogisticated air, which, instead of rising to the surface, may be absorbed by the water, and, before the water is so much impregnated as to suffer any to escape, the plant may cease to vegetate, unless the water is changed. Unless therefore it could be shewn that plants growing in the dark, in water alone, will increase in size, without yielding dephlogisticated air, and without the water becoming more impregnated with it than before, no objection can be drawn from thence.

Mr. Senebier finds, that plants yield much more dephlogisticated air in distilled water impregnated with fixed air, than in plain distilled water, which is perfectly conformable to the abovementioned hypothesis; for as fixed air is a principal constituent part of vegetable substances, it is reasonable to suppose that the work of vegetation will go on better in water containing this substance, than in other water.

There are several memoirs of Mr. Lavoisier published by the Academy of Sciences, in which he intirely discards phlogiston, and explains those phænomena which have been usually attributed to the loss or attraction of that substance, by the absorption or expulsion of dephlogisticated air; and as not only the foregoing experiments, but most other phænomena of nature, seem explicable as well, or nearly as well, upon this as upon the commonly believed principle of phlogiston, it may be proper briefly to mention in what manner I would explain them on this principle, and why I have adhered to the other. In doing this, I shall not conform strictly to his theory, but shall make such additions and alterations as seem to suit it best to the phænomena; the more so, as the foregoing experiments may, perhaps, induce the author himself to think some such additions proper.

According to this hypothesis, we must suppose, that water consists of inflammable air united to dephlogisticated air; that nitrous air, vitriolic acid air, and the phosphoric acid, are also combinations of phlogisticated air, sulphur, and phosphorus, with dephlogisticated air; and that the two former, by a further addition of the same substance, are reduced to the common nitrous and vitriolic acids; that the metallic calces consist of the metals themselves united to the same substance, commonly, however, with a mixture of fixed air; that on exposing the calces of the perfect metals to a sufficient heat, all the dephlogisticated air is driven off, and the calces are restored to their metallic form; but as the calces of the imperfect metals are vitrified by heat, instead of recovering the metallic form, it should seem as if all the dephlogisticated air could not be driven off from them by heat alone. In like manner, according to this hypothesis, the rationale of the production of dephlogisticated air from red precipitate is, that during the solution of the quicksilver in the acid and the subsequent calcination, the acid is decompounded, and quits part of its dephlogisticated air to the quicksilver, whereby it comes over in the form of nitrous air, and leaves the quicksilver behind united to dephlogisticated air, which, by a further increase of heat, is driven off, while the quicksilver re-assumes its metallic form. In procuring dephlogisticated air from nitre, the acid is also decompounded; but with this difference, that it suffers some of its dephlogisticated air to escape, while it remains united to the alkali itself, in the form of phlogisticated nitrous acid. As to the production of dephlogisticated air from plants, it may be said, that vegetable substances consist chiefly of various combinations of three different bases, one of which, when united to dephlogisticated air, forms water, another fixed air, and the third phlogisticated air; and that by means of vegetation each of these substances are decomposed, and yield their dephlogisticated air; and that in burning they again acquire dephlogisticated air, and are restored to their pristine form.

It seems, therefore, from what has been said, as if the phænomena of nature might be explained very well on this principle, without the help of phlogiston; and indeed, as adding dephlogisticated air to a body comes to the same thing as depriving it of its phlogiston and adding water to it, and as there are, perhaps, no bodies entirely destitute of water, and as I know no way by which phlogiston can be transferred from one body to another, without leaving it uncertain whether water is not at the same time transferred, it will be very difficult to determine by experiment which of these opinions is the truest; but as the commonly received principle of phlogiston explains all phænomena, at least as well as Mr. Lavoisier's, I have adhered to that. There is one circumstance also, which though it may appear to many not to have much force, I own has some weight with me; it is, that as plants seem to draw their nourishment almost intirely from water and fixed and phlogisticated air, and are restored back to those substances by burning, it seems reasonable to conclude, that notwithstanding their infinite variety they consist almost intirely of various combinations of water and fixed and phlogisticated air, united according to one of these opinions to phlogiston, and deprived according to the other of dephlogisticated air; so that, according to the latter opinion, the substance of a plant is less compounded than a mixture of those bodies into which it is resolved by burning; and it is more reasonable to look for great variety in the more compound than in the more simple substance.

Another thing which Mr. Lavoisier endeavours to prove is, that dephlogisticated air is the acidifying principle. From what has been explained it appears, that this is no more than saying, that acids lose their acidity by uniting to phlogiston, which with regard to the nitrous, vitriolic, phosphoric, and arsenical acids is certainly true. The same thing, I believe, may be said of the acid of sugar; and Mr. Lavoisier's experiment is a strong confirmation of Bergman's opinion, that none of the spirit of nitre enters into the composition of the acid, but that it only serves to deprive the sugar of part of its phlogiston. But as to the marine acid and acid of tartar, it does not appear that they are capable of losing their acidity by any union with phlogiston. It is to be remarked also, that the acids of sugar and tartar, and in all probability almost all the vegetable and animal acids, are by burning reduced to fixed and phlogisticated air, and water, and therefore contain more phlogiston, or less dephlogisticated air, than those three substances.

PHILOSOPHICAL TRANSACTIONS. VOL. 74, 1784, p. 170

XV. *Answer to Mr.* Kirwan's *Remarks upon the Experiments on Air. By* Henry Cavendish, *Esq., F.R.S. and S.A.*

Read March 4, 1784

In a paper lately read before this Society, containing many experiments on air, I gave my reasons for supposing that the diminution which respirable air suffers by phlogistication, is not owing either to the generation or separation of fixed air from it; but without any arguments of a personal nature, or which related to any one person who espouses the contrary doctrine more than to another. This being contrary to the opinion maintained by Mr. Kirwan, he has written a paper in answer to it, which was read on the fifth of February. As I do not like troubling the Society with controversy, I shall take no notice of the arguments used by him, but shall leave them for the reader to form his own judgement of; much less will I endeavour to point out any inconsistencies or false reasonings, should any such have crept into it; but as there are two or three experiments mentioned there, which may perhaps be considered as disagreeing with my opinion, I beg leave to say a few words concerning them.

Mr. de Lassone found that filings of zinc, digested in a caustic fixed alkali, were partially dissolved with a small effervescence, and that the alkali was rendered in some measure mild. This mildness of the alkali Mr. Kirwan accounts for by supposing, that the inflammable air, which is separated during the solution, and causes the effervescence, unites to the atmospheric air contiguous to it, and thereby generates fixed air, which is absorbed by the alkali. But, in reality, the only circumstance from which Mr. de Lassone judged the alkali to become mild, was its making some effervescence when saturated with acids; and this effervescence is more likely to have proceeded from the expulsion of inflammable air than of fixed air, as it seems likely, that the zinc might be more completely deprived of its phlogiston by the acid than by the alkali.

In the abovementioned paper I say, Dr. Priestley observed, that quicksilver fouled by the addition of lead or tin, deposits a powder by agitation and exposure to the air, which consists in great measure of the calx of the imperfect metal. He found too some powder of this kind to contain fixed air; but it must be observed, that the powder used in this experiment was not prepared on purpose, but was procured from quicksilver fouled by having been used in various experiments, and may therefore have contained other impurities besides the metallic calces. On this Mr. Kirwan remarks, that Dr. Priestley did not at first prepare this powder on purpose, but he afterwards did so prepare it (4 Pr. p. 148 and 149), and obtained a powder exactly of the same sort. It was natural to suppose from this remark, that Dr. Priestley must have obtained fixed air from the powder prepared on purpose, and that I had overlooked the passage; but, on turning to the pages referred to, I was surprised to find that it was otherwise, and that Dr. Priestley not so much as hints that he procured fixed air from the powder thus prepared.

With regard to the calcination of metals it may be proper to remark, that this operation is usually performed over the fire, by methods in which they are exposed to the fumes of the burning fuel, and which are so replete with fixed air, that it is not extraordinary, that the metallic calx should, in a short time, absorb a considerable quantity of it; and in particular red lead, which is the calx on which most experiments have been made, is always so prepared. There is another kind of calcination, however, called rusting, which is performed in the open air; but this is so slow an operation, that the rust may easily imbibe a sufficient quantity of fixed air, notwithstanding the small quantity of it usually contained in the atmosphere.

Mr. Kirwan allows that lime-water is not rendered cloudy by the mixture of nitrous and common air; but contends that this does not prove that fixed air is not generated by the union, as he thinks it may be absorbed by the nitrous selenite produced by the union of the nitrous acid with the lime. This induced me to try how small a quantity of fixed air would be perceived in this experiment. I accordingly repeated it in the same manner as described in my paper, except that I purposely added a little fixed air to the common air, and found that when this addition was $\frac{1}{75}$th of the bulk, or $\frac{1}{50}$th of the weight of the common air, the effect on the lime-water was such as could not possibly have been overlooked in my experiments. But as those who suppose fixed air to be generated by the mixture of nitrous and common air, may object to this manner of trying the experiment, and say, that the quantity of fixed air absorbed by the lime-water was really more than $\frac{1}{75}$th of the bulk of the common air, being equal to that quantity over and above the air generated by the mixture, I made another experiment in a different manner; namely, I filled a bottle with lime-water, previously mixed with as much nitrous acid as is contained in an equal bulk of nitrous air, and having inverted it into a vessel of the same, let up into it, in the same manner as in the above-mentioned experiments, a mixture of common air with $\frac{1}{75}$th of its bulk of fixed air, until it was half full. The event was the same as before; namely, the cloudiness produced in the lime-water was such that I could not possibly have overlooked. It must be observed, that in this experiment no fixed air could be generated, and a still greater proportion of the lime-water was turned into nitrous selenite than in the above-mentioned experiments; so that we may safely conclude, that if any fixed air is generated by the mixture of common and nitrous air, it must be less than $\frac{1}{75}$th of the bulk of the common air.

As for the nitrous selenite, it seems not to make the effect of the fixed air at all less sensible, as I found by filling two bottles with common air mixed with $\frac{1}{100}$dth of its bulk of fixed air, and pouring into each of them equal quantities of diluted lime-water; one of these portions of lime-water being previously diluted with an equal quantity of distilled water, and the other with the same quantity of a diluted solution of nitrous selenite, containing about $\frac{1}{400}$dth of its weight of calcareous earth; when I could not perceive that the latter portion of lime-water was rendered at all less cloudy than the former. Though the nitrous selenite, however, does not make the effect of the fixed air less sensible, yet the dilution of the lime-water, in consequence of some of the lime being absorbed by the acid, does; but, I believe, not in any remarkable degree.

There is an experiment mentioned by Mr. Kirwan which, though it cannot be considered as an argument in favour of the generation of fixed air, as he only supposes, without any proof, that fixed air is produced in it, does yet deserve to be taken notice of as a curious experiment. It is, that, if nitrous and common air be mixed over dry quicksilver, the common air is not at all diminished, that is, the bulk of the mixture will be not less than that of the common air employed, until water is admitted, and the mixture agitated for a few minutes. The reason of this in all probability is, that part of the phlogisticated nitrous acid, into which the nitrous air is converted, remains in the state of vapour until condensed by the addition of water. A proof that this is the real case is, that, in this manner of performing the experiment, the red fumes produced on mixing the airs remain visible for some hours, but immediately disappear on the addition of water and agitation.

The most material experiment alledged by Mr. Kirwan is one of Dr. Priestley's, in which he obtained fixed air from a mixture of red precipitate and iron filings. This at first seems really a strong argument in favour of the generation of fixed air; for though plumbago, which is known to consist chiefly of that substance, has lately been found to be contained in iron, yet one would not have expected it to be decompounded by the red precipitate, especially when the quantity of pure iron in the filings was much more than sufficient to supply the precipitate with phlogiston. The following experiment, however, shews that it was really decompounded; and that the fixed air obtained was not generated, but only separated by means of this decomposition.

500 grains of red precipitate mixed with 1000 of iron filings yielded, by

the assistance of heat, 7800 grain measures of fixed air, besides 2400 of a mixture of dephlogisticated and inflammable air, but chiefly the latter. The same quantity of iron filings, taken from the same parcel, was then dissolved in diluted oil of vitriol, so as to leave only the plumbago and other impurities. These mixed with 500 grains of the same red precipitate, and treated as before, yielded 9200 grain measures of fixed air, and 4200 of dephlogisticated air, of an indifferent quality, but without any sensible mixture of inflammable air. It appears, therefore, that less fixed air was produced when the red precipitate was mixed with the iron filings in substance, than when mixed only with the plumbago and other impurities; which shews, that its production was not owing to the iron itself, which seems to contain no fixed air, but to the plumbago, which contains a great deal. The reason, in all probability, why less fixed air was produced in the first case than [in] the latter is, that in the former more of the plumbago escaped being decompounded by the red precipitate than in the other. It must be observed, however, that the filings used in this experiment were mixed with about $\frac{1}{13}$th of their weight of brass, which was not discovered till they were dissolved in the acid, and which makes the experiment less decisive than it would otherwise be. The quantity of fixed air obtained is also much greater than, according to Mr. Bergman's experiment, could be yielded by the plumbago usually contained in 1000 grains of iron; so that though the experiment seems to shew that the fixed air was only produced by the decomposition of the impurities in the filings, yet it certainly ought to be repeated in a more accurate manner.

Before I conclude this paper, it may be proper to sum up the state of the argument on this subject. There are five methods of phlogistication considered by me in my paper on air; namely, first, the calcination of metals, either by themselves or when amalgamated with quicksilver; secondly, the burning of sulphur or phosphorus; thirdly, the mixture of nitrous air; fourthly, the explosion of inflammable air; and, fifthly, the electric spark; and Mr. Kirwan has not pointed out any other which he considers as unexceptionable. Now the last of these I by no means consider as unexceptionable, as it seems much most likely, that the phlogistication of the air in that experiment is owing to the burning or calcination of some substance contained in the apparatus[1]. It is true, that I have no proof of it; but there is so much probability in the opinion, that till it is proved to be erroneous, no conclusion can be drawn from such experiments in favour of the generation of fixed air. As to the first method, or the calcination of metals, there is not the least proof that any fixed air is generated, though we certainly have no direct proof of the contrary; nor

[1] In the experiment with the litmus I attribute the fixed air to the burning of the litmus, not decomposition, as Mr. Kirwan represents it, which is a sufficient reason why no fixed air should be found when the experiment is tried with air in which bodies will not burn.

did I in my paper insinuate that we had. The same thing may be said of the burning of sulphur and phosphorus. As to the mixture of nitrous air, and the combustion of inflammable air, it is proved, that if any fixed air is generated, it is so small as to elude the nicest test we have. It is certain too, that if it had been so much as $\frac{1}{70}$th of the bulk of the common air employed, it would have been perceived in the first of these methods, and would have been sensible in the second though still less. So that out of the five methods enumerated, it has been shewn, that in two no sensible quantity is generated, and not the least proof has been assigned that any is in two of the others; and as to the last, good reasons have been assigned for thinking it inconclusive; and therefore the conclusion drawn by me in the above-mentioned paper seems sufficiently justified; namely, that though it is not impossible that fixed air may be generated in some chemical processes, yet it seems certain, that it is not the general effect of phlogisticating air, and that the diminution of common air by phlogistication is by no means owing to the generation or separation of fixed air from it.

PHILOSOPHICAL TRANSACTIONS. VOL. 75, 1785, p. 372

XXIII. *Experiments on Air. By* Henry Cavendish, *Esq., F.R.S. and A.S.*

Read June 2, 1785

IN a Paper, printed in the last volume of the *Philosophical Transactions*, in which I gave my reasons for thinking that the diminution produced in atmospheric air by phlogistication is not owing to the generation of fixed air, I said it seemed most likely, that the phlogistication of air by the electric spark was owing to the burning of some inflammable matter in the apparatus; and that the fixed air, supposed to be produced in that process, was only separated from that inflammable matter by the burning. At that time, having made no experiments on the subject myself, I was obliged to form my opinion from those already published; but I now find, that though I was right in supposing the phlogistication of the air does not proceed from phlogiston communicated to it by the electric spark, and that no part of the air is converted into fixed air; yet that the real cause of the diminution is very different from what I suspected, and depends upon the conversion of phlogisticated air into nitrous acid.

The apparatus used in making the experiments was as follows. The air through which the spark was intended to be passed, was confined in a glass tube *M*, bent to an angle, as in fig. 1. (tab. XV.) which, after being filled with quicksilver, was inverted into two glasses of the same fluid, as in the figure. The air to be tried was then introduced by means of a small tube, such as is used for thermometers, bent in the manner represented by *ABC* (fig. 2.) the bent end of which, after being previously filled with quicksilver, was introduced, as in the figure, under the glass *DEF*, inverted into water, and filled with the proper kind of air, the end *C* of the tube being kept stopped by the finger; then, on removing the finger from *C*, the quicksilver in the tube descended in the leg *BC*, and its place was supplied with air from the glass *DEF*. Having thus got the proper quantity of air into the tube *ABC*, it was held with the end *C* uppermost, and stopped with the finger; and the end *A*, made smaller for that purpose, being introduced into one end of the bent tube *M*, (fig. 1.) the air, on removing the finger from *C*, was forced into that tube by the pressure of the quicksilver in the leg *BC*. By these means I was enabled to introduce the exact quantity I pleased of any kind of air into the tube *M*; and, by the same means, I could let up any quantity of soap-lees, or any other liquor which I wanted to be in contact with the air.

In one case, however, in which I wanted to introduce air into the tube many times in the same experiment, I used the apparatus represented in fig. 3. consisting of a tube *AB* of a small bore, a ball *C*, and a tube *DE* of a larger bore. This apparatus was first filled with quicksilver; and then the ball *C*, and the tube *AB*, were filled with air, by introducing the end *A* under a glass inverted into water, which contained the proper kind of air, and drawing out the quicksilver from the leg *ED* by a syphon. After being thus furnished with air, the apparatus was weighed, and the end *A* introduced into one end of the tube *M*, and kept there during the experiment; the way of forcing air out of this apparatus into the tube being by thrusting down the tube *ED* a wooden cylinder of such a size as almost to fill up the whole bore, and by occasionally pouring quicksilver into the same tube, to supply the place of that pushed into the ball *C*. After the experiment was finished, the apparatus was weighed again, which shewed exactly how much air had been forced into the tube *M* during the whole experiment; it being equal in bulk to a quantity of quicksilver, whose weight was equal to the increase of weight of the apparatus.

The bore of the tube *M* used in most of the following experiments, was about one-tenth of an inch; and the length of the column of air, occupying the upper part of the tube, was in general from $1\frac{1}{2}$ to $\frac{3}{4}$ of an inch.

It is scarcely necessary to inform any one used to electrical experiments, that in order to force an electrical spark through the tube, it was necessary, not to make a communication between the tube and the conductor, but to place an insulated ball at such a distance from the conductor as to receive a spark from it, and to make a communication between that ball and the quicksilver in one of the glasses, while the quicksilver in the other glass communicated with the ground.

I now proceed to the experiments.

When the electric spark was made to pass through common air, included between short columns of a solution of litmus, the solution acquired a red colour, and the air was diminished, conformably to what was observed by Dr. Priestley.

When lime-water was used instead of the solution of litmus, and the spark was continued till the air could be no further diminished, not the least cloud could be perceived in the lime-water; but the air was reduced to two-thirds of its original bulk; which is a greater diminution than it could have suffered by mere phlogistication, as that is very little more than one-fifth of the whole.

The experiment was next repeated with some impure dephlogisticated air. The air was very much diminished, but without the least cloud being produced in the lime-water. Neither was any cloud produced when fixed air was let up to it; but on the further addition of a little caustic volatile alkali, a brown sediment was immediately perceived.

Hence we may conclude, that the lime-water was saturated by some acid formed during the operation; as in this case it is evident, that no earth could be precipitated by the fixed air alone, but that caustic volatile alkali, on being added, would absorb the fixed air, and thus becoming mild, would immediately precipitate the earth; whereas, if the earth in the lime-water had not been saturated with an acid, it would have been precipitated by the fixed air. As to the brown colour of the sediment, it most likely proceeded from some of the quicksilver having been dissolved.

It must be observed, that if any fixed air, as well as acid, had been generated in these two experiments with the lime-water, a cloud must have been at first perceived in it, though that cloud would afterwards disappear by the earth being re-dissolved by the acid; for till the acid produced was sufficient to dissolve the whole of the earth, some of the remainder would be precipitated by the fixed air; so that we may safely conclude, that no fixed air was generated in the operation.

When the air is confined by soap-lees, the diminution proceeds rather faster than when it is confined by lime-water; for which reason, as well as on account of their containing so much more alkaline matter in proportion to their bulk, soap-lees seemed better adapted for experiments designed to investigate the nature of this acid, than lime-water. I accordingly made some experiments to determine what degree of purity the air should be of, in order to be diminished most readily, and to the greatest degree; and I found, that, when good dephlogisticated air was used, the diminution was but small; when perfectly phlogisticated air was used, no sensible diminution took place; but when five parts of pure dephlogisticated air were mixed with three parts of common air, almost the whole of the air was made to disappear.

It must be considered, that common air consists of one part of dephlogisticated air, mixed with four of phlogisticated; so that a mixture of five parts of pure dephlogisticated air, and three of common air, is the same thing as a mixture of seven parts of dephlogisticated air with three of phlogisticated.

Having made these previous trials, I introduced into the tube a little soap-lees, and then let up some dephlogisticated and common air, mixed in the above-mentioned proportions, which rising to the top of the tube *M*, divided the soap-lees into its two legs. As fast as the air was diminished by the electric spark, I continued adding more of the same kind, till no further diminution took place: after which a little pure dephlogisticated air, and after that a little common air, were added, in order to see whether the cessation of diminution was not owing to some imperfection in the proportion of the two kinds of air to each other; but without effect[1]. The soap-lees being then poured out of the tube, and separated from the quick-silver, seemed to be perfectly neutralized, as they did not at all discolour paper tinged with the juice of blue flowers. Being evaporated to dryness, they left a small quantity of salt, which was evidently nitre, as appeared by the manner in which paper, impregnated with a solution of it, burned.

For more satisfaction, I tried this experiment over again on a larger scale. About five times the former quantity of soap-lees were now let up into a tube of a larger bore; and a mixture of dephlogisticated and common air, in the same proportions as before, being introduced by the apparatus represented in fig. 3. the spark was continued till no more air could be made to disappear. The liquor, when poured out of the tube, smelled evidently of phlogisticated nitrous acid, and being evaporated to dryness, yielded $1\frac{4}{10}$ gr. of salt, which is pretty exactly equal in weight to the nitre which that quantity of soap-lees would have afforded if saturated with nitrous acid. This salt was found, by the manner in which paper dipped into a solution of it burned, to be true nitre. It appeared, by the test of *terra ponderosa salita*, to contain not more vitriolic acid than the soap-lees themselves contained, which was excessively little; and there is no reason to think that any other acid entered into it, except the nitrous.

A circumstance, however, occurred, which at first seemed to shew, that this salt contained some marine acid; namely, an evident precipitation took place when a solution of silver was added to some of it dissolved in water; though the soap-lees used in its formation were perfectly free from marine acid, and though, to prevent all danger of any precipitate being formed by an excess of alkali in it, some purified nitrous acid had been added to it, previous to the addition of the solution of silver. On consideration, however, I suspected, that this precipitation might arise from the nitrous acid in it being phlogisticated; and therefore I tried whether nitre, much phlogisticated, would precipitate silver from its solution. For this purpose I exposed some nitre to the fire, in an earthen retort, till it had yielded a good deal of dephlogisticated air; and then, having dissolved it in water, and added to it some well-purified spirit of nitre till it was sensibly acid, in order to be certain that the alkali did not predominate, I dropped into it some solution of silver, which immediately made a very copious precipitate. This solution, however, being deprived of some of its phlogiston by evaporation to dryness, and exposure for a few weeks to the

[1] From what follows it appears, that the reason why the air ceased to diminish was, that as the soap-lees were then become neutralized, no alkali remained to absorb the acid formed by the operation, and in consequence scarce any air was turned into acid. The spark, however, was not continued long enough after the apparent cessation of diminution, to determine with certainty, whether it was only that the diminution went on remarkably slower than before, or that it was almost come to a stand, and could not have been carried much further, though I had persisted in passing the sparks.

air, lost the property of precipitating silver from its solution; a proof that this property depended only on its phlogistication, and not on its having absorbed sea-salt from the retort, or by any other means.

Hence it is certain, that nitre, when much phlogisticated, is capable of making a precipitate with a solution of silver; and therefore there is no reason to think, that the precipitate, which our salt occasioned with a solution of silver, proceeded from any other cause than that of its being phlogisticated; especially as it appeared by the smell, both on first taking it out of the tube, and on the addition of the spirit of nitre, previous to dropping in the solution of silver, that the acid in it was much phlogisticated. This property of phlogisticated nitre is worth the attention of chemists; as otherwise they may sometimes be led into mistakes, in investigating the presence of marine acid by a solution of silver.

In the above-mentioned Paper I said, that when nitre is detonated with charcoal, the acid is converted into phlogisticated air; that is, into a substance which, as far as I could perceive, possesses all the properties of the phlogisticated air of our atmosphere; from which I concluded, that phlogisticated air is nothing else than nitrous acid united to phlogiston. According to this conclusion, phlogisticated air ought to be reduced to nitrous acid by being deprived of its phlogiston. But as dephlogisticated air is only water deprived of phlogiston, it is plain, that adding dephlogisticated air to a body, is equivalent to depriving it of phlogiston, and adding water to it; and therefore, phlogisticated air ought also to be reduced to nitrous acid, by being made to unite to, or form a chemical combination with, dephlogisticated air; only the acid formed this way will be more dilute, than if the phlogisticated air was simply deprived of phlogiston.

This being premised, we may safely conclude, that in the present experiments the phlogisticated air was enabled, by means of the electrical spark, to unite to, or form a chemical combination with, the dephlogisticated air, and was thereby reduced to nitrous acid, which united to the soap-lees, and formed a solution of nitre; for in these experiments those two airs actually disappeared, and nitrous acid was actually formed in their room; and as, moreover, it has just been shewn, from other circumstances, that phlogisticated air must form nitrous acid, when combined with dephlogisticated air, the above-mentioned opinion seems to be sufficiently established. A further confirmation of it is, that, as far as I can perceive, no diminution of air is produced when the electric spark is passed either through pure dephlogisticated air, or through perfectly phlogisticated air; which indicates the necessity of a combination of these two airs to produce the acid. Moreover, it was found in the last experiment, that the quantity of nitre procured was the same that the soap-lees would have produced if saturated with nitrous acid; which shews, that the production of the nitre was not owing to any decomposition of the soap-lees.

It may be worth remarking, that whereas in the detonation of nitre with inflammable substances, the acid unites to phlogiston, and forms phlogisticated air, in these experiments the reverse of this process was carried on; namely, the phlogisticated air united to the dephlogisticated air, which is equivalent to being deprived of its phlogiston, and was reduced to nitrous acid.

In the above-mentioned Paper I also gave my reasons for thinking, that the small quantity of nitrous acid, produced by the explosion of dephlogisticated and inflammable air, proceeded from a portion of phlogisticated air mixed with the dephlogisticated, which I supposed was deprived of its phlogiston, and turned into nitrous acid, by the action of the dephlogisticated air on it, assisted by the heat of the explosion. This opinion, as must appear to every one, is confirmed in a remarkable manner by the foregoing experiments; as from them it is evident, that dephlogisticated air is able to deprive phlogisticated air of its phlogiston, and reduce it into acid, when assisted by the electric spark; and therefore it is not extraordinary that it should do so, when assisted by the heat of the explosion.

The soap-lees used in the foregoing experiments were made from salt of tartar, prepared without nitre; and were of such a strength as to yield one-tenth of their weight of nitre when saturated with nitrous acid. The dephlogisticated air also was prepared without nitre, that used in the first experiment with the soap-lees being procured from the black powder formed by the agitation of quicksilver mixed with lead[1], and that used in the latter from turbith mineral. In the first experiment, the quantity of soap-lees used was 35 measures, each of which was equal in bulk to one grain of quicksilver; and that of the air absorbed was 416 such measures of phlogisticated air, and 914 of dephlogisticated. In the second experiment, 178 measures of soap-lees were used, and they absorbed 1920 of phlogisticated air, and 4860 of dephlogisticated. It must be observed, however, that in both experiments some air remained in the tube uncondensed, whose degree of purity I had no way of trying; so that the proportion of each species of air absorbed is not known with much exactness.

As far as the experiments hitherto published extend, we scarcely know more of the nature of the phlogisticated part of our atmosphere, than that it is not diminished by lime-water, caustic alkalies, or nitrous air; that it is unfit to support fire, or maintain life in animals; and that its specific gravity is not much less than that of common air: so that, though the nitrous acid, by being united to phlogiston, is converted into air possessed of these properties, and consequently, though it was reasonable to suppose, that part at least of the phlogisticated air of the atmosphere consists of this acid united to phlogiston, yet it might fairly be doubted whether the whole is of this kind, or whether there are not in reality many different

[1] This air was as pure as any that can be procured by most processes. I propose giving an account of the experiment, in which it was prepared, in a future Paper.

substances confounded together by us under the name of phlogisticated air. I therefore made an experiment to determine, whether the whole of a given portion of the phlogisticated air of the atmosphere could be reduced to nitrous acid, or whether there was not a part of a different nature from the rest, which would refuse to undergo that change. The foregoing experiments indeed in some measure decided this point, as much the greatest part of the air let up into the tube lost its elasticity; yet, as some remained unabsorbed, it did not appear for certain whether that was of the same nature as the rest or not. For this purpose I diminished a similar mixture of dephlogisticated and common air, in the same manner as before, till it was reduced to a small part of its original bulk. I then, in order to decompound as much as I could of the phlogisticated air which remained in the tube, added some dephlogisticated air to it, and continued the spark till no further diminution took place. Having by these means condensed as much as I could of the phlogisticated air, I let up some solution of liver of sulphur to absorb the dephlogisticated air; after which only a small bubble of air remained unabsorbed, which certainly was not more than $\frac{1}{120}$ of the bulk of the phlogisticated air let up into the tube; so that if there is any part of the phlogisticated air of our atmosphere which differs from the rest, and cannot be reduced to nitrous acid, we may safely conclude, that it is not more than $\frac{1}{120}$ part of the whole.

The foregoing experiments shew, that the chief cause of the diminution which common air, or a mixture of common and dephlogisticated air, suffers by the electric spark, is the conversion of the air into nitrous acid; but yet it seemed not unlikely, that when any liquor, containing inflammable matter, was in contact with the air in the tube, some of this matter might be burnt by the spark, and thereby diminish the air, as I supposed in the above-mentioned Paper to be the case. The best way which occurred to me of discovering whether this happened or not, was to pass the spark through dephlogisticated air, included between different liquors: for then, if the diminution proceeded solely from the conversion of air into nitrous acid, it is plain that, when the dephlogisticated air was perfectly pure, no diminution would take place; but when it contained any phlogisticated air, all this phlogisticated air, joined to as much of the dephlogisticated air as must unite to it in order to reduce it into acid, that is, two or three times its bulk, would disappear, and no more; so that the whole diminution could not exceed three or four times the bulk of the phlogisticated air: whereas, if the diminution proceeded from the burning of the inflammable matter, the purer the dephlogisticated air was, the greater and quicker would be the diminution.

The result of the experiments was, that when dephlogisticated air, containing only $\frac{1}{20}$ of its bulk of phlogisticated air (that being the purest air I then had), was confined between short columns of soap-lees, and the spark passed through it till no further diminution could be perceived, the air lost $\frac{43}{200}$ of its bulk; which is not a greater diminution than might very likely proceed from the first-mentioned cause; as the dephlogisticated air might easily be mixed with a little common air while introducing into the tube.

When the same dephlogisticated air was confined between columns of distilled water, the diminution was rather greater than before, and a white powder was formed on the surface of the quicksilver beneath; the reason of which, in all probability, was, that the acid produced in the operation corroded the quicksilver, and formed the white powder; and that the nitrous air, produced by that corrosion, united to the dephlogisticated air, and caused a greater diminution than would otherwise have taken place.

When a solution of litmus was used, instead of distilled water, the solution soon acquired a red colour, which grew paler and paler as the spark was continued, till at last it became quite colourless and transparent. The air was diminished by almost half, and I believe might have been still further diminished, had the spark been continued. When lime-water was let up into the tube, a cloud was formed, and the air was further diminished by about one-fifth. The remaining air was good dephlogisticated air. In this experiment, therefore, the litmus was, if not burnt, at least decompounded, so as to lose entirely its purple colour, and to yield fixed air; so that, though soap-lees cannot be decompounded by this process, yet the solution of litmus can, and so very likely might the solutions of many other combustible substances. But there is nothing, in any of these experiments, which favours the opinion of the air being at all diminished by means of phlogiston communicated to it by the electric spark.

PHILOSOPHICAL TRANSACTIONS. VOL. 76, 1786, p. 241

XIII. *An Account of Experiments made by Mr.* John McNab, *at* Henley House, Hudson's Bay, *relating to freezing Mixtures. By* Henry Cavendish, *Esq.*, *F.R.S. and A.S.*

Read February 23, 1786

IN my observations on Mr. Hutchins's Experiments, printed in the LXXIIId volume of the *Philosophical Transactions*, I gave my opinion concerning the cause of the cold produced by mixing snow with different liquors. As there were some circumstances, however, which seemed to form a difficulty in the way of this opinion, I was desirous of having further experiments made on the subject; and at the same time I thought that, by proper management, a greater degree of cold might be produced than had hitherto been done. On mentioning the experiments I wished to have made to Mr. Hutchins, he very obligingly desired Mr. McNab, Master at Henley-House, to try them; who was so good as to undertake the business, and has executed it in the most satisfactory manner; as he has not only taken great pains, but has shewn the utmost attention and accuracy, in observing and relating all the phænomena which occurred, and has manifested great judgement in frequently adapting the manner of trying the experiments to appearances which occurred in former ones, to which we are indebted for great part of the most curious facts in this paper. His endeavours have also been attended with much success, as he has not only shewn many remarkable circumstances relating to the freezing of the nitrous and vitriolic acids, and the phænomena of freezing mixtures; but has also produced degrees of cold greatly superior to any before known.

1. In the above-mentioned Paper I said, that the cold produced by mixing spirit of nitre with snow, is owing to the melting of the snow; and that in all probability there is a certain degree of cold, in which spirit of nitre is so far from dissolving snow, that it will yield out part of its own water, and suffer that to freeze, as is the case with solutions of common salt; so that if the cold of the materials, before mixing, is equal to this, no additional cold can be produced. A circumstance, however, which at first sight seems repugnant to this opinion, occurred in an experiment of

Fahrenheit's for producing cold by a mixture of spirit of nitre and ice; namely, that the acid, which had been repeatedly cooled by different frigorific mixtures, was found frozen before it was mixed with the ice; notwithstanding which, cold was produced by the mixture. Professor Braun also found, that cold was produced by mixing frozen spirit of nitre with snow. On consideration, however, this appeared by no means inconsistent with the opinion there laid down, as there was great reason to think, that the freezing of the acid was of a different kind from that considered in the above-mentioned Paper, and that it did not proceed from the watery part separating from the rest and freezing; but that the whole acid, or perhaps the more concentrated part, froze; in which case it would not be extraordinary that the acid should dissolve more snow, and produce cold.

2. To clear up this point, I sent to Hudson's Bay a bottle of spirit of nitre, of nearly the same strength as Fahrenheit's; and desired Mr. McNab to expose it to the cold, and, if it froze, to ascertain the temperature, and decant the fluid part into another bottle, and send both home to be examined, as it would thereby be known, whether it was the whole acid, or only the watery part, which froze. For the same purpose also I sent some dephlogisticated spirit of nitre of the same strength, and also some strong oil of vitriol. I also sent some spirit of nitre and spirit of wine, both diluted with so much water, that it was expected, that with the cold of Hudson's Bay they would suffer the first kind of congelation; that is, their watery part would freeze, and thereby make the difference between the two kinds of freezing more apparent.

3. In the same Paper I say,

That on adding snow gradually to some of the spirit of nitre used by Mr. Hutchins, I found, that the addition of a small quantity produced heat instead of cold; and it was not until so much was added as to increase the heat from 28° to 51°, that the addition of more snow began to produce cold; the quantity of snow required for this purpose being pretty exactly one-quarter of the weight of the spirit of nitre, and the heat of the snow and air of the room, as well as the acid, being 28°. The reason of this is, that a great deal of heat is produced by mixing water with spirit of nitre, and the stronger the spirit is, the greater is the heat produced. Now it appears from this experiment, that before the acid was diluted, the heat produced by its union with the water formed from the melted snow was greater than the cold produced by the melting of the snow; and it was not till it was diluted by the addition of one-quarter of its weight of that substance, that the cold generated by the latter cause began to exceed the heat generated by the former. From what has been said, it is evident, that the cold of a freezing mixture, made with the undiluted acid, cannot be quite so great as that made with the same acid, diluted with a quarter of its weight of water, supposing the acid and snow to be both at 28° of heat; and there is no reason to think, that the event will be different if they are colder; for the undiluted acid will not begin to generate cold, until so much snow is dissolved as to increase its heat from 28° to 51°, so that no greater cold will be produced, than would be obtained by mixing the diluted acid heated to 51° with snow of the heat of 28°. This method of adding snow gradually to an acid, is much the best way I know of finding what strength it ought to be of, in order to produce the greatest effect possible.

As it seemed likely that, by following this method, a greater degree of cold might be produced than had been done hitherto, I sent three other bottles of spirit of nitre and oil of vitriol, all three diluted, but not so much so, but that I thought they would require a little further dilution, in order to reduce them to their properest degree of strength. I also sent a bottle of highly rectified spirit of wine, and a mixture of equal quantities of the above-mentioned common spirit of nitre and oil of vitriol; and desired Mr. McNab to find what degree of cold could be produced by mixing them with snow, after having first reduced them, in the above-mentioned manner, to their best degree of strength[1].

He was also desired to ascertain how much snow he added; for as their strength was determined before they were sent out, it would thereby be known what was the best strength of these liquors for frigorific mixtures.

All these bottles were numbered with a diamond; and as I shall sometimes distinguish them by these numbers, and as it may be of use to those who may consult the original, I have added the following list of these bottles, with their contents.

No.	Liquors mentioned in Art. 3	Weight of marble which they dissolve	Specific gravity at 60° of heat
168	Spirit of nitre	,582	1,4371
27	Dephlogisticated spirit of nitre	,53	1,4040
103	Diluted oil of vitriol	,654	1,5596
28	Equal weights of No. 168 and No. 103 ...	—	—
8	Very highly rectified spirit of wine	—	,8195
	Liquors mentioned in Art. 2		
151	Strong oil of vitriol	,98	1,8437
142	Spirit of nitre	,525	1,4043
139	Some of the same diluted with twice its weight of water	—	—
141	Dephlogisticated spirit of nitre	,53	1,4033
143	Some of the same spirit of wine as in No. 8 diluted with 1½ its weight of water ...	—	—
72	Diluted oil of vitriol for comparing the thermometers	,629	—
171	Oil of vitriol of about the usual strength, but the exact strength not known, intended to refresh the former when too weak	—	—

[1] This might have been done at home; but I thought it not unlikely that the strength found this way might differ, in some measure, according to the heat in which the experiment was tried.

4. Professor Braun says, that by mixtures of snow and spirit of nitre he sunk thermometers filled with oil of sassafras, and some other essential oils, to – 100° or – 124°; and that, by the same means, he sunk thermometers filled with the highest rectified spirit of wine to – 148°. Though there seemed great reason to think, from Mr. Hutchins's experiments, that there must be some mistake in this; yet, as it was possible that the essential oils, and even spirit of wine of a strength much different from that with which Mr. Hutchins's thermometers were filled, might follow a considerably different progression in their contraction by great degrees of cold, I sent a thermometer filled with oil of sassafras, and two others with spirits of wine. One of these last was filled with the highest rectified spirits I could procure, its specific gravity at 60° of heat being ,8185; the other was intended to be filled with common spirits, though from circumstances I am inclined to suspect *that* also to have been filled with the best spirits. Besides these, there was sent a mercurial thermometer, accurately adjusted, according to the directions of the Committee of the Royal Society, printed in the LXVIIth volume of the *Transactions*; and also the two spirit thermometers used by Mr. Hutchins, which were filled with spirits whose specific gravity was ,8247.

5. These thermometers were compared together by exposing them to the cold, with their balls immersed in a glass vessel filled with diluted oil of vitriol. They were at times also compared in cold more violent than the natural cold of the climate, by adding snow to the acid in which they were tried, in which case care was taken to keep the mixture frequently stirred. Oil of vitriol was recommended for this purpose, as a fluid which would most likely bear any degree of cold without freezing, and whose natural cold might be much increased by the addition of snow. It seems to have answered the purpose very well, and not to have been attended with any inconvenience.

During the first comparison of these thermometers, a whitish globule, such as those which appear in frozen oil, was observed in the tube of the thermometer filled with oil of sassafras. This appearance of congelation did not much increase; but two days after a large air bubble was found in its ball, which prevented Mr. McNab from making further observations with it.

It is well known, that spirit of wine expands more by a given number of degrees of a mercurial thermometer in warm temperatures than in cold ones; and this inequality, as might be expected, was less in the stronger spirit than in the weaker, but the difference was inconsiderable. The oil of sassafras also had some of this inequality, but much less. It however appears to be by no means a proper fluid for filling thermometers with. No appearance was observed which indicates any considerable irregularity in the contraction of spirits of wine in intense cold, or which renders it probable, that thermometers filled therewith could be sunk by a mixture of snow and spirit of nitre to a degree near approaching to that mentioned by Professor Braun.

6. Mr. McNab in his experiments sometimes used one thermometer and sometimes another; but in the following pages I have reduced all the observations to the same standard; namely, in degrees of cold less than that of freezing mercury I have set down that degree which would have been shewn by the mercurial thermometer in the same circumstances; but as that could not have been done in greater degrees of cold, as the mercurial thermometer then becomes of no use, I found how much lower the mercurial thermometer stood at its freezing point, than each of the spirit thermometers, and increased the cold shewn by the latter by that difference.

On the common and dephlogisticated Acids of Nitre.

The following experiments shew, that both these acids are capable of a kind of congelation, in which the whole, and not merely the watery part, freezes. Their freezing point also differs greatly according to the strength, and varies according to a very unexpected law. Like water too they bear being cooled very much below their freezing point before the congelation begins, and as soon as that takes place, immediately rise up to the freezing point.

7. On the morning of Feb. 1 the common and dephlogisticated spirits of nitre, No. 142 and 141, whose specific gravities were 1,4043 and 1,4033, were found clear and fluid, the cold of the air at that time being $-47°$. They also bore being shook without any alteration; but on taking out their stoppers, both of them in a few minutes began to freeze, the congelation beginning by a white appearance at top, which gradually spread to the bottom; and they became so thick as not to move on inclining the phial. For want of a thermometer whose ball reached far enough below its scale, Mr. McNab was not able to determine their cold while in the bottle; but in somewhat more than an hour's time, the frozen acid had so much subsided as to admit of his pouring a little fluid matter out of each into a glass with a thermometer in it[1]; whereby the cold of the common spirit of nitre was found to be $-31°\frac{1}{2}$, and that of the dephlogisticated acid $-30°$, the temperature of the air being $-41°$. Each of these decanted liquors, at the time their temperature was tried, was full of small *spicula* of ice: they were then put into phials well stopped, and they, as well as the undecanted liquors, sent home to be examined. The decanted part of the common spirit of nitre dissolved ,535 of its weight of marble, and the un-

[1] It may be asked, why it was more possible to decant any liquor at this time than at first, as the acid was all the while exposed to a cold much below the freezing point? The reason in all probability is, not that any part of the ice first formed dissolved, but that the small filaments into which it shot collected together, and in some measure subsided to the bottom.

decanted part ,523; for which reason I shall call the strength of the former ,535, and that of the latter ,523; which mode of reckoning is observed in the remainder of this Paper. The strength of the decanted part of the dephlogisticated acid was ,56, and that of the undecanted part ,528; so that it appears that in each of these acids the unfrozen part was a little stronger than the frozen part. It is remarkable, that in the common spirit of nitre, the decanted part, though stronger than the other, was paler coloured and less fuming.

8. On Dec. 21, the temperature of the air being – 28°, some dephlogisticated spirit of nitre (No. 27), of nearly the same strength as the former acid, was poured into a jar, in order to be diluted with snow, as recommended in Art. 2. Immediately after it was decanted, it began to freeze, in the same manner as before described, except that a less portion of it seems to have congealed: its temperature, tried by dipping a thermometer into it, was – 19°, where it remained stationary for many minutes; it was then diluted with snow, as will be mentioned in Art. 14, whereby its strength was reduced to ,434.

9. On Dec. 29th, this diluted acid was completely melted, and half of it poured into a jar with a ground stopper, and both portions exposed to the air. In the morning they were perfectly fluid; but on taking the stopper out of the jar, and dipping in it a thermometer, the acid immediately froze, beginning by forming a white coat round the ball of the thermometer, which gradually spread through the whole fluid; and at the same time the thermometer rose till it stood stationary at – 5°. The cold of the acid before it began to freeze must have been about – 30°½, that being the temperature of a glass of vitriolic acid standing near it; but the thermometer which was dipped into it was five or six degrees colder, which seems to be the cause of the congelation beginning round the ball.

In the afternoon a thermometer was dipped into the other half of the acid, where, as the weather had grown less cold, it stood above a minute at – 25°, without freezing; then, however, the acid froze, with the same appearance as in the morning, and at the same time the thermometer rose to – 4°, and became stationary.

This acid, being left in the air with the thermometer in it, was found in the evening at – 45°; it however was not intirely frozen, being only thick as an unguent, which shews that the unfrozen part must have been of a different strength from the frozen part; but it does not appear whether stronger or weaker. The next morning it was frozen solid, though the cold was only half a degree greater. On Jan. 16th, this acid was again tried in the same manner; it then suffered a thermometer, whose ball had been previously warmed in the hand, to be dipped into it, and remain there several minutes without freezing, though its temperature was – 35°. But on lifting up the thermometer, a drop fell from its ball into the acid, which immediately set it a freezing, and it rose up to – 4°½.

10. On Dec. 22d, the spirit of nitre (No. 168) which a few days before had been diluted with snow, so as to be reduced to the strength of ,411, was divided into two equal parts, and exposed to the cold. On Dec. 29th, when the temperature of the air was $-17^\circ\frac{1}{2}$, one of these parts was found beginning to freeze; the other was fluid, but began to freeze on dipping in a thermometer; the thermometer in both kept stationary at $-1^\circ\frac{1}{2}$. The latter was twice re-melted and exposed to the cold, and both times the temperature of the frozen acid came out the same as before.

11. The white colour of the ice in these experiments seems owing only to its consisting of very slender filaments; for in some cases, where it froze slower, and where, in consequence, it shot into larger solid masses, they were transparent, and of the same colour as the acid itself. By the continuance of a sufficient cold, the acid, which by hasty freezing put on the white appearance, would become hard solid ice, but yet still retained its white appearance, owing perhaps to the filaments first shot consisting of an acid differing in strength from that which froze afterwards, and filled up the interstices.

In all these experiments, whether the ice was formed into minute filaments or solid masses, still, whenever there was a sufficient quantity of fluid matter to admit of it, they constantly subsided to the bottom; a proof that the frozen part was heavier than the unfrozen. The difference indeed is so great, that in one case where it froze into solid crystals on the surface, these crystals, when detached by agitation, fell with force enough to make a tinkling noise against the bottom of the glass.

These acids contract very much on freezing. Whenever the acid is frozen solid, the surface, instead of being elevated in ridges, like frozen water, is depressed and full of cracks. In one experiment Mr. McNab, after a glass almost full of acid was nearly frozen, filled it to the brim with fresh acid; and then, after it was completely frozen, the surface was visibly depressed, with fissures one-eighth of an inch broad, extending from top to bottom. It is this contraction of the acid in freezing which makes the frozen part subside in the fluid part; as it was found, in the undiluted acid, that the latter consisted of a stronger, and consequently heavier, acid than the former. But still the subsidence of the frozen part shews, that the ice is not mere water, or even a very dilute acid; which indeed was proved by the examination of the liquors sent home.

The ninth and tenth articles shew, that though the acids bear being cooled greatly below the freezing point, without any congelation taking place, yet as soon as they begin to freeze they immediately rise up to their freezing point; and this point is always very nearly, if not exactly, the same in the same acid; for those acids were frozen and melted again three or four times, and were cooled considerably more below the freezing point in one trial than another, and yet as soon as they began to freeze the thermometer immersed in them constantly rose nearly to the same point.

The quantity which these acids will bear being cooled below the freezing point, without freezing, is remarkable. The diluted spirit of nitre, whose freezing point is $-1°\frac{1}{2}$, once bore being cooled to near $-39°$, without freezing, that is, near 37 degrees below its freezing point. The diluted dephlogisticated spirit of nitre, whose freezing point is $-5°$, bore cooling to $-35°$; and the dephlogisticated spirit of nitre (141) whose true freezing point is most likely $-19°$ (*see next article*) bore being cooled to $-49°$: perhaps too they might have born to be cooled considerably lower without freezing, but how much does not appear. It must be observed, however, that the same diluted spirit which at one time bore being cooled to $-39°$, at another froze, without any apparent cause, when its cold was certainly less than $-30°$, and most likely not much below $-18°$.

12. The freezing point differs remarkably, according to the strength of the acid. In the diluted dephlogisticated and common spirit of Art. 7 and 8, the freezing point was $-5°$ and $-1°\frac{1}{2}$. In the dephlogisticated and common spirit of Art. 5 the decanted parts of which were stronger than the foregoing in scarcely so great a proportion as that of four to three, it seemed to be $-30°$ and $-31°\frac{1}{2}$. It may indeed be suspected, that as this point was determined only by pouring a small quantity of the acid into a glass, at a time when the air and glass were much colder than the acid, these decanted liquors might be cooled by the air and glass, and thereby make the freezing point appear lower than it really was: but I do not think this could be the case; for as the decanted liquors were full of small filaments of ice, they could hardly be cooled sensibly below their freezing points without freezing; and any cold, communicated to them by the air or glass, would serve only to convert more of them into ice, without sensibly increasing their cold: so that I think this experiment determines the true freezing point of their decanted part; but it must be observed, that as the decanted part was rather stronger than the rest, it is very possible that the freezing point of the undecanted part might be considerably less cold.

A circumstance which might incline one to think, that the way by which the freezing point was determined in this experiment is defective is, that the freezing point of the dephlogisticated acid No. 27, though nearly of the same strength as that last mentioned, but rather stronger, was much less low, being only $-19°$. But I have little doubt that the true reason of this is, that in the former acid the strength of the decanted part, which is the part whose freezing point was tried, was found to be at least $\frac{1}{20}$ greater than that of the whole mass; whereas in No. 27 the fluid part was in all probability not sensibly stronger than the whole mass; for as No. 27 was cooled only seven degrees below the freezing point, and its temperature was tried soon after its beginning to freeze, not much of the acid could have frozen; whereas the other was cooled 15 degrees below its freezing point, and was exposed for an hour or two to an air not much less cold, in consequence of which a considerable part of the acid must have frozen; so that in all probability the acid, whose freezing point was found to be -30°, was in reality $\frac{1}{20}$ part stronger than that whose freezing point was -19°.

If this reasoning be just, the freezing point of these acids is as follows:

	Strength	Freezing point
Dephlogisticated spirit of nitre, whose strength =	,56	-30°
	,53	-19
	,437	$-4\frac{1}{2}$
Common spirit of nitre, whose strength =	,54	$-31\frac{1}{2}$
	,411	$-1\frac{1}{2}$

On the Phænomena observed on mixing Snow with these Acids.

13. On Dec. 13, snow was added to the spirit of nitre No. 168, as recommended in Art. 2. The snow was put in very gradually, and time was taken to find what effect each addition had on the thermometer and mixture, before more was added. The temperature of the acid before the mixture was -27°, and each addition of snow raised the thermometer a little, till it rose to $-1^\circ\frac{1}{4}$; after which the next addition made it sink to -2°, which shewed that sufficient snow had then been added. The quantity of snow used was pretty exactly $\frac{4}{10}$ of the weight of the acid, the weight of the acid being 13 oz. so that the strength of the diluted acid was reduced to ,411.

The acid before the addition of snow had no signs of freezing, its temperature being in all probability much above its freezing point; yet the snow did not appear to dissolve, but formed thin white cakes, which however did not float on the surface, but fell to the bottom, and when broke by the spatula formed a gritty sediment; so that it appears, that these cakes are not simply undissolved snow, but that the adjoining acid absorbed so much of the snow in contact with it, as to become diluted sufficiently to freeze with that degree of cold, and then congealed into these cakes. The quantity of congealed matter seems to have kept increasing till the end of the experiment.

14. On Dec. 21, an experiment was made in the same manner with the dephlogisticated spirit of nitre No. 27. The acid began to freeze in pouring it into the jar in which the mixture was to be made, and stood stationary there at -19°, as related in Art. 6; so that the liquor at the beginning of the experiment was white and thick, which made the effect of the addition of the snow less sensible. However, the congealed matter constantly subsided to the bottom, and the quantity seems to have continued increasing to the end of the experiment. The heat of the mixture rose to -4° before cold began to be produced, and the quantity of snow added was $\frac{22}{100}$ of that of the acid, so that the strength of the acid was reduced to ,437 by the dilution.

A very remarkable circumstance in this experiment is, that the acid, while the snow was adding, first became of a yellowish, and afterwards of a greenish or bluish hue. This colour did not go off by standing, but continued at least ten days, during which time the acid constantly kept that colour, except when by hasty freezing it shot into small filaments, in which case it put on the white appearance which these acids always assumed under those circumstances; but once that by gradual freezing it shot into transparent ice, this ice was of a bluish colour.

It is difficult to conceive what this colour should proceed from. Spirit of nitre is well known to assume this colour when much phlogisticated and properly diluted; but one does not see why it should become phlogisticated by the addition of the snow, and still less why the dephlogisticated acid should become more phlogisticated thereby than the common acid did; for though it is not extraordinary, that a process not capable of producing any increase of phlogistication in the common acid, should make this as much phlogisticated as that, yet it is very extraordinary that it should make it more so. No notice is taken of any effervescence or discharge of air while it was assuming this colour, nor was it observed that it became more smoking thereby, or that the top of the phial in which it was kept became full of red fumes, as might naturally be expected if it was rendered much phlogisticated. These are circumstances which, considering Mr. McNab's great attention to set down all the phænomena that occurred, I should think would hardly have been omitted if they had really happened.

15. It is remarkable, that in both these experiments the addition of snow produced heat, until it arrived pretty exactly at what was found to be the freezing point of the diluted acid; but that as soon as it arrived at that point, the addition of more snow began to produce cold. This can hardly be owing merely to accident, and to both acids having happened to be of that precise degree of heat before the experiment began, that their heat after dilution should coincide with the freezing point answering to their new strength. The true cause seems to be as follows. It will be shewn in Art. 16 and 17, that the freezing point of these acids, when diluted as in the foregoing experiments, is much less cold than when they are considerably more diluted; and it was before shewn to be much less cold than when not diluted; so that there must be a certain degree of strength, not very different from that to which these acids were reduced by dilution, at which they freeze with a less degree of cold than when they are either stronger or weaker. Now in these experiments, the temperature of the liquors before dilution was below this point of easiest freezing, and a great deal of the acid was in a state of congelation all the time of dilution; the consequence of which is, that when they were diluted to the strength of easiest freezing, they would also be at the heat of easiest freezing; for they could not be below that point, because, if they were, so much of the acid would immediately freeze as would raise them up to it; and they could not

be above it, for, if they were, so much of the congealed acid would dissolve as would sink them down to it. After they were arrived at this strength of easiest freezing, the addition of more snow would produce cold, unless this strength be greater than that at which the addition of a small quantity of snow begins to produce cold; but even were this the case, heat would not be produced, but the temperature of the acids would remain stationary until they were so much diluted that the addition of more snow should produce cold. So that, in either case, the heat of the acids, at the time that the addition of fresh snow began to produce cold, must be that of easiest freezing; and consequently, as this heat was found to coincide very nearly with the freezing point of these acids, after dilution, it follows that their strengths at that time could differ very little from the strength of easiest freezing.

If the temperature of the liquors at the beginning of the experiment had been above the point of easiest freezing, none of the acid would have congealed during the dilution, and nothing could have been learnt from the experiment relating to the point of easiest freezing; but the heat would have kept increasing, till the acid was diluted to that degree of strength at which the cold produced by the dissolving of the snow was just equal to the heat produced by the union of the melted snow with the acid[1]; after which the addition of more snow would begin to produce cold. When I recommended this method of finding the best strength of spirit of nitre for producing cold, by the addition of snow, I was not aware of any impediment from the freezing of the acid, in which case it would have been a very proper method; but on account of this circumstance it can hardly be considered as such, except when the cold of the acid at setting out is less than that of easiest freezing.

In the dephlogisticated spirit of nitre the freezing points answering to the strength of ,434, ,53, and ,56, were said to be $-4^\circ\frac{1}{2}$, -19°, and -30°; and the differences of -30° and -19° from $-4^\circ\frac{1}{2}$ are to each other very nearly in the duplicate ratio of ,126 and ,096, the differences of the corresponding strengths from ,434; which, as ,434 is the strength of easiest freezing, is the proportion that might naturally be expected, and consequently serves in some measure to confirm the reasoning in this and the 12th Article.

16. After Mr. M^c^Nab had diluted these acids as above-mentioned, he divided each of them into two parts, and tried what degree of cold could be produced by mixing them with snow. On January 15th, one of these parts of the common spirit of nitre was tried. It was fluid when the experiment began, though its temperature, as well as that of the snow, was $-21^\circ\frac{1}{2}$; but on adding snow it immediately began to freeze, and grew

[1] In the experiment related in my observations on Mr. Hutchins's Experiments, this strength was rather greater than that of easiest freezing; but whether it is so in degrees of cold exceeding that in which my experiment was tried, does not appear.

thick, and its heat increased to $-2°\frac{1}{2}$; but by the addition of more snow it quickly sunk again, and at last got to $-43°\frac{1}{4}$. During the addition of the snow, the mixture grew thinner, and by the time it arrived at nearly the greatest degree of cold, consisted visibly of three parts: the lowest part, which consisted of frozen acid, was white and felt gritty; the upper part, which occupied about an equal space, was also white, but felt soft, and must have consisted of unmelted snow; the other part, which occupied by much the smallest space, was clear and fluid. The quantity of snow added was about $\frac{9}{13}$ of the weight of the acid, and consequently its strength was reduced to ,243.

Though snow was added to the acid in this experiment as long as, and even longer than, it produced any increase of cold, yet some days after, on adding more snow to the mixture, while it was fluid, and of the temperature of $-40°\frac{3}{4}$, the cold was increased to $-44°\frac{1}{4}$, or 1 degree lower than before. Mr. McNab did not perceive the snow to melt, though in all probability some must have done so, or no cold would have been produced.

The cause of this seems to be, that in the preceding experiment the congealed part of the acid was stronger than the fluid part; so that, though the fluid part was not strong enough to dissolve snow in a cold greater than $-43°\frac{1}{4}$, yet the whole acid together was strong enough to do it in a cold one degree greater.

A circumstance occurred in the last experiment which I cannot at all see the reason of; namely, a small part of the acid being poured into a saucer, before the addition of the snow, it was in an hour's time changed into solid ice, though the cold of the air, at the time the acid was poured out, was only $-41°\frac{1}{4}$, and does not seem to have increased during the experiment.

17. On December 30, the other half of the same acid had been tried in the same manner; at the beginning of the experiment not more than one-ninth part of the acid was fluid, the rest solid clear ice; its temperature was $-34°\frac{1}{2}$, and that of the snow nearly the same; the greatest degree of cold produced was $-42°\frac{3}{4}$; and the quantity of snow employed was about one-eighteenth of the weight of the acid; so that the strength of the mixture was ,38. The freezing point of the acid thus diluted appears to be about $-45°\frac{1}{4}$; for by the increase of warmth during the day-time, most of the congealed matter dissolved; but in the evening it began to freeze again, so as to become thicker, its temperature being then $-45°\frac{1}{4}$; and the next morning it was frozen solid, its cold being one degree greater.

18. On December 12, the diluted spirit of nitre No. 139, whose strength was ,175, was found frozen, its temperature being $-17°$. The fluid part, which was full of thin flakes of clear ice, and was of the consistence of syrup, was decanted into another bottle, and sent back. Its strength was ,21, and was greater than that of the undecanted part in the proportion of ,21 to ,16; so that, as not much of the undecanted part was really con-

gealed, the frozen part of the acid must have been much weaker than the rest, if not mere water. Accordingly, during the melting of the undecanted part, the frozen particles swam at top. Mr. McNab added snow to a little of the decanted liquor, but it did not dissolve, and no increase of cold was produced.

19. From these experiments it appears, that spirit of nitre is subject to two kinds of congelation, which we may call the aqueous and spirituous; as in the first it is chiefly, if not intirely, the watery part which freezes, and in the latter the spirit itself. Accordingly, when the spirit is cooled to the point of aqueous congelation, it has no tendency to dissolve snow and produce cold thereby, but on the contrary is disposed to part with its own water; whereas its tendency to dissolve snow and produce cold, is by no means destroyed by being cooled to the point of spirituous congelation, or even by being actually congealed. When the acid is excessively dilute, the point of aqueous congelation must necessarily be very little below that of freezing water; when the strength is ,21, it is at $-17°$, and at the strength of ,243, it seems, from Art. 16 to be at $-44°\frac{1}{4}$. Spirit of nitre, of the foregoing degrees of strength, is liable only to the aqueous congelation, and it is only in greater strengths that the spirituous congelation can take place. This seems to be performed with the least degree of cold, when the strength is ,411, in which case the freezing point is at $-1°\frac{1}{2}$. When the acid is either stronger or weaker, it requires a greater degree of cold; and in both cases the frozen part seems to approach nearer to the strength of ,411 than the unfrozen part; it certainly does so, when the strength is greater than ,411, and there is little doubt but what it does so in the other case. At the strength of ,54 the point of spirituous congelation is $-31°\frac{1}{2}$, and at ,33 probably $-45°\frac{1}{4}$; at least one kind of congelation takes place at that point, and there is little doubt but that it is of the spirituous kind. In order to present this matter more at one view, I have added the following table of the freezing point of common spirit of nitre answering to different strengths.

Strength	Freezing point °	
,54	$-31\frac{1}{2}$	spirituous congelation
,411	$-1\frac{1}{2}$*	spirituous congelation
,38	$-45\frac{1}{4}$	spirituous congelation
,243	$-44\frac{1}{4}$	aqueous congelation
,21	-17	aqueous congelation

* The point of easiest freezing.

20. In trying the first half of the dephlogisticated spirit of nitre, the cold produced was $-44°\frac{1}{2}$. The acid was fluid before the addition of the snow, and of the temperature of $-30°$, but froze on putting in the thermometer, and rose to $-5°$, as related in Art. 7.

In trying the second part, the acid was about 0° before the addition of the snow, and therefore had no disposition to freeze. The cold produced was − 42°½.

As the quantity of snow added in these experiments was not observed, they do not determine any points of aqueous or spirituous congelation in this acid; but there is reason to think, that these points are nearly the same as those of common spirit of nitre of the same strength, as the cold produced in these experiments was nearly the same as that obtained by the common spirit of nitre.

On the Vitriolic Acid.

21. On December 12, the strong oil of vitriol, No. 151, was found frozen, and was nearly of the colour and consistence of hogs-lard. Its temperature, found by pressing the ball of a thermometer into it, was − 15°, and that of the air nearly the same; but in the night it had been exposed to a cold of − 33°. It dissolved but slowly on being brought into a warm room, and was not completely melted before it had risen to + 20°, and even then was not very fluid, but of a syrupy consistence. During the progress of the melting, the congealed part sunk to the bottom, as in spirit of nitre: and many air bubbles separated from the acid, which, when it was completely melted, formed a little froth on the surface. As soon as it was sufficiently melted to admit of it, which was not till it had risen to the temperature of + 10°, the fluid part was decanted, and both were sent home to be examined.

It is remarkable, that the frozen part did not intirely dissolve until the temperature was so much increased. This would incline one to think, that the frozen part must have differed in some respect from the rest, so as to require much less cold to make it freeze; but yet I could not find that the strength of the decanted part differed sensibly from the rest.

It appeared by another bottle of oil of vitriol, which also froze by the natural cold of the air, that this acid, as well as the nitrous, contracts in freezing.

22. On December 21, when the weather was at − 30°, the vitriolic acid No. 103. was diluted with snow, as directed in Art. 3. The snow dissolved immediately, and no signs of congelation appeared during any part of the process. The temperature of the acid rose only one degree before it began to sink, and the weight of the snow added was only $\frac{10}{122}$ of that of the acid, so that its strength was reduced thereby to ,605; which is therefore the best degree of strength for producing cold by the addition of snow, when the degree of cold set out with is − 30°. This strength is one-fifteenth part less than what I found myself, by a similar experiment, when the temperature of the acid was + 27°; which shews, that the best degree of strength is rather less, when the degree of cold set out with is great than when small, but that it does not differ much.

23. The acid thus diluted was divided into two parts, and the next day Mr. M^c^Nab tried what degree of cold could be produced by adding snow to one of them. The temperature of the air at the time was $-39°$, and the mixture sunk by the process to $-55°\frac{1}{2}$. The snow dissolved readily, and the mixture did not lose much of its fluidity until it had acquired nearly its greatest degree of cold, nor did any congealed matter sink to the bottom in any part of the process. The quantity of snow added was about $\frac{86}{100}$ of the weight of the acid, so that the strength of the mixture was about ,325.

24. On January 1, thin crystals of ice were found diffused all through this mixture, the temperature of the air being $-51°\frac{1}{2}$, but that of the liquor was not tried. As this congelation must have been of the aqueous kind, and seems to have taken place at the temperature of $-51°\frac{1}{2}$, it should follow, that this acid had no power of dissolving snow in a cold of $-51°\frac{1}{2}$; so that it does not at first appear why a cold four degrees greater than that should have been produced in the foregoing experiment. The reason is, that at the time the mixture arrived at $-55°\frac{1}{2}$, it appeared by the diminution of its fluidity to have contained some undissolved snow, and some more was added to it after that time, which before the first of January dissolved and mixed with the acid; so that the acid in the mixture, at the time it sunk to $-55°\frac{1}{2}$, was not quite so much diluted as that which froze on January 1. This is the reverse of what happened in the trial of the nitrous acid in Art. 15. as in that experiment the fluid part, at the time of the greatest cold, was weaker than the whole mixture together; but it must be considered, that *that* mixture contained much congealed acid, as well as undissolved snow, whereas *this* contained only the latter.

25. On January 1, snow was added to the other half of the acid diluted on December 21. The cold produced was much greater than before, namely $-68°\frac{1}{2}$; this seems to have proceeded, partly from the air and materials having been 12 degrees colder in this than in the former experiment, and partly from the snow having been added faster, so that the mixture arrived at its greatest degree of cold in 20′, whereas it before took up 46′. Another reason is, that the former mixture was made in too small a jar, in consequence of which it was poured into a larger before the experiment was completed, whereby some cold was lost. The quantity of snow used in this experiment was less than in the former, so that the strength of the acid after the experiment was about ,343. The mixture also grew much thicker, and had a degree of elasticity resembling jelly; but whether this was owing only to more snow remaining undissolved, or to any other cause, I cannot tell.

26. Great as the foregoing degree of cold is, Mr. M^c^Nab, on February 2, produced one much greater. In hopes of obtaining a greater degree of cold by previously cooling the materials, he cooled about seven ounces of oil of vitriol, whose strength was ,629, that is, rather stronger than the fore-

going, by placing the jar in which it was contained in a freezing mixture of oil of vitriol and snow; the snow intended to be used was also cooled by placing it under the vessel in which the freezing mixture was made. As soon as the acid in the jar was cooled to the temperature of $-57°\frac{1}{2}$, a little of the snow was added, on which it immediately began to freeze, and rose to $-36°$; but in about 40 minutes, as the jar was still kept in the freezing mixture, it sunk to $-48°$; by which time it was grown very thick and gritty, especially at bottom. More of the cooled snow was then added, which in a short time made it sink to $-78°\frac{1}{2}$, and at the same time the thickness and tenacity of the mixture diminished; so that by the time it arrived at the greatest degree of cold, very little thickness remained.

It is worth inquiring, what was the reason of the greater degree of cold produced in this than in the preceding experiment? It could not be owing to the materials being colder; for at the time of the second addition of snow, at which time the experiment may be considered to have begun, the acid was not colder than at the beginning of the preceding experiment, and the snow in all probability not much colder. It could not be owing neither to the jar having been kept in the freezing mixture: for though that mixture was three or four degrees colder than the air in the preceding experiment, yet the acid in the jar, before it acquired much addition of cold, would be robbed of its cold faster by the mixture than it would by air of the same temperature as that in the preceding experiment. Neither could it proceed from any difference in the strength of the acid; for what difference there was must have done more hurt than good. The true reason is, that the acid was in a state of congelation: for as the congealed acid united to the snow and became fluid by the union, it is plain, that cold must have been produced both by the melting of the snow and by that of the acid; whereas, if the acid had been in a fluid state, cold would have been produced only by the first cause, and consequently a greater degree of cold should be produced in this experiment than in the former. The only inconvenience attending the acid being in a state of congelation is, that in all probability it does not unite to the snow so readily as when in a fluid state; but the difference seems not material, as the cold was produced, and the materials melted, in 5 minutes.

27. The day before, Mr. McNab, by adding snow to some of the same acid in the usual manner, when the cold of the materials was $-46°$, produced a cold of only $-66°$.

28. In these four last experiments the acid was reduced, by the addition of the snow, to the strengths of ,325, ,343, ,403, and ,334; and the cold produced in them was before said to be $-55°\frac{1}{2}$, $-68°\frac{1}{2}$, $-78°\frac{1}{2}$, and $-66°$; whence we may conclude, that these are nearly the points of aqueous congelation answering to the foregoing strengths; only it appears, from what was said in Art. 24. that the strengths here set down are all of them rather too small.

Though it is certain that oil of vitriol is capable of the spirituous congelation, and though it appears, both from the foregoing experiments and from some made by the Duc d'Ayen[1] and by M. de Morveau[2], that it freezes with a less degree of cold when strong than when much diluted, it is not certain whether it has any point of easiest freezing, like spirit of nitre, or whether the cold required to freeze it does not continually diminish as the strength increases, without limitation; but the latter opinion is the most probable. For the Duc d'Ayen's and M. de Morveau's acids, which, as they were concentrated on purpose, were most likely stronger than Mr. McNab's, froze with a cold less than zero of Fahrenheit; whereas the freezing point of Mr. McNab's undiluted acid, whose strength was ,98, was $-15°$, and that of the diluted acid, whose strength was ,629, was $-36°$; and when the acid was more diluted, it was found to bear a much greater cold without freezing. It appears also, both from Art. 21. and from M. de Morveau's experiment, that during the congelation of the oil of vitriol, some separation of its parts takes place, so that the congealed part differs in some respect from the rest, in consequence of which it freezes with a less degree of cold; and as there is reason to think from Art. 21. that these two parts do not differ much in strength, it seems as if the difference between them depended on some less obvious quality, and probably on that, whatever it is, which forms the difference between glacial and common oil of vitriol. The oil of vitriol prepared from green vitriol, has sometimes been obtained in such a state as to remain constantly congealed, except when exposed to a heat considerably greater than that of the atmosphere, whence it acquired its name of *glacial*[3]. It is not known indeed upon what this property depends, but it is certainly something else than its strength; for oil of vitriol of this kind is always smoking, and the fumes it emits are particularly oppressive and suffocating, though very different from those of the volatile sulphureous acid. On rectification likewise it yields, with the gentlest heat, a peculiar concrete substance, in the form of saline crystals; and after this volatile part has been driven off, the remainder is no longer smoking, and has lost its glacial quality[4].

On the Mixture of Oil of Vitriol and Spirit of Nitre.

29. This mixture is not so fit for producing cold by the addition of snow, as oil of vitriol alone; for the cold obtained did not exceed $-54\frac{1}{2}°$, in either of the experiments tried with it. The point of spirituous congelation of this mixture, when diluted with somewhat more than one-tenth of

[1] *Diction. de Chym.* par Macquer, 2de édit.

[2] *Nouv. Mém. de l'Académ. de Dijon*, 1782, 1er semestre, p. 68.

[3] *Mém. de l'Académ. des Sc.* 1738, p. 288.

[4] Crell's *Neu. Entdeck. in der Chemie*, Th. 11, p. 100, Th. 12, p. 241, etc., and *Annalen*, 1785, St. 5, p. 438, etc.

its weight of water, is about -- 20°, and is much lower when the acid is considerably more diluted; but as the Society will most likely have less curiosity about the disposition to freeze of this mixture than of the simple acids, I shall spare the particulars.

On the Spirit of Wine.

30. The rectified spirits No. 8. were diluted with snow, in the same manner as the other liquors; but were found not to want any, as the first and only addition of snow produced cold. The quantity added was about $\frac{1}{26}$ of the weight of the spirit.

31. The spirit thus diluted was divided, like the other liquors, into two parts, and each tried separately. The first was at – 45°, before the addition of the snow, and was sunk by the process to – 56°. The snow, even at the first addition, did not dissolve well, so that the spirit immediately became full of white spots[1], and grew thick by the time it arrived at its greatest degree of cold. After standing some hours, the mixture rose to the temperature of – 39°, and was grown clear, but yet was not limpid, but of the consistence of syrup. No cold was produced by adding snow to it in that state, though it appeared that its point of aqueous congelation was at least 6 degrees lower than its temperature at that time[2]; which seems to shew that spirit of wine has scarce any power of dissolving snow when it wants even 6 degrees of its point of aqueous congelation, and therefore is another instance that snow is dissolved much less readily by spirit of wine than by the nitrous and vitriolic acids.

32. In trying the other part of the diluted spirits, the cold produced was only – 47°½, the cold set out with being – 37°.

33. It appeared by the diluted spirit of wine No. 143. which on December 12 froze by the natural cold of the atmosphere, and was treated in the same manner as the diluted spirit of nitre, that when highly rectified spirit of wine, such as No. 8. is diluted with $1\frac{4}{10}$ its weight of water, its point of aqueous congelation will be at – 21°. The congealed part of the spirit was white like diluted milk, and even the decanted part, which was full of thin films of ice, had a milky hue. The fluid part was stronger than the rest, and no increase of cold was produced by adding snow to some of it, both of which are marks of aqueous congelation.

Though the foregoing experiments confirm the truth of what I said, in the account of Mr. Hutchins's experiments, concerning the cause of the

[1] This was not the case during the above-mentioned dilution of the spirits; but the cold was 16 degrees less in that experiment than in this.

[2] On account of the dilution which the spirits suffered by the melting of the snow which remained undissolved at the time of the greatest cold, its point of aqueous congelation was no longer so low as – 56°; but it still was not less than – 45°½, as in the evening it was found at that temperature, without much congealed matter in it.

cold produced by mixing snow with different liquors, and intirely clear up the difficulty relating to it which I mentioned in Art. 1., yet several questions may naturally occur; such as, why the cold produced by the oil of vitriol was so much greater than that obtained by the spirit of nitre, notwithstanding that in warmer climates the nitrous acid seems to produce more cold? and why the cold produced by the nitrous acid, notwithstanding its previous dilution, which might naturally be expected to be of service, was not greater than has been obtained by other persons without that precaution? But as this would lead me into disquisitions of considerable length, without my being able to say any thing very satisfactory on the subject, I shall forbear entering into it. I will only observe, that in most of the foregoing experiments, Mr. McNab would probably have produced more cold, if he had added the snow faster. We ought not, however, to regret that he did not, as its effects on the acids would then have been less sensible.

The natural cold, when these experiments were made, is remarkable; as there were at least nine mornings in which the cold was not less than that of freezing mercury; four in which it was at least eight degrees below that point, or $-47°$; and one in which it was $-50°$. Whereas out of nine winters, during which Mr. Hutchins observed the thermometer at Albany Fort, there were only twelve days in which the cold was equal to that of freezing mercury, and the greatest cold seems to have been $-45°$. I cannot learn whether the last winter was more severe than usual at Hudson's Bay; or whether Henley-House is a colder situation than Albany, which may perhaps be the case; for though it is only 130 miles distant from it, yet it stands inland, and to the W. or S.W. of it, which is the quarter from which the coldest winds blow.

* * *

Mr. McNab's original account of the experiments which furnished the materials of this Paper, having been thought too long to be printed in detail, is deposited in the Archives of the Society.

PHILOSOPHICAL TRANSACTIONS. VOL. 78, 1788, p. 166

XIII. *An Account of Experiments made by Mr.* John McNab, *at* Albany Fort, Hudson's Bay, *relative to the Freezing of Nitrous and Vitriolic Acids. By* Henry Cavendish, *Esq., F.R.S. and A.S.*

Read February 28, 1788

FROM the experiments made by Mr. McNab, of which I gave an account in the LXXVIth Volume of the *Philosophical Transactions*, p. 241. it appeared, that spirit of nitre was subject, not only to what I call the aqueous congelation, namely, that in which it is chiefly, and perhaps intirely, the watery part which freezes, but also to another kind, in which the acid itself freezes, and which I call the spirituous congelation. When its strength is such as not to dissolve so much as $\frac{243}{1000}$ of its weight of marble, or when its strength is less than ,243, as I call it for shortness, it is liable to the aqueous congelation solely; and it is only in greater strengths that the spirituous congelation can take place. This seems to be performed with the least degree of cold when the strength is ,411, in which case the freezing point is at $-1^\circ\frac{1}{2}$. When the acid is either stronger or weaker, it requires a greater degree of cold; and in both cases the frozen part seems to approach nearer to the strength of ,411 than the unfrozen part. The freezing points, answering to different degrees of strength, seemed to be as follows.

Strength	Freezing point °	
,54	$-31\frac{1}{2}$	spirituous congelation
,411	$-1\frac{1}{2}$	
,38	$-45\frac{1}{4}$	
,243	$-44\frac{1}{4}$	aqueous congelation
,21	-17	

As some of these properties, however, were deduced from reasoning not sufficiently easy to strike the generality of readers with much conviction, Mr. McNab was desired to try some more experiments to ascertain the truth of it; which he was so good as to undertake, and has executed them with the same care and accuracy as the former.

For this purpose, I sent him some bottles of spirit of nitre of different strengths, and he was desired to expose each of these liquors to the cold till they froze; then to try their temperature by a thermometer; afterwards to keep them in a warm room till the ice was almost melted, and then again expose them to the cold, and, when a considerable part of the acid had frozen, to try the temperature a second time; then to decant the unfrozen part into another bottle, and send both parts back to England, that their strength might be examined.

The intent of this second exposure to the cold was as follows. Spirit of nitre bears, like other liquors, to be cooled greatly below its freezing point without freezing: then the congelation begins suddenly; the liquor is filled with fine spicula of frozen matter, and the ice becomes so loose and porous, that, if the process be continued long enough for a considerable portion of the acid to congeal, scarce any of the fluid part can be decanted: whereas, if it be heated in this state till the frozen part is almost, but not intirely, melted, and be again exposed to the cold, as the liquor is then in contact with the congealed matter, it begins to freeze as soon as it arrives at the freezing point, and the ice becomes much more solid and compact.

The intent of decanting the fluid part, and sending both parts back, that their strength might be determined, was partly to examine the truth of the supposition laid down in my former Paper, that the strength of the frozen part approaches nearer to ,411 than that of the unfrozen; but it is also a necessary step towards determining the freezing point answering to a given strength of the acid; for as the frozen part is commonly of a different strength from the unfrozen, the strength of the fluid part, and the cold necessary to make it freeze, is continually altering during the progress of the congelation. In consequence of this, the temperature of the liquor is not that with which the frozen part congealed; but it is that necessary to make the remainder, or the fluid part, begin to freeze, or, in other words, it is the freezing point of the fluid part. This is the reason that a thermometer, placed in spirit of nitre, continually sinks during the progress of congelation; which is contrary to what is observed in pure water, and other fluids in which no separation of parts is produced by freezing.

Moreover, from the above-mentioned experiments of Mr. McNab it appeared, that oil of vitriol, as well as spirit of nitre, is subject to the spirituous congelation; but it seemed uncertain, whether, like the latter, it had any point of easiest freezing, or whether it did not uniformly freeze with less cold as the strength increased. For this reason, some bottles of oil of vitriol, of different strengths, were sent, which he was desired to try in the same manner as the former. This point, indeed, has since been determined by Mr. Keir, who has shewn that oil of vitriol has a strength of easiest freezing; and that at that point a remarkably slight degree of cold is sufficient for its congelation.

The result of Mr. McNab's experiments on the nitrous acid is given in the following table.

No.	Decanted part		Undecanted part		Strength of the whole mass	Strength before sent	Freezing point by first method	Freezing point by second method
	Quantity	Strength	Quantity	Strength			°	°
6	—	—	—	—	—	,561	− 41,6	—
7	1410	,445	2137	,435	,439	,437	+ *1,7	− 3,8
8	1658	,390	1940	,422	,407	,408	− 3,5	− 4
9	1368	,353	2438	,416	,393	,391	− 4,5	− 11
10	2206	,343	1920	,373	,357	,357	− 12,5	− 13,8
11	3620	,310	602	,381	,320	,320	− 22,5	− 23
12	2155	,276	1494	,293	,283	,280	− 39,1	− 40,3
13	1618	,241	1961	,235	,238	,238	− 34	− 32

The first column contains the numbers by which Mr. McNab has distinguished the different bottles. The second and third columns contain the quantity and strength of the decanted part of the liquor; and the fourth and fifth shew the quantity and strength of the undecanted part of the liquor. The sixth column gives the strength of both parts put together, or the strength of the whole mass; and the seventh is the strength of the same acid, as it was determined before it was sent to Hudson's Bay. The strengths of the decanted and undecanted parts were found by saturating the liquor returned home with marble; and that of the whole mass was inferred by computation from the quantity and strength of the decanted and undecanted parts; and as the strength thus inferred never differs from that determined before the liquors were sent to Hudson's Bay by more than $\frac{1}{100}$ part of the whole, it is not likely that the strengths of the decanted and undecanted parts here set down should differ from the truth by much more than that quantity.

The eighth column contains the freezing points found in the first method, or the temperature of the liquors after the hasty congelation which took place on exposing them to the cold without any frozen matter in them; and the ninth contains their temperature after the more gradual congelation which took place when they were cooled with some frozen matter in them; and as the unfrozen part of the acid was decanted immediately after the temperature had been observed, it follows, that this column shews the true freezing points of the decanted liquors. In like manner the eighth column shews the freezing points of that part of the liquor which remained fluid in the first manner of trying the experiment; but as the strength of this part was not determined, the precise strengths to which these freezing points correspond are unknown. Thus much, however, is certain, that these points must be below those of the whole mass, and in all probability must be above those of the decanted liquor; as there is great reason to think, that the quantity of frozen matter was always less, and consequently the strength of the fluid part differed less from that of the whole mass, in the first way of trying the experiment than in the second.

Before I draw any conclusions from these experiments, it will be proper to take notice of some particularities which occurred in trying them.

No. 6 was made to congeal by a freezing mixture of snow and diluted oil of vitriol. By the time the acid was cooled to – 42°, icy filaments were formed on the inside of the phial above the acid. Ten minutes after, the acid being cooled one degree more, the phial was taken out and agitated. This mixed the icy filaments with the acid, and made it freeze, which it seems not to have done before, in consequence of which its temperature rose to – 41°½. After having melted the greatest part of these filaments, and again exposed it to the freezing mixture, some snow accidentally fell into the acid, and made an uncertainty in the freezing point, for which reason it is not set down. But as it is evident, that the quantity of congealed matter in the first experiment was excessively small, the strength of the unfrozen part could not differ sensibly from that of the whole mass, and therefore – 41°½ is the true freezing point that answers to the strength of ,561.

It is remarkable, that No. 8 acquired by congelation a bluish colour, not unlike that which the dephlogisticated nitrous acid, in Mr. M^c^Nab's former experiments, acquired by dilution with snow. It is not said, how long the acid retained this colour, but it was intirely gone when the phial arrived in England. I am quite at a loss to account for this phænomenon, and why it happened to this bottle only.

No. 12 when cooled to – 17° seemed to contain many icy particles; but as it afterwards bore to be cooled to – 48°, without their increasing, we may conclude, that they were not frozen spirit of nitre, but only some heterogeneous matter separated from it. A little of the congealed part of No. 8 dropped into it while at this point, made it freeze, and it rose to – 39°.

In all the foregoing acids the ice was heavier than the fluid part, and in consequence subsided to the bottom; a proof that it was the spirituous congelation which had taken place in them: but in No. 13 the frozen part swam at top, which shews, that the congelation was of the aqueous kind.

It may appear remarkable to those who read Mr. M^c^Nab's experiments, that these acids bore to be heated so much above their freezing points before the ice intirely dissolved. No. 6. bore to be heated 18 degrees, No. 7. 13 degrees, and No. 12. 17 degrees above their freezing points, before all the congealed acid had disappeared. But as, in order to dissolve this congealed matter, they were brought into a room in all probability a great many degrees warmer than the points to which they were heated, so that the liquors heated fast; and as during the dissolution the ice would subside to the bottom; it is not extraordinary, that the fluid part in the phial might be many degrees warmer than the frozen part, unless the phials were much agitated during the time, which nothing shews them to have been; especially if we consider the great quantity of heat which, in all probability, must be communicated to the frozen acid in order to melt it; and that, perhaps, the frozen acid may receive and part with its heat but slowly. It must be observed that in No. 6. and 12. the frozen part might very likely be of a considerably different strength, and in consequence its freezing point might be several degrees different from that of the whole mass, so that the temperature to which the fluid was heated, in order to melt the ice, might very likely not differ so much from the freezing point of the ice itself as is here set down. But this could not be the case with No. 7.

It must be observed, that when Mr. McNab wanted to try the temperature of No. 7. after it had frozen in the first manner, the stopper stuck so tight that he was not able to remove it without warming it before the fire. The thermometer was then introduced, and stood several minutes therein at $+ 1^\circ\frac{1}{2}$, or $+ 2^\circ$. As the thermometer remained so long at this point, one might naturally suppose, that this was the true freezing point of the unfrozen acid. But yet, from what has been just said, it seems not improbable that it may be otherwise, and that the true freezing point may be sensibly lower; for which reason it is marked in the table with an asterisk (*) as doubtful.

It was before said, that the temperatures in the ninth column of the foregoing table, are the freezing points answering to the strengths expressed in the third column, and that $- 41^\circ\frac{1}{2}$ is the freezing point answering to the strength of ,561; whence the freezing points determined by these experiments, and their respective strengths, are as follows:

Strength	Freezing point
	°
,561	− 41,6
,445	− 3,8
,390	− 4
,353	− 11
,343	− 13,8
,310	− 23
,276	− 40,3

By interpolation from these *data*, according to Newton's method[1], it appears, that the strength at which the acid freezes with the least cold is ,418, and that the freezing point answering to that strength is $- 2^\circ\frac{4}{10}$.

In order to shew more readily the freezing point answering to any given strength, I have computed, by the same method, the following table, in which the strengths increase in arithmetical progression.

It was before shewn, that the freezing points, found by the first method, ought to be below those of the whole mass, and must, in all probability, be

[1] *Princip. Math.* Lib. III. prop. 40, lem. 5.

above those of the decanted liquor. In order to see how this agrees with observation, I computed in the above-mentioned manner the freezing

Strength	Freezing point	Difference
	°	°
,568	− 45,5	+ 15,4
,538	− 30,1	+ 12
,508	− 18,1	+ 8,7
,478	− 9,4	+ 5,3
,448	− 4,1	+ 1,7
,418	− 2,4	− 1,8
,388	− 4,2	− 5,5
,358	− 9,7	− 8
,328	− 17,7	− 10
,298	− 27,7	

points answering to the strength of the whole mass, and compared them with the observed freezing points. The result is given in the following table.

No.	Strength of the whole mass	Strength of the decanted liquor	Computed freezing point of the whole mass	Observed freezing point: In first method	Observed freezing point: In second method
			°	°	°
7	,439	,445	− 3,2	+ 1,7	− 3,8
8	,407	,390	− 2,6	− 3,5	− 4,
9	,393	,353	− 3,7	− 4,5	− 11,
10	,357	,343	− 10,	− 12,5	− 13,8
11	,320	,310	− 19,9	− 22,5	− 23,
12	,283	,276	− 35,6	− 39,1	− 40,3

It may be observed, that the freezing point of No. 7. tried in the first way, is considerably above that corresponding to the strength of the whole mass; but as this experiment was shewn [p. 218] to be doubtful, and not unlikely to exceed the truth, we may safely reject it as erroneous. All the others, as might be expected, are lower than those corresponding to the strength of the whole mass, and above those observed in the second manner, and therefore serve to confirm the truth of the above determination of the freezing points of spirit of nitre; and also shew, that in this acid the point of spirituous congelation is pretty regular, and does not depend much, if at all, on the rapidity with which the congelation is performed.

The point of aqueous congelation, however, seems liable to considerable irregularity; for No. 13, after having been exposed to the cold, froze on agitation, the congelation, as was before said, being of the aqueous kind, and the thermometer stood stationary therein at − 34°. The ice being then almost melted, it was again exposed to the cold, till a good deal was frozen; but yet its temperature was then no lower than $-32°\frac{1}{4}$, though the quantity of frozen matter must certainly have been much more than in the first trial. The fluid part being then decanted, and the frozen part melted, both were again exposed to the cold. They both were made to congeal by agitation, and the temperature of the undecanted was then found to be $-35°$, and that of the decanted part $-37°$: so that it should seem as if the freezing point found by the hasty congelation was always lower than that found the other way, which may, perhaps, proceed from this cause; namely, that when sufficient time is allowed, the watery part will separate from the rest, and freeze in a degree of cold much less than what is required to produce that effect, when it is performed in a more rapid manner.

These experiments confirm the truth of the conclusions I drew from Mr. McNab's former experiments; for, first, there is a certain degree of strength at which spirit of nitre freezes with a less degree of cold than when it is either stronger or weaker; and when spirit of nitre, of a different strength from that, is made to congeal, the frozen part approaches nearer to the foregoing degree of strength than the unfrozen. Likewise this strength, as well as the freezing point corresponding thereto, and the freezing point answering to the strength of ,54, come out very nearly the same as I concluded from those experiments; for by the present experiments they come out ,418, $-2°\frac{4}{10}$ and $-31°$, and by the former, ,411, $-1°\frac{1}{2}$, and $-31°$. But the freezing point answering to the strength of ,38 is totally different from what I there supposed. This must have been owing to the strength of that acid having been very different from what I thought it; which is not improbable, as its strength was inferred only from the quantity of snow which was added to it in finding the degree of cold produced by its mixture with snow.

After the foregoing experiments were finished, Mr. McNab made some more for determining the freezing points both of the decanted and undecanted part; but for want of a sufficient explanation of the manner in which they were executed, I have not been able to make any use of them. In their present state they shew much appearance of irregularity; but this would very likely have been cleared up, if the circumstances had been more fully detailed.

On the Vitriolic Acid.

An irregularity of a remarkable kind occurred in trying two of these acids; namely, when the undecanted part was melted and again made to congeal, its freezing point was found to be much less cold than that of the decanted part, and the difference was much greater than could be attributed to the difference of strength. This seems to have happened only in the two strongest acids, namely, No. 1. and 2. and in great measure confirms the supposition which I formed from Mr. McNab's former experiments, that the congealed part of oil of vitriol differs from the rest, not merely in strength, but also in some other respect, which I am not acquainted with. It should seem, however, that this property does not extend to weak oil of vitriol.

It perhaps may be suspected, that this property takes place in the nitrous acid also, and was the cause of the slow melting of the ice taken notice of in [p. 217]. But I think it more likely, that that phænomenon proceeded from the causes there assigned.

Some smaller irregularities occurred in trying the vitriolic acid, the cause of which I believe was, that when this acid has been cooled below the freezing point, and begins to freeze, the congelation proceeds but slowly; so that a considerable time elapses before it rises to the true freezing point. Something of the same kind seems to take place in the nitrous acid also, though in a less degree; for the decanted liquors usually continued to freeze and deposit a small quantity of ice, for a few minutes after they were poured off, though their cold, at least in some instances, was found rather to diminish during that time. It must be observed, that small spicula of ice always came over along with the decanted liquor; and to this, in all probability, the new-formed ice attached itself; for otherwise it is likely, that no ice would have been produced.

The following table contains the strength of the acids as determined before they were sent to Hudson's Bay, and the quantity and strength of the decanted and undecanted parts when they arrived at London, and the strength of the whole mass as computed from thence. For the sake of uniformity, I have expressed their strengths, like those of the nitrous acid, by the quantity of marble necessary to saturate them, though I did not find their strength by actually trying how much marble they would dissolve; as that method is too uncertain, on account of the selenite formed in the operation, and which in good measure defends the marble from the action of the acid. The method I used was, to find the weight of the plumbum vitriolatum formed by the addition of sugar of lead, and from thence to compute the strength, on the supposition that a quantity of oil of vitriol, sufficient to produce 100 parts of plumbum vitriolatum, will dissolve 33 of marble; as I found by experiment that so much oil of vitriol would saturate as much fixed alkali as a quantity of nitrous acid sufficient to dissolve 33 of marble. It may be observed, that the quantity of alkali, necessary to saturate a given quantity of acid, can hardly be determined with much accuracy, for which reason the foregoing less direct method was adopted; especially as the precipitation of plumbum vitriolatum shews the proportional strengths, which is the thing principally wanted, with as great accuracy as any method I know.

No.	Strength before sent	Decanted part		Undecanted part		Strength of whole mass
		Quantity	Strength	Quantity	Strength	
1	,977	1375	,967	3460	,963	,964
2	,918	3915	,919	1876	,905	,914
3	,846	88	,777	4915	,850	,849
4	,758	389	,710	{3795	,753	,755
				{547	,803	

The undecanted part of No. 4 was divided into two parts; namely, the less and the more congealable part; and it is the latter whose quantity and strength is given in the last line.

It is well known, that oil of vitriol attracts moisture with great avidity; and some of these acids were much exposed to the air during the experiments made with them, and may therefore be supposed to have attracted so much moisture from the air, as might sensibly diminish their strength; and this seems actually to have been the case with some of them. But as the bottles were well stopped, and as, except in one acid which was the most exposed to the air, the strength of the whole mass comes out not much less than that determined before the liquors were sent to Hudson's Bay, I imagine their strength could not sensibly alter during their voyage home; and consequently their strength, at the time the last observations were made with them, could not differ much from that here set down.

It would be tedious to give the experiments for determining their freezing points in detail; but the result is as follows. The freezing point of No. 1 tried in the first method, was somewhat above + 1°, but it is uncertain how much; that tried in the second manner seemed − 6°½. But the freezing point of the undecanted part, after having been intirely melted, and again exposed to the cold, was + 9°. It must be observed, that though this part was in all probability at first stronger than the decanted part, yet at the time its freezing point was tried, it seems to have become rather weaker than that, owing to its exposure to the air. It was before said, that the freezing point tried in the second manner is that of the decanted liquor; so that the freezing point of the decanted part seems to have been 13 or 14 degrees colder than that of the undecanted part; though the difference of strength, if there was any, must in all probability have tended to produce the contrary effect.

The freezing point of No. 2. tried in the first way, was − 26°; and that tried in the second was − 30°, or − 26°; but yet the freezing point of the undecanted part was 26 or 30 degrees higher, namely, at zero; a difference which could scarcely have proceeded from the difference of strength.

The freezing point of No. 3 could hardly differ much from + 42°; and that of No. 4 was about −.45°.

It should be remarked, that when this last acid, as well as No. 1. and 2. were exposed to a great cold, a sediment formed in them. This must have been of a very different nature from frozen acid, as appeared both from its texture, which was soft and mucilaginous to the feel, instead of being gritty as the frozen acid always was; and also from its being not much increased by an increase of cold; and therefore seems to have been some impurity separated from the acid. The quantity was greatest in No. 4.; but even in this, though it appeared great, it is likely that the real quantity was very small.

Another bottle of acid, whose strength was ,659, was sent; but Mr. McNab was not able to make this freeze.

From these experiments it should seem, that the freezing point of oil of vitriol, answering to different strengths, is nearly as follows:

Strength	Freezing point °
,977	+ 1
,918	− 26
,846	+ 42
,758	− 45

From hence we may conclude, that oil of vitriol has not only a strength of easiest freezing, as Mr. Keir has shewn; but that, at a strength superior to this, it has another point of contrary flexure, beyond which, if the strength be increased, the cold necessary to freeze it again begins to diminish.

The strength answering to this latter point of contrary flexure must, in all probability, be rather more than ,918, as the decanted or unfrozen part of No. 2. seemed rather stronger than the undecanted part; and for a like reason the strength of easiest freezing is rather more than ,846.

Mr. Keir found that oil of vitriol froze, with the least degree of cold, when its specific gravity at 60° of heat was 1,780, and that the freezing point answering to that degree of strength was + 46°; which agrees pretty nearly with these experiments, as the strength of oil of vitriol of that specific gravity is ,848, that is, nearly the same as that of No. 3.

Philosophical Transactions. Vol. 78, 1788, p. 261

XVII. *On the Conversion of a Mixture of dephlogisticated and phlogisticated Air into nitrous Acid, by the electric Spark. By* Henry Cavendish, *Esq., F.R.S. and A.S.*

Read April 17, 1788

In Volume LXXV. of the *Philosophical Transactions*, p. 372. I related an experiment, which shewed, that by passing repeated electric sparks through a mixture of atmospheric and dephlogisticated air, confined in a bent glass tube by columns of soap-lees and quicksilver, the air was converted into nitrous acid, which united to the soap-lees and formed nitre. But as this experiment has since been tried by some persons of distinguished ability in such pursuits without success, I thought it right to take some measures to authenticate the truth of it. For this purpose, I requested Mr. Gilpin, Clerk of the Royal Society, to repeat the experiment, and desired some of the Gentlemen most conversant with these subjects to be present at putting the materials together, and at the examination of the produce.

This laborious experiment Mr. Gilpin was so good as to undertake. It was performed in the same manner, and with the same apparatus, which was used in my own experiments, and which is described in the beginning of the above-mentioned Paper, and is accompanied with a drawing. The method used for introducing air into the bent tube, was that described in the last paragraph of p. [194] in that Paper, by means of the apparatus represented in fig. 3. or the reservoir, as I shall call it. The soap-lees, like those of my own experiments, were prepared from salt of tartar, and were of such strength as to yield $\frac{1}{10}$ of their weight of nitre when saturated with nitrous acid. The dephlogisticated air was prepared from turbith mineral, and seemed by the nitrous test to contain about $\frac{1}{18}$ part of phlogisticated air.

On December 6, 1787, in the presence of Sir Joseph Banks, Dr. Blagden, Dr. Dollfuss, Dr. Fordyce, Dr. J. Hunter, and Mr. Macie, the materials were put together. The quantity of soap-lees, introduced into the bent tube, was 180 measures, each of which contained one grain of quicksilver; and, as the bore of the tube was rather more than one-third of an inch in diameter, it formed a column of five or six-tenths of an inch in length, which, by the introduction of the air, was divided into two parts, one resting on the quicksilver in one leg of the tube, and the other on that in the other leg. The dephlogisticated air was mixed with one-third part of its bulk of atmospheric air of the room in a separate jar, and the reservoir was filled with the mixture; and from thence Mr. Gilpin, as occasion required, forced air into the bent tube, to supply the place of that absorbed by means of the electric spark.

From what has been said, it appears, that the mixture employed contained a less proportion of common air than that used in either of my experiments. This made it necessary for Mr. Gilpin now and then to introduce some common air by means of the bent tube represented in fig. 3. of the above-mentioned paper, whenever from the slowness of the absorption he thought there was too small a proportion of phlogisticated air in the tube.

My reason for this manner of proceeding was, that as my first experiment seemed to shew, that the dephlogisticated air ought to be in a rather greater proportion to the phlogisticated than the latter did, I was somewhat uncertain as to the proper quantities, and doubted whether I could proportion them in such manner as that it should not be necessary, during the course of the experiment, to add either dephlogisticated or common air. I therefore mixed the airs in such proportion, that I was sure there could be no occasion to add the former; since it was much easier, as well as more unexceptionable, to add common air than dephlogisticated air.

On December 24, as the air in the reservoir was almost all used, this apparatus was again filled in the presence of most of the above-mentioned Gentlemen, with a mixture of the same dephlogisticated air and common air, in the same proportions as before; and the same thing was repeated on January 19.

On January 23, the bent tube was, by accident, raised out of one of the glasses of mercury into which it was inverted, by which it was filled with air, and a good deal of the soap-lees were lost; there, however, was enough remaining for examination.

On January 28, and 29, the produce of this experiment was examined in the presence of Sir Joseph Banks, Dr. Blagden, Dr. Dollfuss, Dr. Fordyce, Dr. Heberden, Dr. J. Hunter, Mr. Macie, and Dr. Watson. It appeared that 9290 measures of the mixed air had been forced into the bent tube from the reservoir[1]. Besides this, Mr. Gilpin had at different times introduced 872 measures of common air, which makes in all 10,162 of air, consisting of 6968 of dephlogisticated air, and 3194 of common air. But as there were 900 measures of air remaining in the tube when the accident happened, the quantity absorbed was only 9262; but this is a much greater

[1] The method of ascertaining the quantity of air forced in was by weighing the reservoir, as mentioned in the above-mentioned Paper, p. 374 [p. 188 of this volume].

quantity that [*sic*] what from my own experiments seemed necessary for this quantity of soap-lees.

The soap-lees were poured into a small glass cup, and the tube washed with a little distilled water, in order that as little as possible might be lost. As they were by this means considerably diluted, they were evaporated to dryness; but it was difficult to estimate the quantity of the saline residuum, as it was mixed with a few particles of mercury.

Some vitriolic acid, dropped on a little of this residuum, yielded a smell of nitrous acid, the same as when dropped on nitre phlogisticated by exposure to the fire in a covered crucible; but it was thought less strong. The remainder was dissolved in a small quantity of distilled water, and the following experiments were tried with the solution.

It did not at all discolour paper tinged with the juice of blue flowers.

It left a nauseous taste in the mouth like solutions of mercury, and most other metallic substances.

Paper dipped into it, and dried, burnt with some appearance of deflagration, but not so strongly or uniformly as when dipped in a solution of nitre. The marks of deflagration, however, were stronger than when the Paper was dipped into a solution of mercury in spirit of nitre, but not so strong as when equal parts of this solution and solution of nitre were used.

A solution of fixed vegetable alkali, dropped into some of it diluted, produced a slight reddish-brown precipitate, which afterwards assumed a greenish colour.

A bit of bright copper being dipped into it, acquired an evident whitish colour, though not so white as when dipped into the solution of mercury in spirit of nitre.

From these experiments it appears, that the mixture of the two airs was actually converted into nitrous acid, only the experiment was continued too long, so that the quantity of air absorbed was greater than in my experiments, and the acid produced was sufficient, not only to saturate the soap-lees, but also to dissolve some of the mercury. The truth of the latter part is proved by the metallic taste of the residuum, its not discolouring the blue paper, the precipitate formed by the addition of fixed alkali, and the white colour given to the copper; and the nitrous fumes produced by the addition of oil of vitriol, as well as the manner in which paper impregnated with the residuum burnt, shew as plainly, that the acid produced was of the nitrous kind. It is remarkable, however, that during this experiment there were no signs which shewed when the soap-lees became saturated. The only time when the diminution proceeded much slower than usual was on January 4. It then seemed to go on very slowly; but as the air absorbed at that time was only 4830 measures, which is much less than what seems requisite to saturate the alkali, and as the diminution immediately went on again upon adding more common air, it seems not likely, that the soap-lees were saturated at that time.

On January 10, Mr. Gilpin observed a small quantity of whitish sediment on the surface of the mercury; which seems to shew, that the soap-lees were then saturated, and that the acid was beginning to corrode the mercury. The quantity of air absorbed was also 6840 measures, which is about as much as I expected would be required. However, as I was persuaded, from the event of my own experiments, that the diminution would either intirely cease, or go on very slowly, as soon as the soap-lees were saturated; and as I was unwilling to stop the experiments before that happened, I thought it best to continue the electrification.

On the same morning Mr. Gilpin found, that about 120 measures of the air in the bent tube had been spontaneously absorbed during the night, the quantity therein being so much less than it was the preceding evening, though the electrical machine had not been worked, or anything done to it during the intermediate time. The reason of this in all probability is, that as the acid was then corroding the mercury, the soap-lees became impregnated with nitrous air, which, during the night, united to the dephlogisticated air, and caused the diminution.

Though in reality the event of this experiment was such as to establish the truth of my position, that the mixture of dephlogisticated and phlogisticated air is converted by the electric spark into nitrous acid, as fully as if the experiment had been stopped in proper time; yet, as the event was in some measure different from that of my own experiments, and might afford room for cavil, I was desirous of having it repeated; and as Mr. Gilpin was so obliging as to undertake it again, the materials were, on February 11. put together for a fresh experiment, in the presence of most of the above-mentioned Gentlemen. The soap-lees employed were the same as before, but 183 measures were now introduced. The dephlogisticated air was different, the former parcel being all used. It was prepared, like the former, from turbith mineral, but was rather purer, as it seemed to contain only $\frac{1}{32}$ of phlogisticated air. The proportion in which it was mixed with common air was that of 22 to 10; so that a greater proportion of common air was now used, in consequence of which it was not necessary for Mr. Gilpin to introduce common air so often.

On February 29, the reservoir was again filled with air of the same kind, in presence of some of the same Gentlemen. As it was found by the last experiment that we must not depend on the saturation of the soap-lees being made known by any alteration in the rate of diminution, the process was stopped as soon as the air absorbed was such as from my own experiments I judged sufficient to neutralize the soap-lees. This was effected on the 15th of March. The air remaining in the tube, when Mr. Gilpin left off working, was 600 measures; but at the time the produce was examined, it was reduced to about 120, so much having been absorbed without the help of any electrification, which is a still more remarkable instance of spontaneous absorption than what occurred in the former experiment.

A few days after the experiment began, a black film was formed in one of the legs, which, I suppose, must have been a mercurial *ethiops*; but whether owing to some small degree of foulness in the mercury or tube, or to any other cause, I cannot tell. This foulness seemed not to increase; but on March 10, when the air absorbed was about 5200, a whitish sediment began to appear on the surface of the mercury.

On March 19, the produce was examined in the presence of Dr. Blagden, Dr. Dollfuss, Dr. Fordyce, Dr. Heberden, Dr. J. Hunter, Mr. Macie, and Dr. Watson. The mixed air forced into the bent tube from the reservoir was 6650 measures, besides which Mr. Gilpin had at different times introduced 630 of common air, which makes in all 7280, containing 4570 of dephlogisticated, and 2710 of common air.

The soap-lees were evaporated to dryness as before. The residuum weighed two grains, but there were two or three globules of mercury mixed with it, which might very likely weigh half a grain. This being dissolved in a small quantity of water, the following experiments were made with it.

It did not at all discolour paper tinged with blue flowers.

Slips of paper were dipped into it, and dried; and, by way of comparison, other slips of paper were dipped into a solution both of common nitre and phlogisticated nitre, and also dried. The former burnt in the same manner, and with as strong marks of deflagration, as the latter.

It had a strong taste of nitre, but left also a slight metallic taste on the tongue.

It did not give any white colour to a piece of clean copper put into it.

In order to see whether the whitish sediment, which was before said to be formed in the bent tube, contained any mercury, the remainder of this solution was diluted with some more distilled water, and suffered to stand till the white sediment had subsided. The clear liquor being then poured off, the remainder, containing the sediment, which seemed to amount only to a very small quantity, was put on a piece of bright copper, and dried upon it; a piece of clean gold was then laid over it, and both were exposed to heat. Both metals acquired a whitish colour, especially the gold, but which was very indeterminate.

In order to discover how nice a test of alcalinity the paper tinged with blue flowers was, a saturated solution of common nitre was mixed with $\frac{1}{120}$ of its bulk of the soap-lees; and this mixture was found to turn the paper evidently green; so that, as the solution of nitre contains about twice as much alkali as the soap-lees, it appears, that if the residuum had wanted only $\frac{1}{240}$ part of being saturated, it would have discoloured the paper.

From the foregoing trials it appears, that the mixture of dephlogisticated and common air in this experiment was actually converted into nitrous acid, and was sufficient not only to saturate the soap-lees, but also to dissolve some of the mercury. The quantity dissolved, however, was very small, and not sufficient to diminish sensibly the deflagrating quality of the nitre; so that the proof of the air being converted into nitrous acid was as evident as if no mercury had been dissolved.

In this experiment, as well as the former, no indication of the soap-lees becoming saturated was afforded by any cessation in the diminution of the air; whereas, in my experiments, it was very manifest. I do not know what this difference should be owing to, except to Mr. Gilpin's giving much stronger electrical sparks than I did. In his experiments the metallic knob which received the spark, and conveyed it to the bent tube, was usually placed at about $2\frac{1}{2}$ inches from the conductor, so that the spark jumped through $2\frac{1}{2}$ inches of air, in passing from the conductor to the knob, besides from $1\frac{1}{2}$ to $2\frac{1}{2}$ inches of air in the tube; whereas in my experiments, I believe, the knob was never placed at the distance of more than $1\frac{1}{4}$ inch from the conductor, and the quantity of air in the tube was much less; but the conductor and electrical machine were the same.

Except this, the only difference I know in the manner of conducting the experiment is, first, that Mr. Gilpin usually continued working the machine for half an hour at a time, whereas I seldom worked it more than ten minutes; and, secondly, that in Mr. Gilpin's Experiments the common air in the reservoir bore a less proportion to the dephlogisticated air than in mine; in consequence of which it was necessary for him frequently to introduce common air. On this account, the proportion of the two airs in the bent tube would be considerably different at different times; but on the whole, the common air absorbed bore a greater proportion to the dephlogisticated than in mine.

Though the whole quantity of air absorbed in these experiments is known with considerable precision, yet it is impossible to determine, with any accuracy, how much of each kind was absorbed, on account of our uncertainty about the nature of the air which remained at the end of the experiment. But if in the last experiment we suppose that the air absorbed spontaneously between the 15th and 19th of March was intirely dephlogisticated, and that what remained at the end of that time was of the purity of common air, it will appear, that 4090 of dephlogisticated and 2588 of common air, which is equivalent to 4480 of pure dephlogisticated air and 2198 of phlogisticated air, were absorbed at the time the electrification was stopped, and consequently the dephlogisticated air is $\frac{204}{100}$ of the phlogisticated air; whereas in my first experiment it seemed to be $\frac{220}{100}$, and in my last $\frac{253}{100}$.

But the quantity of acid produced, and consequently, I suppose, the saturation of the soap-lees, depends only on the quantity of phlogisticated air absorbed; and the effect of the greater or less quantity of dephlogisticated air is only to make the nitre produced more or less phlogisticated. Now, in this experiment, the bulk of the phlogisticated air was $12\frac{2}{10}$ that of the soap-lees. In my first experiment it was $11\frac{9}{10}$, and in my last $10\frac{8}{10}$.

As many persons seem to have supposed that the diminution of the air in these experiments is much quicker than it really is, though I do not know any thing in my Paper which should lead to suppose that it was not very slow, it may be proper to say something on this head. As the quickness of the diminution depends so much on the power of the electrical machine, I can only speak as to what happens with the machine used in these experiments. This was one of Mr. Nairne's patent machines, the cylinder of which is 12½ inches long, and 7 in diameter. A conductor of 5 feet long, and 6 inches in diameter, was adapted to it, and the ball which received the spark was placed at two or three inches from another ball, fixed to the end of the conductor. Now, when the machine worked well, Mr. Gilpin supposes he got about two or three hundred sparks a minute, and the diminution of the air during the half hour which he continued working at a time, varied in general from 40 to 120 measures, but was usually greatest when there was most air in the tube, provided the quantity was not so great as to prevent the spark from passing readily.

The only persons I know of, who have endeavoured to repeat this experiment, are, M. Van Marum, assisted by M. Paets Van Trootswyk; M. Lavoisier, in conjunction with M. Hassenfratz; and M. Monge. I am not acquainted with the method which the three latter Gentlemen employed, and am at a loss to conceive what could prevent such able philosophers from succeeding, except want of patience. But M. Van Marum, in his *Premiere Continuation des Expériences, faites par le moyen de la Machine électrique Teylerienne*, p. 182. has described the method employed by him and M. Van Trootswyk. They used a glass tube, the upper end of which was stopped by cork, through which an iron wire was passed, and secured by cement, and the lower end was immersed into mercury; so that the electric spark passed from the iron wire to the soap-lees. After so much of a mixture of five parts of dephlogisticated and three of common air as was equal to twenty-one times the bulk of the soap-lees[1] was absorbed, some paper was moistened with the alkali, which by its burning appeared to contain nitre, but shewed that the alkali was not near saturated. The experiment was then continued with the same soap-lees till more of the air, equal to fifty-six times the bulk of the soap-lees, was absorbed, which is near double the quantity required to saturate them; but yet the diminution went on as fast as ever. It was then tried, by the burning of paper dipped into them, how nearly they were saturated; but they still seemed far from being so.

The circumstance of using the iron wire appears evidently objectionable, on account of the danger of the iron wire being calcined by the electric spark, and absorbing the dephlogisticated air; and when I first read the account, I thought this the most probable cause of the difference in the result of our experiments; but I am now inclined to think that the case was otherwise. From the manner in which M. Van Marum expresses

[1] This is rather more than half of that requisite to saturate the soap-lees.

himself, it seems that the only circumstance, from which they concluded that the alkali was not saturated, was the imperfect marks of deflagration, that the paper dipped into it exhibited in burning; which, as we have seen, might proceed as well from some of the mercury having been dissolved as from the alkali not being saturated. I am much inclined to think, therefore, that, so far from the soap-lees not having been saturated, the quantity of acid produced was in reality much more than sufficient for this purpose, and had dissolved a good deal of the mercury; for the quantity of air absorbed favours this opinion, and the phænomena agree well with Mr. Gilpin's first experiment, in which this was certainly the case; whereas, if the diminution had proceeded chiefly from the dephlogisticated air being absorbed by the iron, the tube towards the end of the experiment would have been filled chiefly with phlogisticated air, which would have made the diminution proceed much slower than before; but we are told, that it went on as fast as ever. It is most likely, therefore, that the apparent disagreement between their experiment and mine proceeded only from their having continued the process too long, and from their not having properly examined the produce.

M. Van Marum then proceeds to say:

Surpris de cette différence de résultat j'envoyai une description exacte de nos expériences à M. Cavendish, le priant en même tems de m'instruire s'il pourroit trouver la cause de cette différence; et comme la seule différence essentielle, par laquelle notre expérience différoit de celle de M. Cavendish, consistoit en ce que nous avons employé de l'air pur produit du précipité rouge ou du minium, au lieu de l'air pur produit de la poudre noire formée par l'agitation du mercure avec le plomb, dont M. Cavendish ne donne pas la maniere de le produire[1], je le priai de me communiquer de quelle maniere il étoit venu a cet air, parceque je desirois de répéter l'expérience avec ce même air: mais comme il ne m'a fourni aucune élucidation sur la cause vraisemblable de la différence du resultat de nos expériences, et qu'il ne lui a pas plu de me communiquer sa maniere de produire l'air pur qu'il avoit employé pour ses expériences, m'écrivant, qu'il s'étoit proposé d'en parler dans un écrit public, la longueur ennuyante de ces expériences nous a fait prendre la resolution de différer leur continuation, pour obtenir une parfaite saturation de la lessive, jusqu'à ce que M. Cavendish ait publié sa maniere de produire l'air pur, dont il s'est servi, nous contentant pour le present d'avoir vu, que l'union du principe d'air pur et de la mofette produit de l'acide nitreux, suivant la découverte de M. Cavendish.

[1] The using the iron wire formed a material difference in our manner of conducting the experiment, and one which may, perhaps, have had great influence on the result; but I do not see how the using some other kind of dephlogisticated air, instead of that prepared from Dr. Priestley's black powder, can in the least degree form an essential difference, as in the same paragraph in which I mention my having used this kind of air in my first experiment, I say, that in my second experiment I used air prepared from turbith mineral.

As I should be sorry to be thought to have refused any necessary information to a Gentleman who was desirous to repeat one of my experiments, and who by his situation was able to do it with less trouble than any one else, I hope the Society will indulge me in adding a copy of my answer, that they may judge whether this is in any degree a fair representation of it.

To M. Van Marum.

Sir,

I received the honour of your letter, in which you inform me of your ill success in trying my experiment on the conversion of air into nitrous acid by the electric spark. It is very difficult to guess why an experiment does not succeed, unless one is present and sees it tried; but if you intend to repeat the experiment, your best way will be to try it with the same kind of apparatus that I described in that Paper. If you do so, and observe the precautions there mentioned, I flatter myself you will find it succeed. The apparatus you used seems objectionable, on account of the danger of the iron being corroded by absorbing the dephlogisticated air.

As to the dephlogisticated air procured from the black powder formed by agitating mercury mixed with lead, as it was foreign to the subject of the Paper, and as I proposed to speak of it in another place, I did not describe my method of procuring it. As far as I can perceive, the success depends intirely on carefully avoiding every thing by which the powder can absorb fixed air, or become mixed with particles of an animal or vegetable nature, or any other inflammable matter: for which reason care should be taken not to change the air in the bottle in which the mercury is shaken, by breathing into it, as Dr. Priestley did, or even by blowing into it with a bellows, as thereby some of the dust from the bellows may be blown into it. The method which I used to change the air was, to suck it out by means of an air-pump, through a tube which entered into the bottle, and did not fill up the mouth so close but what air could enter in from without, to supply the place of that drawn out through the tube.

I am, &c.

With regard to the main experiment, it was not in my power to give him further information than I did; as I pointed out the only circumstance to which, at that time, I could attribute the difference in our results. And with regard to the manner of preparing the dephlogisticated air from the black powder, I have mentioned all the particulars in which my manner of proceeding differed from Dr. Priestley's, and have also explained on what I imagine the success intirely depends; so that, I believe, no one at all conversant in this kind of experiments will think that I did not communicate to him my method of procuring that air.

PHILOSOPHICAL TRANSACTIONS. VOL. 80, 1790, p. 101

X. *On the Height of the Luminous Arch which was seen on* Feb. 23, 1784. *By* Henry Cavendish, *Esq., F.R.S. and A.S.*

Read February 25, 1790

THIS arch was observed, at the same time, at Cambridge by Mr. Wollaston; at Kimbolton in Huntingdonshire, by the Rev. Mr. Hutchinson; and at Blockley near Campden in Gloucestershire, by Mr. Franklin; and is described in letters from those gentlemen read to the Royal Society in December 1786[1].

It has been remarked, that as the arches of the kind described in these Papers have usually but a very slow motion, their height above the surface of the earth may readily be determined, provided they are observed about the same time, at places sufficiently distant; and they seem to be the only meteors of the aurora kind whose height we have any means of ascertaining.

The three places at which this phænomenon was seen are not so well suited for this purpose as might at first be expected from their distance, because they lie too much in the direction of the arch; they however seem sufficient to determine its height within certain limits, and perhaps are as well adapted for it as any observations we are likely to have of such phænomena.

The latitude of Cambridge is 52° 12′ 36″: that of Kimbolton is said by Mr. Hutchinson to be 52° 20′, and, according to the survey of Huntingdonshire, published by Jefferies, is 52° 19′ 50″; so that we may suppose it to be seven geographical miles north of Cambridge, and by the maps it seems to be about 18 such miles west of it: and Blockley is by the map 12 geographical miles south and 72 west of Cambridge.

At Cambridge the observations of its track seem to have been made at about 9 h. 15′ P.M. or 8 h. sidereal time. At Kimbolton, allowing for the difference of meridians, they could hardly have been made more than 5′ sooner; and at Blockley they were most likely made nearly at the same times as at Cambridge.

[1] See pp. 43–46, of this Volume [i.e. Vol. 80 of the *Phil. Trans.*].

At Blockley the arch passed about 7° south of the zenith; but it is unnecessary to determine this point with precision. At Kimbolton it was found by a quadrant to pass 11° to the south of it; and at Cambridge it was observed to pass through δ and ϵ Tauri, β Aurigæ, θ Ursæ majoris, Cor Caroli, and Arcturus. Now, if an arch was drawn through these stars, it must, I think, have appeared sensibly waved to the eye; whereas Mr. Wollaston did not take notice of any crookedness in this part of its course. It is most likely, therefore, that the middle of the arch must have passed to the south of β Aurigæ, and to the north of θ Ursæ; and if a circle is drawn through δ Tauri, Arcturus, and a point one degree north of the zenith, it will differ but little from a great circle, will agree as well with the positions of these stars as any regular line which can be drawn, and will pass 2½ degrees below β Aurigæ, and as much above θ Ursæ; which is not a greater difference from observation than may well have taken place, considering how much care and acquaintance with the fixed stars are required to determine a path by them so nearly.

The direction of the arch here described in that part near the zenith is W. 18° S.; and if a line is drawn through Cambridge in this direction, Kimbolton is 12,8 geographical miles north of it; and therefore, as the arch appeared 12° more south at Kimbolton than at Cambridge, the height of the arch above the surface of the earth must be 61½ geographical or 71 statute miles. If we suppose that the middle of the arch really passed through β Aurigæ, the height comes out 52 statute miles. On the whole, I should think, the height could hardly be less than 52 miles, and is not likely to have much exceeded 71.

The common aurora borealis has been supposed, with great reason, to consist of parallel streams of light shooting upwards, which, by the laws of perspective, appear to converge towards a point; and when any of these streams are over our heads, they appear actually to come to a point, and form a corona. Hence, from analogy, it seems not unlikely, that these luminous arches may consist of parallel streams of light, disposed so as to form a long thin band, pretty broad in its upright direction, and stretched out horizontally to a great length one way, but thin in the opposite direction. If this is the case, they will appear narrow and well-defined to an observer placed in the plane of the band; but to one placed at a little distance from it, they will appear broader, fainter, and less well-defined; and when the observer is removed to a great distance from the plane, they will vanish, or appear only as an obscure ill-defined light in the sky.

There are two circumstances which rather confirm this conjecture: first, that though we have an account of another arch besides this[1] having been seen at great distances in the direction of the arch, we have none of any having been seen in places much distant from each other in the contrary direction; and, secondly, that most of them have passed near the

[1] That of Feb. 15, 1750. *Phil. Trans.* XLVI. pp. 472 and 647.

zenith, whereas otherwise they ought frequently to appear in other situations; for if they appeared near the zenith to an observer in one latitude, they should appear in a very different situation in a latitude much different from that.

I wish it to be understood, however, that I do not offer this as a theory of which I am convinced; but only as an hypothesis which has some probability in it, in hopes that by encouraging people to attend to these arches, it may in time appear whether it is true or not. If it should hereafter be found, that these arches are never seen at places much distant from each other in a direction perpendicular to the arch, it would amount almost to a proof of the truth of the hypothesis; but if they ever are seen at the same time at such places, it would shew that the hypothesis is not true.

Supposing the hypothesis to be well-founded, the height above determined will answer to the middle part of the band, provided the breadth of it was small in respect of its distance from the earth, but otherwise will be considerably below the middle. If the breadth of the band was equal to the distance of its lower edge from the earth, the height of the lower edge would be three-fourths of that above found; and if the breadth was many times greater, would be half of it.

In the common aurora borealis, an arch is frequently seen low down in the northern part of the sky, forming part of a small circle. What this is owing to, I cannot pretend to say; but it is likely that it proceeds from streams of light which appear more condensed when seen in that direction than in any other, and consequently that the streams which form the arch to an observer in one place are different from those which form it to one at a distant place, and consequently that no conclusion as to its height can be drawn from observations of it in different places. Attempts, however, have been made to determine the height of the aurora from such observations, and even from those of the Corona[1]; though the latter method must surely be perfectly fallacious, and most likely the former is so too.

[1] Bergman. *Opusc.* Vol. v.

PHILOSOPHICAL TRANSACTIONS. VOL. 82, 1792, p. 383

XX. *On the Civil Year of the* Hindoos, *and its Divisions; with an Account of three* Hindoo *Almanacs belonging to* Charles Wilkins, *Esq. By* Henry Cavendish, *Esq.*

Read June 21, 1792

THOUGH we have received much information concerning the astronomy of the Hindoos, we know but little of their civil year, and its divisions; and what accounts of it we have received vary much from each other, owing partly, as will be seen, to different methods being used in different parts of India. As it occurred to me, that the best way by which a person in Europe could clear up the difficulties in this subject, would be to examine the *patras*, or almanacs, published by the Hindoos themselves, I applied to Mr. Wilkins, well known for his skill in the Sanskreet language, who was so good as to lend me three such, and assist me in finding out their meaning.

One of them was procured by Mr. Wilkins at Benares, and is computed for that place. The second came from Tanna, in the island of Salsette, near Bombay; but it appears to be the copy of a Benares patra, as it is disposed in the same form as the first, and is adapted to the same latitude and longitude. The third is computed for Nadeea, a town of Bengal, about 50 miles N. of Calcutta, almost as noted for learned men as Benares, and much frequented by students from the coast of Coromandel. The language of all three of them is a corrupt Sanskreet; but the last is written in the common Bengal character.

It appears from these almanacs that the civil year is regulated very differently in different parts of India; but before I speak of this year, it will be proper to mention a few words of the astronomical, which in all parts serves to regulate the civil year.

The astronomical year begins at the instant when the sun comes to the first point of the Hindoo zodiac. In the present year, 1792, it began, according to the principles delivered in the *Surya Siddhanta*[1], on April 9, at 22^h 14′ after midnight of their first meridian, which is about 41′ of time west of Calcutta; but according to Mr. Gentil's account of the Indian

[1] See an account of this in the 2d volume of the *Asiatic Researches*.

astronomy, it began 3^h $24'$ earlier. As this year, however, is longer than ours, its commencement falls continually later in respect of the Julian year by $50'$ $26''$ in four years.

This year is divided into 12 months, each of which corresponds to the time of the sun's stay in some sign, so that they are of different lengths, and seldom begin at the beginning of a day.

The civil day, in all parts of India, begins at sun-rise, and is divided into 60 parts, called dandas, which are again divided into 60 palas.

The only parts of the Benares patras which are of any material use for my purpose, are the names of the months which are set down at the top of each page, and the three first columns, the first of which contains the day of the month, according to the civil account, the next the day of the week, and the third the time at which the lunar *teethee* ends; but as many may like to be informed of the nature of an Hindoo almanac, I shall give an account of the remaining parts at the end of this paper.

In those parts of India in which this almanac is used, the civil year is lunisolar, consisting of 12 lunar months, with an intercalary month inserted between them occasionally. It begins at the day after the new moon next before the beginning of the solar year[1].

The lunar month is divided into thirty parts, called teethees; these are not strictly of the same length, but are equal to the time in which the moon's true motion from the sun is 12°. From the new moon till the moon arrives at 12° distance from the sun, is called the first teethee. From thence till it comes to 24°, is called the second teethee; and so on till the full moon; after which the teethees return in the same order as before.

The civil day is constantly called by the number of that teethee which expires during the course of the day.

[1] My reasons for saying that the civil year begins at the day after the new moon next before the beginning of the solar year, are as follow: 1st. These almanacs begin at this time, and, moreover, the year of Veekramādeetya and Sālavāhana, which is set down at the top of each page, is the same in the first page as in all the following, which would be improper, unless the year began at this time! 2dly. In the calculation of the eclipse of the sun, in Père Patouillet's Memoir, given in Bailly's *Astronomie Indienne*, the computation is made for the new moon preceding the beginning of the solar year, and yet the year of Sālavāhana, and of the cycle of 60, set down in the Memoir, is the same as if the solar year was already begun. 3dly. Père du Champ, in his table of the names of the years of the cycle of 60, given in the same book, had added to some of them the corresponding year of Christ, together with a day of the month. This day, in all of them, is the day next after the new moon, preceding the beginning of the solar year: and though no explanation is given, must evidently be intended for the day on which the year begins. And, 4thly. It is said in the *Ayeen Akbery*, by Abraham Roger, and, I believe, some other authors, that the year begins at this time. To the three last authorities, indeed, it may be objected, that they are taken from places in which we do not know that the Benares almanac is used; but they shew, that in some parts of India the year begins at that time, and if it does so in any place, it most likely does at Benares.

As the teethee is sometimes longer than one day, a day sometimes occurs in which no teethee ends. When this is the case, the day is called by the same number as the following day; so that two successive days go by the same name.

It oftener happens that two teethees end on the same day, in which case the number of the first of them gives name to the day, and there is no day called by the number of the last; so that a gap is made in the order of the days.

In the latter part of the month the days are counted from the full moon, in the same manner as in the former part they are counted from the new moon; only the last day, or that on which the new moon happens, is called the 30th instead of the 15th.

It follows from what has been said, that each half of the month constantly begins on the day after that on which the new or full moon falls; only sometimes the half month begins with the second day, the first being wanting.

The manner of counting the days, as we have seen, is sufficiently intricate; but that of counting the months, is still more so.

The civil year, as was before said, begins at the day after the new moon; and moreover, in the years which have an intercalary month, this month begins at the day after the new moon; but notwithstanding this, the ordinary civil month begins at the day after the full moon. To make their method more intelligible, I will call the time from new moon to new moon, the natural month. The civil month Visākha begins at the day after the full moon of that natural month which commences at the beginning of the civil year, or, in other words, at the day after the full moon of that natural month during which the sun enters the first Hindoo sign. Jyēshtha begins on the day after the full moon of that natural month during which the sun enters the second sign, and so on. The names of the civil months, with the names of the signs which the sun enters during the natural month at the full moon of which the civil month begins, are given in the following table, [p. 239] to which I have also added the day of our month when the sun entered that sign in the latter part of the year 1784, and beginning of 1785, taken from the Benares almanac, the time of the day being counted from sun-rise, and expressed in the Hindoo manner.

It may be observed, that in general, Visākha begins at the day after that full moon which is nearest to the instant at which the sun enters Mesha, whether before or after; however, it is not always accurately the nearest.

The two parts of each month are distinguished in these almanacs by the addition of the syllables *vadee* and *soodha* to the name; thus the first half of Visākha, or that from the day after the full, to the day after the new moon is called Visākha-vadee, and the remainder Visākha-soodha[1];

[1] *Soodha* signifies clear, pure, or complete; but the word *Vadee* is not to be found in any of Mr. Wilkins's dictionaries.

Civil Month	Sign	Day on which the ⊙ enters it			
			1784		
			day.	dan.	pa.
Visākha	Mesha	April	9,	37,	7
Jyēshtha	Vreesha	May	10,	34,	8
Āshāra	Meetoona	June	11,	0,	8
Srāvana	Karkata	July	12,	37,	58
Bhādra	Seengha	August	13,	7,	11
Aswēena	Kanyā	Sept.	13,	7,	36
Kārteeka	Toolā	Octob.	13,	32,	55
Mārgaseersha	Vreescheeka	Nov.	12,	25,	38
Powsha	Dhanoo	Decem.	11,	54,	18
			1785		
Māgha	Makara	Jan.	10,	13,	11
Phālgoona	Koombha	Feb.	8,	40,	21
Chitra	Meena	March	10,	30,	38

but, I believe, the more usual way of distinguishing them is by the words *kreeshna paksha*, or the dark side, and *sookla paksha*, the bright side.

A consequence of this way of counting the months is, that the first half of Chitra falls in one year, and the latter half in the following year.

Whenever the sun enters no sign during a natural month, this month is intercalary, and makes an irregularity, which may best be explained by an example.

In the year 1779, the sun entered into no sign during the natural month which began at the end of the first fortnight of Srāvana; accordingly the whole of this month was intercalary, and the fortnight which preceded it was called Neeja Srāvana vadee, instead of simply Srāvana vadee, as it would otherwise have been named. The first half of the intercalary month was called Adheeka Srāvana soodha, and the latter half Adheeka Srāvana vadee, and the fortnight after the intercalary month, Neeja Srāvana soodha[1].

It appears, therefore, that the two parts of the month where the intercalation takes place, are separated from each other by the interval of the whole intercalary month, and have the word Neeja prefixed to them; and the two parts of the intercalary month are called by the same name, but have the word Adheeka prefixed[2].

[1] *Adheeka* signifies over and above, or intercalary. *Neeja* prefixed to the name of the month signifies that month itself.

[2] What has been here said, agrees perfectly with Mr. Wilkins's almanacs; the only doubt is, whether there may not be some different method of regulating the month, which may also agree with these almanacs, and may be the true one. It is proper, therefore, that I should state my reasons for the account here given. Du Champ, who seems a very accurate writer, says (see Bailly, p. 320) that he was informed by a Hindoo calculator, that whenever the sun enters no sign during a lunar month, that month is doubled. This passage agrees very well with these almanacs, if by month we mean the time between two new moons; but disagrees entirely with

In these almanacs no notice is taken of the solar months, notwithstanding that a column is allotted to the day of the Mahometan calendar, which seems to shew that, in the countries which use the Benares patra, it is not customary to date by the solar month; for it is very unlikely that the computers of these almanacs should have given the days of the Mahometan calendar, and yet have omitted days used in their own.

In those parts of India which use the Nadeea patra, the case is quite different. This almanac contains the name of the solar and lunar month, with the corresponding days of the week and solar month, and the number of the lunar teethee which ends on those days. It begins with the day after that on which the astronomical year commences. This is marked as the first of the month, the next day is called the second, and so on, regularly to the end of the month. In like manner, all the other months begin on the day after the astronomical commencement, and the days are continued regularly to the end, so that the number of days in the month varies from 29 to 32[1].

them if we mean by it the time between two full moons; and moreover, in Mr. Wilkins's almanac it is the period from one new moon to another, which is called Adheeka. It seems certain, therefore, that in this passage the word month must mean what I have called the natural month; and that the rule for intercalation is such as I have mentioned, namely, that it shall take place whenever the sun enters no sign during the natural month. It is certain also that the ordinary civil month begins at the day after the full moon; and granting these two points, I cannot see any way in which the months can be regulated so as to differ in substance from what I have said.

[1] Perhaps I do not express myself accurately in saying that the civil month begins at the day after the commencement of the astronomical. It is true, that in this almanac it is the day after the commencement of the astronomical month, which is marked by the number one; but it must be observed that the Hindoos count by years complete, not by years current: for example, the year 1000 of the Kalee Yug begins at the time when 1000 years are completed from the Kalee Yug; and it is likely that the same manner of counting is adopted with regard to days, so that the day of the month marked one, does not signify the first day, but the day which begins at the expiration of the first day, and consequently that the civil month begins at the sun-rise of the day on which the astronomical month begins. I, however, have chosen to say that it begins at the day after, partly because I am not sure that the foregoing is the true meaning of the Hindoos, and partly because it would have been difficult to express myself in such manner as not to run great risk of being misunderstood, if I had done otherwise. What is here said applies equally to the lunar month in this and the Benares almanacs.

Though it is foreign to the subject of this paper, I cannot refrain from taking notice of an error, which I apprehend many European astronomers have fallen into, from not distinguishing between days current and days complete. It is common to say that the astronomical day begins twelve hours later than the civil day, and the nautical day twelve hours sooner; and it is true that the hour, which, according to the civil account is called one in the afternoon of the first of January, is written by astronomers January $1^{d}1^{h}$, but this, I apprehend, ought not to be read 1^{h} on the first of January, but 1^{d} and 1^{h} from the beginning of January, so that in reality the astronomical and nautical day both begin 12^{h} before the civil. A proof of the truth

The names of the months are the same as those of the lunar months in the Benares patra, Visākha being the first, or that which corresponds with the sign Mesha.

The lunar months begin, not at the full, as in the Benares patra, but at the new moon, and are called by the name of that solar month which ends during the course of them; for example, the lunar month, during which the solar month Visākha ends, is called Chandra (or lunar) Visākha, so that each month begins a fortnight later than by the Benares patra.

The teethees do not recommence at the full moon, but are continued to the end of the month, or to the 30th. In other respects they are counted as in the Benares patra; that is, the same notation is used whenever a day occurs in which no teethee ends, or when two teethees end on the same day.

Unluckily no intercalary month occurred in the year for which this almanac was computed, so that it gives us no information about the method of intercalation; but from analogy we may conclude, that those lunar months in which the sun enters no sign are intercalary, and are called by the name of either the preceding or following month, with the addition of some word to denote that they are intercalary[1].

As the Nadeea almanac begins with the day after the commencement of the solar year, and gives the day of the solar month, which the Benares patra does not, it affords reason to think that the custom of that part of India in which it is used, is to date by the solar month, and begin the year on the next day to the astronomical year; and accordingly Mr. Wilkins informs me, that the Hindoos of Bengal, in all their common transactions, date according to solar time, and use what is commonly called the Bengal era, but in the correspondence of the Brahmins, dating books, and regulating feasts and fasts, they generally note the teethee; and if the year is mentioned, it is often that of Veekramādeetya, sometimes that of Sālavāhana, but more frequently the vulgar Bengal year.

From what has been said, it appears, that the Hindoo civil months, both solar and lunar, consist, neither of a determinate number of days, nor are regulated by any cycle, but depend solely on the motions of the sun and moon, so that a Hindoo has no way of knowing what day of the month it is, but by consulting his almanac; and what is more, the month ought sometimes to begin on different days, in different places, on account of the difference in latitude and longitude, not to mention the difference

of this is, that in astronomical tables the place of the heavenly bodies set down for the beginning of the year, is the place for noon of the last civil day of the preceding year; and moreover, in Halley's tables this place is said to be *annis Julianis ineuntibus*, which shews that he thought that this was not merely a practice used for the sake of convenience, but that the year actually begins at this time.

[1] The Chinese, who, like the Hindoos, consider that lunar month as intercalary in which the sun enters no sign, call it by the same name as the preceding month; and it is likely that the Bengalese do so too.

which may arise from errors in computation. The inconvenience with which this must be attended seemed so great to me, that two or three years ago, by the assistance of Sir Joseph Banks, I proposed a query on the subject to Mr. Davis, author of the very valuable paper, in the Asiatic Researches, on the Hindoo astronomy, inquiring whether any method was taken to avoid the ambiguity, and was favoured with the following answer.

My Pundit, and others with whom I have conversed on the subject, although well aware of the circumstance (that the month may begin on different days in different places) do not think the ambiguity thence arising of much consequence, nor is there any method they know of taken to avoid it. The almanacs in common use are computed at Benares, Tirhut[1], and Nadeea, and three principal seminaries of Hindoo learning in the Company's provinces, whence they are annually dispersed throughout the adjacent country. Every Brahmin in charge of a temple, or whose duty it is to announce the times for the observance of religious ceremonies, is furnished with one of these almanacs; and if he be an astronomer, he makes such corrections in it as the difference of the latitude and longitude render necessary.

The beginning of the solar month falling on different days of the week, is not, as I have observed, regarded; but a disagreement in the computation of the teethee, which sometimes also happens, occasions no small perplexity, because by the teethees, or lunar days, are regulated most of their religious festivals: and I am assured that an instance of this kind, which occurred in Cossim Ally's time, obliged the Rajah of Nadeea to settle by proclamation which of the disputed computations should be regarded as the true one.

To the best of Mr. Wilkins's knowledge, the Nadeea almanac is used all over Bengal, and the Benares all over the upper part of India: and it is likely, therefore, that the Tirhut is used all over Bahar; but of the nature of this almanac I have no information; only to judge from the date of the inscription found at Mongueer[2], it is more likely to agree with the Nadeea than Benares patra.

As one of Mr. Wilkins's Benares patras came from Salsette, we may conclude that this almanac is in use in that part of India. The inscriptions too, found at Salsette and Delhi[3], confirm the opinions that this manner of dating is in use in both those places, as both are dated by the day of the bright side of the moon.

It appears from P. du Champ, and P. Patouillet, and I believe I may add Abraham Roger, that in the part of India from which they write, the civil year begins at the new moon before the beginning of the astronomical year[4]; which seems to shew that the Benares manner of dating is in use in

[1] A district in North Bahar.

[2] *Asiatic Researches*, vol. I. p. 127.

[3] *Asiatic Researches*, vol. I. pp. 363 and 379.

[4] Narsapour, from which P. Patouillet writes, is near the coast, and in the latitude of 16°½ N. Chrisnabouram, from which P. du Champ's Memoir is sent, is in

great part of the coast of Coromandel; but there is some reason to think, that in the neighbourhood of Madras and Pondicherry, they date in a manner different from that used either at Benares or Nadeea: for Mr. Gentil makes the month Chitra or Sitterey, as he spells it, correspond with the sign Mesha, in which he agrees with an alamanac published by an European at Madras, which seems to shew that in those places they date by solar months, but make Chitra correspond with the first sign.

Mr. Wilkins thinks he has heard of one or two places on the east coast of the Peninsula, and in particular Orissa, at which almanacs are computed; but he is not acquainted with the nature of them.

I shall now give a more particular account of the three almanacs. The two Benares patras are preceded by a preface, which begins with an invocation to the Deity, and then gives a whimsical account of the four Yoogas, or ages, and of the inferiority of each succeeding age to that preceding it, and concludes with astrological remarks.

There are no titles to any of the columns of which the almanacs are composed, nor is any explanation of them given in any part of the work; but by a careful examination of the numbers, a person acquainted with astronomical computations may, without much difficulty, find out their meaning.

The calendar part contains one page for each half of the lunar month. At the top of each page is given the year of the eras of Veekramādeetya and Sālavāhana. After this comes the name of the month, and in one almanac is given also the name and number of the month used by the Mahometans.

The part below this consists of eleven columns. The first gives the day of the month, according to the civil reckoning; the next the day of the week; and the two following contain the time of the day, that is the danda and pala at which the lunar teethee ends. The fifth column contains the name of the nakshatra[1] which the moon quits during the course of the day; and the two next shew the time at which she quits it.

The next three columns are very odd; they serve to shew the moon's place in what may be called a moveable zodiac, the first point of which moves backwards with the same velocity with which the sun moves forwards, and coincides with the sun at the beginning and middle of the Hindoo year. This zodiac is divided into twenty-seven equal parts, and the first of these three columns gives the name of the 27th part which the moon quits during the course of the day, and the two others the time at which she quits it. I do not know what use these columns can be applied

nearly the same latitude, but about 2° inland, and Paliacat, where Abraham Roger resided, is on the coast, in the latitude of 13°½, or near ½ a degree N. of Madras. This author, however, has expressed himself so inaccurately, that I am not sure whether they begin the year at that time or not.

[1] Otherwise called the 27 lunar mansions.

to, unless that of astrology. No trace of any thing of the kind has occurred to me in any account of the Hindoo astronomy[1].

In these columns the names of the days of the week, and nakshatras, are expressed by the first syllable of the word.

The last column is the day of the month used by the Mahometans.

As no explanation of these columns is given in the almanacs, it will be proper to mention my reasons for supposing them to be such as I have asserted.

The numbers in the third and fourth columns increase while the moon is near her apogee, and diminish during the rest of the month, which shews that it must be the time at which the moon completes some part of a revolution; and by examining these numbers during twelve revolutions of the moon in anomaly, it appears that the moon moves over 336 of these parts in 330^{d} $41^{dan.}$ $43^{pal.}$ which differs very little from the time answering to 336 teethees, so that there can be no doubt but that these columns shew the time at which the teethee ends. But a further proof of the truth of it is, that the time given in these columns for the end of the last teethee of each half month, agrees pretty nearly with the time of the new and full moon given in the nautical almanac, after allowing for the difference of longitude between Greenwich and Benares, and the time between sun-rise, at the latter place, and noon; which shews also that the time in these columns is reckoned from sun-rise, as might naturally be expected.

In regard to the moon's place in the nakshatras and moveable zodiac, it appears, by examining the fifth and eighth columns, that in each of them are 27 characters, which return constantly in order, except when the regularity is broken, either by the moon quitting two spaces in the same day, or by not quitting any one space in the day. The numbers also, both in the sixth and seventh, and in the ninth and tenth columns, increase when the moon is near the apogee, and diminish when she is near the perigee, which shews that they must be the time at which the moon finishes some 27th part of a revolution of one kind or other; and by examining the alteration of the numbers during twelve revolutions of the moon in anomaly, it appears first, that the moon describes 326 of the spaces given in the fifth column, in 329^{d} $57^{dan.}$ $38^{pal.}$ which is the time in which the moon moves over that number of nakshatras; and secondly, that the moon describes 350 of the spaces given in the eighth column in 329^{d} $36^{dan.}$ $48^{pal.}$ which is the time in which the sum of the mean motions of the moon and sun are equal to 350 27ths of a circle; or in other words, is the time in which the moon's motion in the moveable zodiac is 350 of these 27th parts; and moreover, I cannot find any other 27th of a revolution of the moon which will agree with this time; which is a sufficient proof that the numbers in the ninth and tenth columns are the times at which

[1] From a circumstance not worth mentioning, I find that the place of the moon in this moveable zodiac, is called the Yug.

the moon quits one of these 27th parts in the moveable zodiac. But a thing which more strongly proves the truth of this, and which also shews that the first point of this moveable zodiac coincides with the first point of the fixed zodiac, when the sun also coincides with it, is this: according to my supposition it is evident, that whenever the sun quits a nakshatra at the same time that the moon quits some other nakshatra, the moon must at the same time quit some 27th part of the moveable zodiac; and consequently that the numbers in the ninth and tenth columns should agree with those in the sixth and seventh; and accordingly we find, that on all the days of the year, in which the sun quits a nakshatra, the numbers in these two pairs of columns are nearly alike.

Underneath these eleven columns are tables of the diurnal motion and places of the sun and five planets, and of the moon's node in the Hindoo zodiac, for each week of the year; and between these tables and the eleven columns is set down the day of the month and week, and number of the week for which these places are given, and also the interval at that time between sun-rise and midnight, and the length of the day. The day of the week for which these places are given, is that which is the first in the current solar year, and the number of the week is also counted from the beginning of the solar year. The places are given for midnight.

On the right hand of the eleven principal columns is a space allotted for miscellaneous occurrences. In this is set down the time at which the sun enters each sign, and the beginning and end of eclipses. In these two years no solar eclipses were visible, but the end of the lunar eclipse is denoted by a Sanskreet word, signifying delivery; the meaning of the term used for the beginning is not so clear. The number of digits eclipsed is not set down. The other articles in this space consist chiefly of the time at which the moon and planets come to certain situations. Of this there is not a great deal which I understand, and what I do, is not worth taking notice of. There are also some figures and tables between the preface and calendar, which, as far as I can find, relate only to astrology.

The Nadeea almanac contains, besides the articles above-mentioned, the time of the day at which the lunar teethee ends, the number of the nakshatra and yug (place in the moveable zodiac) which the moon quits on that day, and the time at which she quits them, besides a few occasional remarks. It is disposed in a much coarser manner than the Benares patra, as each page contains as many days as it will hold, so that the month seldom begins at the beginning of a page. It contains no preface, and no explanation of the columns. The days of the week are not denoted by the first syllables of the name, but only by a number, expressing their order in the week, which caused some trouble in finding what day was meant by these numbers; but, by a variety of circumstances, I think it certain that the number 1 must denote Sunday.

PHILOSOPHICAL TRANSACTIONS. VOL. 87, 1797, p. 119

Extract of a Letter from Henry Cavendish, *Esq. to Mr.* Mendoza y Rios, *January*, 1795

[Addition to a Paper by Joseph de Mendoza y Rios, entitled "Recherches sur les principaux Problèmes de l'Astronomie Nautique."]

[Read December 22, 1796]

THE methods in which the whole distance of the moon and star is computed, particularly yours, require fewer operations than those in which the difference of the true and apparent places is found; but yet, as in the former methods, it is necessary either to take proportional parts, or to use very voluminous tables; I am much inclined to prefer the latter. This induced me to try whether a convenient method of the latter kind might not be deduced from the fundamental proposition used in your paper, and I have obtained the following, which has the advantage of requiring only short tables, and wanting only one proportional part to be taken, and I think seems shorter than any of the kind I have met with.

Let h and H be the apparent and true altitude of the star; l and L the apparent and true altitude of the moon, g and G the apparent and true distance of the moon and star. Let the sine and cosine of $g = d$ and δ, the sine and cosine of $l = a$ and α, the sine and cosine of $h = b$ and β; and the sine of the actual and mean horizontal parallax $= p$ and π; and let the sine of $L = a - m + pe$, and its cosine $= \alpha\,(1 + \mu - p\epsilon)$ and let the sine of $H = b - n$, and its cosine $= \beta\,(1 + \nu)$.

Then the cosine of G

$$= \delta\,(1 + \mu - p\epsilon)\,(1 + \nu) + (a - m + pe)\,(b - n) - ab\,(1 + \mu - p\epsilon)\,(1 + \nu),$$

which equals

$$\delta + \delta\mu + \delta\nu - \delta p\epsilon + \delta\mu\nu - \delta p\epsilon\nu + ab - bm + bpe - an + nm - npe$$
$$- ab - ab\mu + abp\epsilon - ab\nu - ab\mu\nu + ab\nu p\epsilon = \delta + \delta\mu + \delta\nu - \delta p\epsilon$$
$$- bm - ba\mu + bpe + bap\epsilon - an - ab\nu + nm - npe - ab\mu\nu$$
$$+ ab\nu p\epsilon + \delta\mu\nu - \delta\pi\epsilon\nu.$$

To make use of this rule, it must be considered that the quantity $\delta\mu\nu - \delta p\epsilon\nu$ is so small that it may safely be disregarded; but

$$nm - npe - ab\mu\nu + ab\nu p\epsilon,$$

if the altitudes are not more than 5°, may amount to about 12″, and therefore ought not to be neglected. The quantity $e + a\epsilon$ also differs very little from one, but it is not quite equal to it. Let therefore a table be made under a double argument, namely, the altitudes of the moon and star, giving the value of

$$nm - n\pi e - ab\mu\nu + ab\nu\pi\epsilon + b\pi e + ba\pi\epsilon - b\pi,$$

answering to different values of these altitudes, which call A. Let a second table be made under a double argument, namely, the altitude of the star and the apparent distance of the moon and star, giving the value of $\delta\nu$, which call D. Let a third table be made with the observed altitude for argument, giving the logarithm of $am + a^2\mu$; and let this quantity, answering to the moon's altitude, be called M, and that answering to the star's altitude, N; observing that the same table will do for the moon and star; but a fourth table should be made for the sun, so as to include its parallax; and, lastly, let a fifth table be made, with the moon's altitude for argument, giving the logarithm of $\frac{\epsilon}{a} - \frac{\mu}{\pi a}$, which call C. Then will

$$\cos . G = \delta - \delta apC - \frac{bM}{a} - \frac{aN}{b} + bp + D - A.$$

It must be observed that $\delta apC = \delta p\epsilon - \frac{\delta\mu p}{\pi}$, whereas it ought to equal $\delta p\epsilon - \delta\mu$; but μ cannot exceed 57″, and the horizontal parallax cannot differ from the mean by more than $\frac{1}{15}$ part of the whole; so that the error arising from thence cannot exceed 3″ or 4″. This small error however may be diminished by giving the quantity C for more than one horizontal parallax.

Addition to the foregoing Letter.

I have procured tables of the above-mentioned kind to be computed, which are intended to be inserted in a work now printing by Mr. Mendoza y Rios. Allowance is made in them for the alteration of the refractive power of the atmosphere, which is done by two new tables, one giving the correction of the logarithms M and N, and the other the sum of the corrections of $\delta\mu$ and $\delta\nu$. Now it must be observed, that the quantities μ and ν vary only from 57″ to 51″; and therefore the corrections of $\delta\mu$ and $\delta\nu$, may, without any material error, be considered as the same at all altitudes; and therefore the sum of the corrections may be comprehended in a table, under a double argument, namely, the refractive power of the atmosphere and the apparent distance.

In order to avoid as much as possible the inconvenience arising from using negative quantities, or giving different cases, the table D is continued to 125° of apparent distance, and the numbers in the table A are increased by 0,0003, so as to make them always positive; and to compensate this, the numbers in D are increased by 0,0002, and those in the correction of $\delta\mu + \delta\nu$ by 0,0001. It was found proper also to give the table C for four different values of horizontal parallax.

The above tables are short, and do not require proportional parts to be taken. The only part of the work in which this is wanted, is in finding the angle answering to the natural cosine of the true distance. In finding the natural cosine of the apparent distance this is avoided, by neglecting the odd seconds in working the problem, and adding them to the result.

PHILOSOPHICAL TRANSACTIONS. VOL. 88, 1798, p. 469

XXI. *Experiments to determine the Density of the Earth. By* Henry Cavendish, *Esq., F.R.S. and A.S.*

Read June 21, 1798

MANY years ago, the late Rev. John Michell, of this Society, contrived a method of determining the density of the earth, by rendering sensible the attraction of small quantities of matter; but, as he was engaged in other pursuits, he did not complete the apparatus till a short time before his death, and did not live to make any experiments with it. After his death, the apparatus came to the Rev. Francis John Hyde Wollaston, Jacksonian Professor at Cambridge, who, not having conveniences for making experiments with it, in the manner he could wish, was so good as to give it to me.

The apparatus is very simple; it consists of a wooden arm, 6 feet long, made so as to unite great strength with little weight. This arm is suspended in an horizontal position, by a slender wire 40 inches long, and to each extremity is hung a leaden ball, about 2 inches in diameter; and the whole is inclosed in a narrow wooden case, to defend it from the wind.

As no more force is required to make this arm turn round on its centre, than what is necessary to twist the suspending wire, it is plain, that if the wire is sufficiently slender, the most minute force, such as the attraction of a leaden weight a few inches in diameter, will be sufficient to draw the arm sensibly aside. The weights which Mr. Michell intended to use were 8 inches diameter. One of these was to be placed on one side the case, opposite to one of the balls, and as near it as could conveniently be done, and the other on the other side, opposite to the other ball, so that the attraction of both these weights would conspire in drawing the arm aside; and, when its position, as affected by these weights, was ascertained, the weights were to be removed to the other side of the case, so as to draw the arm the contrary way, and the position of the arm was to be again determined; and, consequently, half the difference of these positions would shew how much the arm was drawn aside by the attraction of the weights.

In order to determine from hence the density of the earth, it is necessary to ascertain what force is required to draw the arm aside through a given space. This Mr. Michell intended to do, by putting the arm in motion, and observing the time of its vibrations, from which it may easily be computed[1].

Mr. Michell had prepared two wooden stands, on which the leaden weights were to be supported, and pushed forwards, till they came almost in contact with the case; but he seems to have intended to move them by hand.

As the force with which the balls are attracted by these weights is excessively minute, not more than $\frac{1}{50,000,000}$ of their weight, it is plain, that a very minute disturbing force will be sufficient to destroy the success of the experiment; and, from the following experiments it will appear, that the disturbing force most difficult to guard against, is that arising from the variations of heat and cold; for, if one side of the case is warmer than the other, the air in contact with it will be rarefied, and, in consequence, will ascend, while that on the other side will descend, and produce a current which will draw the arm sensibly aside[2].

As I was convinced of the necessity of guarding against this source of error, I resolved to place the apparatus in a room which should remain constantly shut, and to observe the motion of the arm from without, by means of a telescope; and to suspend the leaden weights in such manner, that I could move them without entering into the room. This difference in the manner of observing, rendered it necessary to make some alteration in Mr. Michell's apparatus; and, as there were some parts of it which I thought not so convenient as could be wished, I chose to make the greatest part of it afresh.

Fig. 1. (Tab. XXIII.) is a longitudinal vertical section through the instrument, and the building in which it is placed. *ABCDDCBAEFFE*, is the case; *x* and *x* are the two balls, which are suspended by the wires *hx* from the arm *ghmh*, which is itself suspended by the slender wire *gl*. This arm consists of a slender deal rod *hmh*, strengthened by a silver wire

[1] Mr. Coulomb has, in a variety of cases, used a contrivance of this kind for trying small attractions; but Mr. Michell informed me of his intention of making this experiment, and of the method he intended to use, before the publication of any of Mr. Coulomb's experiments.

[2] M. Cassini, in observing the variation compass placed by him in the Observatory, (which was constructed so as to make very minute changes of position visible, and in which the needle was suspended by a silk thread,) found that standing near the box, in order to observe, drew the needle sensibly aside; which I have no doubt was caused by this current of air. It must be observed, that his compass-box was of metal, which transmits heat faster than wood, and also was many inches deep; both which causes served to increase the current of air. To diminish the effect of this current, it is by all means advisable to make the box, in which the needle plays, not much deeper than is necessary to prevent the needle from striking against the top and bottom.

hgh; by which means it is made strong enough to support the balls, though very light[1].

The case is supported, and set horizontal, by four screws, resting on posts fixed firmly into the ground: two of them are represented in the figure, by *S* and *S*; the two others are not represented, to avoid confusion. *GG* and *GG* are the end walls of the building. *W* and *W* are the leaden weights; which are suspended by the copper rods *RrPrR*, and the wooden bar *rr*, from the centre pin *Pp*. This pin passes through a hole in the beam *HH*, perpendicularly over the centre of the instrument, and turns round in it, being prevented from falling by the plate *p*. *MM* is a pulley, fastened to this pin; and *Mm*, a cord wound round the pulley, and passing through the end wall; by which the observer may turn it round, and thereby move the weights from one situation to the other.

Fig. 2. (Tab. XXIV.) is a plan of the instrument. *AAAA* is the case. *SSSS*, the four screws for supporting it. *hh*, the arm and balls. *W* and *W*, the weights. *MM*, the pulley for moving them. When the weights are in this position, both conspire in drawing the arm in the direction *hW*; but, when they are removed to the situation *w* and *w*, represented by the dotted lines, both conspire in drawing the arm in the contrary direction *hw*. These weights are prevented from striking the instrument, by pieces of wood, which stop them as soon as they come within $\frac{1}{5}$ of an inch of the case. The pieces of wood are fastened to the wall of the building; and I find, that the weights may strike against them with considerable force, without sensibly shaking the instrument.

In order to determine the situation of the arm, slips of ivory are placed within the case, as near to each end of the arm as can be done without danger of touching it, and are divided to 20ths of an inch. Another small slip of ivory is placed at each end of the arm, serving as a vernier, and subdividing these divisions into 5 parts; so that the position of the arm may be observed with ease to 100ths of an inch, and may be estimated to less. These divisions are viewed, by means of the short telescopes *T* and *T*, (fig. 1.) through slits cut in the end of the case, and stopped with glass; they are enlightened by the lamps *L* and *L*, with convex glasses, placed so as to throw the light on the divisions; no other light being admitted into the room.

The divisions on the slips of ivory run in the direction *Ww*, (fig. 2.) so that, when the weights are placed in the positions *w* and *w*, represented by the dotted circles, the arm is drawn aside, in such direction as to make the

[1] Mr. Michell's rod was entirely of wood, and was much stronger and stiffer than this, though not much heavier; but, as it had warped when it came to me, I chose to make another, and preferred this form, partly as being easier to construct and meeting with less resistance from the air, and partly because, from its being of a less complicated form, I could more easily compute how much it was attracted by the weights.

index point to a higher number on the slips of ivory; for which reason, I call this the positive position of the weights.

FK, (fig. 1.) is a wooden rod, which, by means of an endless screw, turns round the support to which the wire *gl* is fastened, and thereby enables the observer to turn round the wire, till the arm settles in the middle of the case, without danger of touching either side. The wire *gl* is fastened to its support at top, and to the centre of the arm at bottom, by brass clips, in which it is pinched by screws.

In these two figures, the different parts are drawn nearly in the proper proportion to each other, and on a scale of one to thirteen.

Before I proceed to the account of the experiments, it will be proper to say something of the manner of observing. Suppose the arm to be at rest, and its position to be observed, let the weights be then moved, the arm will not only be drawn aside thereby, but it will be made to vibrate, and its vibrations will continue a great while; so that, in order to determine how much the arm is drawn aside, it is necessary to observe the extreme points of the vibrations, and from thence to determine the point which it would rest at if its motion was destroyed, or the point of rest, as I shall call it. To do this, I observe three successive extreme points of a vibration, and take the mean between the first and third of these points, as the extreme point of vibration in one direction, and then assume the mean between this and the second extreme, as the point of rest; for, as the vibrations are continually diminishing, it is evident, that the mean between two extreme points will not give the true point of rest.

It may be thought more exact, to observe many extreme points of vibration, so as to find the point of rest by different sets of three extremes, and to take the mean result; but it must be observed, that notwithstanding the pains taken to prevent any disturbing force, the arm will seldom remain perfectly at rest for an hour together; for which reason, it is best to determine the point of rest, from observations made as soon after the motion of the weights as possible.

The next thing to be determined is the time of vibration, which I find in this manner: I observe the two extreme points of a vibration, and also the times at which the arm arrives at two given divisions between these extremes, taking care, as well as I can guess, that these divisions shall be on different sides of the middle point, and not very far from it. I then compute the middle point of the vibration, and, by proportion, find the time at which the arm comes to this middle point. I then, after a number of vibrations, repeat this operation, and divide the interval of time, between the coming of the arm to these two middle points, by the number of vibrations, which gives the time of one vibration. The following example will explain what is here said more clearly. [See p. 253.]

The first column contains the extreme points of the vibrations. The second, the intermediate divisions. The third, the time at which the arm

Extreme points.	Division.	Time.	Point of rest.	Time of middle of vibration.
27,2		h. ′ ″		h. ′ ″
	25	10 23 4	—	10 23 23
	24	57		
22,1	—	—	24,6	
27	—	—	24,7	
22,6	—	—	24,75	
26,8	—	—	24,8	
23	—	—	24,85	
26,6	—	—	24,9	
	25	11 5 22	—	11 5 22
	24	6 48		
23,4				

came to these divisions; and the fourth, the point of rest, which is thus found: the mean between the first and third extreme points is 27,1, and the mean between this and the second extreme point is 24,6, which is the point of rest, as found by the three first extremes. In like manner, the point of rest found by the second, third, and 4th extremes, is 24,7, and so on. The fifth column is the time at which the arm came to the middle point of the vibration, which is thus found: the mean between 27,2 and 22,1 is 24,65, and is the middle point of the first vibration; and, as the arm came to 25 at $10^h\ 23'\ 4''$, and to 24 at $10^h\ 23'\ 57''$, we find, by proportion, that it came to 24,65 at $10^h\ 23'\ 23''$. In like manner, the arm came to the middle of the seventh vibration at $11^h\ 5'\ 22''$; and, therefore, six vibrations were performed in $41'\ 59''$, or one vibration in $7'\ 0''$.

To judge of the propriety of this method, we must consider in what manner the vibration is affected by the resistance of the air, and by the motion of the point of rest.

Let the arm, during the first vibration, move from D to B, (Tab. XXIV. fig. 3.) and, during the second, from B to d; Bd being less than DB, on account of the resistance. Bisect DB in M, and Bd in m, and bisect Mm in n, and let x be any point in the vibration; then, if the resistance is proportional to the square of the velocity, the whole time of a vibration is very little altered; but, if T is taken to be the time of one vibration, as the diameter of a circle to its semicircumference, the time of moving from B to n exceeds $\frac{1}{2}$ a vibration, by $\frac{T \times Dd}{8Bn}$ nearly; and the time of moving from B to m falls short of $\frac{1}{2}$ a vibration, by as much; and the time of moving from B to x, in the second vibration, exceeds that of moving from x to B, in the first, by $\frac{T \times Dd \times . Bx^2}{4Bn^2 \times \sqrt{Bx \times x\delta}}$, supposing Dd to be bisected in δ; so that, if a mean is taken, between the time of the first arrival of the arm at x and its returning back to the same point, this mean will be earlier than the true time of its coming to *B*, by

$$\frac{T \times Dd \times Bx^2}{8Bn^2 \times \sqrt{Bx} \times x\delta}.$$

The effect of motion in the point of rest is, that when the arm is moving in the same direction as the point of rest, the time of moving from one extreme point of vibration to the other is increased, and it is diminished when they are moving in contrary directions; but, if the point of rest moves uniformly, the time of moving from one extreme to the middle point of the vibration, will be equal to that of moving from the middle point to the other extreme, and moreover, the time of two successive vibrations will be very little altered; and, therefore, the time of moving from the middle point of one vibration to the middle point of the next, will also be very little altered.

It appears, therefore, that on account of the resistance of the air, the time at which the arm comes to the middle point of the vibration, is not exactly the mean between the times of its coming to the extreme points, which causes some inaccuracy in my method of finding the time of a vibration. It must be observed, however, that as the time of coming to the middle point is before the middle of the vibration, both in the first and last vibration, and in general is nearly equally so, the error produced from this cause must be inconsiderable; and, on the whole, I see no method of finding the time of a vibration which is liable to less objection.

The time of a vibration may be determined, either by previous trials, or it may be done at each experiment, by ascertaining the time of the vibrations which the arm is actually put into by the motion of the weights; but there is one advantage in the latter method, namely, that if there should be any accidental attraction, such as electricity, in the glass plates through which the motion of the arm is seen, which should increase the force necessary to draw the arm aside, it would also diminish the time of vibration; and, consequently, the error in the result would be much less, when the force required to draw the arm aside was deduced from experiments made at the time, than when it was taken from previous experiments.

Account of the Experiments.

In my first experiments, the wire by which the arm was suspended was $39\frac{1}{4}$ inches long, and was of copper silvered, one foot of which weighed $2\frac{4}{10}$ grains: its stiffness was such, as to make the arm perform a vibration in about 15 minutes. I immediately found, indeed, that it was not stiff enough, as the attraction of the weights drew the balls so much aside, as to make them touch the sides of the case; I, however, chose to make some experiments with it, before I changed it.

In this trial, the rods by which the leaden weights were suspended were of iron; for, as I had taken care that there should be nothing magnetical in the arm, it seemed of no signification whether the rods were magnetical or not; but, for greater security, I took off the leaden weights, and tried what effect the rods would have by themselves. Now I find, by computation, that the attraction of gravity of these rods on the balls, is to that of the weights, nearly as 17 to 2500; so that, as the attraction of the weights appeared, by the foregoing trial, to be sufficient to draw the arm aside by about 15 divisions, the attraction of the rods alone should draw it aside about $\frac{1}{10}$ of a division; and, therefore, the motion of the rods from one near position to the other, should move it about $\frac{1}{5}$ of a division.

The result of the experiment was, that for the first 15 minutes after the rods were removed from one near position to the other, very little motion was produced in the arm, and hardly more than ought to be produced by the action of gravity; but the motion then increased, so that, in about a quarter or half an hour more, it was found to have moved $\frac{1}{2}$ or $1\frac{1}{2}$ division, in the same direction that it ought to have done by the action of gravity. On returning the irons back to their former position, the arm moved backward, in the same manner that it before moved forward.

It must be observed, that the motion of the arm, in these experiments, was hardly more than would sometimes take place without any apparent cause; but yet, as in three experiments which were made with these rods, the motion was constantly of the same kind, though differing in quantity from $\frac{1}{2}$ to $1\frac{1}{2}$ division, there seems great reason to think that it was produced by the rods.

As this effect seemed to me to be owing to magnetism, though it was not such as I should have expected from that cause, I changed the iron rods for copper, and tried them as before; the result was, that there still seemed to be some effect of the same kind, but more irregular, so that I attributed it to some accidental cause, and therefore hung on the leaden weights, and proceeded with the experiments.

It must be observed, that the effect which seemed to be produced by moving the iron rods from one near position to the other, was, at a medium, not more than one division; whereas the effect produced by moving the weight from the midway to the near position, was about 15 divisions; so that, if I had continued to use the iron rods, the error in the result caused thereby, could hardly have exceeded $\frac{1}{30}$ of the whole. [See Tables on p. 256 *et seq.*]

It must be observed, that in this experiment, the attraction of the weights drew the arm from 11,5 to 25,8, so that, if no contrivance had been used to prevent it, the momentum acquired thereby would have carried it to near 40, and would, therefore, have made the balls to strike against the case. To prevent this, after the arm had moved near 15 divisions, I returned the weights to the midway position, and let them remain there, till the arm came nearly to the extent of its vibration, and then again moved them to the positive position, whereby the vibrations were so much

EXPERIMENT I. Aug. 5.

Weights in midway position.

Extreme points.	Divisions.	Time.	Point of rest.	Time of mid. of vibration.	Difference.
		h. ′ ″		h. ′ ″	′ ″
	11,4	9 42 0			
	11,5	55 0			
	11,5	10 5 0	11,5		
At 10h 5′, weights moved to positive position.					
23,4					
27,6	—	—	25,82		
24,7	—	—	26,07		
27,3	—	—	26,1		
25,1	—	—			
At 11h 6′, weights returned back to midway position.					
5,					
	11	0 0 48			
	12	1 30	—	0 1 13	
18,2	—	—	12	—	14 56
	12	16 29			
	11	17 20	—	16 9	
6,6	—	—	11,92	—	14 36
	11	30 24			
	12	31 11	—	30 45	
16,3	—	—	11,72	—	15 13
	12	45 58			
	11	47 4	—	45 58	
7,7					

Motion on moving from midway to pos. = 14,32
pos. to midway = 14,1
Time of one vibration = 14′ 55″

diminished, that the balls did not touch the sides; and it was this which prevented my observing the first extremity of the vibration. A like method was used, when the weights were returned to the midway position, and in the two following experiments.

The vibrations, in moving the weights from the midway to the positive position, were so small, that it was thought not worth while to observe the time of the vibration. When the weights were returned to the midway position, I determined the time of the arm's coming to the middle point of each vibration, in order to see how nearly the times of the different vibrations agreed together. In great part of the following experiments, I contented myself with observing the time of its coming to the middle point of only the first and last vibration.

EXPERIMENT II. Aug. 6.

Weights in midway position.

Extreme points.	Divisions.	Time.	Point of rest.	Time of mid. of vibration.	Difference.
		h. ′ ″		h. ′ ″	′ ″
	11	10 4 0			
	11	11 0			
	11	17 0			
	11	25 0	11,		

Weights moved to positive position.

Extreme points.	Divisions.	Time.	Point of rest.	Time of mid. of vibration.	Difference.
29,3					
24,1	—	—	26,87		
30	—	—	27,57		
26,2	—	—	28,02		
29,7	—	—	28,12		
26,9	—	—	28,05		
28,7	—	—	27,85		
27,1	—	—	27,82		
28,4	—	—			

Weights returned to midway position.

Extreme points.	Divisions.	Time.	Point of rest.	Time of mid. of vibration.	Difference.
6					
	12	1 3 50	—	1 4 1	
	13	4 34			
18,5	—	—	12,37	—	14 52
	13	18 29	—	18 53	
	12	19 18			
6,5	—	—	11,67	—	14 46
	11	33 48	—	33 39	
	12	34 51			
15,2	—	—	11	—	13 46
	·13	45 8	—	47 25	
	12	46 22			
7,1	—	—	10,75	—	15 25
	11	2 3 48	—	2 2 50	
	12	5 18			
13,6					

Motion of arm on moving weights from midway to pos. = 15,87
pos. to midway = 15,45
Time of one vibration = 14′ 42″

EXPERIMENT III. Aug. 7.

The weights being in the positive position, and the arm a little in motion.

Extreme points.	Divisions.	Time.	Point of rest.	Time of mid. of vibration.	Difference.
		h. ′ ″		h. ′ ″	′ ″
31,5					
29	—	—	30,12		
31	—	—	30,02		
29,1					
Weights moved to midway position.					
9					
	14	10 34 18	—	10 34 55	
	15	35 8			
20,5	—	—	14,8	—	14 44
	15	49 31	—	49 39	
	14	50 27			
9,2	—	—	14,07	—	14 38
	14	11 5 7	—	11 4 17	
	15	6 18			
17,4	—	—	13,52	—	14 47
	14	11 18 46	—	11 19 4	
	13	19 58			
10,1	—	—	13,3	—	14 27
	13	33 46	—	33 31	
	14	35 26			
15,6					
Weights moved to positive position.					
32					
	28	0 2 48	—	0 2 59	
	27	3 56			
23,7	—	—	27,8		
31,8	—	—	28,27		
25,8	—	—	28,62		
	27	44 58	—	47 40	
	28	46 50			
31,1					

Motion of the arm on moving weights from pos. to mid. = 15,22
mid. to pos. = 14,5

Time of one vibration, when in mid. position = 14′ 39″
pos. position = 14 54

These experiments are sufficient to shew, that the attraction of the weights on the balls is very sensible, and are also sufficiently regular to determine the quantity of this attraction pretty nearly, as the extreme results do not differ from each other by more than $\frac{1}{10}$ part. But there is a circumstance in them, the reason of which does not readily appear, namely, that the effect of the attraction seems to increase, for half an hour, or an hour, after the motion of the weights; as it may be observed, that in all three experiments, the mean position kept increasing for that time, after moving the weights to the positive position; and kept decreasing, after moving them from the positive to the midway position.

The first cause which occurred to me was, that possibly there might be a want of elasticity, either in the suspending wire, or something it was fastened to, which might make it yield more to a given pressure, after a long continuance of that pressure, than it did at first.

To put this to the trial, I moved the index so much, that the arm, if not prevented by the sides of the case, would have stood at about 50 divisions, so that, as it could not move farther than to 35 divisions, it was kept in a position 15 divisions distant from that which it would naturally have assumed from the stiffness of the wire; or, in other words, the wire was twisted 15 divisions. After having remained two or three hours in this position, the index was moved back, so as to leave the arm at liberty to assume its natural position.

It must be observed, that if a wire is twisted only a little more than its elasticity admits of, then, instead of setting, as it is called, or acquiring a permanent twist all at once, it sets gradually, and, when it is left at liberty, it gradually loses part of that set which it acquired; so that if, in this experiment, the wire, by having been kept twisted for two or three hours, had gradually yielded to this pressure, or had begun to set, it would gradually restore itself, when left at liberty, and the point of rest would gradually move backwards; but, though the experiment was twice repeated, I could not perceive any such effect.

The arm was next suspended by a stiffer wire.

EXPERIMENT IV. Aug. 12.

Weights in midway position.

Extreme points.	Divisions.	Time.	Point of rest.	Time of mid. of vibration.	Difference.
		h. ′ ″		h. ′ ″	′ ″
	21,6	9 30 0			
	21,5	52 0			
	21,5	10 13 0	21,5		
Weights moved from midway to positive position.					
27,2					
22,1	—	—	24,6		

EXPERIMENT IV. Aug. 12 (*cont.*)

Extreme points.	Divisions.	Time.	Point of rest.	Time of mid. of vibration.	Difference.
		h. ′ ″		h. ′ ″	′ ″
27	—	—	24,67		
22,6	—	—	24,75		
26,8	—	—	24,8		
23,0	—	—	24,85		
26,6	—	—	24,9		
23,4					
Weights moved to negative position.					
15					
	17	19 25	—	10 20 31	
	19	20 41			
22,4	—	—	18,72	—	7 0
	20	26 45	—	27 31	
	19	27 22			
15,1	—	—	18,52	—	6 57
	19	35 1	—	34 28	
	20	48			
21,5	—	—	18,35	—	7 23
	20	40 23	—	41 51	
	19	41 18			
15,3	—	—	18,22	—	6 48
	18	48 36	—	48 39	
	19	49 24			
20,8	—	—	18,1	—	6 58
	19	54 45	—	55 37	
	18	55 45			
15,5					
Weights moved to positive position.					
31,3					
	25	11 10 25	—	11 10 40	
	23	11 3			
17,1	—	—	24,02	—	7 3
	22	17 6	—	17 43	
	23	26			
30,6	—	—	24,17	—	7 1
	25	24 33	—	24 44	
	23	25 17			
18,4	—	—	24,32	—	7 5
	23	31 21	—	31 49	
	25	32 9			
29,9	—	—	24,4	—	6 59
	25	38 39	—	38 48	
	23	39 31			

Experiment IV. Aug. 12 (*cont.*)

Extreme points.	Divisions.	Time.	Point of rest.	Time of mid. of vibration.	Difference.
		h. ′ ″		h. ′ ″	′ ″
19,4	—	—	24,5	—	7 6
	23	45 16	—	45 54	
	25	46 12			
29,3					

Motion of arm on moving weights from midway to pos. = 3,1
pos. to neg. = 6,18
neg. to pos. = 5,92

Time of one vibration in neg. position = 7′ 1″
pos. position = 7 3

Experiment V. Aug. 20.

The weights being in the positive position, the arm was made to vibrate, by moving the index.

Extreme points.	Divisions.	Time.	Point of rest.	Time of mid. of vibration.	Difference.
29,6					
21,1	—	—	25,2		
29,	—	—	25,17		
21,6					

Weights moved to negative position.

Extreme points.	Divisions.	Time.	Point of rest.	Time of mid. of vibration.	Difference.
22,6					
	20	10 22 47	—	10 23 11	
	19	23 30			
16,3	—	—	19,27		
21,9	—	—	19,15		
16,5	—	—	19,1		
21,5	—	—	19,07		
16,8	—	—	19,07		
21,2	—	—	19,07		
17,1	—	—	19,05		
20,8	—	—	19,02		
17,4	—	—	19,05		
20,6	—	—	19,02		
	20	11 32 16	—	11 33 53	
	19	33 58			
17,5	—	—	18,97		
					7 13
	19	41 16	—	41 6	
	20	43 0			
20,3					

Weights moved to positive position.

Extreme points.	Divisions.	Time.	Point of rest.	Time of mid. of vibration.	Difference.
20,2					
	24	11 49 10	—	11 40 37	
	26	50 19			

EXPERIMENT V. Aug. 20 (*cont.*)

Extreme points.	Divisions.	Time.	Point of rest.	Time of mid. of vibration.	Difference.
		h. ′ ″		h. ′ ″	′ ″
29,4	—	—	24,95	—	7 7
	26	56 15	—	56 44	
	25	47			
20,8	—	—	24,92		
28,7	—	—	24,87		
21,3	—	—	24,85		
28,1	—	—	24,75		
21,5	—	—	24,67		
27,6	—	—	24,67		
22	—	—	24,7		
	24	0 45 48	—	0 46 21	
	25	46 43			
27,2	—	—	24,7	—	7 1
	25	53 11	—	53 22	
	24	54 9			
22,4					

Motion of arm on moving weights from pos. to neg. = 5,9
neg. to pos. = 5,98
Time of one vibration, when weights are in neg. position = 7′ 5″
pos. position = 7 5

In the fourth experiment, the effect of the weights seemed to increase on standing, in all three motions of the weights, conformably to what was observed with the former wire; but, in the last experiment, the case was different; for though, on moving the weights from positive to negative, the effect seemed to increase on standing, yet, on moving them from negative to positive, it diminished.

My next trials were, to see whether this effect was owing to magnetism. Now, as it happened, the case in which the arm was inclosed, was placed nearly parallel to the magnetic east and west, and therefore, if there was any thing magnetic in the balls and weights, the balls would acquire polarity from the earth; and the weights also, after having remained some time, either in the positive or negative position, would acquire polarity in the same direction, and would attract the balls; but, when the weights were moved to the contrary position, that pole which before pointed to the north, would point to the south, and would repel the ball it was approached to; but yet, as repelling one ball towards the south has the same effect on the arm as attracting the other towards the north, this would have no effect on the position of the arm. After some time, however, the poles of the weight would be reversed, and would begin to attract the balls, and would therefore produce the same kind of effect as was actually observed.

To try whether this was the case, I detached the weights from the upper part of the copper rods by which they were suspended, but still retained the lower joint, namely, that which passed through them; I then fixed them in their positive position, in such manner, that they could turn round on this joint, as a vertical axis. I also made an apparatus, by which I could turn them half way round, on these vertical axes, without opening the door of the room.

Having suffered the apparatus to remain in this manner for a day, I next morning observed the arm, and, having found it to be stationary, turned the weights half way round on their axes, but could not perceive any motion in the arm. Having suffered the weights to remain in this position for about an hour, I turned them back into their former position, but without its having any effect on the arm. This experiment was repeated on two other days, with the same result.

We may be sure, therefore, that the effect in question could not be produced by magnetism in the weights; for, if it was, turning them half round on their axes, would immediately have changed their magnetic attraction into repulsion, and have produced a motion in the arm.

As a further proof of this, I took off the leaden weights, and in their room placed two 10-inch magnets; the apparatus for turning them round being left as it was, and the magnets being placed horizontal, and pointing to the balls, and with their north poles turned to the north; but I could not find that any alteration was produced in the place of the arm, by turning them half round; which not only confirms the deduction drawn from the former experiment, but also seems to shew, that in the experiments with the iron rods, the effect produced could not be owing to magnetism.

The next thing which suggested itself to me was, that possibly the effect might be owing to a difference of temperature between the weights and the case; for it is evident, that if the weights were much warmer than the case, they would warm that side which was next to them, and produce a current of air, which would make the balls approach nearer to the weights. Though I thought it not likely that there should be sufficient difference, between the heat of the weights and case, to have any sensible effect, and though it seemed improbable that, in all the foregoing experiments, the weights should happen to be warmer than the case, I resolved to examine into it, and for this purpose removed the apparatus used in the last experiments, and supported the weights by the copper rods, as before; and, having placed them in the midway position, I put a lamp under each, and placed a thermometer with its ball close to the outside of the case, near that part which one of the weights approached to in its positive position, and in such manner that I could distinguish the divisions by the telescope. Having done this, I shut the door, and some time after moved the weights to the positive position. At first, the arm was drawn aside only in its usual manner; but, in half an hour, the effect was so much increased, that the arm was drawn 14 divisions aside, instead of about three, as it would otherwise have been, and the thermometer was raised near 1°½; namely, from 61° to 62°½. On opening the door, the weights were found to be no more heated, than just to prevent their feeling cool to my fingers.

As the effect of a difference of temperature appeared to be so great, I bored a small hole in one of the weights, about three-quarters of an inch deep, and inserted the ball of a small thermometer, and then covered up the opening with cement. Another small thermometer was placed with its ball close to the case, and as near to that part to which the weight was approached as could be done with safety; the thermometers being so placed, that when the weights were in the negative position, both could be seen through one of the telescopes, by means of light reflected from a concave mirror.

EXPERIMENT VI. Sept. 6.

Extreme points.	Divisions.	Time.	Point of rest.	Thermometer In air.	Thermometer In weight.
		h. ′			
		Weights in midway position.			
	18,9	9 43	—	55,5	
	18,85	10 3	18,85		
		Weights moved to negative position.			
13,1	—	10 12	—	55,5	55,8
18,4	—	18	15 82		
13,4	—	25			
missed					
13,6	—	39	—	55,5	55,8
17,6	—	46	15,65		
13,8	—	53	15,65		
17,4	—	11 0	15,65		
14,0	—	7	15,65		
17,2	—	14	—	55,5	
		Weights moved to positive position.			
25,8	—	23			
17,5	—	30	21,55		
25,4	—	37	21,6		
18,1	—	44	21,65		
25,0	—	51			
missed					
24,7	—	0 5			
19,	—	12	21,77		
24,4	—	19			

Motion of arm on moving weights from midway to – = 3,03

– to + = 5,9

Experiment VII. Sept. 18.

Extreme points.	Divisions.	Time.	Point of rest.	Thermometer In air.	Thermometer In weight.
		h. ′			
Weights in midway position.					
	19,4	8 30	—	56,7	
	19,4	9 32	—	56,6	
Weights moved to negative position.					
13,6	—	40	—	—	57,2
18,8	—	47	16,25		
13,8	—	54			
Eight extreme points missed.					
16,9	—	10 58			
14,5	—	11 5	15,62		
16,6	—	12			
Weights moved to positive position.					
26,4	—	20	—	56,5	
17,2	—	28	21,72		
26,1	—	35			
Four extreme points missed.					
19,3	—	0 10			
25,1	—	17	22,3		
19,7	—	24			

Motion of arm on moving weights from midway to − = 3,15
− to + = 6,1

Experiment VIII. Sept. 23.

Extreme points.	Divisions.	Time.	Point of rest.	Thermometer In air.	Thermometer In weight.
Weights in midway position.					
	19,3	9 46	—	53,1	
	19,2	10 45	19,2	53,1	
Weights moved to negative position.					
13,5	—	56	—	—	53,6
18,6	—	11 3	16,07		
13,6	—	10			
Four extreme points missed.					
17,4	—	44			
14,1	—	51	15,7		
17,2	—	58	—	—	53,6
Weights moved to positive position.					
15,7	—	0 1			
26,7	—	8	21,42		
16,6	—	15	—	53,15	

EXPERIMENT VIII. Sept. 23 (*cont.*)

Two extreme points missed.

Extreme points.	Divisions.	Time.	Point of rest.	Thermometer In air.	Thermometer In weight.
		h. ′			
25,9	—	36			
18,1	—	43	21,9		
25,5	—	50			

Motion of arm on moving weights from midway to − = 3,13
− to + = 5,72

In these three experiments, the effect of the weight appeared to increase from two to five tenths of a division, on standing an hour; and the thermometers shewed, that the weights were three or five tenths of a degree warmer than the air close to the case. In the two last experiments, I put a lamp into the room, over night, in hopes of making the air warmer than the weights, but without effect, as the heat of the weights exceeded that of the air more in these two experiments than in the former.

On the evening of October 17, the weights being placed in the midway position, lamps were put under them, in order to warm them; the door was then shut, and the lamps suffered to burn out. The next morning it was found, on moving the weights to the negative position, that they were 7°½ warmer than the air near the case. After they had continued an hour in that position, they were found to have cooled 1°½, so as to be only 6° warmer than the air. They were then moved to the positive position; and in both positions the arm was drawn aside about four divisions more, after the weights had remained an hour in that position, than it was at first.

May 22, 1798. The experiment was repeated in the same manner, except that the lamps were made so as to burn only a short time, and only two hours were suffered to elapse before the weights were moved. The weights were now found to be scarcely 2° warmer than the case; and the arm was drawn aside about two divisions more, after the weights had remained an hour in the position they were moved to, than it was at first.

On May 23, the experiment was tried in the same manner, except that the weights were cooled by laying ice on them; the ice being confined in its place by tin plates, which, on moving the weights, fell to the ground, so as not to be in the way. On moving the weights to the negative position, they were found to be about 8° colder than the air, and their effect on the arm seemed now to diminish on standing, instead of increasing, as it did before: as the arm was drawn aside about 2½ divisions less, at the end of an hour after the motion of the weights, than it was at first.

It seems sufficiently proved, therefore, that the effect in question is produced, as above explained, by the difference of temperature between the weights and case; for, in the 6th, 8th, and 9th experiments, in which the weights were not much warmer than the case, their effect increased but little on standing; whereas, it increased much, when they were much warmer than the case, and decreased much, when they were much cooler.

It must be observed, that in this apparatus, the box in which the balls play is pretty deep, and the balls hang near the bottom of it, which makes the effect of the current of air more sensible than it would otherwise be, and is a defect which I intend to rectify in some future experiments.

Experiment IX. April 29.

Weights in positive position.

Extreme points.	Divisions.	Time.	Point of rest.	Time of mid. of vibrations.
		h. ′ ″		h. ′ ″
34,7				
35	—	—	34,84	
34,65				
Weights moved to negative position.				
23,8				
	28	11 18 29	—	11 18 43
	29	58		
33,2	—	—	28,52	
	29	25 27	—	25 40
	28	57		
23,9	—	—	28,25	
32	—	—	28,01	
24,15	—	—	27,82	
31	—	—	27,63	
24,4	—	—	27,55	
30,4	—	—	27,47	
	28	0 7 4	—	0 7 26
	27	53		
24,7				

Motion of arm = 6,32
Time of vibration = 6′ 58″

Experiment X. May 5.

Weights in positive position.

Extreme points.	Divisions.	Time.	Point of rest.	Time of mid. of vibrations.
34,5				
33,5	—	—	33,97	
34,4				
Weights moved to negative position.				
22,3				
	28	10 43 42	—	10 43 36
	29	44 6		

EXPERIMENT X. May 5 (*cont.*)

Extreme points.	Divisions.	Time.	Point of rest.	Time of mid. of vibration.	Difference.
		h. ′ ″		h. ′ ″	′ ″
33,2	—	—	27,82	—	7 0
	28	50 33	—	50 36	
	27	51 0			
22,6	—	—	27,72		
32,5	—	—	27,7		
23,2	—	—	27,58		
31,45	—	—	27,4		
23,5	—	—	27,28		
	27	11 25 20	—	11 25 24	
	28	58			
30,7	—	—	27,21	—	7 3
	28	32 0	—	32 27	
	27	32 40			
23,95	—	—	27,21	—	6 56
	27	39 19	—	39 23	
	28	40 2			
30,25					

Motion of arm = 6,15
Time of vibration = 6′ 59″

EXPERIMENT XI. May 6.

Weights in positive position.

Extreme points.	Divisions.	Time.	Point of rest.	Time of mid. of vibration.	Difference.
34,9					
34,1	—	—	34,47		
34,8	—	—	34,49		
34,25					

Weights moved to negative position.

Extreme points.	Divisions.	Time.	Point of rest.	Time of mid. of vibration.	Difference.
23,3					
	28	9 59 59	—	10 0 8	
	29	10 0 27			
33,3	—	—	28,42	7 5	
	29	6 52			
	27	7 51			
23,8	—	—	28,35		
32,5	—	—	28,3		
24,4					
missed					
24,8					
31,3	—	—	28,17		
	29	10 48 37	—	10 49 8	
	28	49 21			

EXPERIMENT XI. May 6 (*cont.*)

Extreme points.	Divisions.	Time.	Point of rest.	Time of mid. of vibration.
		h. ′ ″		h. ′ ″
25,3	—	—	28,2	
	28	56 8	—	56 13
	29	56 56		
30,9				

Motion of arm = 6,07
Time of vibration = 7′ 1″

In the three foregoing experiments, the index was purposely moved so that, before the beginning of the experiment, the balls rested as near the sides of the case as they could, without danger of touching it; for it must be observed, that when the arm is at 35, they begin to touch. In the two following experiments, the index was in its usual position.

EXPERIMENT XII. May 9.

Weights in negative position.

Extreme points.	Divisions.	Time.	Point of rest.	Time of mid. of vibration.
		h. ′ ″		h. ′ ″
	17,4	9 45 0		
	17,4	58 0		
	17,4	10 8 0		
	17,4	10 0	17,4	
28,85				
Weights moved to positive position.				
	24	10 20 50	—	10 20 59
	22	21 46		
18,4	—	—	23,49	
28,3	—	—	23,57	
19,3	—	—	23,67	
27,8	—	—	23,72	
20	—	—	23,8	
27,4	—	—	23,83	
	24	11 3 13	—	11 3 14
	23	54		
20,55	—	—	23,87	
	23	9 45	—	10 18
	24	10 28		
27				

Motion of arm = 6,09
Time of vibration = 7′ 3″

Experiment XIII. May 25.

Extreme points.	Divisions.	Time. h. ′ ″	Point of rest.	Time of mid. of vibrations. h. ′ ″
		Weights in negative position.		
16				
18,3	—	—	17,2	
16,2				
		Weights moved to positive position.		
29,6				
	25	10 22 22	—	10 22 56
	24	0 45		
17,4	—	—	23,32	
	23	29 59	—	30 3
	24	30 23		
28,9	—	—	23,4	
	24	36 58	—	37 7
	23	37 24		
18,4	—	—	23,52	
	23	10 44 3	—	10 44 14
	24	31		
28,4	—	—	23,62	
19,3	—	—	23,7	
27,8	—	—	23,7	
	24	11 5 26	—	11 5 31
	23	6 1		
19,9	—	—	23,72	
	23	12 12	—	12 35
	24	50		
27,3				
		Weights moved to negative position.		
13,5				
21,8	—	—	17,75	
	18	37 34	—	37 39
	17	38 10		
13,9	—	—	17,67	
	17	44 26	—	44 45
	18	45 4		
21,1	—	—	17,62	
14,4	—	—	17,6	
20,5	—	—	17,52	
14,7	—	—	17,47	
20	—	—	17,42	
	18	0 19 57	—	0 20 24
	17	20 52		

Experiment XIII. May 25 (*cont.*)

Extreme points.	Divisions.	Time. h. ′ ″	Point of rest.	Time of mid. of vibration. h. ′ ″
15	—	—	17,37	
	17	27 15	—	27 30
	18	28 15		
19,5				

Motion of the arm on moving weights from – to + = 6,12

+ to – = 5,97

Time of vibration at + = 7′ 6″

– = 7 7

Experiment XIV. May 26.

Weights in negative position.

Extreme points.	Divisions.	Time. h. ′ ″	Point of rest.	Time of mid. of vibration. h. ′ ″
	16,1	9 18 0		
	16,1	24 0		
	16,1	46 0		
	16,1	49 0	16,1	

Weights moved to positive position.

Extreme points.	Divisions.	Time. h. ′ ″	Point of rest.	Time of mid. of vibration. h. ′ ″
27,7				
	23	10 0 46	—	10 1 1
	22	1 16		
17,3	—	—	22,37	
	22	7 58	—	8 5
	23	8 27		
27,2	—	—	22,5	
	23	15 2	—	15 9
	22	32		
18,3	—	—	22,65	
26,8	—	—	22,75	
19,1	—	—	22,85	
26,4	—	—	22,97	
	23	43 40	—	43 32
	22	44 22		
20	—	—	23,15	
	22	49 53	—	50 41
	23	50 37		
26,2				

Weights moved to negative position.

Extreme points.	Divisions.	Time. h. ′ ″	Point of rest.	Time of mid. of vibration. h. ′ ″
12,4				
	16	11 7 53	—	11 8 25
	17	8 27		
21,5	—	—	17,02	
	17	15 30	—	15 27
	16	16 3		

EXPERIMENT XIV. May 26 (*cont.*)

Extreme points.	Divisions.	Time.	Point of rest.	Time of mid. of vibration.
		h. ′ ″		h. ′ ″
12,7	—	—	16,9	
20,7	—	—	16,85	
13,3	—	—	16,82	
20	—	—	16,72	
13,6	—	—	16,67	
	16	11 50 33	—	11 50 58
	17	51 19		
19,5	—	—	16,65	
	17	57 53	—	58 6
	16	58 44		
14				

Motion of arm by moving weights from − to + = 6,27
+ to − = 6,13
Time vibration at + = 7′ 6″
− = 7 6

In the next experiment, the balls, before the motion of the weights, were made to rest as near as possible to the sides of the case, but on the contrary side from what they did in the 9th, 10th and 11th experiments.

EXPERIMENT XV. May 27.

Weights in negative position.

Extreme points.	Divisions.	Time.	Point of rest.	Time of mid. of vibration.
		h. ′ ″		h. ′ ″
3,9				
3,35	—	—	3,61	
3,85	—	—	3,61	
3,4				
	Weights moved to positive position.			
15,4				
	10	10 5 59	—	10 5 56
	9	6 27		
4,8	—	—	9,95	
	9	12 43	—	13 5
	10	13 11		
14,8	—	—	10,07	
	10	20 24	—	20 13
	9	56		
5,9	—	—	10,23	
14,35	—	—	10,35	
6,8	—	—	10,46	

EXPERIMENT XV. May 27 (*cont.*)

Extreme points.	Divisions.	Time. h. ′ ″	Point of rest.	Time of mid. of vibration. h. ′ ″
13,9	—	—	10,52	
	11	48 30	—	48 42
	10	49 11		
7,5	—	—	10,6	
	10	55 26	—	55 48
	11	56 10		
13,5				

Motion of the arm = 6,34
Time of vibration = 7′ 7″

The two following experiments were made by Mr. Gilpin, who was so good as to assist me on the occasion.

EXPERIMENT XVI. May 28.

Weights in negative position.

Extreme points.	Divisions.	Time. h. ′ ″	Point of rest.	Time of mid. of vibration. h. ′ ″
22,55				
8,4	—	—	15,09	
21	—	—	14,9	
9,2				

Weights moved to positive position.

Extreme points.	Divisions.	Time. h. ′ ″	Point of rest.	Time of mid. of vibration. h. ′ ″
26,6				
	22	10 22 53	—	10 23 15
	21	23 20		
15,8	—	—	21	
	20	30 7	—	30 30
	21	36		
25,8	—	—	21,05	
	22	37 23	—	37 45
	21	55		
16,8	—	—	21,11	
	20	44 29	—	45 1
	21	45 4		
25,05	—	—	21,11	
	22	51 54	—	52 20
	21	52 32		
17,57	—	—	21,2	
	21	59 31	—	59 34
	22	11 0 13		
24,6	—	—	21,28	
	22	6 24	—	11 6 49
	21	7 9		
18,3				

Motion of the arm = 6,1
Time of vibration = 7′ 16″

Experiment XVII. May 30.

Weights in negative position.

Extreme points.	Divisions.	Time.	Point of rest.	Time of mid. of vibration.
		h. ′ ″		h. ′ ″
	17,2	10 19 0		
	17,1	25 0		
	17,07	29 0		
	17,15	40 0		
	17,45	49 0		
	17,42	51 0		
	17,42	11 1 0	17,42	

Weights moved to positive position.

Extreme points.	Divisions.	Time.	Point of rest.	Time of mid. of vibration.
28,8				
	24	11 11 23	—	11 11 37
	23	49		
18,1	—	—	23,2	
	22	18 13	—	18 42
	23	43		
27,8	—	—	23,12	
	24	25 19	—	25 40
	23	49		
18,8	—	—	23,2	
	23	32 41	—	32 43
	24	33 13		
27,38	—	—	23,31	
	24	39 28	—	39 44
	23	40 3		
19,7	—	—	23,44	
	23	46 33	—	46 46
	24	47 11		
27	—	—	23,52	
	24	53 36	—	53 48
	23	54 17		
20,4	—	—	23,57	
	23	0 0 34	—	0 0 55
	24	1 18		
26,5	—	—	23,55	
	24	7 34	—	7 50
	23	8 21		
20,8	—	—	23,59	
	23	14 30	—	14 58
	24	15 24		
26,25				

EXPERIMENT XVII. May 30 (*cont.*)

Weights moved to negative position.

Extreme points.	Divisions.	Time. h. ′ ″	Point of rest.	Time of mid. of vibrations. h. ′ ″
13,3				
	17	0 32 19	—	0 32 44
	18	48		
22,4	—	—	17,95	
	18	39 46	—	39 44
	17	40 19		
13,7	—	—	17,85	
	17	46 26	—	46 48
	18	47 0		
21,6	—	—	17,72	
	18	53 43	—	53 50
	17	54 20		
14	—	—	17,6	
	17	1 0 39	—	1 0 55
	18	1 20		
20,8	—	—	17,47	
	18	7 39	—	7 59
	17	8 21		
14,3	—	—	17,37	
	17	14 54	—	15 4
	18	15 42		
20,1	—	—	17,27	
	18	21 32	—	22 5
	17	22 22		
14,6				

Motion of the arm on moving weights from − to + = 5,78
+ to − = 5,64

Time of vibration at + = 7′ 2″
− = 7 3

On the Method of computing the Density of the Earth from these Experiments.

I shall first compute this, on the supposition that the arm and copper rods have no weight, and that the weights exert no sensible attraction, except on the nearest ball; and shall then examine what corrections are necessary, on account of the arm and rods, and some other small causes.

The first thing is, to find the force required to draw the arm aside, which, as was before said, is to be determined by the time of a vibration.

The distance of the centres of the two balls from each other is 73,3 inches, and therefore the distance of each from the centre of motion is 36,65, and the length of a pendulum vibrating seconds, in this climate, is 39,14; therefore, if the stiffness of the wire by which the arm is suspended is such, that the force which must be applied to each ball, in order to draw the arm aside by the angle A, is to the weight of that ball as the arch of A to the radius, the arm will vibrate in the same time as a pendulum whose length is 36,65 inches, that is, in $\sqrt{\frac{36,65}{39,14}}$ seconds; and therefore, if the stiffness of the wire is such as to make it vibrate in N seconds, the force which must be applied to each ball, in order to draw it aside by the angle A, is to the weight of the ball as the arch of $A \times \frac{1}{N^2} \times \frac{36,65}{39,14}$ to the radius. But the ivory scale at the end of the arm is 38,3 inches from the centre of motion, and each division is $\frac{1}{20}$ of an inch, and therefore subtends an angle at the centre, whose arch is $\frac{1}{766}$; and therefore the force which must be applied to each ball, to draw the arm aside by one division, is to the weight of the ball as $\frac{1}{766N^2} \frac{36,65}{39,14}$ to 1, or as $\frac{1}{818N^2}$ to 1.

The next thing is, to find the proportion which the attraction of the weight on the ball bears to that of the earth thereon, supposing the ball to be placed in the middle of the case, that is, to be not nearer to one side than the other. When the weights are approached to the balls, their centres are 8,85 inches from the middle line of the case; but, through inadvertence, the distance, from each other, of the rods which support these weights, was made equal to the distance of the centres of the balls from each other, whereas it ought to have been somewhat greater. In consequence of this, the centres of the weights are not exactly opposite to those of the balls, when they are approached together; and the effect of the weights, in drawing the arm aside, is less than it would otherwise have been, in the triplicate ratio of $\frac{8,85}{36,65}$ to the chord of the angle whose sine is $\frac{8,85}{36,65}$, or in the triplicate ratio of the cosine of $\frac{1}{2}$ this angle to the radius, or in the ratio of ,9779 to 1.

Each of the weights weighs 2439000 grains, and therefore is equal in weight to 10,64 spherical feet of water; and therefore its attraction on a particle placed at the centre of the ball, is to the attraction of a spherical foot of water on an equal particle placed on its surface, as

$$10,64 \times ,9779 \times \left(\frac{6}{8,85}\right)^2 \text{ to } 1.$$

The mean diameter of the earth is 41800000 feet[1]; and therefore, if the mean

[1] In strictness, we ought, instead of the mean diameter of the earth, to take the diameter of that sphere whose attraction is equal to the force of gravity in this climate; but the difference is not worth regarding.

density of the earth is to that of water as D to one, the attraction of the leaden weight on the ball will be to that of the earth thereon, as

$$10{,}64 \times {,}9779 \times \left(\frac{6}{8{,}85}\right)^2 \text{ to } 41800000D :: 1 \text{ to } 8739000D.$$

It is shewn, therefore, that the force which must be applied to each ball, in order to draw the arm one division out of its natural position, is $\frac{1}{818N^2}$ of the weight of the ball; and, if the mean density of the earth is to that of water as D to 1, the attraction of the weight on the ball is $\frac{1}{8739000D}$ of the weight of that ball; and therefore the attraction will be able to draw the arm out of its natural position by $\frac{818N^2}{8739000D}$ or $\frac{N^2}{10683D}$ divisions; and therefore, if on moving the weights from the midway to a near position the arm is found to move B divisions, or if it moves $2B$ divisions on moving the weights from one near position to the other, it follows that the density of the earth, or D, is $\frac{N^2}{10683B}$.

We must now consider the corrections which must be applied to this result; first, for the effect which the resistance of the arm to motion has on the time of the vibration: 2d, for the attraction of the weights on the arm: 3d, for their attraction on the farther ball: 4th, for the attraction of the copper rods on the balls and arm: 5th, for the attraction of the case on the balls and arm: and 6th, for the alteration of the attraction of the weights on the balls, according to the position of the arm, and the effect which that has on the time of vibration. None of these corrections, indeed, except the last, are of much signification, but they ought not entirely to be neglected.

As to the first, it must be considered, that during the vibrations of the arm and balls, part of the force is spent in accelerating the arm; and therefore, in order to find the force required to draw them out of their natural position, we must find the proportion which the forces spent in accelerating the arm and balls bear to each other.

Let *EDCedc* (fig. 4) be the arm. *B* and *b* the balls. *Cs* the suspending wire. The arm consists of 4 parts; first, a deal rod *Dcd*, 73,3 inches long; 2d, the silver wire *DCd*, weighing 170 grains; 3d, the end pieces *DE* and *ed*, to which the ivory vernier is fastened, each of which weighs 45 grains; and 4th, some brass work *Cc*, at the centre. The deal rod, when dry, weighs 2320 grains, but when very damp, as it commonly was during the experiments, weighs 2400; the transverse section is of the shape represented in fig. 5; the thickness *BA*, and the dimensions of the part *DEed*, being the same in all parts; but the breadth *Bb* diminishes gradually, from the middle to the ends. The area of this section is ,33 of a square inch at the middle, and ,146 at the end; and therefore, if any point x (fig. 4.) is taken in cd, and $\frac{cx}{cd}$ is called x, this rod weighs $\frac{2400 \times ,33}{73,3 \times ,238}$ per inch at the middle; $\frac{2400 \times ,146}{73,3 \times ,238}$ at the end, and $\frac{2400}{73,3} \times \frac{,33 - ,184x}{,238} = \frac{3320 - 1848x}{73,3}$ at x; and therefore, as the weight of the wire is $\frac{170}{73,3}$ per inch, the deal rod and wire together may be considered as a rod whose weight at x

$$= \frac{3490 - 1848x}{73,3} \text{ per inch.}$$

But the force required to accelerate any quantity of matter placed at x, is proportional to x^2; that is, it is to the force required to accelerate the same quantity of matter placed at d as x^2 to 1; and therefore, if cd is called l, and x is supposed to flow, the fluxion of the force required to accelerate the deal rod and wire is proportional to $\frac{x^2 l \dot{x} \times 3490 - 1848x}{73,3}$, the fluent of which, generated while x flows from c to d, $= \frac{l}{73,3} \times \frac{3490}{3} - \frac{1848}{4} = 350$; so that the force required to accelerate each half of the deal rod and wire, is the same as is required to accelerate 350 grains placed at d.

The resistance to motion of each of the pieces de, is equal to that of 48 grains placed at d; as the distance of their centres of gravity from C is 38 inches. The resistance of the brass work at the centre may be disregarded; and therefore the whole force required to accelerate the arm, is the same as that required to accelerate 398 grains placed at each of the points D and d.

Each of the balls weighs 11262 grains, and they are placed at the same distance from the centre as D and d; and therefore, the force required to accelerate the balls and arm together, is the same as if each ball weighed 11660, and the arm had no weight; and therefore, supposing the time of a vibration to be given, the force required to draw the arm aside, is greater than if the arm had no weight, in the proportion of 11660 to 11262, or of 1,0353 to 1.

To find the attraction of the weights on the arm, through d draw the vertical plane dwb perpendicular to Dd, and let w be the centre of the weight, which, though not accurately in this plane, may, without sensible error, be considered as placed therein, and let b be the centre of the ball; then wb is horizontal and $= 8,85$, and db is vertical and $= 5,5$; let $wd = a$, $wb = b$, and let $\frac{dx}{dc}$, or $1 - x = z$; then the attraction of the weight on a particle of matter at x, in the direction dw, is to its attraction on the same particle placed at b :: $b^3 : (a^2 + z^2 l^2)^{\frac{3}{2}}$, or is proportional to $\frac{b^3}{(a^2 + z^2 l^2)^{\frac{3}{2}}}$, and the force of that attraction to move the arm, is proportional to $\frac{b^3 \times \overline{1 - z}}{(a^2 + z^2l^2)^{\frac{3}{2}}}$, and the weight of the deal rod and wire at the point x, was before said to be $\frac{3490 - 1848x}{73{,}3} = \frac{1642 + 1848z}{73{,}3}$ per inch; and therefore, if dx flows, the fluxion of the power to move the arm

$$= l\dot{z} \times \frac{1642 + 1848z}{73{,}3} + \frac{b^3 \times \overline{1 - z}}{(a^2 + z^2l^2)^{\frac{3}{2}}} = \dot{z} \times (821 + 924z) \times \frac{b^3 \times \overline{1 - z}}{(a^2 + l^2z^2)^{\frac{3}{2}}}$$

$$= \frac{b^3\dot{z} \times \overline{821 + 103z - 924z^2}}{(a^2 + l^2z^2)^{\frac{3}{2}}} = \frac{b^3z \times \overline{821 + 103z + \frac{924a^2}{l^2}}}{(a^2 + l^2z^2)^{\frac{3}{2}}} - \frac{924b^3\dot{z} \times \overline{\frac{a^2}{l^2} + z^2}}{(a^2 + l^2z^2)^{\frac{3}{2}}};$$

which, as $\frac{a^2}{l^2} = {,}08 = \frac{b^3\dot{z} \times \overline{895 + 103z}}{(a^2 + l^2z^2)^{\frac{3}{2}}} - \frac{924b^3\dot{z}}{l^2\sqrt{a^2 + l^2z^2}}$

The fluent of this

$$= \frac{895b^3z}{a^2\sqrt{a^2 + l^2z^2}} - \frac{103b^3}{l^2\sqrt{a^2 + l^2z^2}} + \frac{103b^3}{l^2a} - \frac{924b^3}{l^3}\log.\frac{lz + \sqrt{a^2 + l^2z^2}}{a},$$

and the force with which the attraction of the weight, on the nearest half of the deal rod and wire, tends to move the arm, is proportional to this fluent generated while z flows from 0 to 1, that is, to 128 grains.

The force with which the attraction of the weight on the end-piece de tends to move the arm, is proportional to $47 \times \frac{b^3}{a^3}$, or 29 grains; and therefore the whole power of the weight to move the arm, by means of its attraction on the nearest part thereof, is equal to its attraction on 157 grains placed at b, which is $\frac{157}{11262}$, or ,0139 of its attraction on the ball.

It must be observed, that the effect of the attraction of the weight on the whole arm is rather less than this, as its attraction on the farther half draws it the contrary way; but, as the attraction on this is small, in comparison of its attraction on the nearer half, it may be disregarded.

The attraction of the weight on the furthest ball, in the direction bw, is to its attraction on the nearest ball :: $wd^3 : wD^3$ :: ,0017 : 1; and therefore the effect of the attraction of the weight on both balls, is to that of its attraction on the nearest ball :: ,9983 : 1.

To find the attraction of the copper rod on the nearest ball, let b and w (fig. 6.) be the centres of the ball and weight, and ea the perpendicular part of the copper rod, which consists of two parts, ad and de. ad weighs 22000 grains, and is 16 inches long, and is nearly bisected by w. de weighs 41000, and is 46 inches long. wb is 8,85 inches, and is perpendicular to ew. Now, the attraction of a line ew, of uniform thickness, on b, in the direction bw, is to that of the same quantity of matter placed at w :: $bw : eb$; and therefore the attraction of the part da equals that of $\frac{22000 \times wb}{db}$, or 16300, placed at w; and the attraction of de equals that of

$$\overline{}\,41000 \times \frac{\overline{ew}}{ed} \times \frac{\overline{bw}}{be} - \overline{}\,41000 \times \frac{\overline{dw}}{ed} \times \frac{\overline{bw}}{bd},$$

or 2500, placed at the same point; so that the attraction of the perpendicular part of the copper rod on b, is to that of the weight thereon, as 18800 : 2439000, or as ,00771 to 1. As for the attraction of the inclined part of the rod and wooden bar, marked Pr and rr in fig. 1, it may safely be neglected, and so may the attraction of the whole rod on the arm and farthest ball; and therefore the attraction of the weight and copper rod, on the arm and both balls together, exceeds the attraction of the weight on the nearest ball, in the proportion of ,9983 + ,0139 + ,0077 to one, or of 1,0199 to 1.

The next thing to be considered, is the attraction of the mahogany case. Now it is evident, that when the arm stands at the middle division, the attractions of the opposite sides of the case balance each other, and have no power to draw the arm either way. When the arm is removed from this division, it is attracted a little towards the nearest side, so that the force required to draw the arm aside is rather less than it would otherwise be; but yet, if this force is proportional to the distance of the arm from the middle division, it makes no error in the result; for, though the attraction will draw the arm aside more than it would otherwise do, yet, as the accelerating force by which the arm is made to vibrate is diminished in the same proportion, the square of the time of a vibration will be increased in the same proportion as the space by which the arm is drawn aside, and therefore the result will be the same as if the case exerted no attraction; but, if the attraction of the case is not proportional to the distance of the arm from the middle point, the ratio in which the accelerating force is diminished is different in different parts of the vibration, and the square of the time of a vibration will not be increased in the same proportion as the quantity by which the arm is drawn aside, and therefore the result will be altered thereby.

On computation, I find that the force by which the attraction draws the arm from the centre is far from being proportional to the distance, but the whole force is so small as not to be worth regarding; for, in no position of the arm does the attraction of the case on the balls exceed that of $\frac{1}{5}$th of a spheric inch of water, placed at the distance of 1 inch from the centre of the balls; and the attraction of the leaden weight equals that of 10,0 spheric feet of water placed at 8,85 inches, or of 234 spheric inches placed at 1 inch distance; so that the attraction of the case on the balls can in no position of the arm exceed $\frac{1}{1170}$ of that of the weight. The computation is given in the Appendix.

It has been shewn, therefore, that the force required to draw the arm aside one division, is greater than it would be if the arm had no weight, in the ratio of 1,0353 to 1, and therefore $= \frac{1,0353}{818N^2}$ of the weight of the ball; and moreover, the attraction of the weight and copper rod on the arm and both balls together, exceeds the attraction of the weight on the nearest ball, in the ratio of 1,0199 to 1, and therefore $= \frac{1,0199}{8739000D}$ of the weight of the ball; consequently D is really equal to

$$\frac{818N^2}{1,0353} \times \frac{1,0199}{8739000B}, \text{ or } \frac{N^2}{10844B}, \text{ instead of } \frac{N^2}{10683B},$$

as by the former computation. It remains to be considered how much this is affected by the position of the arm.

Suppose the weights to be approached to the balls; let W (fig. 7.) be the centre of one of the weights; let M be the centre of the nearest ball at its mean position, as when the arm is at 20 divisions; let B be the point which it actually rests at; and let A be the point which it would rest at, if the weight was removed; consequently, AB is the space by which it is drawn aside by means of the attraction; and let $M\beta$ be the space by which it would be drawn aside, if the attraction on it was the same as when it is at M. But the attraction at B is greater than at M, in the proportion of $WM^2 : WB^2$; and therefore,

$$AB = M\beta \times \frac{WM^2}{WB^2} = M\beta \times 1 + \frac{2MB}{MW}, \text{ very nearly.}$$

Let now the weights be moved to the contrary near position, and let w be now the centre of the nearest weight, and b the point of rest of the centre of the ball; then

$$Ab = M\beta \times 1 + \frac{2Mb}{MW}, \text{ and } Bb = M\beta \times 2 + \frac{2Mb}{MW} + \frac{2MB}{MW} = 2M\beta \times 1 + \frac{Bb}{MW};$$

so that the whole motion Bb is greater than it would be if the attraction on the ball was the same in all places as it is at M, in the ratio of $1 + \frac{Bb}{MW}$ to one; and, therefore, does not depend sensibly on the place of the arm, in either position of the weights, but only on the quantity of its motion, by moving them.

This variation in the attraction of the weight, affects also the time of vibration; for, suppose the weights to be approached to the balls, let W be the centre of the nearest weight; let B and A represent the same things as before; and let x be the centre of the ball, at any point of its vibration; let AB represent the force with which the ball, when placed at B, is drawn towards A by the stiffness of the wire; then, as B is the point of rest, the attraction of the weight thereon will also equal AB; and, when the ball is at x, the force with which it is drawn towards A, by the stiffness of the wire, $= Ax$, and that with which it is drawn in the contrary direction, by the attraction, $= AB \times \frac{WB^2}{Wx^2}$; so that the actual force by which it is drawn towards A

$$= Ax - \frac{AB \times WB^2}{Wx^2} = AB + Bx - \overline{AB \times 1 + \frac{2Bx}{WB}} = Bx - \frac{2Bx \times AB}{WB},$$

very nearly. So that the actual force with which the ball is drawn towards the middle point of the vibration, is less than it would be if the weights were removed, in the ratio of $1 - \frac{2AB}{WB}$ to one, and the square of the time of a vibration is increased in the ratio of 1 to $1 - \frac{2AB}{WB}$; which differs very little from that of $1 + \frac{Bb}{MW}$ to 1, which is the ratio in which the motion of the arm, by moving the weights from one near position to the other, is increased.

The motion of the ball answering to one division of the arm $= \frac{36,35}{20 \times 38,3}$; and, if mB is the motion of the ball answering to d divisions on the arm, $\frac{MB}{WM} = \frac{36,35d}{20 \times 38,3 \times 8,85} = \frac{d}{185}$; and therefore, the time of vibration, and motion of the arm, must be corrected as follows:

If the time of vibration is determined by an experiment in which the weights are in the near position, and the motion of the arm, by moving the weights from the near to the midway position, is d divisions, the observed time must be diminished in the subduplicate ratio of $1 - \frac{2d}{185}$ to 1, that is, in the ratio of $1 - \frac{d}{185}$ to 1; but, when it is determined by an experiment in which the weights are in the midway position, no correction must be applied.

To correct the motion of the arm caused by moving the weights from a near to the midway position, or the reverse, observe how much the position of the arm differs from 20 divisions, when the weights are in the near position: let this be n divisions, then, if the arm at that time is on the same side of the division of 20 as the weight, the observed motion must be diminished by the $\frac{2n}{185}$ part of the whole; but, otherwise, it must be as much increased.

If the weights are moved from one near position to the other, and the motion of the arm is $2d$ divisions, the observed motion must be diminished by the $\frac{2d}{185}$ part of the whole.

If the weights are moved from one near position to the other, and the time of vibration is determined while the weights are in one of those positions, there is no need of correcting either the motion of the arm, or the time of vibration.

CONCLUSION

The following Table contains the Result of the Experiments.

Exper.	Mot. weight	Mot. arm.	Do. corr.	Time vib.	Do. corr.	Density.
1	m. to +	14,32	13,42		—	5,5
	+ to m.	14,1	13,17	14′,55″	—	5,61
2	m. to +	15,87	14,69	—	—	4,88
	+ to m.	15,45	14,14	14,42	—	5,07
3	+ to m.	15,22	13,56	14,39	—	5,26
	m. to +	14,5	13,28	14,54	—	5,55
4	m. to +	3,1	2,95		6,54	5,36
	+ to −	6,18	—	7,1	—	5,29
	− to +	5,92	—	7,3	—	5,58
5	+ to −	5,9	—	7,5	—	5,65
	− to +	5,98	—	7,5	—	5,57
6	m. to −	3,03	2,9	—	—	5,53
	− to +	5,9	5,71		—	5,62
7	m. to −	3,15	3,03	7,4		5,29
	− to +	6,1	5,9	by mean	6,57	5,44
8	m. to −	3,13	3,00		—	5,34
	− to +	5,72	5,54	—	—	5,79
9	+ to −	6,32	—	6,58	—	5,1
10	+ to −	6,15	—	6,59	—	5,27
11	+ to −	6,07	—	7,1	—	5,39
12	− to +	6,09	—	7,3	—	5,42
13	− to +	6,12	—	7,6	—	5,47
	+ to −	5,97	—	7,7	—	5,63
14	− to +	6,27	—	7,6	—	5,34
	+ to −	6,13	—	7,6	—	5,46
15	− to +	6,34	—	7,7	—	5,3
16	− to +	6,1	—	7,16	—	5,75
17	− to +	5,78	—	7,2	—	5,68
	+ to −	5,64	—	7,3	—	5,85

From this table it appears, that though the experiments agree pretty well together, yet the difference between them, both in the quantity of motion of the arm and in the time of vibration, is greater than can proceed merely from the error of observation. As to the difference in the motion of the arm, it may very well be accounted for, from the current of air produced by the difference of temperature; but, whether this can account for the difference in the time of vibration, is doubtful. If the current of air was regular, and of the same swiftness in all parts of the vibration of the ball, I think it could not; but, as there will most likely be much irregularity in the current, it may very likely be sufficient to account for the difference.

By a mean of the experiments made with the wire first used, the density of the earth comes out 5,48 times greater than that of water; and by a mean of those made with the latter wire, it comes out the same; and the extreme difference of the results of the 23 observations made with this wire, is only ,75; so that the extreme results do not differ from the mean by more than ,38, or $\frac{1}{14}$ of the whole, and therefore the density should seem to be determined hereby, to great exactness. It, indeed, may be objected, that as the result appears to be influenced by the current of air, or some other cause, the laws of which we are not well acquainted with, this cause may perhaps act always, or commonly, in the same direction, and thereby make a considerable error in the result. But yet, as the experiments were tried in various weathers, and with considerable variety in the difference of temperature of the weights and air, and with the arm resting at different distances from the sides of the case, it seems very unlikely that this cause should act so uniformly in the same way, as to make the error of the mean result nearly equal to the difference between this and the extreme; and, therefore, it seems very unlikely that the density of the earth should differ from 5,48 by so much as $\frac{1}{14}$ of the whole.

Another objection, perhaps, may be made to these experiments, namely, that it is uncertain whether, in these small distances, the force of gravity follows exactly the same law as in greater distances. There is no reason, however, to think that any irregularity of this kind takes place, until the bodies come within the action of what is called the attraction of cohesion, and which seems to extend only to very minute distances. With a view to see whether the result could be affected by this attraction, I made the 9th, 10th, 11th, and 15th experiments, in which the balls were made to rest as close to the sides of the case as they could; but there is no difference to be depended on, between the results under that circumstance, and when the balls are placed in any other part of the case.

According to the experiments made by Dr. Maskelyne, on the attraction of the hill Schehallien, the density of the earth is $4\frac{1}{2}$ times that of water; which differs rather more from the preceding determination than I should have expected. But I forbear entering into any consideration of which determination is most to be depended on, till I have examined more carefully how much the preceding determination is affected by irregularities whose quantity I cannot measure.

APPENDIX

On the Attraction of the Mahogany Case on the Balls.

The first thing is, to find the attraction of the rectangular plane $ck\beta b$ (fig. 8.) on the point a, placed in the line ac perpendicular to this plane.

Let $ac = a$, $ck = b$, $cb = x$, and let $\frac{a^2}{a^2+x^2} = w^2$, and $\frac{b^2}{a^2+x^2} = v^2$, then the attraction of the line $b\beta$ on a, in the direction ab, $= \frac{b\beta}{ab \times a\beta}$; and therefore, if cb flows, the fluxion of the attraction of the plane on the point a, in the direction cb,

$$= \frac{b\dot{x}}{\sqrt{a^2+x^2} \times \sqrt{a^2+b^2+x^2}} \times \frac{x}{\sqrt{a^2+x^2}} = \frac{-b\dot{w}}{w\sqrt{b^2+\frac{a^2}{w^2}}} = \frac{-b\dot{w}}{\sqrt{b^2w^2+a^2}} = \frac{-\dot{v}}{\sqrt{1+v^2}},$$

the variable part of the fluent of which $= -\log. v + \sqrt{1+v^2}$, and therefore the whole attraction $= \log. \left(\frac{ck+ak}{ac} \times \frac{ab}{b\beta + a\beta}\right)$; so that the attraction of the plane, in the direction cb, is found readily by logarithms, but I know no way of finding its attraction in the direction ac, except by an infinite series.

The two most convenient series I know, are the following:

First series. Let $\frac{b}{a} = \pi$, and let A = arc whose tang. is π,

$$B = A - \pi,\ C = B + \frac{\pi^3}{3},\ D = C - \frac{\pi^5}{5},\ \&c.$$

then the attraction in the direction ac

$$= \sqrt{1-w^2} \times A + \frac{Bw^2}{2} + \frac{3Cw^4}{2.4} + \frac{3.5Dw^6}{2.4.6},\ \&c.$$

For the second series, let A = arc whose tang.

$$= \frac{1}{\pi},\ B = A - \frac{1}{\pi},\ C = B + \frac{1}{3\pi^3},\ D = C - \frac{1}{5\pi^5},\ \&c.$$

then the attraction

$$= \text{arc}.\ 90^\circ - \sqrt{(1+v^2)} \times A - \frac{Bv^2}{2} + \frac{3Cv^4}{2.4} - \frac{3.5Dv^6}{2.4.6},\ \&c.$$

It must be observed, that the first series fails when π is greater than unity, and the second, when it is less; but, if b is taken equal to the least of the two lines ck and cb, there is no case in which one or the other of them may not be used conveniently.

By the help of these series, I computed the following table [p. 286].

Find in this table, with the argument $\frac{ck}{ak}$ at top, and the argument $\frac{cb}{ab}$ in the left-hand column, the corresponding logarithm; then add together this logarithm, the logarithm of $\frac{ck}{ak}$, and the logarithm of $\frac{cb}{ab}$; the sum is logarithm of the attraction.

	,1962	,3714	,5145	,6248	,7071	,7808	,8575	,9285	,9815	
,1962	,00001									
,3714	,00039	00148								
,5145	,00074	00277	00521							
,6248	00110	00406	00778	01183						
,7071	00140	00522	01008	01525	02002					
,7808	00171	00637	01245	01896	02405	03247				
,8575	00207	00772	01522	02339	03116	03964	05057			
,9285	00244	00910	01810	02807	03778	04867	06319	08119		
,9815	00271	01019	02084	03193	04368	05639	07478	09931	12849	
1,	00284	01054	02135	03347	04560	05975	07978	10789	14632	1[illegible]

To compute from hence the attraction of the case on the ball, let the box $DCBA$, (fig. 1.) in which the ball plays, be divided into two parts, by a vertical section, perpendicular to the length of the case, and passing through the centre of the ball; and, in fig. 9, let the parallelopiped $ABDEabde$ be one of these parts, $ABDE$ being the abovementioned vertical section; let x be the centre of the ball, and draw the parallelogram $\beta npm\delta x$ parallel to $BbdD$, and $xgrp$ parallel to βBbn, and bisect $\beta\delta$ in c. Now, the dimensions of the box, on the inside, are $Bb = 1{,}75$; $BD = 3{,}6$; $B\beta = 1{,}75$; and $\beta A = 5$; whence I find, that if xc and βx are taken as in the two upper lines of the following table, the attractions of the different parts are as set down below.

xc	,75	,5	,25
βx	1,05	1,3	1,55
Excess of attract. of $Ddrg$ above $Bbrg$ ...	,2374	,1614	,0813
,, ,, $mdrp$ above $nbrp$ ...	,2374	,1614	,0813
,, ,, $mesp$ above $nasp$ ...	,3705	,2516	,1271
Sum of these ...	,8453	,5744*	,2897
Excess of attract. of $Bbn\beta$ above $Ddm\delta$ ...	,5007	,3271	,1606
,, ,, $Aan\beta$ above $Eem\delta$ ...	,4677	,3079	,1525
Whole attraction of the inside surface of the half box	,1231	,0606	,0234

It appears, therefore, that the attraction of the box on x increases faster than in proportion to the distance xc.

The specific gravity of the wood used in this case is ,61, and its thickness is $\frac{3}{4}$ of an inch; and therefore, if the attraction of the outside surface of the box was the same as that of the inside, the whole attraction of the box on the ball, when $cx = {,}75$, would be equal to $2 \times {,}1231 \times {,}61 \times \frac{3}{4}$ cubic inches, or ,201 spheric inches of water, placed at the distance of one inch from the centre of the ball. In reality, it can never be so great as this, as the attraction of the outside surface is rather less than that of the inside; and, moreover, the distance of x from c can never be quite so great as ,75 of an inch, as the greatest motion of the arm is only $1\frac{1}{2}$ inch.

PHILOSOPHICAL TRANSACTIONS. VOL. FOR 1809,
page 221.

XIII. *On an Improvement in the Manner of dividing astronomical Instruments. By* Henry Cavendish, *Esq., F.R.S.*

THE great inconvenience and difficulty in the common method of dividing, arises from the danger of bruising the divisions by putting the point of the compass into them, and from the difficulty of placing that point mid-way, between two scratches very near together, without its slipping towards one of them; and it is this imperfection in the common process, which appears to have deterred Mr. Troughton from using it, and thereby gave rise to the ingenious method of dividing described in the preceding part of this volume [Troughton, *Phil. Trans.* 1809, 105]. This induced me to consider, whether the abovementioned inconvenience might not be removed, by using a beam compass with only one point, and a microscope instead of the other; and I find, that in the following manner of proceeding, we have no need of ever setting the point of the compass into a division, and consequently that the great objection to the old method of dividing is entirely removed.

In this method, it is necessary to have a convenient support for the beam compass: and the following seems to me to be as convenient as any. Let *CC* (Fig. 1.) be the circle to be divided, *BBB* a frame resting steadily on its face, and made to slide round on it with an adjusting motion to bring it to any required point: $d\delta$ is the beam compass, having a point near δ, and a microscope *m* made to slide from one end to the other. This beam compass is supported at *d*, in such manner as to turn round on this point as a center, without shake or tottering; and at the end δ it rests on another support, which can readily be lowered, so as either to let the point rest on the circle, or to prevent its touching it. It must be observed, however, that as the distance of *d* from the center of the circle must be varied, according to the magnitude of the arch to be divided, the piece on which *d* is supported had best be made to slide nearer to, or further from, the center; but the frame must be made to bear constantly against the edge of the circle to be divided, so that the distance of *d* from the center of this circle, shall not alter by sliding the frame.

This being premised, we will first consider the manner of dividing by continued bisection. Let *F* and *f* be two points on this limb which are to be bisected in ϕ. Take the distance of the microscope from the point nearly equal to the chord of $f\phi$, and place *d* so that the point and the axis of the microscope shall both be in the circle in which the divisions are to be cut. Then slide the frame *BBB* till the wire of the microscope bisects the point *F*; and having lowered the support at δ, make a faint scratch with the point.

Having done this, turn the beam compass round on the center *d* till the point comes to *D*, where it must rest on a support similar to that at δ; and having slid the frame till the wire of the microscope bisects the point *f*, make another faint scratch with the point, which, if the distance of the microscope from the point has been well taken, will be very near the former scratch; and the point mid-way between them will be the accurate bisection of the arch *Ff*; but it is unnecessary, and better not to attempt to place a point between these two scratches.

Having by these means determined the bisection at ϕ, we must bisect the arches $F\phi$ and $f\phi$ in just the same manner as before, except that the wire of the microscope must be made to bisect the interval between the two faint scratches, instead of bisecting a point.

It must be observed that when the arch to be bisected is small, it will be necessary to use a bent point, as otherwise it could not be brought near enough to the axis of the microscope; and then part of the rays, which form the image of the object seen by the microscope, will be intercepted by the point; but I believe, that by proper management this may be done without either making the point too weak, or making the image indistinct; but if this cannot be done, we may have recourse to Mr. Troughton's expedient of bisecting an odd number of contiguous divisions.

It must be observed too, that in the bisections of all the arches of the same magnitude, the position of the point *d* on the frame remains unaltered; but its position must be altered every time the magnitude of the arch is altered.

It is scarcely necessary to say, that the bisections thus made are not intended as the real divisions, but only as marks from which they are to be cut. In order to make the real divisions, the microscope must be placed near the point, and the support *d* must be placed so that $d\delta$ shall be a tangent to the circle at δ. The wire of the microscope must then be made to bisect one of these marks, and a point or division cut with the point, and the process continued till the divisions are all made.

It is plain that in this way, without some further precaution, we must depend on the microscope not altering its position in respect of the point during the operation; for which reason I should prefer placing the axis of the microscope at exactly the same distance from the center of motion *d*, as the point; but removed from it sideways, by nearly the semi-diameter of the object glass; so that having made the division, we may move the beam compass till the division comes within the field of the microscope, and then see whether it is bisected by the wire, and consequently see whether the microscope has altered its place.

In the operation of bisection, as above described, it may be observed, that if the two scratches are placed so near together, that in making the second the point of the compass runs into the burr raised by the first, there seems to be some danger that the point may be a little deflected from its true course; though in Bird's account of his method, I do not find that he apprehends any inconvenience from it. One way of obviating this inconvenience, if it does exist, would be to set the beam compass not so exactly to the true length, as that one scratch should run into the burr of the other; but as this would make it more difficult to judge of the true point of bisection, perhaps it might be better to make one scratch extend from the circle towards the center, and the other from it.

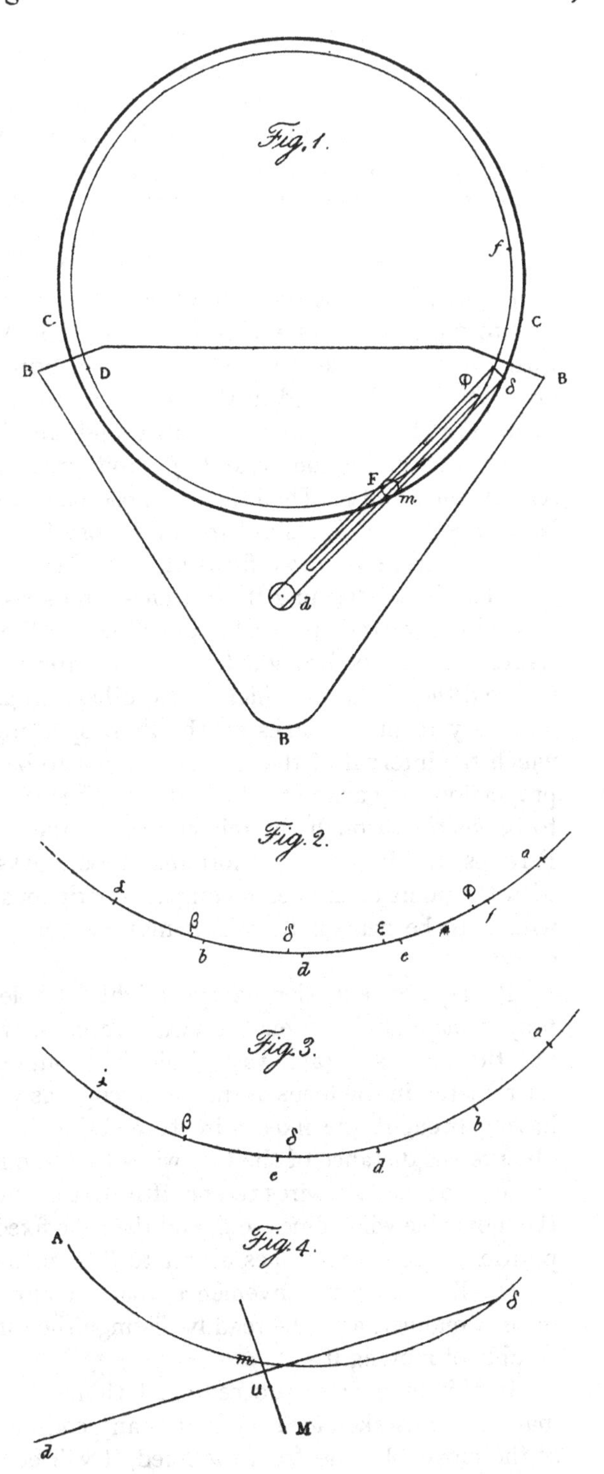

It is clear, that the entire arc of a circle cannot be divided to degrees, without trisection and quinquesection; and I do not know whether our artists have recourse to this operation, or whether they avoid it by some contrivance similar to Bird's, namely, that of laying down an arch capable of continued bisection; but if the method of quinquesection is preferred, it may be performed by either of the three following methods:

First Method.

Let αa (Fig. 2) be the arch to be quinquesected. Open the beam compass to the chord of one fifth of this arch; bring the microscope to a, and with the point make the scratch f; then bring the microscope to f, and draw the scratch e; and in the same manner make the scratches d and b. Then turn the beam compass half round, and having brought the microscope to α, make the scratch β; and proceeding as before, make the scratches δ, ϵ and ϕ. Then the true position of the first quinquesection will be between b and β, distant from β by one fifth of $b\beta$; and the second will be distant from δ by two fifths of $d\delta$, and so on.

Then, in subdividing these arches, and striking the true divisions, the wire of the microscope, instead of bisecting the interval between the two scratches, must be brought four times nearer to β than to b. But in order to avoid the confusion which would otherwise proceed from this, it will be necessary to place marks on the limb opposite to all those divisions, in which the interval of the scratches is not to be bisected, shewing in what proportion they are to be divided; and these marks should be placed so as to be visible through the microscope, at the same time as the scratches. Perhaps, the best way of forming these marks, would be to make dots with the point of the beam compass contiguous to that scratch which the wire is to be nearest to, which may be done at the time the scratch is drawn.

Perhaps an experienced eye might be able to place the wire in the proper manner, between the two scratches, without further assistance; but the most accurate way would be to have a moveable wire with a micrometer, in the focus of the microscope, as well as a fixed one; and then having brought the fixed wire to b, bring the moveable one to β, and observe the distance of the two wires by the micrometer; then reduce the distance of the two wires to one fifth part of this, and move the frame till the moveable wire comes to β, and then the fixed wire will be in the proper position, that is four times nearer to β than to b.

It will be a great convenience, that the moveable wire should be made in such manner, as to be readily distinguished from the fixed, without the trouble of moving it.

In this manner of proceeding, I think a careful operator can hardly make any mistake: for if he makes any considerable error in the distance of the moveable wire from the fixed, it will be detected by the fixed wire not appearing in the right position, in respect of the two scratches; and as the mark is seen through the microscope, at the same time as the scratches, there is no danger of his mistaking which scratch it is to be nearest to, or at what distance it is to be placed from it.

To judge of the comparative accuracy of this method with that of bisection, it must be considered that the arches $\alpha\beta$, $\beta\delta$, &c. though made with the same opening of the compass, will not be exactly alike, owing partly to irregularities in the brass, and partly to other causes. Let us suppose, therefore, that in dividing the arch *aa* into five parts, the beam compass is opened to the exact length, but that from the above-mentioned irregularities the arches $\alpha\beta$, $\beta\delta$, $\delta\epsilon$, and $\epsilon\phi$ are all too long by the small quantity ϵ, and that the arches *af*, *fe*, *ed*, and *db* are all too short by the same quantity, which is the supposition the most unfavourable of any to the exactness of the operation; then the error in the position of $\beta = \epsilon$, and the point *b* errs 4ϵ in the same direction, and therefore the point assumed as the true point of quinquesection, will be at the distance of $\frac{3\epsilon}{5}$ from β, and the error in the position of this point $= \epsilon \times 1\frac{3}{5}$.

By the same way of reasoning, the error in the position of the point taken between *d* and $\delta = \epsilon \times 2\frac{2}{5}$.

In trisecting the error of each point $= \epsilon \times 1\frac{1}{3}$; and in bisecting, the error $= \epsilon$; and in quadrisecting, the error of the middle point $= 2\epsilon$.

It appears therefore that in trisecting, the greatest error we are liable to does not exceed that of bisection in a greater proportion than that of 4 to 3; but in quinquesecting the error of the two middle points is $2\frac{2}{5}$ times greater than in bisecting. It must be considered, however, that in the method of continued bisection, the two opposite points must be found by quadrisection; and the error of quinquesection exceeds that of quadrisection in no greater proportion than that of six to five; so that we may fairly say, that if we begin with quinquesection, this method of dividing is not greatly inferior, in point of accuracy, to that by continued bisection.

Second Method.

This differs from the foregoing, in placing dots or scratches in the true points of quinquesection and trisection, before we begin to subdivide. For this purpose, we must have a microscope placed as in page [287] first par. at the same distance from the center of motion as the point is; and this microscope must be furnished with a moveable wire and micrometer, as in page [290]; and then having first made the fixed wire of this microscope correspond exactly with the point, we must draw the scratches *b* and β, *d* and δ, &c. as before, and bring the fixed wire to the true point of quinquesection between *b* and β, in the manner directed in page [290], and with the point strike the scratch or dot; and if we please, we may, for further security, as soon as this is done, examine, by means of the moveable wire, whether this intermediate scratch or dot is well placed.

The advantage of this method is, that when this is done, we may subdivide and cut the true divisions, by making the wire of the microscope bisect the intermediate scratches, instead of being obliged to use the more troublesome operation of placing it in the proper proportion of distance between the two extremes.

This method certainly requires less attention than the former, and on the whole seems to be attended with considerably less trouble; but it is not quite so exact, as we are liable to the double error of placing the intermediate point and of subdividing from it.

As in this method the intermediate points are placed by means of the micrometer, there is no inconvenience in placing the extreme scratches *b* and β, &c. at such a distance from each other, that the intermediate one shall be in no danger of running into the bur raised by the extremes.

Third Method.

Let $a\alpha$ (Fig. 3) be the arch to be quinquesected; lay down the arches *ab*, *bd*, and *de*, as in the first method; then turn the beam compass half round, and lay down the arches $\alpha\beta$ and $\beta\delta$; then, without altering the frame, move the moveable wire of the microscope till it is four times nearer to δ than to *e*, and, having first rubbed out the former scratches, lay them down again with the compass thus altered; but as this method possesses not much, if any, advantage over the second, in point of ease, and is certainly inferior to it in exactness, it is not worth while saying anything further about it.

It was before said[1], that the center of motion of the beam compass is to be placed, so that the point and axis of the microscope shall both be in the circle in which the divisions are made; but it is necessary to consider this more accurately. Let $A\delta$ (Fig. 4) be the circle in which the scratches are to be made, δ the point of the beam compass, which we will suppose to be exactly in this circle, *d* the center on which it turns, and *Mm* the wire in the focus of the microscope, and let *m* be that point in which it is cut by the circle; and let us suppose that this point is not exactly in the line $d\delta$, then, when the beam compass is turned round, the circle will cut the wire in a different point μ, placed as much on one side of $d\delta$, as *m* is on the other so that if the wire is not perpendicular to $d\delta$, the arch set off by the beam compass, after being turned round, will not be the same as before; but if it is perpendicular, there will be no difference; for which reason, care should be taken to make the wire exactly perpendicular to $d\delta$, which is easily examined by observing whether a point appears to run along it, while the beam compass is turned a little on its center. It is also necessary to take care that the point δ is in the arc of the circle, while the bisection

[1] Page 288.

is observed by the microscope, which may most conveniently be obtained, by placing a stop on the support on which that end of the beam compass rests. If proper care, however, is taken in placing the wire perpendicular, no great nicety is required either in this or in the position of *d*.

Another thing to be attended to, in making the wire bisect two scratches, is to take care that it bisects them in the part where they cut the circle; for as the wire is not perpendicular to the circle, except in very small arches, it is plain, that if it bisects the scratches at the circle, it will not bisect them at a distance from it.

There are many particulars in which my description of the apparatus to be employed will appear incomplete; but as there is nothing in it which seems attended with difficulty, I thought it best not to enter further into particulars, than was necessary to explain the principle, and to leave the rest to any artist who may choose to try it.

It is difficult to form a proper judgment of the conveniences or inconveniences of this method, without experience; but, as far as I can judge, it must have much advantage, both in point of accuracy and ease, over that of dividing by the common beam compasses; but it very likely may be thought that Mr. Troughton's method is better than either. Whether it is or is not, must be left for determination to experience and the judgment of artists. Thus much, however, may be observed, that this, as well as his, is free from the difficulty and inaccuracy of setting the point of a compass exactly in the center of a division. It also requires much less apparatus than his, and is free from any danger of error, from the slipping or irregularity in the motion of a roller; in which respect his method, notwithstanding the precautions used by him, is perhaps not entirely free from objection; and what with some artists may be thought a considerable advantage, it is free from the danger of mistakes in computing a table of errors, and in adjusting a sector according to the numbers of that table.

UNPUBLISHED PAPERS

FROM THE

ORIGINAL MANUSCRIPTS

IN THE POSSESSION OF

THE DUKE OF DEVONSHIRE,

K.G., LL.D., F.R.S.

WITH EXPLANATORY NOTES BY THE EDITOR

UNPUBLISHED PAPERS

AMONG the manuscripts preserved at Chatsworth not only are there the "minutes," as they are termed by Cavendish, of much of the experimental matter of his published work, arranged more or less systematically, and paged and indexed by himself, together with some of the rough drafts of certain of his memoirs and reports, with odds and ends of calculations, notes and memoranda, abstracts of foreign notices, and a few letters; but there are also a variety of papers relating to experimental inquiries, some of which must have required a considerable expenditure of time and labour, and although certain of these papers are put together as if for the press, or for perusal by a friend, they were, for reasons which are not apparent, withheld from publication. In a few cases these unpublished papers were obviously intended as sections of memoirs which were eventually communicated to the Royal Society and are printed in the *Philosophical Transactions*. In others, they are in the form of short notes on side-issues, details of experiments made to settle queries which occurred in the course of the main inquiry, but which even when settled, and in spite of their occasional interest and novelty, he refrained from publishing. The manuscripts deal with a wide range of subjects, and with many departments of physical science—mathematics, astronomy, geodesy, geology, mineralogy, chemistry, heat, electricity, terrestrial magnetism and meteorology. Some small portion of this material has already seen the light. The Rev. William Vernon Harcourt, who examined the papers in connection with the Water Controversy, printed one or two of them as a postscript to his address in 1839 as President of the British Association, and, as already stated, Professor Clerk Maxwell has sifted and published the notes of experimental work relating to electricity. Of the remainder, there is, of course, much that it is unnecessary to reproduce, in spite of its historical interest. Some of it is merely the detail of observations already incorporated in the published memoirs. Other portions are too fragmentary and detached, and their meaning is not always apparent. But there are certain of the papers, more especially those dealing with chemical and physical subjects which unquestionably are of value and interest, and should find a place in any account of Cavendish's work which aims at being reasonably complete. Accordingly when drawn up by Cavendish, as if for publication, they will be printed *in extenso* in what follows. When still in the form of "minutes" an abstract of their contents, as far as possible in the original words, will be given.

It is not possible in all cases to associate definite dates with the papers; this can only be surmised from such internal evidence as they afford. They will be dealt with in sections, and, as far as practicable, in what is presumed to be their chronological sequence.

EXPERIMENTS ON ARSENIC

The parcel of unpublished papers under this title contains: (1) notes of the details of the experiments in question; (2) a rough draft of a systematic account of them; (3) a fair copy of this draft. It is from the last-named that the account has now been printed. Apparently it was written out for the information of a friend, whose name is not mentioned, but who seems to have witnessed some of the experiments. From a date among the notes it would appear that the work was done in or about 1764. It consists of a study of the action of alkalis and acids upon arsenious oxide (As_4O_6); the preparation of properties of arsenic acid and of potassium and other arsenates, with "conjectures" concerning the nature of arsenic acid and its relations to arsenious oxide. In the outset, it was to a large extent a review of Macquer's work, published originally in the memoirs of the French Academy, but contains many original observations. It describes accurately the preparation and behaviour of solutions of potassium arsenite, obtained by dissolving arsenious oxide in a boiling solution of potassium carbonate. Cavendish found that "the greatest quantity of arsenic [arsenious oxide] which a solution of f. Alkali can retain properly dissolved, is about 2½ times of the dry alcaline salt contained in the solution." The theoretical ratio of potassium carbonate to arsenious oxide is 2·8. He accurately notes the action of the mineral acids on solutions of this salt.

He prepares what he terms "neutral arsenical salt" (potassium arsenate) by Macquer's method of heating a mixture of equal weights of nitre and arsenious oxide. If the deflagrated mass is dissolved in a proper quantity of hot water

it readily shoots on cooling into crystals, which do not at all grow moist in the air, and require about 3½ times their weight of water to dissolve them. A solution of these crystals scarcely alters the colour of syrop of violets; if anything they give it a reddish cast; they turn tournsol paper of a brownish red.

This is the first accurate description of acid potassium arsenate (KH_2AsO_4). Cavendish points out that, strictly speaking, it is not a neutral salt, as Macquer supposed, since it dissolves, with effervescence, the carbonates of potassium and lime. He also correctly describes its behaviour with solutions of copper and iron salts. In preparing potassium arsenate by Macquer's process, he collected the red fumes evolved in a solution of pearl-ash, and incidentally obtained potassium nitrite, and notes its behaviour with the mineral acids, and with acetic acid. He assumes that the nitrous acid so formed is a modification of nitric acid "so much altered by this process as to have a less affinity to f. Alk. than the marine acid [hydrochloric acid] though not so small, I suppose, as distilled vinegar."

By heating strong oil of vitriol with arsenious oxide he appears to have obtained the compound described by Richter and analysed by Reich (*J. pr. Chem.* 90. 176) and subsequently examined by Mr R. H. Adie (*Chem. Soc. Journ.* 55. 1899, 157). Cavendish found that 1 part of arsenious oxide yielded 1·5 parts of "an irregular crystallised mass" after solution in strong oil of vitriol. Theoretically 1 part of arsenious oxide forms 1·4 parts of $As_4O_6 . (SO_3)_2$.

The action of strong nitric acid upon arsenious oxide is described at length, as the result was very different from that observed in the case of the other mineral acids. "The arsen. by being dissolved in this acid was found to have undergone the change necessary to enable it to form the neut. arsen. salt when united to f. Alk." In other words, it was transformed into arsenic acid which, when combined with potash, formed potassium arsenate.

The saturated solution yielded on evaporation crystals of nitre mixed with other crystals of a different shape which proved to be neut. arsen. salt. The success of this experiment induced me to try whether by dissolving arsen. [arsenious oxide] in aq. fort. and driving off the acid by heat, I could not procure the arsen. which had sufferd the above-mentioned change (or the arsenical acid, if you will allow me to call it by that name) by itself.

The experiment is described at length. 4 oz. of arsenious oxide were used: the "Caput mortuum" in the retort, after the excess of acid had been driven off, and the residue heated to "almost as great a heat...as the furnace would admit of" was found to "weigh 4 oz. 13 dwt. 6 gra. *id est* about $\frac{1}{6}$ part more than arsen. from which it was made." This increase corresponds almost exactly with the theoretical amount. "It attracted the moisture of the air though but slowly. It requires very little water to dissolve it; I believe scarcely more than $\frac{1}{2}$ its weight; but it does not dissolve fast without the assistance of heat." The substance was arsenic pentoxide, and its properties are accurately described. He proved that it contained no nitric acid, and that it yielded potassium hydrogen arsenate, identical with that formed by Macquer's process.

It also seems to possess all the properties of an acid (unless perhaps it should fail in respect of taste which I have not thought proper to try) since it effervesces with, and neutralizes the fixed and volatile alcalies and calcarious Earths and magnesia, and turns syrop of violets red and also unites to the Earth of alum.

The excess of the weight of the cap. mort. above that of the arsen. it was made from, must be owing, I suppose, to its retaining some of the matter of the aq. fort. used in making it.

How near he was to discovering what that particular matter was, is indicated by his next experiment in which he roasts an intimate mixture of arsenious oxide and potassium carbonate

in a broad shallow earthen pan, care being taken to keep it frequently stirrd. The heat was as great as the matter could bear without caking together. Some of it was taken out now and then, and dissolved in water, and tried with solut. silver; the colour of the precipitate formed thereby changed gradually the more the matter was Calcined, from a pale yellow [silver arsenite], which it was at 1st to a purplish red [silver arsenate] the same as that made by neut. arsen. salt [potassium arsenate].

The mass was then dissolved in water and carefully neutralised with hydrochloric acid.

It was then evap. There 1st shot some crystals resembling neut. arsen. salt, and afterwards some crystals of Sal Sylvii [potassium chloride]. Some of the crystals resembling neut. arsen. salt were dissolved in water: the solution efferv'd with whiting and f. alkali; reddened the colour of blue papers; made the same colourd precip. with solut. silver and blue vitr. as the neut. arsen. salt; in a word I could perceive no difference between that and the neut. arsen. salt made in the common manner.

He then theorises as to the rationale of the change, and as to the nature of the difference between arsenious oxide and arsenic oxide, and, like a true phlogistian, he is oblivious to the significance of the increase of weight he had found to occur when common or white arsenic passes into the "arsenical acid." He says:

I think these experiments shew pretty plainly that the only difference between plain arsenic and the arsenical acid is that the latter is more thoroughly deprived of its Phlogiston than the former. For all the ways I know of making arsenical acid or neut. arsen. salt are such as may reasonably be supposed to deprive the arsen. of its Phlogiston as, for example, in making arsen. acid by solution in aq. fort. the nitrous acid [nitric acid] is known to have a great disposition to lay hold of Phlogiston, and there are strong reasons for thinking that the dissolving of metallic substances in that acid is a very powerful method of depriving them of it, as I shall take notice of by and by.

It is not necessary to follow Cavendish into the maze in which he now enters. His reasoning is consistent, and from his point of view, ingenious and thoroughly sound. The paper is full of acute and accurate observations, many of which contain the germs of future discoveries as, for example, the evolution of chlorine by the action of strong hydrochloric acid on arsenic acid, and the formation of the brown coloured compound formed by the action of green vitriol upon solutions of nitrites etc.

Why Cavendish refrained from publishing these results can only be surmised. He must have known that they were largely original. Had he done so at or about the time they were obtained he would have anticipated Scheele, who is usually credited with the discovery of arsenic acid, by some nine or ten years. Scheele's well-known memoir, which appeared in 1775, contains a great number of observations on arsenic acid, but his method of procuring it was not so simple or direct as that discovered by Cavendish, which is the one now in use, and there is not that sense of quantitative accuracy in Scheele's work which seems to pervade all Cavendish's attempts.

EXPERIMENTS ON TARTAR

The account of these experiments is written, apparently for publication, on small sheets of paper ($6\frac{3}{8} \times 4\frac{1}{8}$ in.). There is no indication of the time at which they were performed, but they seem to have been made at two different periods as the description is divided into two sections entitled, respectively, "old experiments on tartar" and "new experiments on tartar." Although practically unpunctuated, and plentifully interspersed with contractions, the account is easily legible and its meaning is clear. Nothing is stated as to the object of the inquiry, but it eventually resolved itself into an attempt to determine the amount of alkali, respectively, in cream of tartar (potassium hydrogen tartrate, $C_4H_5O_6K$) and the more readily soluble normal potassium tartrate ($C_4H_4O_6K_2 . \frac{1}{2}H_2O$).

No reference is made to any prior workers on the chemical nature of tartar, but Cavendish was probably familiar with all that had been published on the subject up to his time, although the *Phil. Trans.*, even down to the end of the eighteenth century, contains no reference to tartar nor are any memoranda or notes from foreign literature to be found among his papers. Wine-lees or argol, the *tartarum*, or *tartarus* (Arabic *tartir*), of the iatro-chemists was originally considered to be an acid, and its solution was termed *aqua dissolvens*. It was known to the Romans that it yielded an alkali on incineration, but until late in the eighteenth century it was considered that this substance was formed during the burning, and was not a real constituent of tartar, in spite of the observations of Kunkel in 1677 and of Duhamel and Grosse in 1732. The existence of the alkali may be said to have been first definitely established by Marggraf in 1764, but the precise nature of the action of acids and alkalis upon tartar was not clearly understood, and it was apparently this matter that Cavendish set himself to elucidate. It was probably one of his earliest attempts at chemical inquiry, and may have been begun shortly after the publication of Marggraf's work.

The "old experiments on tartar" consisted of a study of the action of nitric and sulphuric acids on cream of tartar. In the case of nitric acid Cavendish recognised the formation of nitre, proving that the alkali present in tartar is identical with that in nitre. He further noticed that as the solution in nitric acid was gradually neutralised by lime, cream of tartar was again precipitated, but he failed to recognise the formation of calcium tartrate.

The action of sulphuric acid liberated tartaric acid which was isolated by evaporating the solution on the water-bath. The solution neutralised with lime caused a precipitate which was held to be selenite (calcium sulphate). This was repeatedly boiled with water and filtered; "the solution seemed by evaporation to consist of selenite, cr. tart and vitr. tart [potassium sulphate] but it was not examined accurately." The experiments with the two acids were repeated, but no additional observations of importance were noted. Although, as in all Cavendish's experiments, the work was quantitative, it is impossible to deduce any numerical results from the figures given. At this stage the work was temporarily abandoned. When it was resumed we have no means of knowing.

In 1769 Scheele communicated his memorable paper on tartaric acid to the Swedish Academy of Science, and an account of it subsequently appeared in Crell's *Chemical Journal*. This may have induced Cavendish to take up the inquiry again. On the other hand, he may have resumed it before the appearance of Scheele's memoir, and it may have been this latter circumstance that caused him to withhold his own paper from publication. But in any case Cavendish's "new experiments on tartar" furnished far more definite information than his previous ones. He began by studying the action of chalk upon a hot solution of cream of tartar. He notes the evolution of gas and is aware of the nature of the change, viz. that "soluble tartar" ($C_4H_4O_6K_2$) goes into the solution, and that "tartareous selenite" (calcium tartrate, $C_4H_4O_6Ca$) is precipitated, and can be separated by filtration.

To a portion of the solution of the "soluble tartar" he added some f. alkali which caused no precipitate: hence no lime was in the solution. To a second and known portion of the solution he added nitric acid so long as a precipitate was formed. This consisted of cream of tartar, "which being washed and dried weighed 231 grains." It can hardly be expected from the nature of the operations that the quantitative results would be of a high degree of accuracy, but it may be interesting to see how far the numbers given by Cavendish accord with theory. To begin with, he operated upon 2370 grains of cream of tartar. This would be decomposed by the chalk as follows:

$$\underset{\text{cream of tartar}}{2C_4H_5O_6K} + \underset{\text{chalk}}{CaCO_3} = \underset{\text{sol. tartar}}{C_4H_4O_6K_2} + \underset{\text{tart. selenite}}{C_4H_4O_6Ca} + H_2O + \underset{\text{fixed air}}{CO_2}$$

The action of the nitric acid on the "soluble tartar" is:

$$\underset{\text{sol. tartar}}{C_4H_4O_6K_2} + \underset{\text{aq. fort.}}{HNO_3} = \underset{\text{cream of tartar}}{C_4H_5O_6K} + \underset{\text{nitre}}{KNO_3}$$

In other words, 2 mols of the original cream of tartar would give 1 mol. of reprecipitated cream of tartar and 1 mol. of nitre. Now from $\frac{22}{100}$ of the solution of the "soluble tartar" Cavendish obtained 231 grains of cream of tartar by the addition of nitric acid and 126 grains of true nitre; that is, the amount of cream of tartar was 1·83 times that of the nitre; theoretically he should have obtained 260·7 grains of cream of tartar and 140 grains of nitre, or 1·86 times the weight of the nitre. Having regard to the conditions, imperfect decomposition, washing and separation, and the fact that the original cream of tartar might not be pure, the result is in fair accordance with the theory of the reactions as we now know them.

One half of the calcium tartrate ("tartareous selenite") was then triturated with dilute oil of vitriol. The precipitate evidently contained free chalk as the mixture effervesced on the addition of the acid. The liquid, which was very acid, was separated from the gypsum by filtration, evaporated to a small bulk, again filtered from a further sediment (of calcium sulphate) and further evaporated to a thin syrup when tartaric acid "shot on cooling." The crystals weighed approximately 245 grains. The mother liquor was again concentrated when a further crop of crystals was formed. These last were purified by re-solution, added to what remained of the mother liquor, and the tartaric acid in solution precipitated by salt of tartar (potassium carbonate). In all 73 grains of cream of tartar were thus obtained. The total weight of tartaric acid obtained from this portion of the tartareous selenite was thus about 303 grains as against a theoretical yield of 473 grains.

The 245 grains of tartaric acid were then converted by a solution of barilla (sodium carbonate) into the acid sodium tartrate, and into the neutral salt "which had a good deal of the mawkish bitter of Glauber's salt."

By triturating the tartareous selenite with a solution of volatile sal ammoniac (ammonium carbonate), "fixed air" (carbon dioxide) was evolved, and ammoniacal tartar (the neutral ammonium tartrate) passed into solution. The clear liquor on concentration crystallised "pretty readily" and the crystals of ammonium tartrate were found to be soluble in about twice their weight of water.

The ammonium tartrate was then subjected to destructive distillation, with a view, apparently, of preparing the *spiritus tartari* of Lully (pyruvic acid) but the distillate was not examined. Cavendish seems, however, to have noticed the formation of pyrotartaric acid, and tartaric anhydride, but although the behaviour of the salt on heating is accurately described, the account of the results is too vague to enable any certain conclusions to be drawn.

Cavendish then enters upon some calculations as to the distribution of the tartaric acid in the "soluble tartar" (normal potassium tartrate) and the "tartareous selenite" and the equivalent amounts of cream of tartar. He found that the "soluble tartar" ($C_4H_4O_6K_2$) obtained from 2370 grains of cream of tartar was capable of yielding rather more than 1050 grains of cream of tartar which, as it existed in solution, contains in addition rather more alkali than is contained in 573 grains of nitre. Cavendish thus clearly recognised that the "soluble tartar"—the normal potassium tartrate—contained considerably more alkali than cream of tartar, the acid potassium tartrate. 2370 grains of cream of tartar should furnish 1185 grains of cream of tartar by the action of nitric acid upon the normal salt: Cavendish found 1050. This amount (1050) to be converted into the normal salt he found would require as much alkali as is contained in 573 of nitre, that is 221 grains (K)—making in all 1271 grains: the quantity of the normal tartrate equivalent to 1050 grains of the acid tartrate is 1262 grains.

The calculations based upon the amount of cream of tartar derivable from the "tartareous selenite" are vitiated by the fact that the calcium tartrate was not homogeneous. But the general conclusion drawn is that the amount of alkali required to completely saturate tartaric acid, that is to convert it into the normal tartrate, is at least double that contained in cream of tartar.

Cavendish proved that tartaric acid contains no alkali as a normal constituent—a fact of importance in relation to the vague ideas then current as to the mutual relations of cream of tartar and tartaric acid. In fact, certain pharmacopœias confused the two substances as late as nearly the end of the eighteenth century. He further found that the whole of the tartaric acid in calcium tartrate could be completely "dislodged" by oil of vitriol.

He also noted the "stiff gluey" character of calcium tartrate when precipitated from alkaline solutions and speaks of the difficulty in filtering it, and of the fact that its solutions become turbid on warming, and that the salt is soluble in "sope leys" (potash solution).

The paper concludes with the details of calculations of the amount of alkali contained in cream of tartar or required to completely saturate it, and of the equivalent quantity of marble and pearl ash. The statement that the acid of 1 part of tartar is saturated by ·525 of whiting is substantially correct: the theoretical amount is ·531.

It will thus be seen that Cavendish's work on tartar was remarkably accurate, and was a notable contribution to the chemical history of the subject. Whether it was done independently of Scheele's work, or withheld from publication on the ground that he had been anticipated, it is impossible to say.

But even in the latter case its publication would have been of service as tending to clear up much that was vague concerning the nature of tartar, the production from it of tartaric acid, and the properties of the tartrates. But Cavendish was never in a hurry to publish his work: his interest in it was largely satisfied when he had satisfied the questioning of his own intelligence.

ON THE SOLUTION OF METALS IN ACIDS

Digression to paper on Inflammable Air.

[This "digression" was intended to be added to the section on Inflammable Air in Cavendish's paper on Factitious Air, published in 1766. For some reason it was omitted, possibly because, on reflection, he considered as, he says, that he had not "made sufficient experiments to speak quite positively as to this point," i.e. of the solution of metals in acids. The "digression," however, is interesting for several reasons. It serves to show what were Cavendish's views concerning the nature of the action of acids upon metals in general, and the reasons for the difference in their mutual behaviour. These views, as far as can be gathered, were consistently held by him to the last, at least to the extent that phlogiston was concerned in the phenomena. It will be noticed that the arguments, and to a great extent the language, are identical with those in the paper on "Arsenic," which seems additional evidence that Cavendish's experiments on that substance were made prior to 1766 and that therefore he anticipated Scheele in the discovery of free arsenic acid by 9 or 10 years.]

If it is not digressing too much I should be glad to make some observations concerning the solution of metals in acids. There seems to be only the 3 above-mentioned metallic substances which dissolves easily in spt. of salt or the diluted vitriolic acid. I have not indeed made sufficient experiments to speak quite positively as to this point, but I will relate what I have tried relating to copper, which is always looked upon as one of the most soluble of the metallic substances. Some clean copper wire seemed not at all acted on by oil of vitriol diluted with an equal weight of water while cold, though it was kept in the acid several days; it gave no signs of solution neither though assisted by a heat almost sufficient to make the acid boil. Copper does not discharge any air-bubbles neither, when put into strong spt. of salt while cold, though if kept a great while in the acid, especially if exposed to the open air, it does dissolve slowly; but with the assistance of heat it makes a considerable effervescence, and discharges vapours which, as will be shown hereafter are not inflammable; but though it makes so much effervescence yet it dissolves extremely slowly. [Cf. p. 11.]

All metallic substances except gold and platina unite, readily with the assistance of heat, to the concentrated acid of vitriol. The union is performed with a great effervescence and discharge of vapours smelling strongly of the volatile sulphureous acid. All metallic substances also except gold and platina dissolve readily in the nitrous acid [nitric acid]. They dissolve with great effervescence and discharge plenty of vapours, which appear plainly by the smell to contain a great deal of the acid; but which are of a more penetrating smell, more volatile, and in general are of a deeper colour than the fumes of the plain nitrous [nitric] acid; and which seem therefore to be composed of the nitrous acid united to the phlogiston of the metal. It is remarkable, too, that, though in general the nitrous acid has the least affinity to metallic substances of any of the mineral acids, yet it dissolves them all the readiest of any acid.

The reason of these phenomena seems to be as follows. It is well known that no metallic substance, the perfect ones excepted, can dissolve in acids without being deprived of its phlogiston. This seems to form the principal impediment towards their solution in acids. Zinc, iron and tin seem to have a greater affinity to the vitriolic and marine acids than they have to their phlogiston; whence they dissolve without much difficulty in either of these acids. I do not at all know indeed why they show so much less disposition to unite to the concentrated vitriolic acid than to the diluted. But all the other metallic substances seem to have a greater affinity to their phlogiston than they have to either of the mineral acids. In all probability the reason why notwithstanding this they unite so readily, with the assistance of heat, to the concentrated vitriolic acid is, that this acid when heated to a certain degree has a great disposition to unite to phlogiston, the affinity of the phlogiston to the acid counteracting its own affinity to the metal, whereby part of the acid unites to the metal, while the remainder unites to the phlogiston. The volatile sulphureous smell produced during the solution is a certain proof that the acid does actually unite to the phlogiston of the metal, and is a strong reason to suppose that it is in good measure owing to the affinity of the phlogiston to the acid that the metal is enabled to unite to the acid. In all probability the reason why metals dissolve so easily in the nitrous acid is owing to a like cause, namely, the affinity of the phlogiston of the metals to the nitrous acid; the fumes produced in dissolving metals in this acid seeming to be no more than the acid united to the phlogiston of the metals. The diluted vitriolic acid and marine acid seem to have very little disposition to unite to phlogiston; which is most likely the reason why metals dissolve so difficultly in those acids. The experiment which will be mentiond hereafter, concerning the solution of copper in spt. of salt, looks, however, as if spt. of salt had some small disposition to unite to phlogiston. It seems not unlikely that the reason why the other metallic substances will not furnish inflammable air, as well as zinc, iron and tin, is that their phlogiston will not fly off in close vessels without uniting to the acid whereby it is separated, and thereby losing its inflammable quality.

As the precipitates from the solutions of mercury and the perfect metals in acids are reducible without the help of inflammable fluxes, it has usually been thought that they are not deprived of their phlogiston by solution in acids. But yet the volatile sulphureous acid produced by dissolving silver and mercury in oil of vitriol is a strong proof that these

2 metallic substances are deprived of their phlogiston by solution in that acid at least. I should imagine therefore that mercury and the perfect metals were deprived of their phlogiston by solution in acids as well as the imperfect ones; but that by reason of their great affinity to phlogiston they acquired it again from the matter which must be added to separate the acid from them, when assisted by the heat necessary to reduce them; since there seems no reason to think that the purest fixed alcali, or even lime, is quite free from phlogiston. The effervescence and elastic vapours produced during their solution in aqua fortis or aqua regia (which are seemingly just of the same nature as those which attend the solution of the imperfect metals in these acids) agree very well with this hypothesis; whereas it is likely that if they were not deprived of their phlogiston thereby they would dissolve quietly and without effervescence; for the effervescence can proceed only from the separation of some elastic fluid either from the metal or the acid. If this hypothesis is true it will account very well for gold not being soluble in any simple acid, but only in a mixture of the nitrous and marine acids. Gold, I imagine, has little or no affinity to the nitrous acid, but only to spt. of salt; but its affinity to that acid alone is not sufficient to deprive it of its phlogiston. It therefore requires the united efforts of the nitrous and marine acids, the nitrous to absorb the phlogiston and the marine to dissolve the metal.

That gold has little or no affinity to the nitrous acid seems likely from what Dr. Lewis says, that gold when by particular management made to dissolve in the nitrous acid is precipitated again only by exposure to the air, and that upon committing a solution of gold in aqua regia to distillation the nitrous acid flies off, leaving the gold united to the spt. of salt.

EXPERIMENTS ON FACTITIOUS AIR

Part IV

Containing experiments on the air produced from vegetable and animal substances by distillation.

[This paper was evidently intended to form a continuation of Cavendish's first communication to the *Philosophical Transactions*, entitled "Three papers containing Experiments on Factitious Air," and published in 1766. There is nothing to show why it was withheld from publication. The experiments were probably made not later than 1767. They are interesting as early attempts to gain an insight into the nature of the inflammable air obtained from wood and charcoal, and hence are of importance in connection with the Water Controversy.]

I received the air produced from these substances in inverted bottles of water nearly in the same manner as in the former experiments read to the Society, by means of the apparatus represented in the annexed drawing: where *A* represents a brass pot, in which are placed the materials

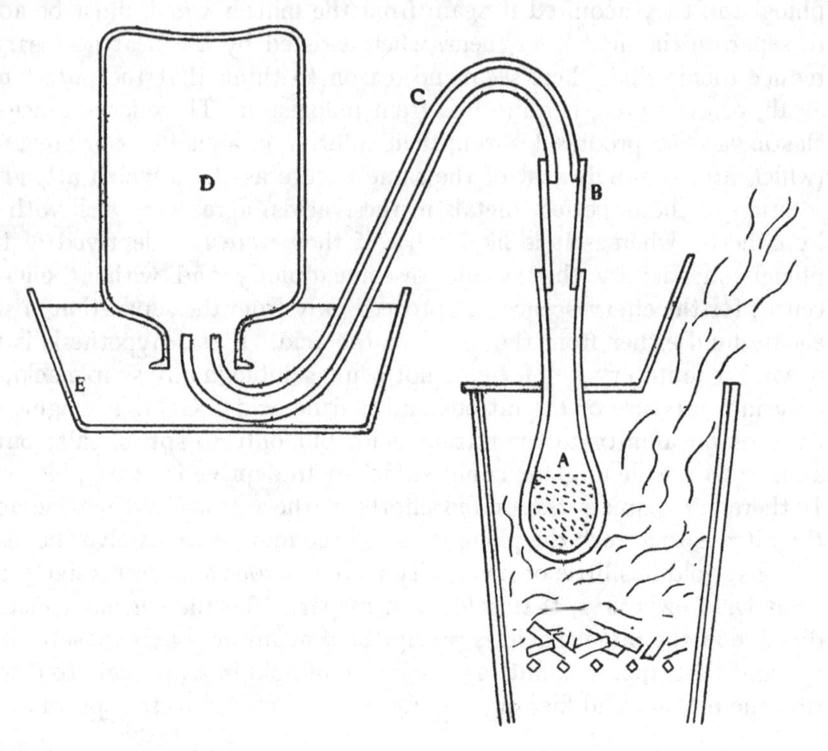

for distillation: *B* is a brass pipe fitted on to it by a cement composed of ochre and drying oil: *C* is a bent glass tube fitted to the brass pipe; and *D* is the bottle in which the air is to be received, filled with water, and inverted into the vessel of water *E*, over the end of the glass tube.

I used the cement above-mentioned in preference to that used in the former experiments, as it will bear a greater heat; and the reason why I placed the brass pipe *B* between the glass tube and distilling vessel, is that I expected the cement would bake so hard by the heat, that I could not get a glass tube out without breaking it. The joints were so well secured by this means that extreamely little air seemed to escape, though the brass pot was heated pretty strongly red hot. The pot was made of such a shape that I could easily clean the inside by scraping it with a bent piece of iron.

Exp. 1. 400 grains of raspings of Norway oak, called wainscot by the Carpenters, were distilled in the above-mentioned manner, till no more air would rise with a heat just sufficient to make the distilling vessel obscurely red hot. The bottle in which the air was received was then removed, and another put in its place, and the distillation compleated with a pretty strong red heat. By this means that part of the air which requires a red heat to disengage it was procured separate from that which rises with a less heat. Each of these parcels of air were then brought in contact with sope leys in the manner described in my experiments on Rathbone-place water, in order to see whether they contained any fixed air, and to free them from it if there was any. The first parcel of air, namely that which rose first in distillation, measured 22100 grains when first made, and was reduced by the sope leys to 12700. The 2nd parcel measured 34600 grains, and was reduced by the same means to 30700 grains.

The quantity of common air containd in the distilling apparatus, allowing for the room occupied by the wood, was about 1700 grains; all of which must have been forced into the inverted bottle along with the first distilled parcel of air, and would not be absorbed by the sope leys. 1700 is about $\frac{2}{15}$ of 12700; so that the first distilled air, when reduced by the sope leys, contains about $\frac{2}{15}$ of its bulk of common air, or is a mixture of above 13 parts of pure factitious air to 2 of common air. The last distilled parcel must have been intirely free from common air.

All that air which was absorbed by the sope leys may, I think, be fairly supposed to be fixed air. The remaining air of each parcel was inflammable, but required a much greater quantity of common air to make it explode than the inflammable air from metals does: for a vial holding near 1200 grain measures being filled with 1 part of the first distilled air with $2\frac{1}{2}$ of common air, the mixture caught fire on applying a lighted candle to the mouth of the vial and went off with a small puff; but when the vial was filled with 1 part of the same air to 2 of common air it would not catch fire. In like manner a mixture of 1 part of the 2nd distilled air with 3 of common air went off with a puff; but 1 part of the same air with $2\frac{1}{2}$ of common air would not. So that the first distilled air required to be mixed with not less than between 2 and $2\frac{1}{2}$ times its bulk of common air, and the 2nd distilled air with between $2\frac{1}{2}$ and 3 times its bulk of common air, before it would explode; whereas the air from metals, when tried the same way, would explode though mixed with only $\frac{1}{2}$ its bulk of common air.

I next tried which of these parcels of air would explode with most force when mixed with considerably more common air than what was sufficient to enable them to catch fire. For this purpose I mixed some of each of these parcels of air, and also some inflammable air from zinc, with 4 times their bulk of common air, and tried them in the same bottle. The first distilled air went off with the least noise. As for the 2 others I was uncertain which made most: but the air from zinc went off with a sharper sound than the other, and no light could be seen in the bottle; whereas in the trials of each of the distilled airs a small light was seen.

The experiment was then repeated with mixtures of each of these airs with 5 times their bulk of common air. The first distilled air took fire, but with scarce any noise. The 2 others went off as near as I could judge with the same degree of noise, the distilled air with a small light visible in the bottle and a duller sound; the air from zinc without any light and a sharper sound.

It should seem therefore as if the second distilled air contained about as much phlogiston as the air from zinc, but that the first did not contain so much; for when the quantity of common air is considerably more than sufficient to consume the whole of the inflammable air, it seems likely that the loudness of the explosion should be in proportion to the quantity of phlogiston contaïnd in the mixture.

In all these experiments the air was measured in a cylindrical glass with divisions on its sides, in such manner that I think I could not well err more than 5 grains or a 240th part of the whole mixture. The vial in which the explosions were made had a glass tube about an inch and a $\frac{1}{2}$ long and $\frac{4}{10}$ of an inch in bore fitted to its mouth, by way of contracting the orifice.

I also tried the specific gravity of each of these parcels of distilled air in my usual manner. 1000 grains of the first distilled air being forced into a bladder, which held 48000 grains and had a brass cock fitted to it, the bladder increased $\frac{3}{4}$ of a grain in weight on pressing out the air. So that, supposing common air to be 800 times lighter than water, this air, which was before said to contain $\frac{2}{15}$ of its bulk of common air, should be about $\frac{1}{17}$th part lighter than common air; and the pure factitious air without any mixture of common air should be $\frac{1}{14}$th or $\frac{1}{15}$th part lighter than common air, or near $6\frac{1}{2}$ times heavier than the inflammable air from metals.

21100 grain measures of the last distilled air being forced into the same bladder, there was an increase of 12 grains on pressing it out; whence this air appears to be lighter than common air in the proportion of 11 to 6, or near 4 times heavier than the air from metals.

The *caput mortuum* or matter remaining in the brass pot after the distillation was compleated, consisting of the wood reduced to charcoal, weighd 134 grains.

On the whole, the 400 grains of wainscot yielded with a heat less than sufficient to make it red hot, 9400 gra. measures of fixed air, whose specific gravity was before found to be about $1\frac{1}{2}$ times greater than that of common air, and 12700 of an inflammable air, and which was about $\frac{1}{15}$th part lighter than common air, and which required to be mixed with more than 2^{ce} [twice] its bulk of common air to make it explode. With a greater heat than that, it yielded 5800 grains of fixed air, and 30700 of an inflammable air, which required to be mixed with above $2\frac{1}{2}$ times its bulk of common air to make it explode, and whose density was $\frac{6}{11}$ of that of common air. The weight of all this air together is 64 grains *id est* $\frac{16}{100}$

of the wood it was produced from, or near $\frac{1}{4}$ of the loss of weight which it suffered in distillation. It must however be observed that there was most likely more fixed air discharged than is here set down; as in all probability some of it must have been absorbed by the water.

As this inflammable distilled air is much heavier than that from metals, and requires to be mixed with a much greater proportion of common air to make it explode, I at first imagined it might consist of an inflammable air exactly of the same kind as that from metals, mixed with a good deal of air, heavier than it, and which had a power of extinguishing flame like fixed air; as I hinted before with regard to the air produced from meat by putrefaction; but on consideration, I fancy this air must really be of a different kind from that of metals; for if it had been only a compound of that air with some of a different kind, then a mixture of that compound with common air must necessarily I think have exploded with less noise than a mixture of pure inflammable air with the same proportion of common air, as it contains less inflammable matter than the latter mixture, and that compounded with a substance which should rather diminish than increase the explosion. Whereas the last distilled air was found to make as great an explosion as the air from metals, when both were mixed with 4 times their bulk of common air.

Exp. 2nd. In another trial made in the same manner, except that the whole of the distilled air was received together, without changing the bottle, the like quantity of wainscot yielded 19200 grain measures of fixed and 42700 of inflammable air. The inflammable air required to be mixed with more than 2^{ce} its bulk of common air to make it explode, and its density was less than that of common air in the proportion of 1·52 to 1. The weight of the whole of this air is 71 grains, *id est* near $\frac{18}{100}$ of the wainscot it was produced from. This experiment is exactly consistent with the former, except that the quantity of fixed air was greater, as might be expected, since the distillation was performed in much less time, and consequently much less fixed air could be absorbed by the water.

Exp. 3. I made another experiment with the same quantity of wainscot, the distilling pot being this time placed in oil, that I might see what would be the nature of the air which would rise with no greater heat than that of boiling oil. The oil caught fire which prevented me from compleating the experiment; I however, got 11500 grain measures of air, 5400 of which were fixed air, the remaining 6100 were inflammable, requiring somewhat more than 2^{ce} their bulk of common air to make them explode. Their density allowing for the common air in the distilling vessel was about $\frac{1}{18}$ part greater than that of common air.

Exp. 4. I also examined the air produced from tartar by distillation, though not in so careful a manner as the wainscot. It yielded more fixed, and less inflammable air than wainscot; 400 grains of it yielding 46600 grains of fixed air and 23500 of inflammable air. The inflammable air required to be mixed with more than 4 times its bulk of common air to make it explode, and was about $\frac{1}{11}$ part heavier than common air.

Exp. 5. 900 grains of Hartshorn shavings were distilled exactly in the same manner as the wainscot in the first experiment, except that the heat was raised to a rather greater degree before the bottle was changed. The first distilled parcel of air measured 33600 grains, and was reduced by sope leys to 20400. The common air left in the distilling vessel was 1630 grains; so that this air when reduced by the sope leys contained $\frac{4}{50}$ of its bulk of common air. The last distilled parcel measured 9400 grains, and was reduced by sope leys to 8900.

Each of these parcels of air, when thus reduced, was found to be inflammable. The first distilled air, tried in the same bottle as was used for similar experiments on the air from wainscot, caught fire on applying a lighted candle when mixed with 5 times its bulk of common air, but would not when mixed with only 4 times its bulk. The 2nd parcel caught fire when mixed with $2\frac{1}{2}$ times its bulk of common air, but would not with 2ce its bulk. I then compared the loudness of the explosion made by each of these parcels of air and of some air from zinc, when mixed with 6 times their bulk of common air. I could perceive very little difference between the 2 parcels of distilled air: but both of them seemed to make rather more noise than the air from zinc. The same difference in the manner of explosion between the distilled air and air from zinc might be observed with these, as with that from wainscot; namely that the distilled airs went off with the duller sound, and exhibited a light in the bottle, which was not visible with the air from zinc.

18240 grain measures of the first distilled air being forced into a bladder holding about 21600, there was an increase of weight of $5\frac{3}{4}$ grains on pressing out the air, so that allowing for the common air mixed with it the pure factitious air is lighter than water [air] in the proportion of 137 to 100.

8160 grain measures of the 2nd distilled air being forced into a bladder holding near 14000, it increased $4\frac{1}{4}$ grains on pressing out the air, whence it appears to be lighter than common air in the proportion of 171 to 100.

The *caput mortuum* consisting of the hartshorn burnt to a coal weighed 623 grains. The weight of all the air discharged appears from what has been said to be 51 grains, *id est* $\frac{1}{18}$ part of the weight of the hartshorn, or about $\frac{2}{11}$ of the loss of weight which it sufferd in distillation.

We have examined therefore 3 substances of very different natures, namely the first a simple wood, the 2nd a vegetable substance of a saline nature, and the 3rd an animal substance of the nature of bones. Each of them agreed in furnishing some fixed and some inflammable air; but the proportions of these airs were considerably different, the nature of the inflammable air was not quite the same in each, but yet hardly differing more than that produced from the same substance at different periods of the distillation; so that there should seem to be a considerable resemblance between the air produced by distillation from all animal and vegetable substances.

Dr Hales in his *Vegetable Statics* has given the quantity of air produced by distillation from a great variety of different substances. He observed that the air produced from some of them was inflammable, but does not mention whether he found any which was not. One of the methods which he used for measuring the air did not differ essentially from that used in these experiments.

In the first and 2nd experiments we have an examination of all the air which can be procured from wainscot by distillation in close vessels; but this is by no means all the air which it contains; for the *caput mortuum*, which, as was before said, consists of the wood burnt to charcoal, seems to contain a very remarkable quantity of fixed air.

The alcali produced by deflagrating nitre with charcoal is well known to effervesce with acids, and consequently to contain fixed air; which air I think can proceed only from the charcoal; for when nitre is alcalized by metals in their metallic form, which contain no fixed air[1] the alcali makes no effervescence with acids, as I know by experience; and I think it seems very unlikely that the nitre should furnish fixed air when deflagrated by charcoal, and not produce any when deflagrated by metals. This induced me to make the following experiment.

Exp. 6. 150 grains of the *caput mortuum* remaining after the distillation of wainscot in the first and 2nd experiments, well dried, were ground with 5 times their weight of nitre and about 130 grains of water, and when the whole was thought to be perfectly mixed, was deflagrated by little and little in an iron ladle. The intention of the water, was to make the matter deflagrate with less violence; whereby there was less danger of any fixed air being dissipated by the heat. The deflagrated matter was put into water to dissolve the alcali. The insoluble matter, consisting partly of the ashes of the *caput mortuum* and partly of some of the *caput mortuum* which had escaped the fire, weighed, when well dried, 38 grains; so that the loss of weight which the *caput mortuum* sufferd in

[1] The late Dr Hadley found that the volatile alcali produced by distilling sal ammoniac with red lead made a very considerable effervescence with acids; whereas that procured by distilling sal ammoniac with some metal in its metallic form (I believe it was copper) made none at all; which shews that though metals themselves contain no fixed air, yet some metallic calces contain a great deal; and probably all those do which are exposed during their calcination to the fumes of the burning fewel, and thereby have an opportunity of absorbing fixed air; for those fumes contain a great deal. It is doubtless owing to the fixed air it absorbs that lead increases in weight by being converted into minium.

deflagration was 112 grains. In order to find the quantity of fixed air in the alcaline solution, ½ of it was saturated with the vitriolic acid, and the loss of weight which it suffered in effervescence observed with the same precautions as were used for finding the quantity of fixed air in pearl ashes in the 2nd part of these experiments [see *Phil. Trans.* 1766, also p. 91 *et seq.*]: it appeared to contain 62 grains. As the experiment makes the quantity of fixed air produced from the *caput mortuum* appear to be greater than the loss of weight which it sufferd in deflagration, which is impossible, I took another method to find the quantity in the remaining ½ of the alcaline solution; namely, I mixed it with a sufficient quantity of lime water, whereby all the fixed air therein was transferred into the lime, which was thereby precipitated. I then found the quantity of fixed air in this precipitate; it appeard to be 59 grains which is only 3 grains less than it appeard to be the other way. By a mean of these experiments the quantity of fixed air separated from the 150 grains of *caput mortuum* should be 121 grains which is 9 grains more than the loss of weight which it sufferd in deflagration.

By a like experiment made with some more *caput mortuum* of the same kind the quantity of fixed air seemed still greater.

As it is impossible that the quantity of fixed air separated from the *caput mortuum* should exceed the loss which it suffers in deflagration, I must either be mistaken in supposing that all the fixed air in the alcali proceeded from the *caput mortuum*, and not from the nitre, or else some moisture must have flown off along with the fixed air in saturating the alcali with the acid, which would make the quantity of fixed air therein appear greater than it really is. This last supposition seems much the most probable.

In the 10th experiment of my 2nd paper on Factitious Air, in which the fixed air produced by dissolving marble in spirit of salt was made to pass through a glass cylinder filled with filtering paper, very little moisture was found to be condensed in the paper; from whence I concluded that very little moisture could fly off along with the fixed air in effervescence, as thinking that the greatest part of what did fly off must have been condensed in the filtering paper. But this conclusion was too hasty; as it seems not improbable, that a considerable quantity of moisture might fly off along with the fixed air, but which might adhere to it too strongly to be absorbed from it by the filtering paper. Perhaps some moisture may be necessary to enable the fixed air to assume an elastic form.

If we suppose, as I think seems much most probable, that all the fixed air in the alcali proceeded from the *caput mortuum*, it follows that this substance contains a remarkably greater proportion of fixed air than any other substance we have examined; though we cannot determine the exact quantity. I should think it very likely though, that, excepting the small quantity of ashes which it leaves on burning, it consists almost intirely of fixed air. It is almost needless to say that, according to this supposition, the determinations of the quantity of fixed air in alcaline substances, in the latter part of my 2nd paper on Factitious Air, cannot be depended on as to the exact quantity of it in any one substance; but I see no reason why this should incline one to think there is any fallacy in the determination of the proportion which the quantity of it in one substance bears to that in another.

Exp. 7. I made also an experiment of the same kind with common charcoal. It appeared to contain a great deal of fixed air, though not so much as the *caput mortuum*, namely $\frac{93}{100}$ of the loss of weight which it sufferd in deflagration. This difference may very likely be owing, partly to the charcoal not being dried before I weighed it, for charcoal contains a good deal of moisture though kept in a dry room, a circumstance that I did not attend to when I made the experiment, and partly perhaps to its not being well burnt; for I hardly imagine there can be any difference between the *caput mortuum* and charcoal perfectly burnt.

Exp. 8. The *caput mortuum* remaining after the distillation of Hartshorn shavings, when deflagrated with nitre is changed into a white calx, which in the experiments I made seemd not at all inferior in weight to the matter unburnt. So that as it seems to suffer very little loss of weight in deflagration it cannot be expected to furnish much fixed air. However, 382 grains of this matter well dried, being ground with $\frac{3}{5}$ of their weight of nitre and deflagrated, the fixed alcali produced thereby appeard to contain 22 grains of fixed air. The ashes washed and dried weighd not at all less than the *caput mortuum* they were produced from; so that it did not appear from this experiment that the *caput mortuum* sufferd any loss by deflagration. This in all probability must be owing either to some of the fixed alcali adhering to the ashes in such manner that it was not separated from thence by washing, or else to the ashes containing more moisture than the *caput* they were made from, though they were dried with a pretty considerable heat: for it is certain that the *caput mortuum* must have lost some weight by deflagration, though most likely it was not much. The ashes dissolved readily in spt. of salt and appeard to contain 30 grains of fixed air.

The event was very nearly the same in another experiment made with some more *caput mortuum* of the same kind.

[From the circumstance that Cavendish refrained from communicating this paper we may infer that he was not altogether satisfied with the results of his experiments, or that he was unable to explain them as fully as he might wish, and this would seem to be confirmed by the fact that he returned to the subject at several subsequent periods, as his laboratory notes show. The inflammable gas, freed from carbon dioxide, was a variable mixture of carbonic oxide, marsh-gas and hydrogen, in relative amounts depending upon the stage of the distillation, temperature, presence of moisture, etc. Doubtless it was this variable nature, affecting the physical characters of the "inflammable air," its specific gravity, ignition point, explosibility etc. which, in the absence of analytical means to prove that it was non-homogeneous, baffled Cavendish. Had he been able to procure either carbon monoxide or marsh-gas in a separate state, the inquiry would, in all probability, have been greatly facilitated. How near he was to the discovery and isolation of carbonic oxide on several occasions is obvious from his notes. He shows that after freeing the gas prepared from charcoal from carbonic acid, it at once killed a bird. "I plunged a burning candle in this air freed of its fixed air: it inflamed with a slight explosion: the flame was blue, in colour like that of burning sulphur." Unlike carbonic acid it was not affected by tincture of litmus. "The investigation of this shall be my employment for another memoir." Although he returned to the subject again and again, he got little beyond the stage of clearly recognising that the inflammable air from metals and from charcoal were not the same things, in which respect he was in advance of Priestley. Had Cavendish published Part IV of his paper on "Factitious Air" in or about 1767, it might have expedited the more general recognition of this fact, as, in spite of uncertainties, his determination of the physical characteristics of the inflammable air from wood, charcoal, etc., left little room for doubt that it was essentially different from that from zinc and iron. The delay may be said to have occasioned Priestley's error, and thus indirectly led to the controversy concerning Cavendish's claim to be the first to establish and announce the fact of the compound nature of water.]

THE LABORATORY NOTES OF "EXPERIMENTS ON AIR"

[The laboratory notes in connection with the "Experiments on Air' have been preserved, and are among the Chatsworth papers, as a separat parcel, paged and indexed by Cavendish. They are written on smal detached sheets, $6\frac{3}{8} \times 4\frac{1}{8}$ inches, about 400 in number, and are mostl in the form of short memoranda and jottings of experimental data, interspersed with calculations of results. They rarely contain any deductions any account of these was presumably reserved for the draft of the memoi in case it should be published. It is, however, not difficult to follo and interpret the greater portion of the notes in the light of the paper in the *Phil. Trans.* But many observations are recorded which are no included in the published memoirs, possibly for the reason that they wer not directly relevant to the subject-matter, or that Cavendish was no satisfied with the numerical results. The whole of the work, which extend over several years and includes many hundreds of measurements of on kind or another, is almost entirely quantitative in character. It seems t have been impossible for Cavendish to work otherwise than by weight and measure.

If he has to prepare oxygen he invariably states the amount of red precipitate or turbith mineral he uses and the volume of the gas obtained. If he has occasion to use "sope-lees" he notes its strength by the amount of nitre the solution is capable of yielding on neutralisation with nitric acid. This custom is frequently of service in unravelling the real nature of the phenomena he observes, and which he sets down with remarkable fidelity. In the light of present knowledge we are now able to perceive many things which were obscure to him, and to see how frequently he was on the verge of discoveries which are credited to later workers.

A few extracts from these notes of hitherto unpublished observations are given—as showing Cavendish's manner of work and the patient care and thoroughness with which he investigated all sides of a subject on which he was engaged. For example, he has occasion to mix two gases which have no chemical action on each other. He inquires if the resultant volume is the sum of the volumes, or if there is "any penetration of parts." Do gases of different densities mix perfectly or does the heavier one subside? In connection with the working of his "new eudiometer" he is concerned to know whether there is any difference in the rate of movement of nitrogen and of common air. He is anxious to obtain some numerical estimate of the relative violence of the detonation of explosive gaseous mixtures, and contrives a kind of dynamometer for the purpose. The notes reveal with what care he investigated the different modes of preparing nitric oxide in order to ensure uniformity in its character in view of its use in the eudiometer, and they show the patience with which he established the conditions upon which depended the formation of the nitric acid in his synthesis of water. He was the first to show that a fairly accurate determination of the amount of oxygen in air could be made by the combustion of phosphorus. He performed an approximately accurate analysis of carbon dioxide and made repeated but unavailing attempts to establish the composition of carbon monoxide.]

Is there any penetration of parts on mixing gases? "It was tried whether there was any penetration of parts on mixing common and inflammable air by means of the eudiometer. For this purpose 1 measure common air was let up by the cock into small bottle containing least measure of inflamm. air [that is, the least measure required to combine with the oxygen in it]. The diminution of bulk appeard to be ·002. One measure of infl. air being let up that way into the small bott. with least meas. comm. air, the diminution of bulk appeard to be ·004. Conseq. the diminution of bulk cannot exceed $\frac{3}{2250}$ of the whole."

Do the gases of a mixture separate out in the order of their densities? "It was also tried whether these airs mix perfectly or whether the com. air subsided to bottom. For this reason some of these were mixed in oblong spheroidical bottle and placed over the syphon in apparatus so that the syphon reachd to the top of bott. After standing some hours the air was drawn off by pouring water gently into the vessel so as not to

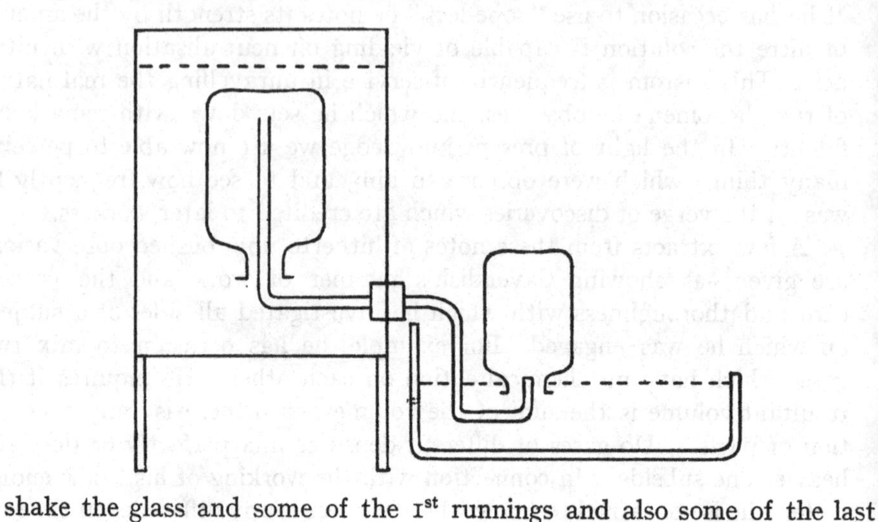

shake the glass and some of the 1st runnings and also some of the last caught in separate bottles, and the test of these bottles tried by new eudiom. For the 1st exper. test of $\left.\begin{matrix}\text{1st}\\\text{last}\end{matrix}\right\}$ runnings was $\left\{\begin{matrix}\cdot 472\\\cdot 482\end{matrix}\right.$. But the quant. of last runnings tried was only ·783 which might make the dimin. greater than it ought to be. In the 2nd exper. the test of $\left.\begin{matrix}\text{1st}\\\text{last}\end{matrix}\right\}$ runnings was $\left\{\begin{matrix}\cdot 537\\\cdot 540\end{matrix}\right.$ so that the last runnings appear to contain very little if at all more com. air than the first."

A "Measurer of explosions of inflam. air." "The strength of the explosion was tried by the machine rep^{r} in fig. *AB* is the brass cyl. for

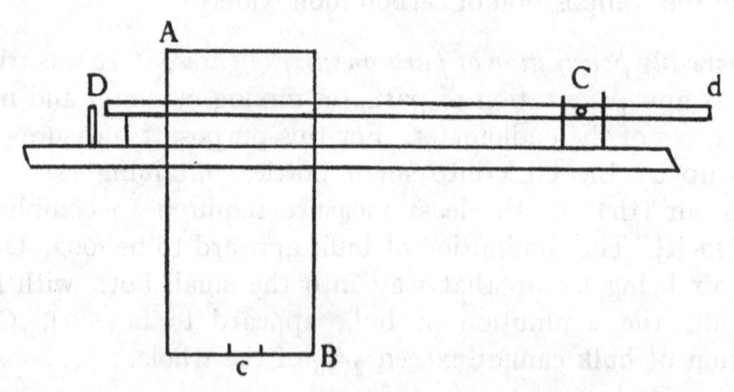

the air with a hole *c* at bottom. This is fixed tight to a board *Dd* turning on center at *C* and more or less weight is laid on *Dd*. In using it the cyl.

was first filled with water, the water entering by hole at bott. and the air escaping by cock at top, the app. for firing the air by el[ectricity] being placed in recess so as not to be wetted thereby. A proper quantity of mixt air was then pourd in by hole at bott. (but not near suffic. to fill the whole cyl.) and fired and the height which the cyl. sprung up thereby measured by an index. The force with which the cyl. was pressed down was found by hanging it to the end of pair of scales while in its place by a wire fastend to the middle of the top of the cyl.

The hole was exactly ½ inc. in diam. the cyl. was about 6 inc. long and 1·8 in diam. and held 4300 gra. [measures]. The recess in which the app. for firing held about 12 gra.

1713 [grain measures] of a mixt of 1 part of inf. [hydrogen] and 2 of common air being put into this machine and fired, it sprung up 4·23 [inches?] when weight lying on it was 32400 [grains] and 3·43 [inches] when the weight was 37300.

As it was suspected that the water driven out from the hole by the explosion was resisted by the water below and thereby was pressed against the bott. of the cyl. and made the force appear too great the machine was alterd in this manner.

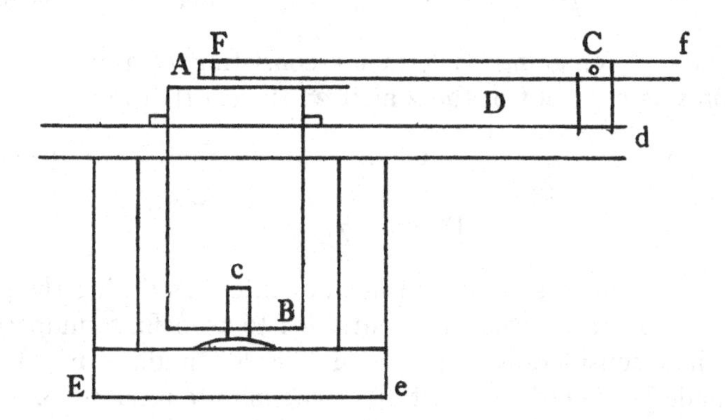

AB is the cylinder passed through a round hole in the fixed board *Dd* so as to rise and fall in it freely and resting thereon by a projecting piece of brass *c* is a cylindrical piece of brass filling up the whole [hole] almost intirely and fixed upon the piece of wood *Ee* fastened to the bottom of the board *Dd*. *Ff* is a board turning on a center at *C* and resting on the brass cyl. intended to carry weight. Mixt air is put into the cyl. and fired as before and the height which it rises measured by index as before.

As the bottom of the cyl. almost touchd the wood *Ee* it was suspected that the small quant. water issuing between the plug and the sides of the hole might have partly the same effect as was apprehended in the former machine. A small piece of wood was in some trials placed between

Ee and the bott. cyl. so as to keep it raised about ¼ inch above it; but the effect was the contrary to what was expected as it always sprung up more when this wood was placed under than without it.

Some of the obs. made with it are given below:

Index before firing	D° after	Weights lying on board	Dist. nearest end from end board	Weight with which cyl. is pressed
·0	·0	1 + 2 + 3	14·6	97000
·72	·77	......	...	D°
·0	·0	1 + 3 + 4	14·6	84000
·73	·82	......		D°
·0	·16	1	14·6	68000
·74	1·27	......		D°
·0	·38	1	13·6	62000
·75	1·63	......		D°
·0	1·38	1	11·5	50000
·77	2·67	......		D°

In the next the bit of wood was placed between *Dd* and the projecting brass

·93	1·15	1	14·6	68000

In all the foregoing 1713 of air consisting of 1 part inf. and 2 of common was used but in the 2 next 2[ce] [twice] that quant. was used:

·0	·03	1	14·6	68000
·0	·13	1	13·6	62000
·0	·10	D° with 1713		"

[The experiments were continued at intervals during the greater part of August to Sept. 9, but apparently led to no definite numerical results, although a considerable number of "trials" under variable conditions were made.] "There seemed to be some minute interval of time between the electric spark and the explosion."..."It does not seem as if there was any connection between the strength of the spark and of the explosion."

Rate of efflux of gases. "It was tried whether the *vis inertia* of phlogisticated air was the same in proport. to its weight as that of common air by noting the time in which a given quantity passed through a given hole when urged by a given pressure by means of the following apparatus. *A* is a tin vessel 8½ inches in diam. and 10 deep with a small hole in the diaphragm *a*. This vessel is suspended over a vessel of water by the rod *B* turning on a center near the middle point and partly ballanced by a weight at the other end and sufferd to descend as the air runs through the hole. The time in which it descended a given space (about 7½ inches) was found by observing the time in which the knob *b* moved from one mark to another. The force with which the vessel was pressed down was about 10½ oz. the rest of the weight of the tin vessel being taken off by the counterpoise. The way by which it was filled with air was by holding

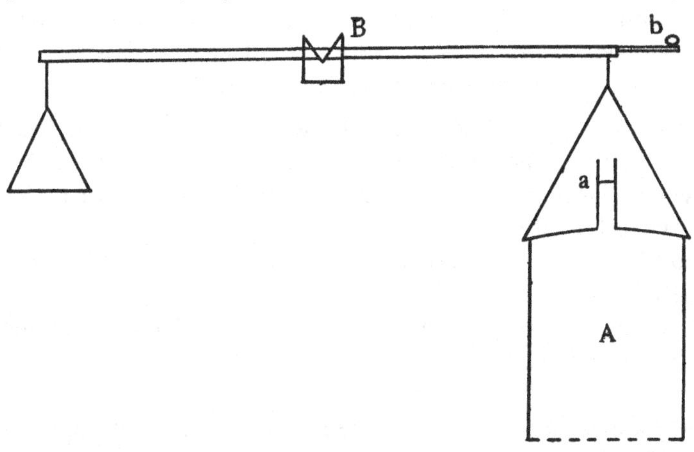

it under water till all the air was run out, then stopping pipe with thumb, raising up the vessel till the bottom was near the surface and pouring in the air [i.e. the gas to be experimented upon].

The event was as follows:

Oct. 28. 1780.

With com. air was	2′ 15″ running out
A 2nd time	2 12½
With air phlogisticated by liver of sulph.	2 7
With com. air	2 9 ″

[Common air phlogisticated with liver of sulphur would be practically nitrogen: its specific gravity was found to be $\frac{1}{47}$ less than that of common air. Apparently Cavendish was not satisfied with the performance of the apparatus, as no experiments with other gases were tried.]

Diminution of common air by phosphorus [Cavendish's title]. "July 30. Supposed th.[ermometer] 70° Bar. 30·09. A bit of dry phosph. was put into inverted bott. In a day's time it [the air] was a good deal diminished, but a drop of water having fallen on the phosph. prevented its acting; it was therefore fired by burning glass which was done with much smoke but no sensible flame. The phosphorus seemed intirely consumed. The original quantity of air was 3772 and the diminution was 651, bar. being then 30·02 and thermometer 71°. Its test was ·336."

[Neglecting the corrections for the slight change in temperature and pressure, this result would give 17·3 as the percentage volume of oxygen in air instead of 20·9. As all the phosphorus was consumed, its amount was probably insufficient to combine with all the oxygen.]

"Aug. 1. Some phosphorus was fired by burning glass in bott. with some common air in it. The original quantity of air was 3951, and, after the firing, was reduced to 3595. On burning another piece it was reduced to 3258. A little more phosph. was then put in and fired. It went out immed. and produced not much smoke. Next day it [the air] was found to be reduced to 3234. Allowance in all these trials being made for alteration of th. and bar. which was not great."

[This method of conducting the experiment was not likely to give accurate results. The data given point to the air containing 18·1 per cent. of oxygen. Nothing is stated as to how the successive pieces of phosphorus were introduced.

Phosphorus was an expensive article in 1780 and Cavendish was naturally very sparing of its use.

These are the only instances in the notes of its employment in the analysis of air.]

Air from plants. "Oct. 24, 1782. Sunshine. Some pond weed from Shepherd well was put in inverted jar and some more in large bott. Some chickweed also with some other water plants were put in a bott. In the evening the air was separated and the test tried in 2nd meth. was as follows:

Pond weed in jar	2·356
D° in bott.	2·528 [?]
Chickweed	2·278

The three parcels put together measured 11650 [grain measures] and lost 675 by washing with lime water. The next morning cloudy but now and then a little ⊙ [sun]. The 2 last were exposed again to ⊙ with fresh water and sufferd to remain all night and next day.

Oct. 26. Sunshine. Some more pond weed was exposed in 2 large jars and in the evening the air from them was collected and mixed together, and at the same time that from the old plants was collected and mixed together and all the botts and jars filled with fresh water and exposed. The test of the air from the $\left.\begin{matrix}\text{old}\\\text{new}\end{matrix}\right\}$ plants was $\left\{\begin{matrix}2{\cdot}79\\2{\cdot}547\end{matrix}\right.$. They did not lose much by washing with lime water.

Oct. 30. Some sunshine. Oct. 31. A good deal. In the evening the air was collected and the bottles filled with fresh water and exposed. The test of the air from the $\left.\begin{matrix}\text{old}\\\text{new}\end{matrix}\right\}$ plants was $\left\{\begin{matrix}2{\cdot}97\\2{\cdot}79\end{matrix}\right.$.

Nov. 4. The air was taken from the 2 bottles: its test was 2·553. The botts were then filled with beccæbunga[1] [?] the jars remaining as before.

Nov. 8. After 2 or 3 sunshiny cold days the air was taken. The test of that from the jars with pond weed was 2·670; that from the beccæbunga was 2·690.

[1] Veronica Beccæbunga.

Nov. 19. The plants had yielded a good deal of air but almost all that from the Beccæbun. was absorbed again. The test of that from the $\left.\begin{matrix}\text{Becca}\\ \text{pond weed}\end{matrix}\right\}$ was $\left\{\begin{matrix}\cdot 70\\ 2\cdot 706\end{matrix}\right.$.

All the above-mentioned except those got before Oct. 31, were mixed together and washed with milk of lime.

The test in $\left.\begin{matrix}1^{st}\\ 2^{nd}\end{matrix}\right\}$ method was $\left\{\begin{matrix}3\cdot 595\\ 2\cdot 710\end{matrix}\right.$, therefore its standard should be $\left\{\begin{matrix}3\cdot 48\\ 3\cdot 623\end{matrix}\right.$."

[This "air" was mixed with hydrogen and exploded. "21 [grains] of water was condensed which were very slightly acid to the taste but turned paper tinged with blue flowers evidently red." In a second experiment the "air" was mixed with a larger volume of hydrogen. 18 grains of water was formed. "It did not taste at all acid, nor turned blue paper at all red even when much diminished by evaporation."]

Air from mines. "Very bad air sent Jan. 22, 1783. 6700 [grain measures] of *B* 5 [probably the number of the bottle in which the sample was collected] was dimin. 90 or $\frac{1}{74}$ by lime water [1·34 per cent. of carbonic acid][1]. This air tested with 1st method was ·94. Standard ·86. Common air 1·08.

Jan. 25. Another bott. was tried and came out as follows $\left\{\begin{matrix}\cdot 923\\ \cdot 923\end{matrix}\right.$."

[Cavendish then compared the mine air with that in which a candle would no longer burn.]

		Test in 1st method	Standard
Air in which wax candle burnt out in receivers holding common air	*200000	·907	·827
	80000	·883	·803
	9500	·668	·581

* N.B. "The candle in the least jar burnt 10 secs., in the 2nd 70 secs. and in the largest either 2 or 3 mins., I am not sure which but I believe 2 mins."

Alteration in air by breathing. "Feby. 5, 1786. Some air was breathed as long as I well could by means of tin pipe from a glass jar holding 60000 gra. inverted into water. 57000 of this breathed air was drawn out into an exhausted globe and well washed with lime water by which 3880

[1] [This air, although undoubtedly very bad, is not as bad as that found occasionally by Angus Smith in mines. The largest amount of carbonic acid, 2·5 per cent., was found in a Cornish mine: the average of 339 analyses was ·785 per cent. by volume of carbonic acid (R. Angus Smith, *Air and Rain*, 8).]

[A candle will not burn with less than 18 per cent. of oxygen when there is 3 per cent. of carbonic acid present at the same time (*loc. cit.*).]

or $\frac{1}{15}$ of the whole was absorbed[1]. Its test after this in 2nd method was ·444; therefore its standard was ·40 and therefore the dephlog. air destroyed was $\frac{\cdot 6}{5} = \frac{1}{8 \cdot 3}$ and therefore the fixed air is $\frac{83}{150} = \frac{166}{300} = \cdot 55$ of the dephl. air destroyed."

CAVENDISH ON CHEMICAL NOMENCLATURE

[Among the miscellaneous papers of the Chatsworth MSS. is a rough draft of a letter in answer to one from Blagden, dated Sept. 16, 1787. Blagden, who at that period was Secretary of the Royal Society, and intimately associated with Cavendish, had written from Dover on his way to France. He sends a short account of General Roy's operations at Dover in establishing a trigonometrical connexion between the observatories of Paris and Greenwich in order to determine the difference of longitude, a work in which the Royal Society was officially interested, and which they had initiated in 1784, but which had been delayed on account of the dilatoriness of Ramsden in supplying the instruments. General Roy had expressed a hope that he might have a visit from Cavendish. In the letter Blagden states that he had brought with him Sir Joseph Banks' presentation copy from Lavoisier of the *Nomenclature Chymique*, which he commends to Cavendish's attention.]

As the weather seems likely to be wet, I have given up all thoughts of coming to you at present, but be so good as to tell Gen. Roy that if the weather grows fair I shall very likely accept his offer and pay him a visit while he is about his operations, especially if the base is going on at the same time.

I was mistaken in supposing the angles could be reduced in Dr. Maskelyne's manner without taking into consideration the elevation or depression of the objects above the horizon. My mistake arose from supposing that if the objects were reduced to the level of the sea by lines drawn parallel to the direction of gravity at those places the reduced objects would subtend the same horizontal angle at the place of observation as the objects themselves which is true in the sphere but not in the spheroid.

Mr. Le Voisier has sent me the Nomenclature. I do not know whether you have seen the sequel of Saussures journey. The most remarkable circumstance is the effect of the rarity of the air on them which was such even after the fatigue of climbing was over that I wonder the French

[1] [This would be equivalent to 6·8 per cent. of carbonic acid. Air expired from the lungs contains from 3·3 to 5·5 per cent. by volume of carbonic acid. On the average, 4·78 vols. per cent. of oxygen is inhaled and 4·38 vols. of carbon dioxide is expired The "respiratory quotient" obtained by Cavendish is too small.]

astronomers at Peru did not observe it, unless you suppose that this effect was made remarkably more sensible by their preceding fatigue.

He computed the height from his observations of the barom. both according to De Luc's and Trembley's rule; according to one it came out a little greater and by the other a little less than according to Sir G. Shuckburgh's measurement[1].

I have been reading La V.'s preface. It has only served the more to convince me of the impropriety of systematic names in chemistry and the great mischief which will follow from his scheme if it should come into use. He says very justly that the only way to avoid false opinions is to suppress as reasoning as much as possible unless of the most simple kind and reduce it perpetually to the test of experiment and can anything tend more to rivet a theory in the minds of learners than to form all the names which they are to use upon that theory.

But the great inconvenience is the confusion which will arise from the different hypotheses entertained by different people and the different notions which must be expected [to] arise from the improvements continually making. If the giving systematic names becomes the fashion it must be expected that other chemists who differ from these in theory will give other names agreeing with their particular theories, so that we shall have as many different sets of names as there are theories and in order to understand the meaning of the names a person employs, it will be necessary first to inform yourself what theories he adopts. An equal inconvenience, too, will arise from the necessity of altering the terms as often as new experiments point out inaccuracies in our notions or give us further knowledge of the composition of bodies. But to shew the ill consequences of what they are about, let them only consider what would be the present confusion if it had formerly been the fashion to give systematic names and that those names had been continually alterd as people's opinions alterd. The great inconvenience is the fashion which so much prevails among philosophers of giving new names whenever they think the old ones improper as they call it. If a name is in use and its meaning well ascertained no inconvenience arises from its conveying an improper idea of its nature and the attempting to alter it serves only to make it more difficult to understand people's meaning.

With regard to distinguishing the neutral salts of less common use by names expressive of the substances they are composed of the case is different, for their number is so great that it would be endless to attempt to distinguish them otherwise; but as to those in common use, or which are found naturally existing, I think it would be better retaining the old

[1] [Observations made in Savoy, to Ascertain the Height of Mountains by means of the Barometer; being an Examination of Mr De Luc's Rules, delivered in his Recherches sur les Modifications de l'Atmosphere. By Sir George Shuckburgh, Bart., F.R.S. *Phil. Trans.* 67, 1777, 513.]

names. But with regard to salts whose properties alter according to the manner of preparing them, such as corrosive sublimate, calomel, etc. I should in particular think it very wrong to attempt to give them names expressive of their composition.

As I think this attempt a very mischievous one it has provoked me to go out of my usual way and give you a long sermon. I do not imagine indeed that their nomenclature will ever come into use, but I am much afraid it will do mischief by setting people's minds afloat and increasing the present rage of name-making.

CAVENDISH'S PAPERS ON HEAT

[On leaving Cambridge in 1753, Cavendish, after a short tour on the Continent with his brother Frederick, appears, from a few fragmentary notes and observations to be found among his papers, to have assisted his father, with whom he resided, in the physical and meteorological inquiries which are known to have engaged Lord Charles Cavendish's attention. It was probably this circumstance which first led the son to engage in experimental investigation. As we have seen, Lord Charles, who seems to have possessed what at the time was a pretty extensive physical cabinet, was interested in thermometry, and occasional mention is made by Cavendish of the instruments which were used by him from time to time. He was thus induced, almost at the very outset of his career as an experimenter, to take up the study of heat, and to pursue the calorimetric inquiries which are to be found, more or less systematically summarised and described, among his unpublished manuscripts. The evidence contained in these papers leaves no room for doubt that Cavendish embarked upon his investigations at about the same period as Black, or possibly somewhat later, and that he worked independently, and to a large extent in ignorance of Black's observations and theories. When they were actually begun is impossible to determine, but it is reasonably certain, from the few dates scattered among the notes, that much of the experimental work was done in or before 1764. Black's earliest observations on heat were made in 1758, but nothing was published concerning them, or his subsequent work, except through his University Lectures at Glasgow and Edinburgh, until much later; and it is certain that Cavendish could have had no knowledge of Black's work through any printed source. How far it was known to scientific and university circles in England at about the period referred to is difficult to determine. Cavendish maintained no active connection with Cambridge, and although he was elected into the Royal Society in 1760 he published nothing until six years later; his talents were then unrecognised, and his shy retiring nature, taciturnity and reluctance to enter into conversation, must have largely prevented him from obtaining information orally even if it were available at the period.

Everything, therefore, goes to show that his work on heat was original and independent, and that he discovered for himself the facts concerning latent and specific heat, thermal expansion, melting points, heat of chemical combination, etc., contained in his manuscript remains. Some of these early observations were published many years subsequently, as incidental and confirmatory results, as, for example, in the paper on the freezing of mercury printed in 1783, and there are occasional allusions to them under such phrases "as I know by experience," but with no details.

It is probably useless to surmise why the young experimenter should have refrained from making known the results of his independent observations, at the time he made them. But Cavendish was no ordinary man, and ordinary conventions and rules of conduct are inapplicable to him.

"In truth," as Wilson points out, "with Cavendish, publication was the exception, not the rule, and he has left so many completed researches unpublished, that no special hypothesis is needed to account for those upon heat having remained in manuscript." "Perhaps," he adds, "a reluctance to enter into even the appearance of rivalry with Black, prevented him from publishing researches which might be thought to trespass upon ground which the latter had marked off for himself, and pre-occupied."

Be this as it may, there can be no question that had Cavendish published his observations at, or near, the time he made them he would have anticipated much of the credit which belongs to Irvine, Crawford and especially Wilcke.]

EXPERIMENTS ON HEAT

It seems reasonable to suppose that on mixing hot and cold water the quantity of heat in the liquors taken together should be the same after the mixing as before; or that the hot water should communicate as much heat to the cold water as it lost itself; so that if the expansion of the mercury in the thermometer is proportional to the increase of heat the difference of the heat of the mixture and of the cold water as measured by the thermometer multiplied by the weight of the cold water should be equal to the difference of heats of the hot water and mixture multiplied by the weight of the hot water, or the excess of the heat of the mixture above that of the cold water should be to the difference of heats of the hot and cold water as the weight of the hot water [is] to that of the mixture.

The following experiments were made with an intent to see whether the excess of the heats of the hot water and the mixture above the cold water really bore that proportion to each other or not.

The apparatus used in these experiments is such as is represented in the annexed figure.

ABCD is a tin cylindrical vessel about 10 inches in diameter and as much in depth holding about 400 oz. of water with a cover of the same metal fitted to it.

M is a slip of tin plate about $1\frac{1}{2}$ inch broad, fastend to a bent piece of wire, passed through a small hole in the cover, serving to stirr the water in the pan.

EFGH is a cylindrical tin funnel, the pipe of which *NP* is about $\frac{1}{2}$ inch in diameter and enters into the pan *ABCD* through a hole made in the cover. This funnel also is furnished with a cover and a stirrer of the same kind as that to the pan. Each cover also has a hole in it to put a thermometer in.

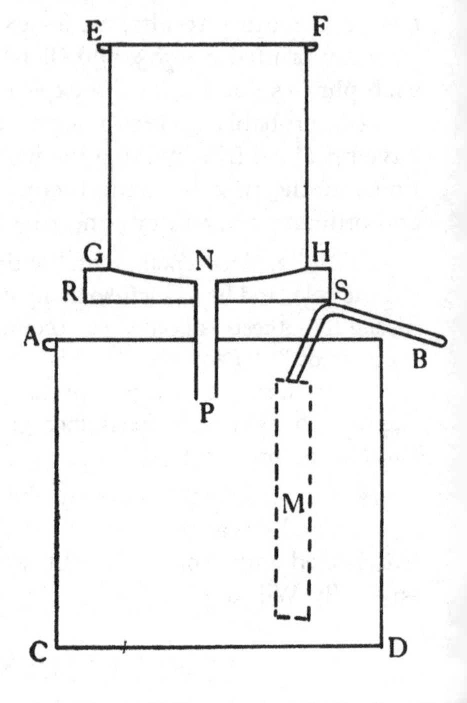

GHRS is a round piece of wood placed under the bottom of the funnel and kept by short legs at about an inch distance from the cover of the pan in order to prevent the water in the pan from being heated by the hot water in the funnel.

The pipe of the funnel is stopt up with a cork, which by means of a piece of wire fastend to it, and passed through a hole in the cover of the funnel, can be pulled out without taking off the cover. In trying the experiment, the cold water was put in the pan and the weight set down after its heat was ascertaind, which, as it differed but little from the temper of the air alterd very slowly. The thermometer was taken out and the funnel with the pipe corked up was put in its place, some hot water was then put into it, the cover put on, and the thermometer put in. When the heat of the water and its rate of cooling was sufficiently ascertaind, the water being kept stirrd all the time, the cork was drawn out of the pipe and the hot water let into the pan, the funnel was then taken off, the thermometer put into the pan and the heat of the mixture found after which, by weighing the mixture, the quantity of hot water put in was known. By this method of proceeding the hot water was put into the pan without being exposed to the open air, and consequently without any evaporation, so that it was coold very little in passing into the pan and I was able in some measure to estimate how much that little was, whereas if it had been pourd in without this precaution it would most likely have been coold considerably by the evaporation and contact of the air, and I should not have been able to form any guess how much it was coold

Experiment 1. At 5^{hr} $55'$ the cold water in the pan was found to be at $48\frac{2}{3}°$ of heat and was found to increase about 1° in 50 minutes.

At $6^{hr.}$	$12^{min.}$	$10^{sec.}$	the hot water in funnel was at	197°
	12	50	,, ,, ,,	196
	13	25	,, ,, ,,	195
At 6	16	0	the mixture was at	99
	20	30	,, ,,	$98\frac{1}{2}$

The cork was pulled out of the pipe of the funnel at $6^{hr.}$ $13^{m.}$ $25^{sec.}$ and the hot water took up near $30^{sec.}$ to run intirely out of the funnel but run almost out in 15 or 20^{secs}. The weight of the cold water was $249\frac{8}{10}$ oz.; that of the mixture $383\frac{8}{10}$ whence the weight of the hot water was 134 oz.

But before we proceed to make this experiment with the rule it was intended to examine, it is necessary to make some corrections. First, as neither the heat of the hot or cold water or mixture were found at the precise time of making the mixture, there must be some corrections made to the heats shewn by the thermometer. Let us assume, therefore, $6^{hr.}$ $13^{m.}$ $40^{sec.}$, or $15^{sec.}$ after the cork was pulled out, as the time of making the mixture, and find what would be their heats at that time according to their observed rates of heating and cooling. The hot water coold 2° in about $1^{m.}$ $15^{secs.}$ or $\frac{4}{10}$ of a degree in $15^{secs.}$; so that its heat at $6^{hr.}$ $13^{m.}$ $40^{sec.}$ would be 194°·6. According to the same method of proceeding, the heat of the cold water at the same time would be 49°, and that of the mixture $99\frac{2}{10}$; so that $194\frac{6}{10}°$, 49° and $99\frac{2}{10}°$ may be looked upon as the true heats of the hot water, cold water, and mixture. Secondly, it must be observed that part of the effect of the hot water was spent in heating the pan, and therefore an allowance must be made on that account. The weight of the pan was 31·25 oz.; that part of the stirrer which was within the pan was 2·3 oz., and the cover weighed $9\frac{9}{10}$ oz. As the inside surface of the pan would be heated much faster by the water than the outside surface would be coold by the air, I believe we may suppose the tin pan would be heated almost as hot as the mixture within it, but that the cover being in contact with the air on both sides would be heated only to a mean heat between the cold water and mixture. Therefore 33·55 oz. of tin plate were heated to the heat of the mixture, and 9·6 oz. were heated $\frac{1}{2}$ as much, which comes to the same thing as if 38·35 oz. were heated to the full heat of the mixture. Whence we may conclude from the experiment that the heat of a mixture of 134 oz. of water whose heat is 194°·6 with 249·8 oz. of water whose heat is 49°, and 38·35 oz. of tin plate of the same degree of heat, will be 99°·2. But it will appear from an experiment, which will be mentiond hereafter, that cold iron filings added to hot water, cool it no more than the addition of $\frac{1}{8\frac{1}{2}}$ that quantity of water would do, and consequently the 38·35 oz. of tin plate in the foregoing experiment has no more effect in cooling the hot water than 4·5 oz. of water. Wherefore adding this to the weight of the cold water I think we may conclude that the true heat of a mixture of 134 oz. of water whose heat = 194°·6 with 254·3 oz. of water whose heat = 49° is 99°·2, which is exactly the same that it ought to be according to the forementiond rule that the difference of heat of the cold water and mixture is [to] the difference of heats of the hot and cold water as the weight of the hot water to the weight of the mixture.

It is plain that the allowance to be made for the effect of the pan and cover is very uncertain, as one can give but a very uncertain guess how much they will be heated, but if their effect had been intirely neglected the computed heat of the mixture, or its heat according to the above mentiond rule, should be but $\frac{6}{10}$ of a degree greater than I made, so that it should seem as if no error to signify could proceed from thence.

In trying the heat of water-mixture in this and all the following experiments of this kind, care was taken that so little of the scale of the thermometer should be out of the pan that no sensible error could proceed from thence.

The experiment was repeated in the same manner with nearly the same quantities of water. The heat of the mixture was $\frac{1}{10}$ of a degree less than it should be by computation.

Experiment 2. With about equal quantities of hot and cold water.

At 7.24.0 the cold water was at	48°
At 7.37.0 the water in funnel was at	193
38.30 ,, ,, ,,	192
40.0 ,, ,, ,,	191
At 7.42.0 the mixture was at	116¾
43.40 ,, ,,	116½
46.30 ,, ,,	116
The weight of the cold water	196·7 oz.
,, ,, mixture	387·2

The heats corrected as before are:

Cold water	48°·3
hot water	190·8
mixture	117·2

Making the same allowance as before for the effect of the pan, the heat of the mixture by computation should be 117°·6; consequently the heat of the mixture was $\frac{4}{10}$ of a degree less than it should be by computation.

By another experiment made with nearly the same quantities of water the heat of the mixture was exactly the same as by computation.

Experiment 3. With about 2 parts of hot water to one of cold.

At 6. 4.0 the cold water was at	47°
6.16.0 ,, ,,	47½
At 6.23.50 the water in the funnel was at	203
25.0 ,, ,, ,,	202
At 6.28.20 the mixture was at	148
31.45 ,, ,,	147
The weight of the cold water	131·5 oz.
,, mixture	392·9 oz.
The corrected heat of the hot water	201·8°
,, ,, cold water	47·9
,, ,, mixture	149·2

The heat of the mixture by computation should be 149°·1; therefore the heat of the mixture is $\frac{1}{10}$ of a degree greater than it should be by computation.

Experiment 4. The experiment was repeated with about equal quantities of hot and cold water in a rather different manner from before, namely, the hot water was put into the pan and the cold water into the funnel, the pipe of the funnel being made to enter into the cover of the pan near one side so that there was room to keep the thermometer in the pan and to stir the water therein while the funnel was in its place.

At $6^{hr.}$ $58^{min.}$ $30^{sec.}$, the water in the pan was at 198°·5: at the same time the water in the funnel was at 45°½. The cork was then immediately taken out and the cold water let into the pan. At 7.1.20 the mixture was at 123° and at 7.7.10 at 122°. The weight of the cold water was 195 oz. and that of the mixture 394½.

As the heats of the hot and cold water were observed at the same time and immediately before the cork was drawn out, there is no need to correct their heats, but only to correct the heat of the mixtures by finding what would have been its heat at the time of drawing out the cork according to its observed rate of cooling, wherefore 123°·5 is to be looked upon as the true heat of the mixture.

If the same allowance is made for the effect of the pan as in the former experiment the heat of the mixture by computation is 123°·7.

But in reality the allowance ought to be rather different. Let us suppose that before the cold water was put in, that part of the pan which was in contact with the hot water was of the same heat as the water; and that the remainder of the pan and the cover was about a mean between the heat of the hot water and the air of the room, *id est* about the heat of the mixture; and let us suppose that after the mixture was made (as the pan was then near full) that the whole pan was of the same heat as the mixture, and the cover about a mean between the heat of the mixture and of the air. Therefore that part of the pan which was not in contact with the hot water was not coold at all by pouring in the colc water. The rest of the pan and the cover was coold in the same proportior as it was heated in the former experiment. As the pan was about ½ ful of hot water, that part which was not in contact with the water was abou ⅖ of the whole; whence the allowance for the effect of the pan, etc. come out about 3 oz. and therefore the heat of the mixture by computation i 123°½ which is exactly the same as the observed heat.

By another experiment tried in the same way with nearly the sam quantities the heat of the mixture was ½ a degree less than it should b by computation.

In all the foregoing experiments the observed heats of the mixtur differ from the computed by not more than the different experiments d from each other, so that the above-mentiond rule seems perfectly con formable to experiment.

It must be observed that in the first method of trying the experimen the greater the allowance I made for the effect of the pan, the less woul the computed heat come out, whereas in the 2nd way of trying, th greater allowance I made the greater would the computed heat tur out. Consequently as the computed heats agree equally well with th observed in both methods of trying the experiment, it appears that th allowance I made could not be sufficiently wrong to be of any signification

Section 2. One would naturally imagine that if cold ☿ [mercury] o any other substance is added to hot water the heat of the mixture woul be the same as if an equal quantity of water of the same degree of hea had been added; or, in other words, that all bodies heat and cool each othe when mixed together equally in proportion to their weights. The followin experiment, however, will show that this is very far from being the case.

The glass bottle used for making the mixtures in the following exper was of the shape marked in the annexed figure and was blown thin and very regular the better to bear sudden alterations of heat and cold.

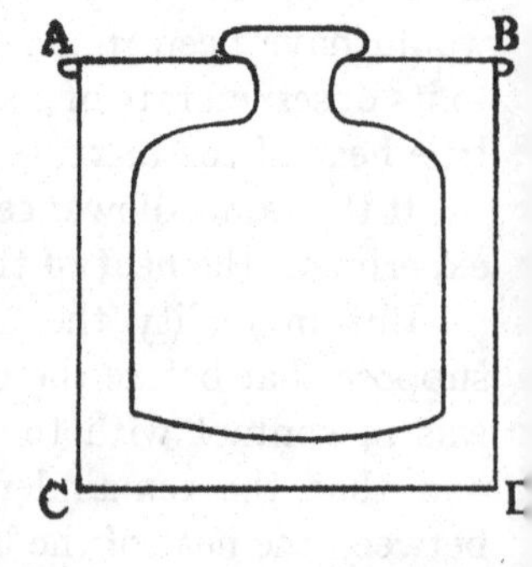

In order to prevent the liquor in the bottle from cooling so fast as it would otherwise do, the bottle was inclosed in a tin pan with a cover to it, *ABCD*, in such manner that only the mouth of the bottle was exposed to the air, all the rest being within the pan. The bottle was kept steady in the middle of the pan by tin rings and the spaces between the bottle and sides of the pa were filled with wool. In trying the experiment the hot water or othe liquor was put into the bottle and when its heat and rate of cooling wa sufficiently ascertaind, the cold (whose heat as being nearly of the tempe of the air alterd very slowly) was put. The heat of the liquor in the bottle was found by putting the ball of the thermometer into it and keeping it there till it was come to the full heat of the liquor, the bottle being all the while gently shaken, after which the thermometer was taken out, the bottle corked up close and kept shaking till the heat of the liquor was tried again. The time during which it was requisite to keep the thermometer in the liquor before it acquired the full heat was different according to the nature of the liquor: in water it was requisite to keep it about [1]; in ☿ rather less; in spirits of wine somewhat more.

The bottle held rather more than 19 oz. of water, and weighed $9^{oz.}$ $18^{dwt.}$ 0^{grains}: the tin pan and wool weighd 11.15.4 [$11^{oz.}$ $15^{dwt.}$ 4^{grains}].

Exp. 5. Mixture of hot and cold water.

At 20 minutes the water in the bottle was at					158½°
23	,,	,,	,,	,,	155
26	,,	,,	,,	,,	152

The cold water was then immediately put in.

At 29 minutes the mixture was at				106½
32	,,	,,	,,	106
35	,,	,,	,,	105½

The weight of the cold water put in was $5^{oz.}$ $11^{dw.}$ $17^{gra.}$ and its heat was 51°½. The weight of the mixture was 11.3.19 and therefore the weight of the hot water 5.12.2.

Exp. 6. Mixture of cold ☿ with hot water.

At 43 minutes the water in the bottle was at					161¼°
46	,,	,,	,,	,,	158
49	,,	,,	,,	,,	155

The ☿ was then immediately put in.

At 52 minutes the mixture was at				108¾
55	,,	,,	,,	108¾
58	,,	,,	,,	108

The weight of the ☿ put in was 171.12.0 and its heat was 54°½: the weight of the mixture was 177.4.3 and therefore the weight of the hot water 5.12.3.

The difference of heat of the hot and cold water in the 5th exp. was the same as the difference of heat of the hot water and ☿ in the 6th exp. and the weight of the hot water was the same in both these experiments, and yet the difference of heat of the hot water and mixture was near as great in the 5th exp. as the 6th; so that the hot water was coold near as much by the addition of 5.11.17 of water as by that of 171.12.0 of ☿; that is, hot water is coold near as much by the addition of 1 part of cold water as by that of 30 parts of ☿ of the same heat.

[1] [Time not stated.]

In order to compare the effects of ☿ and water more exactly, let us assume 26½ in the 5th exp. or ½ a minute after the last time of observing the heat of the hot water, for the time of making the mixture, and find the heats of hot water and mixture at that time, as in the former experiments. By this means the corrected heats of the hot water and mixture come out 151°·5 and 106°·7. The quantity of cold water added was 5oz. 11dwt. 7gr. and its heat was 51°·5. But by the above-mentiond rule 106°·8 is the heat of a mixture of 5.11.7 of water of 51°·5 of heat with 6.14.4 of water of 151°·5 of heat; so that the effect is the same as if the weight of the hot water had been 6.14.4 and the effect of the bottle and pan had been nothing, and consequently the effect of the bottle and pan is equal to that of 1.6.2 of water.

According to an experiment which will be related hereafter, that pounded glass added to cold water cools it as much as $\frac{1}{8{\cdot}073}$ its weight of water of the same degree of heat; consequently supposing the glass bottle to be heated to the same degree as the liquor within, which I think it must necessarily be, its effect must be equal to that of 1.4.12 of water; consequently there remains 2.2 for the effect of the pan, etc.

The heats of the hot water and mixture in the 6th exp. corrected as above are 154°½ and 109°·1, and as the same bottle and pan was used in this exp. as in the last, and as the quantity of hot water also was the same, though I do not know whether that is of any signification, it is plain that the quantity to be allowed for the effect of the bottle and pan must be the same in this exp. as in the last. Therefore the heat of the mixture is the same as if instead of the ☿, 5.15.21 of water of the same degree of heat had been added, for 54°·3, the difference of heats of the mixture and ☿ is to 45°·7, the difference of heats of the hot water and ☿, as 6.18.5. to 5.15.21; consequently cold water added to hot water cools it as much as the addition of 29·61 its weight of ☿ of the same degree of heat will do; or, in other words, the effect of water in cooling hot water is 29·61 times greater than that of ☿ of the same heat.

The experiment was repeated in the same manner, with different quantities of hot water.

Exp. 7. Mixture of cold water with hot.

At 2 minutes the hot water was at			151°
5	,,	,,	149
8	,,	,,	147

The cold water was then put in.

At 11 minutes the mixture was at			119
14	,,	,,	118½
17	,,	,,	117¾

The weight of the cold water added was 3·971 oz. and its heat was 51°¼. The weight of the mixture was 12·715 and consequently the weight of the hot water 8·744.

Exp. 8. Mixture of ☿ with hot water.

At 22 minutes the hot water was at			157½°
25	,,	,,	155
28	,,	,,	152¾

The ☿ was then put in.

At 31 minutes the mixture was at			123
34	,,	,,	122½
37	,,	,,	121½

The weight of the ☿ was 120·212 [oz.] and its heat 51°. The weight of the hot water was 8·719 oz. Proceeding as before, the corrected heats of the hot water and mixture in the 7th exp. are 146°·7 and 119°·5 and the effect of the bottle and pan is equal to that of 1·074 oz. of water. In the 8th exper. the corrected heats of the hot water and mixture are 152°·3 and 123°·6. Therefore the 120·212 oz. of ☿ coold the water as much [as] 3·871 oz. of water would do and therefore the effect of water in cooling hot water is 31·05 times greater than that of ☿.

Having thus found the proportional effect of cold water and ☿ in cooling hot water I next tried whether their effects in cooling hot ☿ would bear the same proportion to each other.

Exp. 9. Mixture of cold ☿ with hot.

At 7 minutes the ☿ in the bottle was at				167¼°
10	,,	,,	,,	163½
13	,,	..	,,	160

The cold ☿ was then put in.

At 16 [min.] the mixture was at			113
19	..	,,	113¼
22	,,	,,	111½

The weight of the cold ☿ was 125·437 [oz.] and its heat 54°⅓. The weight of the hot ☿ was 125·45.

Exp. 10. Cold water added to hot ☿.

At 42′ the ☿ in the bottle was at			168
45	,,	,,	164
48	,,	,,	160¼

The water was then put in.

At 51′ the mixture was at			113
54	,,	,,	111¾
57	,,	,,	110¾

The weight of the cold water was 4·115 [oz.] and its heat 52°¾. The weight of the hot ☿ is 125·425.

In the 9th exp. the corrected heats of the hot ☿ and mixture are 159°·4 and 114°·4; and the effect of the bottle is equal to that of 42·08 of ☿. In the 10th exp. the corrected heats of the hot ☿ and mixture are 159°·6 and 113°·9: wherefore the 4·115 of water coold the ☿ as much as 125·18 of ☿ and therefore water cools hot ☿ as much as 30·42 its weight of ☿ can do, or the effect of water in cooling hot ☿ is 30·42 times greater than that of ☿. By a mean of all 3 experiments the effect of water in cooling hot water or ☿ seems to be 30·1 times greater than that of ☿.

The effect of water in cooling hot water appeard to be 29·61 times greater than that of ☿ by one experiment and 31·05 by the other. Whence it appears that the effects of water and ☿ in cooling hot ☿ bear the same proportion to each other as their effects in cooling hot water.

I then tried whether the effects of hot water and hot ☿ in heating cold water bore the same proportion to each other as those of cold water and ☿ do in cooling hot water.

Exp. 11. Hot water added to cold.

The cold water, which was nearly of the temper of the air, was put in the bottle inclosed in the pan, and the hot water put in a separate bottle wrapt up in flannel to prevent its cooling so fast as it would otherwise do. This bottle of water with a thermometer immersed in the liquor was gently agitated till the thermometer sunk to 138°. The thermometer was then taken out and the hot water immediately poured into the bottle of cold water.

At 1 minute the hot water was pourd in.
At 4 minutes the mixture was at 91°
6 ,, ,, 90½

The weight of the cold water was 5·606 and its heat 54½: the weight of the hot water was 5·59.

Exp. 12. Hot ☿ added to cold water.

This exper. was tried in the same way as the former.

At 48′ the ☿ was pourd in.
At 51 the mixture was at 92½
54 ,, ,, 92
57 ,, ,, 91¾

The weight of the cold water was 5·606 and its heat 56°½.
The weight of the hot ☿ was 171·6 [oz.] and its heat 138°.

In these 2 exper. let us suppose the heats of the liquors the same as they were found by the thermometer, and let us correct the heats of the mixture by finding what their heat ought to have been according to their observed rate of cooling at the time of putting in the hot liquor.

Therefore in the 11th exper. 91°·5 was the corrected heat of the mixture; the effect of the bottle is equal to that of 1·419 of water. In the 12th exper. the corrected heat of the mixture is 93° and the ☿ heated the water as much as 5·7 of water would; and therefore the effect of water in heating water is 30·11 times greater than that of ☿.

Hence, I think, we may conclude that the effects of hot water and hot ☿ in heating cold water bear the same proportion to each other as their effects when cold do in cooling hot water.

This conclusion might have been deduced from the former experiments, provided it is granted that the rule which was shewed to hold good with regard to hot and cold water, holds also good with regard to mixtures of hot and cold ☿; namely, that the difference of heats of the hot ☿ and mixture is to the difference of heats of the mixture and cold ☿ as the weight of the cold ☿ to the weight of the hot. For from these experiments it was deduced that the effects of cold water and ☿ in cooling hot ☿ bore the same proportion to each other as their effects in cooling hot water. Let the effect of cold water in cooling hot ☿ or water be n times greater than the effect of cold ☿; then will the heat of a mixture of one part of cold water with any number of parts (as A) of hot ☿ be equal to the heat of a mixture of n parts of cold ☿ with A parts of hot, which by the postulation is equal to the heat of a mixture of n parts of cold water with A parts of hot, or of 1 part of cold water with $\frac{A}{n}$ of hot. Therefore $\frac{A}{n}$ parts of hot water have the same effect in heating one part of cold water as A parts of hot ☿, or 1 part of hot water has the same effect as n parts of hot ☿, which is the thing to be proven.

On the other hand, if it is granted that the effects of hot water and hot ☿ in heating cold water bear the same proportion to each other as their effects when cold do in cooling hot water and hot ☿, it follows that the above mentiond rule does also hold good with regard to ☿; and also that the effects of hot water and ☿ in heating cold ☿ bear the same proportion to each other as their effects when cold do in cooling hot water and ☿.

Exp. 13. Cold spirits of wine added to hot ☿.

At 44′ the ☿ in the bottle and pan was at	148°¼
47 ,, ,, ,,	145
50 ,, ,, ,,	142

The spirits were then put in.

At 53′ the mixture was at	103°¼
56 ,, ,, ...	102¼
59 ,, ,, ...	101½

The weight of the hot ☿ was 125·45 [oz.]: the weight of the spts. of wine 5·275 and its heat 54°.

The corrected heats of the ☿ and mixture are 141°·5 and 104°; whence, by comparing this exp. with exp. 9th in which the quantity of hot ☿ was exactly the same, it appears that the spirits of wine coold the ☿ as much as 123·62 of ☿ would have done, and therefore the effect of spts. of wine in cooling hot ☿ is 23·43 times greater than that of ☿.

By another exper. of the same kind tried with the same quantity of ☿ and 7·85 of spts., the effect of spts. of wine came out 21·65 greater than that of ☿; and by another experiment with the same quantity of ☿ and 5·852 of spts. it came out 22·82 times greater.

I then tried whether their effects in cooling hot spts. bore the same proportion to each other.

Exp. 14. Cold spts. added to hot.

At 55 minutes the spts. were at			142°½
58	,,	,,	140
61	,,	,,	137½

The cold spts. were then put in.

At 64 minutes the mixture was at			101°¾
67	,,	,,	101
70	,,	,,	100½

The weight of the cold spts. was 6·404 [oz.] and their heat 56° and the weight of the hot spirits was 6·412 [oz.].

Exp. 15. Cold ☿ added to hot spts.

At 50′ the spirits were at		...	143°¾
53	,,	...	140¾
56	,,	...	138

The ☿ then added.

At 59 minutes the mixture was at				101°¾
62	,,	,,	,,	101¼
65	,,	,,	,,	100¾

The weight of the ☿ was 147·562 [oz.] and its heat 55°¾: the weight of the spts. was 6·406 [oz.].

In the 14th exp. the corrected heats of the hot spts. and mixture are 137°·1 and 102°·3 and the effect of the bottle and pan is equal to 2·1 of spts. In the 15th exp. the corrected heats of the hot spt. and mixture are 137°·54 and 102°·17; wherefore the ☿ coold the spt. as much as 6·481 of spts. would have done; whence the effect of spts. in cooling hot spts. is 22·77 times greater than that of ☿.

Hence, I think we may conclude that the effects of spts. and ☿ in cooling hot spts. bear the same proportion to each other as their effects in cooling hot ☿.

By a mean of all the experiments the effect of spts. of wine in cooling hot ☿ and spts. seems to be 22·7 times greater than that of ☿, and, consequently, 1·326 times less than that of water.

I also made an exp. to see whether the effect of spts. cooling hot water bore the same proportions or not to that of ☿. But it must be observed that there is heat generated by mixing spts. of wine with water; *id est*, when water and spts. of wine of the same degree of heat are mixed together the heat of the mixture will be greater than that of the liquors before mixing. Wherefore it is necessary to find how much the heat generated is, which renders the expt. more complicated and less exact.

Exp. 16. Cold spts. of wine added to hot water.

At 39′ water at ... ... $153^{\circ}\frac{3}{4}$
42 ,, ... $151\frac{1}{2}$
45 ,, ... $149\frac{1}{4}$

[The spts. of wine were then put in.]

At 49′ the mixture was at 123
52 ,, ,, ... $122\frac{1}{2}$
55 ,, ,, ... $121\frac{1}{2}$

The weight of the hot water was 9·088: the weight of the spt. was 7·631 and its heat 60°.

Exp. 17. Some warm spts. of wine were put in the glass bottle inclosed in the tin pan and some warm water was kept ready in another bottle wrapt in flannel with a thermometer in it.

At 52′ the spts. were at ... ... $114^{\circ}\frac{3}{4}$
55 ,, ,, ... 113

The warm water was then immediately put in.

At 59′ the mixture was at ... ... $121^{\circ}\frac{1}{2}$
62 ,, ,, ... $120\frac{1}{4}$
65 ,, ,, ... $119\frac{1}{3}$

The weight of the spts. was 7·587 [oz.]. The weight of the water added was 9·175 [oz.] and its heat at 55 minutes was $117^{\circ}\frac{3}{4}$.

The corrected heat of the spt. in this last expt. is 112°·71, and, supposing the warm water to have coold as fast as the spts. its corrected heat is 117°·46, the corrected heat of the mixture is 122°·81.

If we suppose that the effect of warm water in heating spts. is 1·326 times greater than that of spts., the heat of the mixture would be 112°·71 if no heat was generated by the mixing, and as the difference of heat of the 2 liquors is so small, namely, only 4 degrees, the heat would not be much different though the proportion of the effects of water and spts. in heating spts. was considerably different from what we suppose: therefore the mixture was 7°·46 hotter than it would be if no heat was generated.

In the 16th experiment the corrected heats of the hot water and mixture are 148°·88 and 123°·88, and as the quantity of water and spts. is very nearly the same in this exp. as [in] the other, we may suppose that the heat of the mixture in this expt. as much exceeded what it would be if no heat was generated as it did in the other. Therefore the heat of the mixture would have been 116°·42 if no heat had been generated. This heat is the same as the heat of the mixture would have been if instead of the spts., 5·948 of water had been added. Therefore, allowing for the heat generated by the mixture, 7·631 of spt. cool water as much as 5·948 of water and therefore the effect of spts. in cooling water is 1·283 times less than that of water or 23·46 times greater than that of ☿. Therefore as far as can be determined by this exper. the effects of spt. and ☿ in cooling hot water bear the same proportion to each other as their effects in cooling spts. or ☿.

It should seem, therefore, to be a constant rule that when the effects of any 2 bodies in cooling one substance are found to bear a certain proportion to each other that their effects in heating or cooling any other substance will bear the same proportion to each other.

If this rule is true it is plain that in the foregoing experiments the effect of the bottle and pan in heating the cold liquor put in should be the same (or should be equal to that of the same quantity of water) of whatever nature the hot liquor in the bottle was. This, as far as can be expected, is conformable to experiment, for according to the 5th exper. the quantity to be allowed for the effect of the bottle and pan appeard to be 1·269 oz. of water; by the 7th exper. 1·241 oz. and by the 12th exper. it seemd 1·419 oz. According to the 9th exp. it appeard to be 39·21 oz. of ☿ or 1·303 of water; and by the 14th it appeard to be 2·099 of spts. or 1·583 of water. Though there is a considerable difference between the quantity of this allowance as found by the 7th and 14th exps., yet it does not seem greater than may be owing to the error of the exper., as this difference may be accounted for without supposing so great an error in the experiments as what is necessary to reconcile the 3 experiments in which cold spts. were added to hot ☿.

The true explanation of these phenomena seems to be that it requires a greater quantity of heat to raise the heat of some bodies a given number of degrees by the thermometer than it does to raise other bodies the same number of degrees.

I made some experiments also to find the effect of 2 or 3 other liquors and also of several solid bodies in heating or cooling other substances. The experiments upon the liquors were tried in the same glass bottle used for the former experiments by putting hot ☿ in the bottle and adding these liquors to it. Those upon solid substances were tried in a tin bottle nearly of the same shape as the glass one, holding 22·902 oz. of water and weighing 3·9 oz. and inclosed in the same tin pan as the glass bottle. The experiments were tried by putting hot water in the bottle and adding these substances to it. The heat of the solid substances was determined by keeping the bottle in which they were containd in a pail of water till it was supposed that they were arrived at the same heat as the water. In all other respects the experiments were tried in the same manner as the former. In order to find the allowance to be made for the effect of the bottle I made 3 experiments, the result of which is as follows:

	Weights of		Corrected heats of		Heat of cold water	Allowance for effect of bottle and pan
	hot water	cold water	hot water	mixture		
			°	°	°	
1st expt.	12·935	9·137	153·75	108·5	39·5	1·
2nd expt.	15·892	6·112	149·75	120·5	40·5	·85
3rd. expt.	17·171	4·035	155·25	134·5	41·5	·694

The following tables contain the results of all the experiments I have made on this subject as well as those made with the glass bottle as with the tin one.

Experiments tried with glass bottle.

Name of liquor added	Weights of		Corrected heats of		Heat of cold liquor	Effect of these liquors in heating and cooling others: the effect of water = 1
	hot ☿ oz.	cold liquor oz.	hot ☿	mixture		
			°	°	°	°
Saturated solution of sea-salt	125·46	5·506	145·46	105·37	57	$\frac{1}{1{\cdot}21}$
Solution of pearl-ashes in 2ce their weight of water	125·46	5·5	154·13	111·12	54·5	$\frac{1}{1{\cdot}32}$
Oil of vitriol mixed with an equal weight of water	125·46	5·5	156·63	115·58	57	$\frac{1}{1{\cdot}43}$

In all these exper. I took care not to put so much of the solid matter, but what there was water enough above it to intirely cover the ball of the thermometer. These experiments, however, are much less exact than those with liquids as there was much more time spent in pouring in the solid substances than the liquids, and the thermometer was much slower in arriving at the heat of the liquor as I could not agitate the liquor about it so much; besides that, perhaps, that part of the liquor in which the thermometer was immersed was not exactly of the same temper as the rest of the water and the solid matters.

I made an attempt to find the effect of air in cooling water by blowing air from a smith's bellows through the worm of a still, and seeing how the water in the worm-tub was coold thereby. The nozzle of the bellows was fastend to the lower end of the worm and the ball of the thermometer was put into the upper end of it to see how much the air was heated in

Experiments tried with tin bottle.

Name of substance tried	Weights of hot water	Weights of solid substance	Corrected heats of hot water	Corrected heats of mixture	Heat of solid substance	Effect of substance in heating or cooling others
Iron filings	14·965	42	153°	128·5°	53·75°	$\frac{1}{8 \cdot 12}$
The same tried again	15·023	40·581	152¼	130½	52·75	$\frac{1}{9 \cdot 15}$
Lead shot	13·2	102·1	152	133·25	47·5	$\frac{1}{33 \cdot 6}$
The same tried again	13·354	102·03	156	136·5	50	$\frac{1}{32 \cdot 2}$
Tin reduced to lumps of about the size of peas	13·929	57·471	154·5	135·5	48·5	$\frac{1}{18}$
Silver sand	12·954	24·25	148	122·5	48·5	$\frac{1}{5 \cdot 1}$
White glass	11·904	28·892	148·25	126·75	50·75	$\frac{1}{8 \cdot 07}$
White marble	13·435	21·604	157	131·5	50·25	$\frac{1}{4 \cdot 856}$
Brimstone	12·756	17·502	150	131·25	50·75	$\frac{1}{5 \cdot 59}$
Newcastle coal	13·071	11·502	153·25	133	50	$\frac{1}{3 \cdot 42}$
Charcoal	13·583	3·987	156·25	148·5	49·5	$\frac{1}{3 \cdot 57}$

passing through it. The lower board of the bellows was confind in such manner that it could move only a given space so that the same quantity of air was forced into the upper partition of the bellows at each stroke. A mark was placed by the side of the upper board of the bellows, and as soon as the board sunk below that mark, a stroke of the lower board was given so as to keep the upper board always equally elevated. It was found by measuring the surface of the lower board, and the space which it was made to move through, that 283 cubic inches of air was forced into the upper partition at each stroke, supposing that no air escapes through the valve of the lower board. It was found also that only $\frac{955}{1000}$ of the quantity of air forced into the upper partition passed through the worm, the rest escaping by the valve of the bellows and through the pores of the leather as I found that in trying the experiment I was forced to give 1600 strokes of the bellows in 57·0 minutes so as to keep the upper board at the right elevation; whereas when the upper orifice of the worm was stopt up so that no air could pass through the worm, I was obliged to give 645 strokes in the same time.

The tub had a wooden cover to it, with a hole in it to put a thermometer in, and was furnished with a piece of wood by which I could stir the water without taking off the cover or taking out the thermometer. The exper is as follows:

Time hr.	min.	secs.	Number of strokes given	Thermom. in tub	Do. in worm
				°	°
8	17	20	0	119½	118
	28	25	300	118½	117½
	46	20	800	117	116
9	14	30	1600	114¾	114

The water in the worm-tub was 2250 oz. and it was found by experiment that 323 oz. are to be allowed for the effect of the tub and cover and worm. The heat of the air in the room near the bellows was 52°. It was found that the water in the tub would have coold about 3°·79 in the 57 minutes that the experiment lasted if no air had been blown through the worm; so that it seemed as if the water was coold ·96° by blowing the air through.

The quantity of air blown through the worm was 216½ oz. and was heated 63° in passing through. Therefore it requires as great a quantity of heat to raise 216½ oz. of air 63° by the thermometer as to raise 2573 oz. of water ·96° and therefore the effect of air in heating and cooling other bodies is 5·51 times less than that of water.

By another exper. made in the same manner, its effect seemd 9·2 times less than that of water, but the quantity which the water was coold by blowing the air through the worm was so small in both these experiments that one can give but a very imperfect guess at how much its effect is.

Part II

As far as I can perceive it seems a constant rule in nature that all bodies in changing from a solid state to a fluid state or from a non-elastic state to the state of an elastic fluid generate cold, and by the contrary change they generate heat.

I shall first consider those cases in which bodies are changed from a non-elastic to an elastic state or from an elastic to a non-elastic state, and afterwards those in which they are changed from a solid to a fluid state or the contrary.

The reason of this phenomenon seems to be that it requires a greater quantity of heat to make bodies shew the same heat by the thermometer when in a fluid than in a solid state, and when in an elastic state than in a non-elastic state. It is plain that according to this explanation all bodies should generate as much cold in changing from a $\propto$ st. as they generate heat by the contrary change, which as far as I can perceive seems to be the case. There are 2 different ways by which fluids evaporate or are changed into the state of an elastic fluid, namely, first that species of evaporation which is performed with a less heat than that which is sufficient to make them boil, and which is owing to their being absorbed by the air; and, secondly, that species which we call boiling and which is performed independently of the air.

Dr. Cullen has sufficiently proved that most if not all fluids generate cold by the first species of evaporation. There is also a circumstance daily before our eyes which shews that water generates cold by the 2nd species of evaporation.

It is well known that water as soon as it begins to boil continues exactly at the same heat till the whole is boiled away, which takes up a very considerable time. No reason, however, can be assigned why the fire should not continually communicate as much or nearly as much heat to it after it begins to boil as it did when it wanted not many degrees of boiling, and yet during all this time it does not grow at all hotter. This, I think, shews that there is as much heat lost, or, in other words, as much cold generated by the evaporation as there is heat communicated to it by the fire. Thus, when the water is heated to the boiling-point, then as fast as it receives heat from the fire there is immediately so much of the water turned into steam as is sufficient to produce as much cold from the fire, so that the water is prevented from growing hotter, and, moreover, will not be intirely evaporated till it has received as much heat from the fire as there is lost by turning the whole of the water into steam. Whereas if no cold was produced by the evaporation, the water should either grow hotter and hotter the longer it boiled, or else it should be intirely converted into steam immediately after it arrived at the boiling-point.

Perhaps it may be said that the cause of this phenomenon is that the steam is hotter than the boiling water, and by that means carries off the heat communicated to it by the fire. This, however, is by no means the case, as I have found by experiment that the steam is not at all hotter than the water it proceeds from. This I tried by putting a thermometer into a vessel of water inclosed on all sides, except a chimney to carry off the steam, the thermometer being placed so that very little of the ☿ appeard out of the vessel, when I found that the ☿ rose very nearly, but not quite to so great a height when the quantity of water was not sufficient to reach up to the ball of the thermometer as when it rose a little above it. But when the water was enough to rise a great way above the ball then the ☿ rose sensibly higher.

To understand this, it must be observed that the water near the bottom of the vessel will require more heat to make it boil than that above it, as being pressed by a greater weight. Now in all probability the water in all parts of the vessel is heated precisely to that degree which is required to make it boil, for it cannot be heated higher as in that case it would be instantly turnd into steam and if the water boils pretty fast it is not likely that it should anywhere be of a less heat than that. Consequently that part of the water adjoining to the ball of the thermometer would be a little hotter when the water rose a good deal higher than the ball than when it rose but a little above it. As the thermometer rose very near as high when exposed to the steam as when the water rose very little above it the steam seems to be precisely of the same heat as the surface of the boiling water.

For the same reason we may conclude that all other liquors and also all solid bodies which are capable of being volatilized generate cold by being changed into an elastic fluid or vapor, for there is no substance but what takes up a great deal of time after it begins to distil strongly before it is all driven over.

I made some exper. on this principle, namely by heating some water over a lamp and finding how fast the water heated before boiling and how long it was a given quantity, for determining the quantity of cold produced by water in boiling all away (*sic*).

The apparatus is represented in the annexed figure, where *A* is a tin bottle for boiling the water in, the same that was used in the foregoing experiments. *B* is a spirit lamp furnished with seven small wicks: *bbb* are the wicks. *CD* is a brass plate, the use of which is to prevent the spirits in the lamp from being so much heated by the flame as they would otherwise be and *EFGH* is a tin frame surrounding the bottle and serving to keep in the heat. A thermometer was kept in the water all the while it was heating, and the space left between the mouth of the bottle and stem of the thermom. stopped up to prevent evaporation. When the water was near boiling the therm. was taken out and the mouth stopt. up again, all but a sufficient hole to suffer the steam to escape.

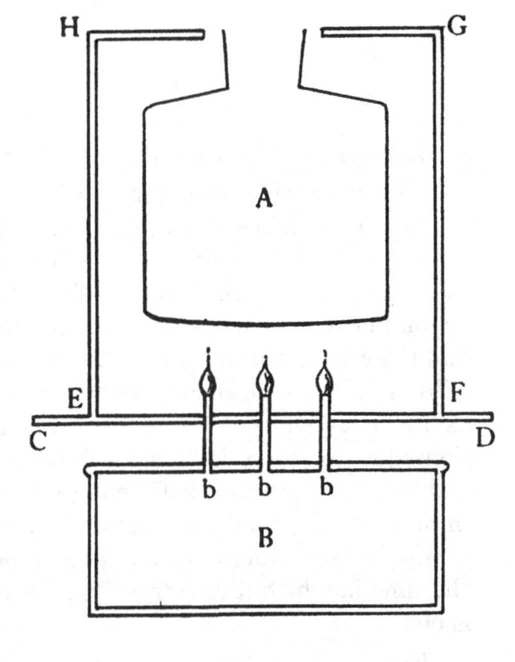

The results of the exper. are as follows. The bottle heated by 4 wicks of the lamp. [See Table on following page.]

The weight of the water before heating was 15·696 oz., it was taken off the lamp at $1^{hr.}\ 40^{m.}\ 15^{s.}$ and was then found to have lost 1·469 in boiling.

It must be observed that on account of the length of the tube of the therm. which was out of the bottle the heat which was called 195° was in reality about [1]96°·1; that which was called 165° was 165°·7 and 95° was really what it appeard to be. Therefore, if we judge by the time in which the water rose from 95° to 195°, the quantity of heat communicated

EXPERIMENT I.

Heats shown by thermom.	Time hrs.	mins.	secs.	Difference secs.	Heats shown by thermom.	Time hrs.	mins.	secs.	Difference secs.
95°	0	55	2		155°	1	7	48	
				63					64
100		56	5		60		8	52	
				65					65
105		57	10		65		9	57	
				65					63
10		58	15		70		11	0	
				63					68
15		59	18		75		12	8	
				62					70
20		60	20		80		13	18	
				66					65
25	1	1	26		85		14	23	
				61					70
30		2	27		90		15	33	
				63					68
35		3	30		95		16	41	
				64					68
40		4	34		200		17	49	
				64					66
45		5	38		205		18	55	
				65					65
50		6	43		210		20	0	

to the water in   seconds was just sufficient to raise it 1°; if we judge by the time in which it rose from 165° to 195° the heat communicated to it in   was just sufficient to raise it 1°. From the rate at which the water heated, I think we may conclude that it began to boil at 1. 20. 22 consequently it continued boiling 19^m. 53^s.. If we suppose that the lamp communicated as much heat to the water while boiling as it did before, and if we estimate that from the time in which it rose from 95° to 195°, it follows that the heat lost by converting 1·469 of water into steam was as much as would raise 15·696 of water   degrees by the thermom. Consequently there is as much heat lost by converting any quantity of water into steam as is sufficient to raise that quantity of water 982°, or, in other words, there are 982° of cold generated by converting water into steam. If we estimate the quantity of heat communicated to the water by the time in which it rose from 165° to 195° the quantity of cold produced is 960°.

Exp. 2. Repeated in the same manner.

Exp. 3. Tried in the same manner except that the water was heated by 7 wicks.

Exp. 4. Tried in the same manner except that the water was heated by only 2 wicks.

Exp. 4 [5?]. The water being heated with 4 wicks but the cover *EFGH* being taken off.

I have been informed that Dr. Black has observed that in distilling water, the water in the worm-tube is heated thereby much more than it would be by mixing with it a quantity of boiling water equal to that which passes through the worm. Upon this principle I made some experiments to determine how much heat is generated by converting water from the state of an elastic to that of a non-elastic fluid.

The still used in these experiments was of copper of the usual shape and held about oz. or gallons. The worm-tub was coverd with a wooden cover made to fit close by means of list: it had a hole in it by which I could let down a thermom. into the water and had a piece of wood fastend to it by which I could stir the water without taking off the cover. The tub held upwards of 2300 oz. and weighd 712·2: the worm weighd 269·5, and the cover and stirrer 156. The thermom. was kept in the water all the while the experiment was trying and care was taken to immerge it to such a depth that very little of the ☿ in the tube should reach above the cover.

SKETCH OF THE OTHER EXPERIMENTS

The heat produced by the condensation of the vapours of boiling water by a mean of several experiments tried in the foregoing manner was about 920°. So that it seems likely that there is just as much heat produced by the condensation of steam into water as there is cold by the changing of water into steam.

An attempt was made to find whether any cold was generated by the emission of fixed air in dissolving alcaline substances in acids. The way I tried it was by finding how much more heat was produced by saturating sope-leys [potash solution], spirits of sal ammoniac made with lime [ammonia solution] and lime slaked with water (all which substances contain no fixed air) with spirit of salt than by saturating the same substances saturated with fixed air; that is solutions of pearl ashes [potassium carbonate], the mild spirits of sal ammoniac [ammonium carbonate] and whiting [calcium carbonate] mixed with water in the same acid. By a comparison of the experiments it seemed that the cold generated by the emission of the fixed air was sufficient to heat a quantity of water equal in weight to the fixed air emitted about 1000 or 1700 degrees.

EXPERIMENTS TO SHOW THAT BODIES IN CHANGING FROM A SOLID STATE TO A FLUID STATE PRODUCE COLD AND IN CHANGING FROM A FLUID TO A SOLID STATE PRODUCE HEAT

With regard to water and ice this may be proved by the same kind of argument as was used in page 34 [of the MS.] from the long time required to thaw ice or to freeze water. There is also a very curious experiment (I believe of Mairan's[1]) which shews that heat is produced by the freezing of water. Put a thermometer into a vessel of water, shut it up close from the air, and expose it to the cold. It will bear cooling some degrees below the freezing point without freezing: then on agitating the water it will begin to freeze and the thermometer in it will immediately rise to the freezing point.

It is evident according to this hypothesis that on the water beginning to freeze, the thermometer in it ought to rise and should continue to rise to the freezing point unless the water was coold so much below the freezing point as to harden into solid ice before it was heated up to the freezing point, and it is evident that it ought not to rise higher than the freezing point for when it is come to that point the water will cease to freeze.

What is the cause that water bears to be coold below the freezing point and then immediately begins to freeze on agitating it or dropping a little bit of ice into it, etc.? I do not at all know, but there are many other instances in nature of things of a like kind.

I made an experiment to determine the quantity of cold produced by the changing of snow into water. It was done by dissolving a given quantity of snow in warm water. The cold produced seemed to be about 170 degrees[2]. There seemed no difference between the cold produced by snow and by the same quantity of ice.

It is well known that on mixing salt or many other substances with snow or ice the snow dissolves, and a great increase of cold is produced. There can be no doubt that this increase of cold is owing to the melting of the snow. I made some experiments to determine the quantity of cold produced by mixing snow with the following substances; namely, a solution of sea-salt, a solution of pearl ashes, spirit of wine and *aqua fortis*. The

[1] [M. de Mairan was Secretary of the French Academy and wrote a number of philosophical treatises; among them Dissertations on Ice, on Phosphori, and on the Aurora Borealis. He died in 1771, aged 93.]

[2] [This is a fair approximation to the truth.]

quantity of cold generated was not very different from that produced by dissolving snow in warm water.

I find also that cold is generated by the melting and heat by the hardening of spermaceti. The cold produced by the melting of spermaceti is sufficient to cool a quantity of water equal to it in weight about 70 degrees, and nearly the same degree of heat is produced by the hardening of spermaceti. It was tried by putting cold spermaceti into hot water and hot spermaceti into cold water. Spermaceti in cooling loses its fluidity at the heat of about 115°[1]. There is very little difference between the heat at which it ceases to be perfectly fluid and that at which it begins to be not at all fluid.

[Addendum.] Heat with which bees-wax melts. Therm. being immersed in melted bees-wax and kept constantly stirred about, the time[s] by the watch at which it arrived at different heats were as follows: heat of room supposed near 60°.

	min.	secs.		
	39	15	200°	
1.30	40	45	190	
2.0	42	45	180	
2.25	45	10	170	
2.45	47	55	160	
1.20	49	15	155	
1.40	50	55	150	Began to harden round edges
1.45	52	40	145	
1.50	54	30	143	
2.0	56	30	142+	A great deal hardend, remainder
8.30	65	0	142−	rather of a syropy consistence
12.0	77	0	141	
8.0	85	0	140	Much the greatest part hardend, re-
7.0	92	0	139[2]	mainder of consist. of very thick syrop

Some Tin and Lead were melted separately in a crucible and a thermometer put into them and sufferd to remain there till they were cold. The thermometer coold pretty fast till the metal began to harden round the edges of the pot. It then remained perfectly stationary till it was all congealed, which took up a considerable time. It then began to sink again. On heating the metal with the thermometer in it, as soon as the metal began to melt round the sides, the thermometer became stationary as near as I could tell at the same point that it did in cooling and remaind so till it was intirely melted.

On putting a thermometer into melted bismuth, the phenomena were the same, except that the thermometer did not become stationary till a good deal of the metal was hardend, unless I took care to keep the thermo-

[1] [Spermaceti melts between 106° and 120° F., depending on its purity.]

[2] [Beeswax melts between 144° and 151° and solidifies between 142° and 146° F. Cavendish's observations are therefore accurate.]

meter constantly stirring about. It then remaind stationary till it was almost all hardend. I do not know what this difference between Bismuth and the 2 other metals should be owing to except to its not transmitting heat so fast as them. I forbear to use the word conducting as I know you have an aversion to the word, but perhaps you will say the word I use is as bad as that I forbear. I did not suffer the thermometer to remain in the bismuth till it was hardend.

The heat at which {Lead, Bismuth, Tin} loses its fluidity is {612°, 510, 443[1]}.

All the following mixtures except the first differ considerably from the 3 simple metals in the manner in which they harden in cooling, as they begin to abate of their fluidity in a heat considerably greater than that in which they grow hard; whereas in the simple metals I could not perceive any difference between the heat in which they ceased to be perfectly fluid and that in which they hardend. In the trials I made the thermometer was kept constantly stirring about from the time the metal began to abate of its fluidity till there was a great deal hardend round the sides of the pot, and the rest had acquired as great a degree of stiffness as would suffer me to stir the thermometer about without breaking it, when I took it out. As soon as the metal began to abate of its fluidity the thermometer began to sink extremely slow in comparison of what it did before, and continued to do so till it was taken out, so that I think there can be very little doubt but what these metallic substances generate heat in hardening as well as the simple metals.

I think it seems likely that the reason why these mixtures begin to abate of their fluidity in a greater heat than that in which they harden is that the metals of which the mixture is composed begin to separate as soon as the heat is not sufficient to keep the mixture quite fluid. This is confirmd by the following experiment. The mixture of equal quantities of lead and tin was melted over again and sufferd to remain quiet till cold. It was then cut in two: the specific gravity of the upper piece was 8·801 and that of the lower 9·031, so that the upper piece appears to contain much less lead than the lower.

Mixtures	Heat at which it abates of its fluidity	Heat at which it grows rather stiff
lead 2, tin 3	351°	
„ 1 „ 1	379	362°
lead 1, bismuth 1	278	258
tin 1, bismuth 1	289	275
tin 1, bismuth 1, lead 1	275	251
„ 3 „ 5 „ 2	236	211

[1] [The true melting points of lead, bismuth and tin are variously stated by different authorities. Modern observations are, respectively, 617° F., 507° F. and 450° F (Heycock and Neville). Both the melting and solidifying points are considerably affected by slight traces of foreign metals.]

THOUGHTS CONCERNING THE ABOVE MENTIOND PHENOMENA

There are several of the above mentiond experiments which at first seemd to me very difficult to reconcile with Newton's theory of heat, but on further consideration they seem by no means to be so. But to understand this you must read the following proposition.

[THE END]

BOILING POINT OF WATER

At the Royal Society, April 18, 1766

Experiments made to determine how much the height of the boiling point is affected by the water boiling fast or slow, or by the bulb being immersed in the water or only exposed to the steam when the whole thermometer, except a small part of the tube near the boiling point, is inclosed in a vessel of water shut up in such a manner as to leave no more passage than what is necessary to carry off the steam and consequently when almost all the ☿ in the tube is nearly at the same heat as that in the ball. The experiments tried first by a very quick thermometer with a cylindrical bulb about 2½ inches long, the top of the bulb about 6 inches below the boiling point. The brass scale to it being divided to 20ths of an inch and furnishd with a nonius. 1·175 inches on the scale answering to 3° on the thermometer, or 1 inch being equal to 2·55 degrees.

		Division on scale	Difference in degrees
The water reaching not quite to the bottom of the cylinder and boiling	very gently	1·43	0
	pretty fast	1·425	+ ·013
The top of the cylindrical bulb barely coverd with water boiling	gently	1·22	+ ·54
	faster[1]	1·255	+ ·45

The first column is the division on the scale which the thermometer stood at; the 2nd is the difference of height expressed in degrees, the mark + signifying that it stood higher. For example, it stood $\frac{54}{100}$ of a degree higher when just immersed in the water and boiling gently than when not at all immersed and boiling gently.

Experiments of the same kind made with another thermometer, the ball of which is 15 inches or about 250 degrees below the boiling point. A small

[1] It is not owing to any mistake in writing it down that the ☿ appears to stand a little lower when the water boild fast than when slow, as it was taken notice of at the time as extraordinary.

brass scale divided to 40ths of an inch being fastend to the tube to shew the alterations of the height of the boiling point. 10 degrees of the thermometer are equal to 22 divisions on the scale.

	Division on scale	Difference in degrees
The ball dipt a little way into the water in an open vessel filled almost to the top and boiling moderately, the tube held inclined so as to be as little heated as possible by the steam of the water	from 3 to $4\frac{1}{2}$	− 3·86 − 3·18
Tried in a close vessel like that in which the former thermometer was tried, only deeper, namely 23 inches, with only 3 or 4 inches of water in it, so that the ball was out of water	$10\frac{3}{4}$	− ·34
The water rising 3 inches above the ball and boiling	gently $11\frac{1}{2}$ faster $11\frac{5}{8}$	0 + ·05
The water rising about 13 inches above the ball and boiling	gently $12\frac{1}{2}$ faster 13	+ ·45 + ·68

The first of these thermometers was tried by me on April 16 in the same manner as above related.

Tried in the pot 23 inches deep water about even with the top of the cylinder of the thermometer or 19 inches deep and boiling	gently 1·32 faster 1·30	+ ·13 + ·18
Tried in the same pot with little water in it and consequently the cylinder a great height above the water boiling	gently 1·545 faster D°	− ·45
In the shallow pot or that in which it was tried on April 18. Water just covering cylinder. Water boiling	gently 1·37 faster 1·345	0 + ·06
In the same pot with the cylinder intirely out of the water: water boiling	gently 1·53 fast D°	− ·41

N.B. If the chimney was uncoverd the thermometer immediately sunk several degrees and rose to the same point as before on putting it on again.

Tried again the deep pot water rising a little above the top of the cylinder and boiling fast	1·32	+ ·13
The same pot: water rising $2\frac{1}{2}$ inches above the top of the cylinder and boiling fast	1·32	+ ·13

The 2nd of these thermometers had also been tried on April 17. A piece of paper with divisions on it answering to degrees being pasted on the tube. Tried in the deep-pot.

	Degrees on paper	Difference in degrees
Water not rising so high as the ball and boiling	gently 1·4 fast D°	− ·55
Water rising a little above the ball and boiling	gently 1·95 fast D°	0

		Degrees on paper	Difference in degrees
Water rising 12 inches above the ball and boiling	gently	2·5	+ ·55
	fast[1]	2·5 or 2·7	+ ·55 or ·57

It appears from hence that if the vessel in which the thermometer is kept is sufficiently close:

First, there is scarce any sensible difference in the height whether the water boils fast or slow.

2nd. The thermometer stands about ½ a degree lower if the ball is exposed only to the steam than if it is immersed a little way in the water.

3rd. When the ball is exposed only to the steam there is no sensible difference whether it is raised a great way above the surface of the water or but a little.

4th. It stands about ½ a degree higher when the ball is 12 inches below the surface of the water than when it is very little.

5th. It stands rather higher if there is a great depth of water below the ball than if there is but little supposing the ball to be immersed in the water to the same depth in both cases.

Trials of boiling point of different thermometers at Royal Society

	Length	Degree at ball	
A thermometer of Bird marked *S*	...	− 40	213
,, of Adams	...	...	213¾
,, of Nairne	6	− 10	213 +
,, of Bird marked *SS*	9½	− 20	212
Large one of Ramsden's unfinished largest ball bott. mark	...	...	213 +
Bird's without mark	11¼	− 30	213½ +
Nairne's No. 5	15	− 70	215¼
— No. 3	16½	− 50	213½
Ramsden's Mr C.	12·5	0	212⅓

Bar. in N[orfolk] L[ibrary] at	begin	30·12	
	end	30·11	mean 213·1
Therm. in D°		55	
Bar. in garden supposed		= 30·1	
Therm. in D°		= 69	

By means of all thermometers height 212° is same as if it had been adjusted in steam when bar. was 29·41.

[This little paper is of interest, not only from the intrinsic importance of the subject, which at that period in the history of thermometry was, of course, considerable, but also as an illustration of a custom which was then rapidly dying out, namely for the Fellows to make experiments in common

[1] The ☿ was unsteady, dancing slowly up and down.

in the Apartments of the Royal Society—one of their number being appointed to undertake the manipulative part, the operator in this particular case being Cavendish. In the early days of the Society this experimenting in common was the usual practice, more importance being attached to it than to the communication and reading of papers.

The "trials" in this case had their origin, probably, in an attempt to elucidate the cause of the difference in the indications of thermometers made by the best "artists" of that time, and which was suspected with good reason to be due to a diversity of practice in determining the upper fixed-point.]

THEORY OF BOILING

There are 2 species of evaporation: 1st that which is performed with a less heat than that of boiling water; and 2ndly that which is called boiling.

The first species of evaporation is owing intirely to the action of the air, and has been very well explained by Mr. Le Roy, who has shewn that air is capable of dissolving a certain quantity of water, just as water does salt; and that when it has acquired that quantity, it is incapable of dissolving more; and that the quantity which it can dissolve is different according to its heat. It follows from hence that water must evaporate very slowly in vessels almost closed or communicating with the outer air only by a long narrow pipe, unless it is heated enough to boil. For when the air within the vessel has absorbed as much water as it can dissolve, no more can evaporate till some more of that air is changed for fresh.

The 2nd species of evaporation or boiling may be performed without any assistance from the air. Its phenomena seem to depend on 4 principles. First, that water as soon as it is heated ever so little above that degree of heat which is acquired by the steam of water boiling in vessels closed as in the experiments tried at the Royal Society, is immediately turned into steam, provided it is in contact either with steam or air: this degree I shall call the boiling heat, or boiling point. It is evidently different according to the pressure of the atmosphere, or more properly to the pressure acting on the water.

But 2ndly, if the water is not in contact with steam or air, it will bear a much greater heat without being changed into steam, namely that which Mr. De Luc calls the heat of ebullition.

3rdly, steam not mixed with air as soon as it is coold ever so little below the degree of heat acquired by the steam of water boiling in vessels closed as above-mentiond is immediately turned back into water.

4th. There is a great quantity of heat lost by the changing of water into steam; and a great quantity of heat acquired by the condensing of steam into water.

My father many years ago made some experiments which seem not only to prove the truth of the 1st and 3rd principles, but also to show the heat of the boiling point answering to much smaller pressures of the atmosphere than can be found otherwise. Namely, he took a barometer with a ball at top and filled it well with ☿; he then introduced into it a small quantity of water well purged of air; on which the surface of the ☿ immediately stood considerably lower than it did before; though on inclining the tube so that the ☿ rose into the ball, the bubble of air left was very little greater than before the introduction of the water. The surface of the ☿ was not sensibly more depressed when the quantity of water introduced was large than when it was just visible, after making allowance for the weight of the water, and was always the same in the same heat of the air, but was very different in different heats; namely, when the air was at 78° it was 0·92 of an inch; when at 30° only 0·16[1].

It appears from hence that when the pressure of the atmosphere on water is diminished to a certain degree (which is different according to the heat of the water) it is immediately turned into steam, provided its continuity is broke, that is, provided it is in contact with steam or air; and is immediately reduced back to water on restoring or increasing the pressure. Or, in other words, that whenever the heat of the water is increased ever so little above a certain degree (which degree is different according to the pressure of the atmosphere) it is immediately turned into steam; and is immediately restored back to its former shape on diminishing the heat. My father never tried this experiment with greater heats than those of the

[1] [Among the Chatsworth MSS. is an interpolation table calculated by Cavendish, from the results of measurements made in conjunction with his father on the Tension of Aqueous Vapour, and which are referred to above. They appear to have been made about 1757 and are based upon a number of observations over a considerable range of atmospheric temperature and probably, therefore, at various seasons of the year. The following excerpt from the table shows the values as compared with the observations of Dalton and Regnault at the corresponding temperatures, and affords evidence of their degree of accuracy.]

Comparison of Lord Charles Cavendish's observations on Vapour Tension of Aqueous Vapour with those of Dalton and Regnault.

	Tension in inches of Mercury		
	Lord C. Cavendish	Dalton	Regnault
°F.	—	—	—
75	·84	·85	·87
70	·70	·72	·73
65	·58	·62	·62
60	·49	·52	·52
55	·41	·44	·43
50	·33	·37	·36
45	·28	·32	·30
40	·24	·26	·25
35	·20	·22	·20

atmosphere; so that though there was the utmost reason to think that this heat at which water was turned into steam was the same as that acquired by the steam of water boiling as above mentiond, yet there was no very direct proof of it. But since the writing of this paper I have made an experiment, which I think puts the matter out of doubt.

AbBG is a bent glass tube of the size of an ordinary barometer tube, with two balls *A* and *B*, each about $1\frac{1}{2}$ inch in diameter. The ball *A* and the tube as far as *b* is filled with ☿, with a little water in the ball *A*. This glass is exposed to the steam of boiling water, or to the water itself in the glass vessel *CED*, coverd with the tin cover *CD*, with a chimney *F*; the tube *BG* being passed through a hole in the cover and secured with cork; and some woollen cloth being placed between the cover and the top of the glass vessel to make it fit the closer.

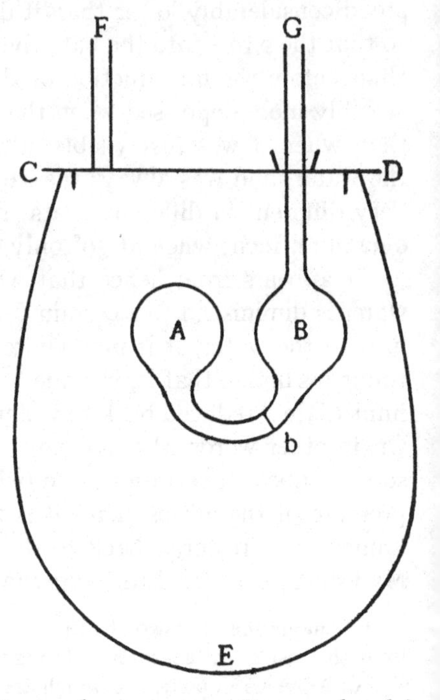

The event was that when the balls were raised above the surface of the water, so as to be exposed only to the steam, that then on giving the apparatus a little shake, so as to break the continuity of the water in the ball, the ☿ descended pretty quickly but not instantaneously, till it stood pretty exactly on a level in the 2 balls, where it remaind as long as the water continued boiling. But if the balls were sunk below the surface of the water, the surface of the ☿ in the inner ball *A* stood $\frac{6}{10}$ or $\frac{7}{10}$ below that in the ball *B*.

The circumstance of the gradual sinking of the ☿ in this experiment seems at first to disagree with that part of the first principle, which says that as soon as the water is heated above the boiling point it is *immediately* turned into steam. But in reality it does not; for as soon as any part of the water is turned into steam, the remainder must (by the 4th principle) be coold thereby; and consequently no more can be turned into steam till it has received so much heat from the steam or water surrounding the ball *A* as to make up that loss.

This experiment shows also, that the heat at which in my father's experiments the ☿ in the barometer is depressed a given depth, is the true heat which the steam of water boiling in vessels closed as above-mentiond, would acquire when pressed by that weight of the atmosphere. Suppose, for example, that when the heat is 61°, the ☿ in a barometer with a little water at top stands $\frac{1}{2}$ an inch lower than in one well filled, as is really the case; it follows that if it was possible to try the heat of boiling water in a place where the pressure of the atmosphere was only $\frac{1}{2}$ an inch, the steam would be found (supposing the vessels to be closed as above mentiond) to acquire a heat of only 61°.

In the above-mentiond experiment the water in the ball *A* was partly, but not imperfectly, purged of air; so that though a small shake was sufficient to break its continuity, yet as the quantity of water in the ball was small, and as that was deprived of great part of its air, the quantity of air in it bore an extreamly small proportion to the capacity of the ball; I think I may safely say not more than $\frac{1}{6000}$ part. This I found by putting the ball into cold water as soon as it was taken out of the steam, so as to cool it as expeditiously as possible, and consequently so as to leave as little time for the air to be absorbed as possible, and then before it was quite cold, measuring the diameter of the bubble.

But I made another experiment with a much greater quantity of air in the ball, namely such that its bulk when cold was about $\frac{1}{24}$ part of the space occupied by the vapour in the ball. I then found that when the apparatus was exposed to the steam of the boiling water, the ☿ in the ball *A* was depressed $1\frac{7}{10}$ inch below the level. It appears, therefore, that in this case the change of the water into steam was so much assisted by the absorption of the air that it was turned into steam, though the pressure was too great by $1\frac{7}{10}$ inch to have sufferd it to acquire that form without the assistance of the air.

In this experiment the quantity of air was found in a more exact manner; namely I pourd in ☿ till it rose above the ball *B* into the tube *BG*; and then found how much the surface of the ☿ in the tube was depressed by extricating the air.

The truth of the 3rd principle is confirmed by the experiment mentiond in the report of the Committee, p. 823 for year 1777[1]; namely that the heat of the steam in a pot 27 inches deep and $4\frac{1}{4}$ in diameter was not sensibly greater near the surface of the water, than near the top of the pot, that is at more than 18 inches above the surface. For if the steam could bear a less heat than that shewn in this experiment without being condensed into water, the steam at the top of the pot could hardly fail of being cooler than that in the lower part; whereas if this principle is true, this circumstance is not at all extraordinary; for while there is any steam remaining uncondensed, it must be as hot [as] at the boiling point.

As steam appears to be incapable of sustaining a degree of heat at all less than what I call the boiling point, or, in other words, as it is condensed

[1] [The Report of the Committee appointed by the Royal Society to consider of the Best Method of Adjusting the Fixed Points of Thermometers; and of the Precautions necessary to be used in making Experiments with those Instruments. *Phil. Trans.* Vol. LXVII. 1777, 816.]

into water the instant it is coold ever so little below that point; so in like manner it seems reasonable to suppose that water whose continuity is broke is incapable of sustaining a heat at all greater than the boiling point, or that it is turned into steam the instant it is heated at all above that point.

The truth of the 2nd principle has been sufficiently established by Mr. De Luc.

As to the 4th principle Dr. Cullen has sufficiently proved that heat is lost or cold generated by the first species of evaporation; and there is a circumstance daily before our eyes which shews that a vast deal of heat is lost by the 2nd species of evaporation. It is well known that water as soon as it begins to boil continues exactly or very nearly at the same heat till the whole is evaporated, which takes up a very considerable time. No reason, however, can be assigned why the fire should not continually communicate nearly as much heat to it after it begins to boil as when it wanted not many degrees of boiling; and yet during all this time it does not grow at all hotter. From hence we may conclude that there is a great deal of heat lost by the evaporation to compensate that communicated to it by the fire. For if no heat was lost by the evaporation, the water should either grow hotter and hotter the longer it boild, or else it should be intirely converted into steam immediately after it arrived at the boiling point.

On this principle I made some experiments to determine the quantity of heat lost by evaporation; namely by heating water in a metal vessel, in the form of a bottle, over a spirit lamp made so as to give as uniform a heat as possible and finding in what time it acquired a given number of degrees of heat before boiling and how much of the weight was lost by boiling a given time. It appeard that the quantity of heat lost by evaporation was 900 or 1000 degrees of Fahrenheit; that is as much heat was lost by evaporation as was sufficient to cool 100 times the weight of water evaporated 9 or 10 degrees.

Dr. Black found that the water in the worm tub of a still is heated much more than it would be by the addition of a quantity of boiling water equal to that distilled; and from thence computed the quantity of heat generated by the condensation of steam. I am not acquainted with the result of his experiment; but I repeated it myself and found the quantity of heat generated to be about 900°. So that there seems to be as much heat generated by the condensation of steam as there is cold by the production of it, and both to be about 900°.

From some experiments I have made the quantity of heat lost by the first species of evaporation seems to be much the same as by the 2nd species.

Mr. De Luc has also given a very clear proof that heat is lost by the 2nd species of evaporation in Art. 1062.

By the help of these principles the chief phenomena of boiling water may be readily explaind. When water is set on the fire and begins to boil, the lamina of water in contact with the bottom of the pot is heated till either small particles of air are detached from it, or till bubbles of steam are produced by ebullition. As these particles or bubbles ascend, the water in contact with them, if at all hotter than the boiling point is immediately turned into steam.

2ndly. These bubbles during their ascent through the water can hardly be hotter than the boiling point; for so much of the water which is in contact with them must instantly be turned into steam that by means of the production of cold thereby, the coat of water next to the evaporated coat of water, and which thereby comes in contact with the bubbles, is no hotter than the boiling point; so that the bubbles during their ascent are continually in contact with water heated only to the boiling point.

3rdly. Though the bubbles of steam can not be hotter than the boiling point, yet the water in general may be considerably hotter and most likely almost always is so in a small degree. For though the coat of water immediately in contact with the bubbles during their passage is not hotter than the boiling point, yet the rest of the water has not time to communicate much of its heat to that coat before the bubble is past. For this reason when the water boils with a vast number of small bubbles its heat ought in general to exceed the boiling heat less than when it boils with large bubbles succeeding one another slowly.

The excess of the heat of the water above the boiling point is influenced by a great variety of circumstances. The quantity of air in the water has a very great influence; for the more air it contains, the less heat will the water in contact with the bottom be capable of receiving, and the greater number of bubbles will be discharged. It is this which seems to be the reason of the difference between water beginning to boil and long boild, and between pump water and rain water.

Rain water contains only a small quantity of air and that common air, much the greatest part of which is discharged before the water begins to boil. But pump water besides this contains a calcareous earth suspended [dissolved] by the means of fixed air; and the calcareous earth detains the fixed air in such manner that it cannot be all separated without a vast deal of boiling.

It seems likely, I think, that the excess of the heat of the boiling water above the boiling point should be greater when the heat is less violent and applied to a greater surface than when more violent and applied to a less surface; and also when the application of the heat is more uniform, like that of oil, than when irregular like that of a common fire, that is acting with greater intensity on one point than another.

4th. It was before said that the bubbles of steam at their issuing from the water can hardly be hotter than the boiling point, and that if the vessel is properly closed, as in the experiments at the Royal Society, the steam can nowhere be colder than that point, and therefore the heat of the steam must be the same in all parts of the vessel.

5th. From what has been said it should seem that steam must afford a considerably more exact method of adjusting the boiling point than water. I do not see indeed any cause which should produce an alteration in the heat of the steam except the vessel being not shut sufficiently close and there being more or less air discharged from the water. To understand this it must be considerd that air is capable of dissolving a certain quantity of water, which is greater the greater is its heat. Now it seems likely that if the air is heated almost to the boiling point, it may be capable of absorbing a weight of water many times greater than itself. Suppose, for example, that air heated to within $\frac{1}{8}$ of a degree of the boiling point is capable of dissolving 100 times its weight of water; and suppose that the quantity of air discharged from the boiling water is $\frac{1}{100}$ part of the weight of the steam; or, in other words, that the vapours discharged consist of 1 part of air to 100 of steam, I think it seems likely that so much more of the water in contact with the bubbles will be turned into steam than would be if no air was discharged, that the bubbles and coat of water in contact with them will be coold $\frac{1}{8}$ of a degree below the boiling point, instead of being exactly of that heat, as they would otherwise be; and consequently the heat of the steam in all parts of the pot will be $\frac{1}{8}$ of a degree below the boiling point[1].

As to the heat of the water, it will be less on 2 accounts: 1st as it will approach nearer to the heat of the bubbles as was before said; and 2ndly as the heat of the bubbles will be less.

If the cover does not fit close, but lets in a little air, and the vapours within the pot thereby consist of one part of air to 100 of steam, I think it seems likely that the heat of the steam should be $\frac{1}{8}$ of a degree below the boiling point, as before. For the steam will bear being coold to that degree without being condensed; and in all probability the evaporation from the surface of the water will be so much increased as to cool the steam to that point. But the heat of the water ought in all probability to be very little affected thereby, except close to the surface.

But if the vessel is quite open, the surface of the water will most likely be many degrees cooler than the boiling point; and therefore it is likely that the water may be sensibly cooller than the boiling point, even to a considerable depth below the surface.

It should seem, therefore, that the heat of the boiling water should be considerably more regular in close vessels than in open ones. For the vessel being close or open can affect the lower parts of the water no otherwise than by the effect which it has on the water at the surface. Now in close vessels the water at the surface must in all cases be exactly at the boiling point; but in open vessels it will be much cooller, and its heat will be very different according as the surface is more or less exposed to the air, and as the water boils faster or slower.

[1] What is said in this paragraph is confirmed by the experiment in which a good deal of air was left in the ball *A*.

The hypothesis which I said I had about the cause of the difference of the heat of boiling and ebullition is as follows.

I suppose that within a certain very minute distance the particles of water repel each other; and that beyond that distance they attract each other; and that the repulsive force increases and diminishes as the heat increases and diminishes; but that the attractive force is either the same in all heats, or that it diminishes as the heat increases, or at least that it increases in a much slower proportion than the repulsive force; and also that the distance to which the attraction extends is many times greater than that to which the repulsion extends.

Thus let *PAap* (fig. 2) represent the section of a flat plate of water. Let *Nn*, *Mm* etc. represent parallel planes. Let *MG* and *GF* be equal to the distance to which the repulsion extends; and let *GN* and *MeE* equal that to which the attraction extends; that is, let a particle be repelled by any particle whose distance from it is less than *GM* or *GF*, and attracted by any whose distance is between *GM* and *GN* or between

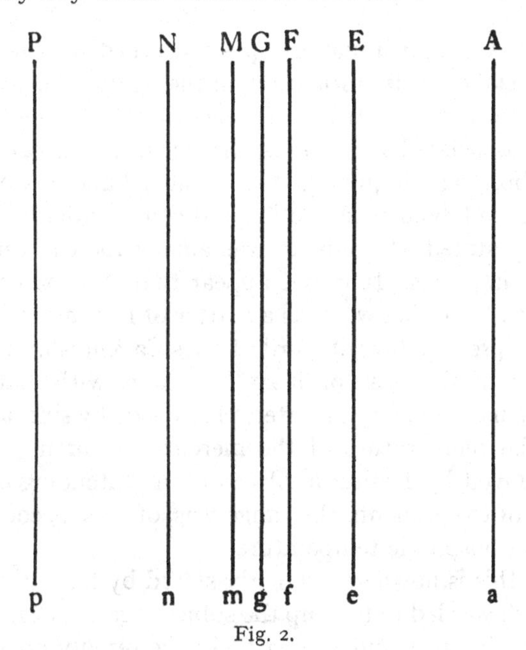

Fig. 2.

GF or *ME*. Then does the repulsion of the thin lamina *GFfg* on *GMmg* tend to separate the part *GAag* from *GPpg*; but it is counteracted by the pressure of the atmosphere on *Aa* and *Pp*; and also by the attraction of *GEeg* on *MNnm*. Suppose now the part *FAaf* to be taken away, so that the plane *Ff* may now become the outside surface of the water. Then is the force which tends to separate the lamina *GFfg* from *GPpg* the same as before; but the force which counteracts it is less; for the counteracting force is only the attraction of *GFfg* on *MNnm*, added to the weight of the atmosphere; and the attraction of *GFfg* on *MNnm* is evidently less than that of *GEeg* on the same. Whence it appears that the repulsion of *GFfg* and *GMmg* on each other may be sufficient to separate *GFfg* from *GPpg*, when *Ff* is the outside surface of the water, though it is not when there is a sufficient thickness of water on each side of it; and consequently that the repulsive force of the particles may be sufficient to separate a thin lamina from the outside of the water, and turn it into steam, though it is not sufficient to separate the water in the middle.

The phenomena of capillary tubes shew that the attraction of water to glass, and in all probability to metals also, is greater than that of water to water; so that it should require a still greater heat to produce a separation in that part of the water which is in contact with the sides of the pot, than in that which is at a little distance from it.

TENSION OF AQUEOUS VAPOUR

Among the Cavendish manuscripts concerned with heat is an account of a long and elaborate investigation on the relation between the temperature and elastic force of steam. The account is complete, and, together with certain associated matters, is written out in great detail on about 85 quarto folios. The inquiry involved many hundreds of measurements and considerable labour in reduction and computation.

As already stated, the subject was among the earliest of Cavendish's experimental inquiries. It would appear that the observations which he associates with his father were, as a matter of fact, made by himself under the direction, presumably, of Lord Charles Cavendish. They consisted in comparing the height of an ordinary barometer with that of a barometer containing a small quantity of water, placed side by side, and noting at the same time the temperature of the mercurial column—a method subsequently employed by Kämtz in observations extending over some years. In both sets of experiments the range was, of course, confined within the extremes of atmospheric temperature.

Although this is nowhere explicitly stated by him, it may be surmised that Cavendish was led to take up the subject again, not merely on account of its scientific interest, but by reason of the rapidly growing importance of the steam engine, and the desirability of obtaining accurate data concerning its theory. As a historical fact the problem was then engaging the attention of many experimenters, although, in all probability, Cavendish was unaffected by that circumstance, even if he was aware of it. Watt had attacked it as far back as 1764 and had made a series of observations in the winter of 1773–74 but nothing was publicly known of his work, and no account of it appeared until much later when he communicated the results to Robison who published them in his *Mechanical Philosophy*,

Vol. II. p. 29, *et seq.* Robison's own results were first stated in his article on Steam in the *Encyclopædia Britannica.* Zeigler's observations made with a Papin's digester, and published at Basle in 1759, were probably unknown to Cavendish who would certainly have compared the results with his own had he been aware of them. Betancourt's memoir *Sur la Force Expansive de la Vapeur* was not printed until 1792, and an account of Southern's experiments, made by direction of Watt, first appeared in Brewster's edition of Robison's Works.

There can be no question, therefore, that Cavendish's inquiry was wholly original and independent, and altogether uninfluenced by any previous or contemporary work on the subject. He seems to have begun the new inquiry in the spring of 1777. It occupied him during much of the following year and until the early part of 1779. The experiments are described under the two main sections of observations below and above 212° F., as the methods employed in the two series are dissimilar to a slight extent.

The observations above 212° F. are first dealt with. The arrangement used in this series is illustrated by the figure on [p. 356].

It is thus described: *AbBG* is a bent glass tube of the size of an ordinary barometer tube with two balls *A* and *B* each about 1½ in. in diameter. The ball *A* and the tube as far as *b* is filled with ☿ [mercury] with a little water at the top of the ball. This glass is exposed to the steam of boiling water, or to the water itself, in the glass *CED*, covered with the tin cover *CD*, with a chimney *F*, the tube *BG* being passed through a hole in the cover, and secured with cork, and some woollen cloth being placed between the cover and the top of the glass *CED* to make it fit the closer. A hole was also made in the cover through which one of the short thermometers used in the experiments for the boiling-point [p. 356] could be passed.

The capacities of the bulbs and tube were ascertained by calibration with known weights of mercury, and the results are embodied in a table showing the capacity of the tube at successive intervals of an inch from the top of the outer tube.

The calibration was necessary as the amount of mercury in the apparatus varied during the course of the experiments, more or less being added, or withdrawn from time to time, as required, in order to keep the mercury in the shorter or inner limb at approximately the same level, and also below the top of the outer tube.

Some preliminary experiments were made to determine the effect of small or varying proportions of air to vapour upon the tension, the observations being made at the boiling point of the water in the "pot," which was heated, in this case directly over the fire, or in a sand tray. The difference of level in the limbs was estimated by means of "a level ruler, with small sliding brass rule." The volume of the air and its ratio to the known content of the bulb were attempted to be ascertained by measuring the size of the bubble left after the apparatus was cooled by immersion in cold water, according to "the rule given in 'Experiments on Barometers'[1] for finding the bulk of bubbles."

Cavendish was aware that the method might not be strictly applicable to this case. He says: "But it must be observed that that rule was found by experiments in which no water was in the ball, and therefore can not hold good in experiments in which there is."

The bulk of residual air would be affected also to a small extent by its solubility in the cooled water in the bulb, but as the quantity of water was very small, its effect on the bulk was probably not greater than the unavoidable error of measurement. A number of experiments were made and the results, although only approximate, when combined with the known effects of determinate volumes of air in depressing the barometer, afforded some information as to the influence of varying amounts of air

[1] The rule here alluded to is given as follows in a short paper in MS. headed "Barometers."

Let the figure represent the section of the ball and bubble. The curvature of *egb* may be supposed nearly the same whatever is the diameter of the globe provided *eb* is small in respect of *ad*. The content of *eabf* is to that of globe as $3fb^4 : ad^4$.

If ratio *efgb* to 1 glob. inch be called r

$$fg = \frac{r}{3fb^2}.$$

1 glob. inch ☿ = 1800 grains.

Hence it appears that when diameter of bubble $= \left\{\begin{array}{l}\cdot 66\\ \cdot 44\\ \cdot 245\\ \cdot 16\end{array}\right.$

$$efba = \left\{\begin{array}{r}53\cdot 5\\ 10\cdot 6\\ 1\cdot 0\\ \cdot 185\end{array}\right\} \text{ grains and } efgb = \left\{\begin{array}{r}17\cdot 4\\ 10\cdot 2\\ 2\cdot 53\\ \cdot 833\end{array}\right\} \text{ grains}$$

$$fg = \left\{\begin{array}{l}\cdot 03\\ \cdot 039\\ \cdot 0315\\ \cdot 0241\end{array}\right\} \text{ inch.}$$

Hence I conclude that the value of *gf* answering to bubbles of different diameters is as follows:

diameter bubble	·10	·13	·16	·2	·25
gf	·018	·021	·024	·028	·032
diameter bubble	·3	·4	·5	·6	·7
	·037	·039	·039	·034	·028

The number of grains of ☿ equal in bulk to bubble

$$= 1350 \times fg + eb^2 + 337 \times \frac{eb^4}{ad},$$

eb and *ad* being expressed in inches; and the bulk of the bubble is to that of the ball as

$$\frac{3fb^4}{ad^4} + \frac{3fb^2 \times fg}{ad^3} \text{ to } 1.$$

on the apparent elastic force of steam. It will be seen later that Cavendish subsequently detected how the effect of the air could be determined, thereby anticipating a discovery we associate with Dalton.

The observations for measurement of vapour tension were now begun. A small quantity of water, "well purged of air" by boiling, was introduced into the bulb *A*, and the apparatus immersed in a solution of pearl-ash, contained in the vessel *CED*. "The heat was estimated by the standard thermometer *S* of Nairne." It had been standardised by the method prescribed by Cavendish's Committee of the Royal Society already referred to. The readings of the temperature were also corrected, when necessary, for that portion of the stem of a lower temperature than that of the bulb, by the method introduced by Cavendish and adopted in the Report of the Committee (*Phil. Trans.* 1777, LXVII. 816).

"The different height of the ☿ [mercury] in the tube and in the ball *A* being estimated by the weight of ☿, by means of the foregoing measures," i.e., the measurements embodied in the calibration table before mentioned.

The following excerpts from the tables of details of the first series of observations will serve to illustrate their characte1.

May 6, 1778. Barometer = 29·93.

	Weight of ☿ in apparatus	Heat, °F.	Level of ☿ in tube below top	Level of ☿ in ball below top of tube (1)	Force with which vapour is compressed (2)
	16·968 oz.	229·6	17·1	29·854	42·559
	"	230	16·8	·872	42·877
tried again	18 oz.	229·1	17·2	·710	42·315
	"	231·8	15·6	·803	44·008
	"	232	15·35	·819	44·274

N.B. (1) Expansion of ☿ by heat is allowed for in this computation.

(2) About 7½ inches of the column of ☿ by which the vapour was compresse was heated to 212°, in consequence of which the pressure of that column was diminished about ·125 inch, according to De Luc's rule, which is allowed for in this computation.

July 4 [1778]. The same experiment repeated with a stronger solution of pearl-ash. Bar. = 30·08.

	Weight of ☿ in apparatus	Heat °F.	Level of ☿ in tube below top	Force with which vapour is compressed
	24·812 oz.	247·4	·95	58·59
		247·7	·64	58·91
		248	·4	59·16
The solution in *CED* weakened by addition of water		246	2·5	56·95
		246·3	2·05	57·43
		246·5	1·85	57·64

The whole quantity of ☿ in the tube was much heated: perhaps to 120°.

For the higher temperatures an oil-bath was employed. The oil was contained in a copper vessel provided with a cover.

"The oil," we read, "was stirred by a semi-circular horizontal tin-plate, filling near ½ the area of the pot, and raised and sunk perpendicularly. It was heated by a spirit lamp of 9 wicks, kept burning during the observations, more or less of the wicks being lighted so as to make the oil heat and cool slowly. The pot [*CED*] was encompassed with a piece of thin brass, covered with flannel and paper, forming a kind of flue. The 2 standard thermometers A. and Br. for 300° were placed in the vessel on different sides of the ball of the apparatus, their balls being on a level with the center of the ball of apparatus and near 4 inches below the cover of the pot. 2 thermometers were placed, the ball of one near the middle of that part of the column of ☿ in the standard thermometers which was out of the pot, the ball of the other near the middle of that part of the column of ☿ of apparatus which was out of the pot."

As the radiation from the heated "pot" would doubtless affect the volatility of the spirit of wine in the lamp below, which might prove troublesome and even dangerous, it was diluted with water so as to minimise this risk. "The specific gravity of the spirits used for the lamps when the thermometer was at 63° was ·8803 [circa 66 per cent. of alcohol by weight], the flame being found not to increase much by the heating of the spirits, when they were no stronger than that."

The following table shows the general character of the measurements. Readings of the thermometers and of the levels of mercury in the apparatus were taken as the temperature slowly rose and again as it fell on cooling and the mean taken, these being subsequently corrected for the lower temperature of the emergent columns.

Sat. Aug. 1, 1778. Bar. at 4.40 p.m. = 30·32, glass ☿ = 11 oz. 13 dwt. 15 gr., heat tube thermom. = 81°, heat column ☿ = 71°.

Time	Thermometers		☿ below top	Pressure
	A.	Br.		
hr. m.	°	°	°	
5.32 P.	230·7	230·5	15·75	44·050
33	1·3	1·	15·6	44·209
34	1·8	1·5	15·25	44·789
36	2·3	2	14·8	45·054
37	2·7	2·5	14·4	45·476
38	3·2	3	14	45·898
40	3·2	3	13·65	46·268
42	2·8	2·5	13·95	45·950
43	2·3	2	14·35	45·528
44	1·8	1·5	14·75	45·106

Heats 2°·5 in 6 min. = 1° in 2·4 mins.
Cools 1°·5 in 4 min. = 1° in 2·7 ,,

	A.	Br.
	°	°
Corr. for heat tube thermom.	1·94	2·14
Supposed true heat corrected	2·17	2·12

For the higher temperatures and pressures a similar apparatus was constructed with a much longer outer tube. It is distinguished in the account as the 11 foot tube and was subsequently employed in an inquiry on the thermal expansibility of air. It, together with the ball, was carefully calibrated by mercury as before, the details being set out at length in the account.

With this apparatus, we read,

> the experiments were made just in the same manner as the foregoing, except that a tin cover was placed...about 3·8 inches above the top of the pot to prevent danger in case the glass [tube] was to break. A thermometer was placed to show the heat of that part of the thermometer and tube which was between the cover and pot, and another to shew the heat of that above the cover.

In all 17 series of observations were made involving several hundred readings of the thermometers and levels of the mercury. It is unnecessary to give further examples of their character: they were exactly similar to those already given. In each case a series of readings was taken at about the particular point determined upon first as the temperature slowly rose and again as it fell, a mean of the whole being taken. Thermostats were unknown in Cavendish's time[1]: his method of controlling the temperature of the bath was to extinguish one or more of the wicks of the spirit lamp. The methods of reduction and computation, and the details of corrections for inequalities of temperature, expansion of mercury, etc. are described in full and the final values are given in the form of synoptical tables.

The apparatus "for trying the force of steam at heats less than 212°" is shown in the figure on p. 368. It is thus described:

> *Bbβ* is a ball and bent tube with a brass plate *Cc* cemented on it to which may be fastend a tin pot *Aa*, which, to make the containd water keep its heat longer, is inclosed in a wooden box with wool between. The same stirrer is used as in former experiments, and 2 thermometers are placed in it, their balls being on a level with the center of ball [*B*]. The tin pot is supplied with hot water from another vessel with a cock, the pipe of which enters into *Aa*. The vessel *Aa* also is furnished with a cock and a pipe to carry off the overflowings. *Dd* is a graduated brass frame with a slider *E* with a nonius and a brass cock *G*, to observe the height of the ☿ in the leg *bβ* or outer leg. In order to observe the ☿ in the inner leg, an additional cock *Ff* was used lying upon the former. To see whether this additional cock[2] was horizontal, a box *Mm* with a board *N* floating in ☿ was used with 2 cocks *P* and *p* fastend upon it. It was observed what division the nonius stood at when each of the cocks *P* and *p* were even with the top of the

[1] [The word "thermostat" appears to have been introduced into the literature by Heeren in 1834 and was applied to an arrangement devised by him for regulating the heat applied to crucibles, beakers, etc. over a spirit lamp. Its use in the sense of an arrangement to maintain a constant temperature is of much later date.]

[2] [Cock—the style or gnomon of a dial: the needle or index of a balance: a pointer or indicator, e.g. weather-cock.]

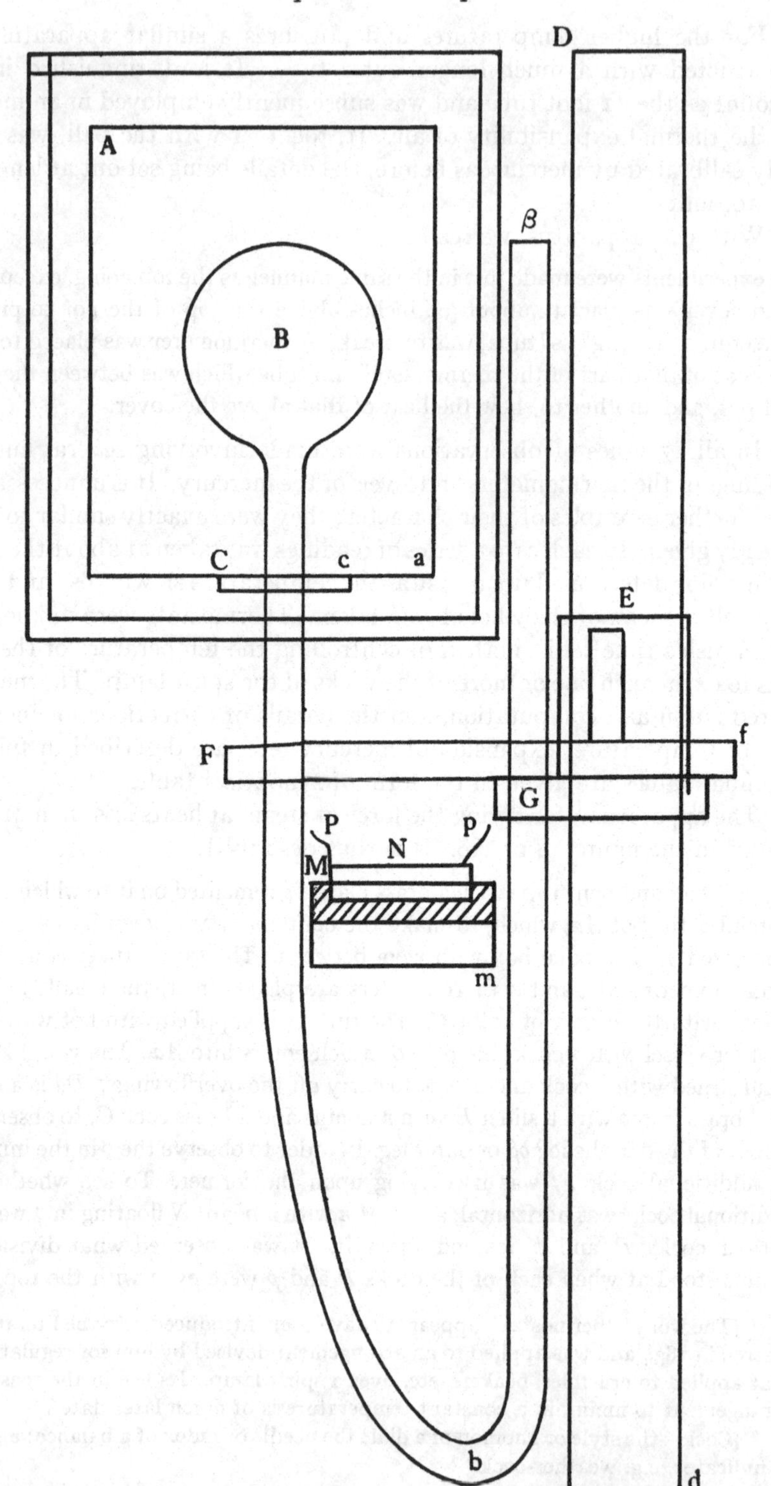
D
A
B
C
c
a
β
E
F
f
G
P
p
N
M
m
b
d

additional cock; the box was then turned end for end and the observations repeated by which means it was found how much one of the floating cocks was higher than the other, and how much that part of the additional cock, answering to one floating cock was higher than that answering to the other.

A slight scratch was made on the inner leg and its position on the graduated scale when the apparatus was in position was ascertained: its distance from a similar scratch near the top of the outer tube was thereby determined.

The bulb and tube were calibrated, as before, by mercury

"The degree of $\left\{\begin{matrix}52\\50\end{matrix}\right.$ on thermometers $\left\{\begin{matrix}L\\S\end{matrix}\right.$ [the thermometers employed] was even with the tin cover and 78 even with the top of the wooden cover. It is supposed, therefore that each thermometer was of the full heat as far as 65 [degree mark], and from thence of the temper of the air." When the apparatus was set up and "the ball and tube well filled [with mercury] without any water in them on Dec. 9, the ball and tube was measured [calibrated] and at the same time it was tried how well it was filled. The heat during the experiment was 50½. The height of the barometer did not alter sensibly during the time."

By comparisons with the barometer, the depression from any vapour [or air] contained in the ball in one experiment "cannot much exceed ·0011 and in the latter ·0034."

After the introduction of the boiled water, observations were made, *de die in diem*, during the latter part of December, 1778, and the early part of January, 1779. Neither Christmas Day, nor New Year's Day, nor Sundays were allowed to interrupt the work. For some reason, which is not apparent, the greater number of the observations were made in the late afternoon and evening, and often far into the night, so that the readings must have been taken by candle- or lamp-light. At that period Cavendish's laboratory was over the stables of his father's house in Great Marlborough Street, and it is possible that the apparatus was so placed that the comparatively feeble light of a mid-winter's day was insufficient for the purpose.

As an example of the observations an extract may be given from the table of measurements made on Dec. 25th. A column is given showing the results of Lord Charles Cavendish's measurements (L.C.C.) at the corresponding temperatures.

Friday, Dec. 25, 1778.

Time	Thermometers		☿ in outer leg	Barometer	Tension	
	L	*S*			H.C.	L.C.C.
	°	°				
6.57 p.m.	71·6	71·7	4·690	30·781	·740	·747
7.2 ,,	71·3	71·5	·680	·782	·731	·739
8 ,,	62·3	62·3	·485	·783	·530	·532
13 ,,	62·2	62·3	·480	·784	·526	·530
27 ,,	52·3	52	·325	·785	·364	·368
32 ,,	52·3	—	·323	·786	·363	·368

As in the first series, the whole of the observations are set out in the form of tables, with all the details of the computations and corrections, so that each step of the work may be followed.

The following table, compiled from the final results as stated by Cavendish, shows the corrected values for temperature and tension, expressed, of course, in Fahrenheit degrees and inches of mercury. As an indication of their accuracy they are compared with Regnault's values.

Tables showing the tension of Aqueous Vapour between 11° *and* 308°, *expressed in degrees Fahrenheit and inches of mercury, as observed by Cavendish, compared with the corresponding values determined by Regnault.*

° F.	Cavendish	Regnault	° F.	Cavendish	Regnault
11·0	·065	·071	122	3·573	3·621
12·0	·074	·074	132	4·690	4·755
22·2	·118	·119	142	6·074	6·183
26·8	·140	·145	152	7·816	7·925
31·0	·174	·174	162	9·934	10·07
36·7	·195	·216	172	12·531	12·74
44·8	·284	·288	182	15·647	15·92
52·0	·384	·388	224·2	37·56	38·19
62·0	·548	·556	229·4	41·30	41·90
71·6	·760	·762	234·0	45·03	45·76
82·0	1·067	1·094	239·3	49·57	50·29
92·0	1·467	1·500	244·3	54·21	55·16
102·0	1·998	2·039	248·6	58·37	59·42
112	2·683	2·725			

Tension of Aqueous Vapour at temperatures above 212° F.

"Experiments made with 11 foot tube. Observed heats (corrected) answering to different pressures: taken from a mean of the observations themselves" (Cavendish).

° F.	Cavendish	Regnault
239·05	50·20	50·08
248·72	59·64	59·52
258·31	70·87	70·31
268·38	83·02	83·32
277·97	97·14	97·40
287·88	113·37	113·91
298·22	132·22	133·33
308·07	153·25	154·40

The first approximately accurate measurements of the tension of aqueous vapour published in England were made by Dalton, by a method "recommended by an elegant simplicity." An account of them appeared in 1805 in Vol. v. of the *Memoirs of the Literary and Philosophical Society of Manchester*. They extend from the freezing to the boiling point of water under ordinary atmospheric pressure. From the rate of increase in tension between these limits Dalton computed the values at temperatures above 212° F., but, as Young from theoretical considerations, and Ure, from the results of direct observation, pointed out, Dalton's method of extrapolation was wholly erroneous, furnishing numbers some 25 or 30 per cent. too low at 50° or 60° above the ordinary boiling point. Cavendish's method, on the other hand, although far more laborious than that of Dalton, and involving many more measurements and much time and trouble in evolving the necessary corrections, afforded in his hands, as the comparison with Regnault's values proves, much more accurate results. This is especially true at the higher temperatures where the values were obtained by direct experiment, and are therefore independent of all extrapolation on hypothetical assumptions.

But, in the absence of other data, Dalton's *Tables of the Force of Aqueous Vapour*, in spite of the criticisms of Biot, were long regarded as authoritative, especially in this country, and as such appeared in practically every text book during the first third of the 19th century, when they were superseded by the work of Dulong and Arago. It is to be regretted therefore, that Cavendish should have refrained from publishing the results of his labours of 1777–1779 on this subject. "Erroneous observations," said Darwin, "are in the highest degree injurious to the progress of Science, since they often persist for a long time. But erroneous theories, when they are supported by facts, do little harm, since everyone takes a healthy pleasure in proving their falsity."

It may be doubted, however, whether this last remark is applicable to Cavendish. Nowhere does he manifest "a healthy pleasure" in dealing with a false theory or unsound hypothesis. He studiously avoided controversy; he seemed, in fact, almost nervously afraid of it. In this, as indeed in other respects, he resembled Newton, who in consequence of the objections which were raised to his theory of light and colours, confessed that he had been imprudent in publishing it, since by catching, as he said, at the shadow he had lost the substance, namely his own peace and quiet. Throughout his long life there was only a single occasion on which Cavendish was, for a brief time, led into controversy when he was provoked to reply to certain strictures by Kirwan; and he is quite apologetic to the Royal Society for the necessity of troubling it with polemical matters. If we are driven to seek a reason for his withholding from publication the results of his long and laborious work on aqueous vapour tension, it may possibly be found in the fact that it would inevitably bring him into conflict with Deluc. As the manuscript shows, the direction of the investigation was to some extent affected by Deluc's theory of the thermometric scale, and the reduction of the observations is encumbered with calculations which prove the unsoundness of Deluc's views. Deluc's "rules" led to a progression of elasticity and a scale of temperature which were found to be wholly inconsistent with observation. Cavendish sets out the numbers establishing this fact, but he nowhere comments on the disparity between theory and experiment. He simply leaves the numbers to speak for themselves.

Deluc, as is well known, held that a mixture of equal weights of hot and cold water had a temperature lower than the arithmetic mean of the initial temperatures. Thus, according to Deluc, a mixture of equal weights of water at 212° F. and 32° F. had the temperature of 119° instead of 122°. Cavendish, for years past, was aware that such a statement was erroneous. His early experiments, made with all the precision and scrupulous care of modern methods, would have enabled him to refute it. But so far as can be gathered, he took no steps to make known the facts. He contented himself with satisfying his own curiosity in the matter. As it was, the error persisted down to the time of Dalton, who not only shared it, but sought to support it by the assumption of a false hypothesis concerning the law of the thermal expansion of liquids.

Accompanying the account of these experiments, and paged up with it, are descriptions of two other inquiries which, it may be presumed, Cavendish considered as connected with, or supplementary to, the main investigation, and which had this been published would, in all probability, have appeared at the same time. The descriptions are headed "Compressibility of Air" and "Expansion of Air." The experiments on the compressibility of air appear to have been made during the last ten days of November and first week of December, 1778, before the final set of experiments on aqueous vapour tension at temperatures below 212° were undertaken. The object of the inquiry, apparently, was to ascertain how far air saturated with moisture obeyed Boyle's law, and whether the partial pressures of the air and aqueous vapour could be differentiated.

The experiments on the thermal expansion of air were not made until the December of the following year.

COMPRESSIBILITY OF AIR

After former experiments with 11 foot tube were finished it was washed with water and on Nov. 20 [1778] was measured while wet. Thermometer about 57°. Thermom. about 57°.

Quant. ☿ [oz. dwt. gr.]			Weight of ☿ which by former measures would fill ball and tube to that height [oz. dwt. gr.]
9. 13. 22	in short leg, 2·52 from bott.		
10. 10. 16	to a point 2·00 from last		
11. 19. 18	in long leg 3·09 from bott.	= 127·51 from top	12. 0. 2
64. 12. 14	,, ,, ,,	= 9·13 ,,	64. 13. 13
67. 11. 18	,, ,, ,,	= 2·00 ,,	67. 11. 18

Top tube = 130·6 from bottom, bottom = 3·75 from mark.

On Nov. 21 about 7.30 P. all the ☿ was pourd out of tube and 1. 0. 17 of ☿ pourd in

surface water	1·09 above bottom
surface ☿ in short leg was	·89 ,,
surface ☿ in long leg	1·71 ,,
surface water in long leg	5·44 ,,

It was then set in large glass of water and at 11.0 P.

$$\begin{array}{r} \text{Surf. ☿ in long leg was} \\ \text{Thermometer in water \quad ,, \quad ,,} \end{array} \left\{ \begin{array}{l} \text{1·72 above bottom.} \\ \text{5·50 \quad ,,} \end{array} \right.$$

Water being at that time 48°·3. The surface ☿ and water in short leg could not be measured but we may conclude from former measure that they were $\left\{\begin{array}{l}1·08\\ ·83\end{array}\right.$ above bottom.

64 oz. 8 dwt. 9 gr. more ☿ was then pourd in making in all 64. 9. 2; and at 11.20 P. thermom. in water same as before.

$\left.\begin{array}{l}\text{Surface ☿}\\ \text{,, \quad water}\end{array}\right\}$ was $\begin{array}{c}4·00\\ ·61\end{array}$ below top, and at 11.36 P. $\left\{\begin{array}{l}4·12\\ ·60\end{array}\right.$ below top; thermom. being still the same as before.

Bar. at 12.10 P. (?A.) was 30·02 and next morning 29·98, heat ☿ therein being about 50°.

In the exp. with short col. empty space was = 10·298 oz. ☿: pressure was = press. atm. + ·84 + $\frac{3·58}{13·5}$ = 1·105. Height Bar. reduced to heat 57° was 30·04 and allowing for 10½ feet greater pressure 30·052. Therefore total pressure = 31·157 inches. Depression of barom. by water at heat 48°·3 = ·315. Therefore if all the moisture was separated from air it would have sustained pressure of 30·842, when of that bulk ☿ filling ball and tube to 4·12 from top = 66·707 oz. ☿ actually in = 64·454. Bulk water in short leg = ·081: therefore empty space in last trial = 2·17. The surface ☿ in ball when there is that quantity of empty space was found by trial to be 4·09 above bottom. Therefore pressure = pressure atm. + 126·48 + $\frac{3·52}{13\frac{1}{2}}$ − 4·09 = 152·751, the press. atm. being 30·04, and therefore pressure which it would have sustained if perfectly dry 152·44. Therefore the empty spaces are to each other as 4·746 to 1, and the pressures as 4·943 to 1.

Dec. 1, 1778, the experiment was repeated.

$\left.\begin{array}{l}\text{Surface water}\\ \text{,, \quad ☿}\end{array}\right\}$ in short leg = $\left.\begin{array}{c}1·6\\ 1·3\end{array}\right\}$ above bottom.

Surface ☿ in long leg = 1·68 above bottom, being then set in water surf. $\left.\begin{array}{c}\text{☿}\\ \text{water}\end{array}\right\}$ in long leg = $\left.\begin{array}{c}1·63\\ 2·04\end{array}\right\}$ above bottom and consequently $\left.\begin{array}{c}\text{water}\\ \text{☿}\end{array}\right\}$ in short leg = $\left.\begin{matrix}1\cdot65\\1\cdot35\end{matrix}\right\}$ above bottom, thermom. being 44° and bar. = 29·94. The ☿ in app. being then made = 59. 9. 17 $\left.\begin{matrix}☿\\\text{water}\end{matrix}\right\}$ in long leg was $\left\{\begin{matrix}16\cdot28\\15\cdot87\end{matrix}\right.$ below top, thermom. being the same. The air was then got out of the ball without losing any of the ☿, the ☿ in apparatus being then made = 63. 6. 13 surf. water was 11·85 below top.

In the 1st experiment, empty space was 10·060 oz. Pressure was 30·23 ☿ filling ball and tube to 16·28 from top or 110·57 from mark is by 1st measure = 61·763, bulk water in short leg = ·084. Therefore empty space = 2·195. By expelling the air the empty space appeard = 2·203. The pressure was 140·20. The depression of the barometer is ·27; therefore the pressures which same quantity dry air would have sustained are as 139·93 to 29·96 or as 4·666 to 1 or as 10·060 to 2·156.

EXPANSION OF AIR

Dec. 19, 1779.

A bent glass like the figure was filled with ☿ by being placed in tin vessel with the capillary *A* and the end of leg *B* passing through holes in the cover. The air in the vessel was heated to about 90°. The ☿ was heated considerably hotter and was pourd in through a capillary funnel entering into *B*. When it was filled the capillary *A* was sealed and the greatest part of the ☿ in the outer leg pourd out. The glass was then fastend to the brass scale used in the preceding experiment [Tension of Aqueous Vapour], and the outer leg measured by pouring in ☿.

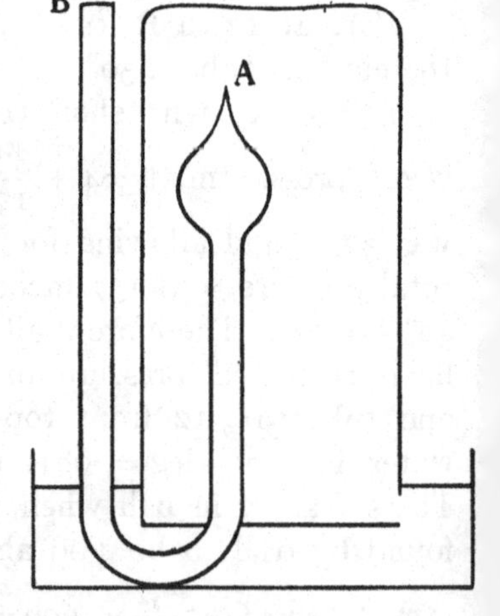

Glass empty = 7 oz. 8 dwt. 5 gr.

Division on scale answering to ☿ in outer leg	Glass filled with ☿ to that mark	Do. in decim. oz.	Diff. for 1 inch	Log. do.
2·895	29. 14. 22	29·746		
			·343	9·5351
8·41	31. 12. 18	31·637		
			·334	·5237
13·97	33. 9. 21	33·494		
			·321	·5061
20·59	35. 12. 8	35·617		

Heat [of] air about 50°.

Dec. 22. Some air was let into the ball in this manner: The ball was placed under a jar inverted to glass of water as in fig. the air being at 32° and the barom. at 29·13, and kept so for about an hour that the whole might be of the temper of the air. The glass jar was then taken off, the end of the capillary broke off[1] and the jar put on again. After it had staid so for about 15′ some ☿ was drawn out of outer leg by syphon and thereby some air let into ball. The jar was then immediately taken off and the capillary sealed. Measures were taken on different days with the apparatus fastend to the scale as before.

The height of ☿ in ball below the top of the capillary was taken by an additional cock lying on the other in the same manner as in the preceding experiment [on tension of aqueous vapour]. The height of the top of the capillary was found by resting a level ruler on the cock used for finding the height of the ☿ in the outer leg, and moving this cock till the ruler touchd the top of the capillary.

Day 1779	Glass with ☿ oz. dwt. gr.	Height ☿ in out. leg	Surf. ☿ in ball below capill.	Height top capill.	Heat
Dec. 23	16. 6. 16	6·76	2·41	18·225	38°
„ 24	20. 15. 13	11·435	2·22	18·223	36½
„ 25	22. 11. 0	13·535	2·155	18·225	32½
Jan. 1	24. 8. 4	16·26	2·095	18·216	37
„ 2	14. 5. 4	4·14	2·515	18·212	37
„ 6	22. 11. 6	13·71	2·14	18·20	37
„ 7				18·20	

Experiment made with this apparatus immersed in glass vessel of water 21½ inch. deep and 9 in. diam., the appar. being immersed to such a depth that the top of the ball was 2 or 3 inches under water. The surface of ☿ in outer leg and the divisions on scale being seen through the glass. A tin cover was fitted to the glass and the standard thermom. *S* let down through a hole in the cover so that its ball was on a level with the ball of the appar. The water was stirred in the same manner as in the preceding exper. and the glass was supplied in the same manner with hot water; the water in the glass was let off by a syphon. The brass scale was fastend firmly to a board by that part which reached above the vessel of water, and the vessel of water was put up without disturbing that fastening, so that the inclination of the brass scale to the vertical line (if any) was the same as before it was immersed in the glass vessel, and consequently the point of the scale on a level with the top of the capillary was the same as before it was immersed in the glass vessel.

[The actual measurements of the thermal expansion of the air were begun in the late afternoon of Dec. 25, 1779, and continued on that day until late at night. They were resumed on the following morning, and were continued on successive days until January 5th.

In principle the method was similar to that adopted in the measurements of vapour tension. It involved reading the heights of the mercury in the two limbs of the apparatus, and noting the height of the barometer

[1] The height of the ☿ in the outer leg was such that no air enterd into the ball thereby, but on the contrary some run out.

at each reading, together with the temperature of the water in which the apparatus was immersed, known weights of mercury being added or withdrawn when necessary, so as to keep the levels in the two limbs at convenient heights for measurement. The range of temperature was between 33° and 150°. All the details of the observations are given, together with those of their computation and reduction. Cavendish's object was not merely to ascertain the thermal expansion of air, but to determine the effect of aqueous vapour on its expansibility. Accordingly on January 6, after it had been ascertained that the weight of mercury in the apparatus agreed with the calculated amount to within 4 grains, some water "purged from air" was introduced into the apparatus containing the air, and on the evening of January 7, the observations were resumed and continued day after day until January 12, when the weight of mercury in the apparatus was checked as before. In this second series the range of temperature was from about 32° to 127°.

The results of these observations showed, to quote Cavendish's words, that "the expansion of air by 1° is $\frac{1}{500}$ of its bulk at 38°$\frac{11}{12}$ or $\frac{1}{481}$ of bulk at 0° or $\frac{1}{511}$ of bulk at 50°."

He further found

> that the increase of pressure which perfectly dry air will sustain by the addition of water, supposing the bulk to remain unaltered, is equal to the depression of the barometer by water [vapour]. The depression of the barometer by water [vapour] at 32° = ·174. So that the increase of pressure which the air bore in this experiment, on the addition of water, fell short of the excess of the depression of the barometer at the observed heats above that at 32 by from ·04 to ·07, so that it should seem as if the moisture in the apparatus [and in the assumed dry air] before the addition of the water was sufficient to depress it from 4 to 7–100ths of an inch.

Of course, it need hardly be stated, attempts to determine the expansi bility of air by heat had been made prior to the time of Cavendish. Th necessity of knowing its amount and rate was perceived long ago, originally no doubt, on account of the use of air as a thermoscopic agent, as als from the recognition of its bearing on astronomical refraction and on th determination of heights by barometric measurements. These attempt date from the end of the 17th century and are connected with the names amongst others, of Amontons, Nuguet and La Hire in France, Hawksbe Shuckburgh and Priestley in England, Musschenbroek in Holland, and D Luc in Switzerland. But the results obtained by these observers wer widely different, partly from the imperfection of the methods employe and partly from the unsuspected influence of moisture in the air or th apparatus. Cavendish's results, although only approximately accurate, a certainly nearer the truth than those of the majority of his predecessor and compare not unfavourably with those of Dalton and Gay Lussac mad nearly 30 years afterwards. Cavendish's values were doubtless affected by the circumstance that the air, although cooled to 32° F. before its introduction into the apparatus was not wholly free from moisture. This was recognised by him, as he clearly states, and he gives an approximate estimate of the amount of moisture which was probably present. He had no means of thoroughly drying the air, as desiccating agents other than dry pearl-ash and filter paper were unknown, or at least unused, at that period.

But the interesting fact clearly appears that he definitely recognised that the increase in volume and pressure of a given bulk of air, which may be due to the moisture contained in it, is equivalent to the tension of aqueous vapour at the corresponding temperature, as measured by the depression of the barometer, whence the volume of the dry air may be ascertained. He thereby anticipated by nearly a third of a century, a discovery which we associate with Dalton.]

Expansion of different airs by heat. The app. consists of a barometer, ball [bulb], and tube about   inc. long[1] with a piece of a thermom. scale [of wood or ivory] fastend to it and inverted into a glass of ☿ [mercury] which is placed in jar of water in which the water by means of syphon is kept always at same height.

[He begins by assuring himself that the thermometer scale is accurately divided, by measuring lengths of successive intervals of 10°.]

Measure of divisions.

		Diff.
40° to 50°	·67 inc.	
		·66
60	1·33 ,,	
		·65
70	1·98 ,,	
		·65
80	2·63 ,,	
		·64
90	3·27 ,,	
		·63
100	3·90 ,,	

☿ [mercury] required to fill ball and tube to a given division.

	☿	Div.	Diff. of ☿	Diff. of div.
	6644	97·8		
	6828	73	184	24·8
	7070	41	242	32
Do. repeated.	☿			
	6648	98		
	6854	70		
	7076	40		

[1] The length is not stated. If measured, it was omitted to be inserted in the space left vacant for it.

Whence the capacity [in grain measures] answering to even divisions is as follows:

		Diff.			Diff.
100°	6634		60°	6928	74
		73			
90	6707		50	7002	74
		73			
80	6780		40	7076	74
		74			
70	6854				

These exp. agree with the first to less than 6 gra.

By mean, 1 inch of tube holds 111 of ☿. Diam. glass into which it is inverted varies from 2·63 to 2·65: by mean 2·64; and 1 cyl. inc. ☿ of that diam. = 18800, and therefore 1 inc. tube holds as much as ·0059 of the glass and 10 div. hold as much as ·00383 of glass.

When ☿ in tube is at 40, ☿ in glass is 1·03 below that divis. and the surface of the water in the large glass [jar] is 8 in. above the ☿ in the glass, and the pressure of this 8 inches = ·59 in. of ☿; and therefore when ☿ in tube is at 40, the press. on air is less than height barom. by 1·03 − ·59, or ·44 inc. and therefore pressure at different divisions is the less than height barom. by quant. in following table:

Div.	Quant.	Diff.	Div.	Quant.	Diff.
40	·44				66
		67	80	3·09	
50	1·11				64
		67	90	3·73	
60	1·78				63
		65	100	4·36	
70	2·43				

[Based on the means of the differences respectively in capacities and pressures, the foregoing tables are then expanded as follows:]

Div.	Capacity	Prop. parts	
100	6634	1	7·4
		2	14·8
90	6707	3	22·2
80	6780	4	29·6
		5	37·0
70	6854	6	44·4
60	6928	7	51·8
		8	59·2
50	7002	9	66·6
40	7076	10	74·0

Div.	Press.	Div.	Press.	Prop. parts	
40	·44	75	2·760	1	0·065
				2	·13
45˙	·775	80	3·090	3	·195
50	1·110	85	3·410	4	·26
				5	·325
55	1·445	90	3·730	6	·390
60	1·780	95	4·045	7	·455
				8	·520
65	2·105	100	4·360	9	·585
70	2·430			10	·650

[The "airs" experimented upon were: common air; nitrous air (nitric oxide); fixed air (carbon dioxide); heavy inflammable air (presumably gas from charcoal); dephlogisticated air (oxygen); phlogisticated air (presumably nitrogen); inflammable air (presumably hydrogen). No details are given concerning the preparation of these "airs," nor how they were introduced into the apparatus; presumably they were collected as prepared over water in the customary manner and were therefore saturated with moisture. Nothing is stated of the way in which the "airs" were heated nor are any details given of the mode in which the observations were made. Presumably the heating was effected by pouring hot water into the jar in which the thermometer was suspended and adjusting the constant level of 8 ins. from the level of the reservoir of mercury by the syphon. The experiments seem to have begun in the early part of December (year not stated) and carried on, with interruptions, into the following January. The following tables are transcribed exactly as they appear in the notes.

The results are uncorrected for the expansion of the glass "ball" (bulb), and of the mercury at the different temperatures. It is unlikely that these corrections were overlooked by Cavendish but he probably considered their influence was negligible in view of the much larger errors of observations.]

Exp. with common air.

	Th.	Div.	Bar.	Capac.	Press.	Log. capac.	Log. press.	Sum. log.	Log. expans.	Expans.	Diff.	Do for ·1°o	Exp. for 1°
c. 5	80[1]	40	28·85	7076	28·41	8498	4535	3033	·0324	1·078			
	66	50	—	7002	27·74	8452	4431	2883	·0174	1·041	37	26	
	52	60	—	6928	27·07	8406	4325	2731	·0022	1·005	36	26	1/390
	41	68	28·87	6869	26·57	8369	4244	2613	·9904	·978	27	25	

[1] The number is set down 85 in the minutes but must certainly be a mistake for 80.

Nitrous air.

	Th.	Div.	Bar.	Capac.	Press.	Log. capac.	Log. press.	Sum. log.	Log. expans.	Expans.	Do. if unif.	Expan. for 1°
c. 7	83·5	44·5	29·37	7043	28·63	8470	4568	3046	·0059	1·014	1·017	
30 P.	60·7	62·5	—	6910	27·43	8395	4382	2777	·9790	·953	·959	
c. 8	54·2	67·6	29·52	6872	27·25	8371	4354	2725	·9738	·942	·942	
50 A.	43	76·5	—	6806	26·67	8329	4260	2589	·9602	·913	·913	1/388
	60·7	64·3	—	6896	27·46	8386	4387	2773	·9786	·952	·950	
	68·3	58·5	—	6939	27·84	8413	4447	2860	·9873	·971	·977	
	89·5	40·5	—	7072	29·05	8495	4631	3126	·0139	1·033	1·033	
	68·3	58·5										

With fixed air.

	Th.	Div.	Bar.	Capac.	Press.	Log. capac.	Log. press.	Sum. log.	Log. expans.	Expans.	Do. if unif.	Expan. for 1°
	86·5	40·2	29·54	7075	29·09	8497	4637	3134	·0387	1·093	1·094	
. 8	66·4	57	—	6950	27·97	8420	4467	2887	·0140	1·033	1·038	
30 P.	46·3	71·8	29·56	6841	27·01	8351	4315	2666	·9919	·982	·982	1/358
P.	68	56	—	6958	28·05	8425	4479	2904	·0157	1·037	1·042	
	86·8	40	—	7076	29·12	8498	4642	3140	·0393	1·095	1·095	

Heavy inflammable air.

	Th.	Div.	Bar.	Capac.	Press.	Log. capac.	Log. press.	Sum. log.	Log. expans.	Expans.	Do. if unif.	Expan. for 1°
. 8	44	70	—	6854	27·14	8359	4336	2695	·9886	·974	·974	
P.	65·8	54·5	—	6969	28·16	8432	4496	2928	·0119	1·028	1·032	1/375
	84	40	29·57	7076	29·13	8498	4643	3141	·0332	1·080	1·080	

Dephlogisticated air.

	Th.	Div.	Bar.	Capac.	Press.	Log. capac.	Log. press.	Sum. log.	Log. expan.	Expan.	Do. if unif.	Ex fo
Dec. 31	94	40	29·9	7076	29·46	8498	4692	3190	·0526	1·129	1·129	
2.0 P.	93	40·8	29·9	7070	29·41	8494	4685	3179	·0515	1·126	1·126	
	68	63·5	29·93	6902	27·92	8390	4459	2849	·0185	1·044	1·054	
5.0 P.	45	80	30·00	6780	26·91	8312	4299	2611	·9947	·988	·988	
	68·8	63·3	30·00	6904	28·01	8391	4473	2864	·0200	1·047	1·056	
	70·2	62·4	30·00	6910	28·06	8395	4481	2876	·0212	1·050	1·060	
	94·5	40·6	30·01	7072	29·53	8495	4702	3197	·0533	1·131	1·130	

Nitrous air.

	Th.	Div.	Bar.	Capac.	Press.	Log. capac.	Log. press.	Sum. log.	Log. expan.	Expan.	Do. if unif.
7.40 P.	91·3	40·5	—	7042	29·57	8477	4709	3186	·0485	1.118	1·121
	69·5	59·5	30·04	6932	28·29	8409	4516	2925	·0224	1·053	1·059
	44	78·3	—	6793	27·07	8321	4325	2646	·9945	·987	·987
	44·5	78	—	6795	27·09	8322	4328	2650	·9949	·988	·988
	67·5	61·5	—	6914	28·16	8397	4496	2893	·0192	1·045	1·054
	92·5	40	—	7076	29·60	8498	4713	3211	·0510	1·125	1·125
	92	40·3	30·04	7074	29·58	8497	4710	3207	·0506	1·124	1·123
Jan. 1	67·5	58·7	29·79	6936	28·11	8411	4489	2900	·0197	1·047	1·054
10 A.	67	59	—	6935	28·10	8410	4487	2897	·0196	1·046	1·052

Phlogist. air.

	Th.	Div.	Bar.	Capac.	Press.	Log. capac.	Log. press.	Sum. log.	Log. expan.	Expan.	Do. if unif.
Jan. 7	86	40	29·76	7076	29·32	8498	4672	3170	·0413	1·099	1·099
0.10	65	58·2	29·72	6941	28·08	8414	4484	2898	·0141	1·033	1·042
1.20 P.	46·5	69·5.	29·69	6858	27·29	8362	4360	2722	·9965	·992	·992
	65	56·5	29·66	6954	28·13	8422	4492	2914	·0157	1·037	1·042
	85·3	40	29·66	7076	29·22	8498	4657	3155	·0398	1·096	1·097

Inflammable air.

	Th.	Div.	Bar.	Capac.	Press.	Log. capac.	Log. press.	Sum. log.	Log. expan.	Expan.	Do. if unif.
Jan. 1	67·5	40·3	—	7074	29·39	8497	4682	3179	·0164	1·039	—
9.30 P.	52·5	51·5	29·85	6991	28·64	8445	4570	3015			

More air taken out.

	Th.	Div.	Bar.	Capac.	Press.	Log. capac.	Log. press.	Sum. log.	Log. expan.	Expan.	Do. if unif.
Jan. 2	59·8	61·5	30·01	6917	28·13	8399	4492	2891	·0040	1·009	1·014
9.30 A.	42·6	73·3	30·01	6830	27·37	8344	4373	2717	·9866	·970	·969
	43	73·2	30·01	6830	27·37	8344	4373	2717	·9866	·970	·970
	60	61·5	30·02	6917	28·14	8399	4493	2892	·0041	1·040	1·015
	86	40·5	30·03	7072	29·56	8495	4706	3201	·0350	1·084	1·084

[All the coefficients of expansion are much too large, but the results must have at least served to indicate, as indeed Cavendish would appear to have recognised, that the rate between the limits he observed is practically uniform. Moreover, he was justified in inferring that it is sensibly the same for all gases, the difference between the several gases not being greater than that found on a repetition of the experiments on the same gas. With that meticulous sense of accuracy which was characteristic of him, he refrains, however, from explicitly stating this conclusion, but his claim to have demonstrated it is at least as valid as that of "Citizen" Charles, who also never published the results of his experiments, which seemed, if we may judge from Guy Lussac's statement, to have been made subsequently to those of Cavendish.]

EXPERIMENT PROPOSED FOR DETERMINING THE DEGREE OF COLD AT WHICH ☿ BEGINS TO FREEZE[1]

(Shewn to Sir J. BANKS in 1781.)

The way I would propose, is to fill the cylindrical glass of one of [the] thermometers with the ivory scales, to the top of the swelled part, so as to cover the ball of the inclosed thermometer and keep it in a freezing mixture till almost all the ☿ in the cylinder is frozen; and observe what heat is shewn by the inclosed thermometer. This must evidently be the cold with which ☿ begins to freeze; for as in this case, the ball of the thermometer will be surrounded for some time with ☿, part of which is actually frozen, it seems impossible that the thermometer should be sensibly above the point of freezing ☿; and while any of the ☿ in the cylinder is unfrozen, it is impossible that it should sink sensibly below that point.

From the Petersburg experiments of freezing ☿ I think it appears clearly, that ☿ contracts in the act of freezing, or that ☿ takes up less room in a solid than in a fluid state, and that the very low degrees to which the thermometers sunk in that experiment, was owing to this contraction of the ☿ during freezing and not that they produced a degree of cold anything near equal to that shewn by the thermometer; and from some circumstances of the experiment, I have little doubt but that the degree at [which] ☿ freezes is less than 200° below nothing; for which reason I thought it unnecessary making the ivory thermometers reach more than 2 or 300 degrees below nothing.

If a thermometer is put into a glass of water, and exposed to a cold sufficient to freeze it, it will remain perfectly stationary from the time that the water begins to freeze, till it is intirely frozen; and will then begin to sink again. In like manner if a thermometer is dipt into melted tin or lead, it will remain perfectly stationary (as I know by experience) from the time that the metal begins to harden round the edges of the pot, till it is all hardend, when it will begin to descend again; and there seems no reason to doubt but what the same thing will obtain with regard to ☿.

The method which I would use in trying the experiment is as follows. Put the ivory thermometer prepared as abovementiond, into the freezing mixture, along with one of the thermometers with wooden scales; taking care that none of the mixture gets into the cylinder. Then as soon as the ☿ in the cylinder begins to freeze, the ivory thermometer will become

[1] [Cf. p. 57 and p. 145.]

stationary, but the wooden one will still continue to descend, on account of the contraction of the ☿ in it by freezing. Write down this stationary degree of the ivory thermometer, which as was before said, is the heat at which the ☿ begins to freeze; and keep the thermometers in the mixture till either the ivory thermometer again begins to descend, or till the wooden one becomes stationary. If the ivory thermometer begins to descend before the wooden one becomes stationary, take it out and see whether the ☿ in the cylinder is frozen. But if the wooden thermometer should become stationary before the ivory thermometer ceases to be so, as will most likely be the case, it will be necessary to refresh the freezing mixture or put the thermometers into a fresh mixture. For the wooden thermometer becoming stationary shows that the mixture is no longer cold enough to freeze ☿. In this manner you must proceed till either the ivory thermometer again begins to descend, or till you have reason to think from the time which it has continued stationary, and the very low degree to which the wooden one has sunk, that a great part of the ☿ in the cylinder is frozen.

It is not impossible but what the ivory thermometer may rise suddenly some degrees on the ☿ in the cylinder beginning to freeze, and then remain stationary. At least this is what happens on exposing a thermometer in a glass of water to the cold. But this would cause no inconvenience in the experiment as the point at which it remaind stationary would still be the freezing point.

The only circumstance I can think of which can prevent the success of the experiment (except that of not being able to produce a sufficient cold) is that possibly the ☿ in the ivory thermometer (as being in vacuo) may freeze with a less degree of cold than that in the cylinder, which is compressed with the whole weight of the atmosphere. I have no reason to think that this will be the case; but if it was it would intirely destroy the success of the experiment; but it would not lead to error; as the ☿ in the ivory thermometer would never become stationary. If this was found to be the case, it would be proper to break off the top of the stem of the ivory thermometer, so as to let the air into it; and try it over again.

Experiment 2. The above experiment gives the degree of cold, as shewn by a ☿ thermometer, at which ☿ begins to freeze. But it is not unlikely that it may require a much less alteration of cold to make a ☿ thermometer sink a given number of degrees, when coold almost to freezing, than when it is of a less degree of cold; and consequently that the real degree of cold at which ☿ freezes may be much less than that shewn by the thermometer. The best way I know of finding whether this is the case or not, will be as follows:

Take a thermometer with a ball of a very large diameter; and keep it in a freezing mixture till it is coold very near to that degree at which it is found by the former experiment that ☿ begins to freeze; then take it out, wipe off slightly, and as quick as you can, so much of the mixture as adheres to it, read off the division; and plunge it into a glass of ☿ of about the temper of the air in the room, and whose heat must be previously found; and keep it stirring about till you have reason to think that the thermometer and glass of ☿ are both of the same heat; and write down this heat, or the heat of the mixture as I shall call it. Next repeat the experiment in the same manner; only making the thermometer about as much warmer than the glass of ☿, as it before was colder; and find the heat of the mixture as before. Then if by this means you find that the difference of heat of the mixture and glass of ☿, bears a less proportion to the difference of heat of the thermometer and glass of ☿ in the 1st experiment, than in the 2nd, it will shew that ☿ expands more by a given alteration of heat, in a very great degree of cold than in a more moderate one. Suppose, for example, that in the $\left\{\begin{matrix}\text{1st}\\\text{2nd}\end{matrix}\right.$ trial, the heat of the ☿ in the glass was $\left\{\begin{matrix}-50^\circ\\-52^\circ\end{matrix}\right.$; the heat of the thermometer before plunging in $\left\{\begin{matrix}-153^\circ\\+31^\circ\end{matrix}\right.$; and the heat of the mixture $\left\{\begin{matrix}-70^\circ\\-30^\circ\end{matrix}\right.$; consequently the difference of the heat of the thermometer and glass of ☿ is $\left\{\begin{matrix}103^\circ\\83^\circ\end{matrix}\right.$; and the difference of heat of the mixture and glass of ☿ $\left\{\begin{matrix}20^\circ\\22^\circ\end{matrix}\right.$. Then the real difference of heat of the thermometer and glass of ☿, in the 1st experiment is to that in the 2nd as 20 : 22; that is as its effect in altering the heat of the mixture. Therefore if we suppose that the differences of heat shewn by the thermometer in heats greater than -53° are proportional to the real differences of heat, the real difference of heat of the thermometer and glass of ☿ in the 1st experiment is $83 \times \frac{20}{22} = 75^\circ\frac{1}{2}$; and therefore the true heat of the thermometer in the 1st experiment is $-50 - 75\frac{1}{2} = -125^\circ\frac{1}{2}$, instead of 153° as shewn by the thermometer.

N.B. It will be proper to repeat this experiment with very different heats, both of the thermometer and glass of ☿, in order to judge by the agreement or disagreement of the experiments the degree of accuracy which can be expected from them.

In degrees of heat between freezing and boiling water, the differences of heat shewn by a ☿ thermometer seem to be actually in proportion to the real differences. For if equal quantities of hot and cold water are mixt together, I am convinced that the heat of the mixture is very exactly the mean between the heats of the hot and cold water. Mr De Luc, indeed, thinks otherwise; but in several experiments which I made with as much accuracy as I could, and in different manners of trying the experiment, the event was always as I mention; and Mr Smeaton has also tried the experiment in a different manner, and with the same event.

[This statement appears to have been drawn up by Cavendish for the information of the President of the Royal Society in connection with a request from the Society that the Hudson's Bay Company might permit one of their servants to make the suggested observations.]

COLD PRODUCED BY RAREFACTION OF AIR

Feb. 25, 1783.

Over the cork of condensing glass of air pump was screwed a brass cap *A* as in fig., the bore of the hollow cylinder being not much more big enough to receive the ball of the thermometer *G* having a small hole at bottom by which the condensed air escaped on opening the cock and blew on the ball of the thermometer.

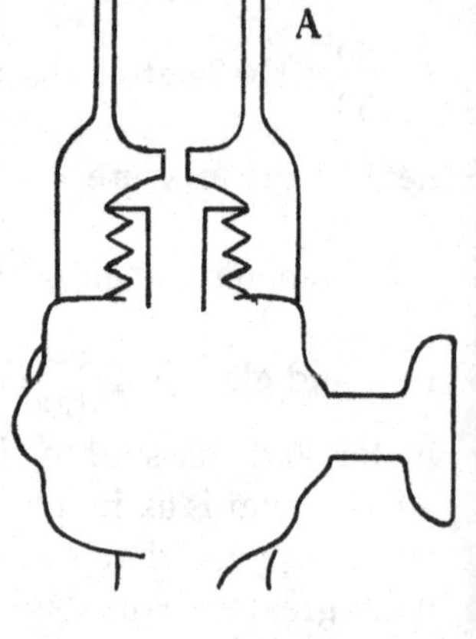

Having then forced one additional atmosphere into the receiver, the thermometer sunk 7 or 8° on letting out but began to rise again before the air was all run out.

A copper globe was prepared 12 inches in diameter to be used instead of condensing glass, and also a wooden cap like *A*.

Apr. 14. With small thermometer and brass cap *A* globe condensed to 30 inches, thermometer sunk 14°½, and a 2nd time 15°½. It was found that the air ran ½ out in 33″.

Apr. 15. Thermometer *G* sunk 14°, air being let out immediately after it was condensed, but being tried again after waiting 10′ before it was let out it sunk 15°½. Air run ½ out in 25″.

With small thermometer, sunk 15°½, waiting 10′: air run ½ out in 24″.

In order to see whether thermometer *G* or small thermometer coold fastest, they were both heated in hand to 88° and exposed to air [of] room which was about [1]. *G* cooled to 70° in 70″ and did the same in a 2nd trial: small thermometer coold as much in 50″ in 1 trial and in 60″ in another.

[1] [Not stated.]

HEAT AND COLD BY EXHAUSTING AND CONDENSING OF AIR

Tried with Nairne's thermometer: ball = ·28 inch in diameter: inclosed in condensing glass.

Time, March 10	Heat air	Heat therm. before	Do. after
8.40 P.		55°	49 on exhausting
9.30	55½	55	59 letting in air
9.40	56	55½	49 exhausting
11th 9.30 A.	48	48	52 letting in air

N.B. It rose about ½° out of this 4° after air was let in.

2.40 P.	55¼	56	62¼ condensing 1 atm.
3.10 P.	55½	57	51 letting out air

Tried with same thermometer in a thinner glass intended only for exhausting.

8.30 P.	55	53½	58 letting in air after having been long exhausted
9.35 P.	53	53	49½ on exhausting

N.B. It sunk near ½° after the exhaustion was finished.

11.15 P.	54½	53	57½ letting in air
12th 8.30 A.	47½	47½	44½ on exhausting
9.30 A.	50	49½	51½ gage sunk to 16¼
10.30 A.	51	50½	53 letting in remainder

Tried in same glass with Thermometer *G* ·25 in. diameter.

11.40	52	52	47 on exhausting

Being exhausted to within ·53.

0.0	52	52½	not altered on depressing gage by 1·4
0.6 P.	52	52½	54½ gage sunk ½ way
0.15 P.	52½	52½	55 letting in remainder of air

The same glass with same thermometer in it was exhausted and thermometer in it stood long stationary at 52°½. The large copper globe heated before fire was then screwed on, and a small thermometer placed in the side hole till it became stationary which it did at 95°. This thermom. was then taken out, the hole screwed up, and the air let out of the globe into the exhausted glass so that it was filled with air of 95° hot. The included thermometer rose to 57°½.

March 12. The same receiver being exhausted, with same thermom. in it, and kept so till it remained stationary at 49° and then air let in in common manner, thermom. rose 5°.

The mouth of one of the large globular glasses was stopt with a cork through which a piece of barometer tube passed reaching to near bottom of glass; a thermometer was also fastend to the cork so that the bulb was near the center of the glass. This glass was heated before the fire and then suspended over the receiver of the air pump in such manner that the tube enterd into a hole in a piece of wood screwed to the cock of the receiver and fitted tight into it. The receiver had been previously exhausted, and the inclosed thermometer stood stationary at 49°½; the warm air from the glass was then let into it, the thermometer in which was at 110°, on which the thermometer in the receiver rose to 54°½.

2 glass tubes were prepared with brass caps with a kind of gimbals fastend to them by which they could be stretchd. Both tubes were about ·22 in. bore, the glass of one being about ·11 and the other about ·03 thick. A thermometer was also prepared with a ball ·21 in. diam. and a scale like Ramsden so that it would go within these tubes. The thick tube with this thermom. inclosed in it, and each end stopt loosely by a cork, was then stretchd with a weight of $\frac{50{,}000 \times 31}{4{\cdot}18} = 375{,}000$ grains which is about as much as a globe of glass of that thickness 6 inches in diam. would be stretchd by 3 atmospheres. I could not perceive the thermometer to be at all affected by that stretching or taking away the stretch. No alteration neither could be perceivd by stretching the thin tube with the same weight.

The same thermometer was inclosed in a tube of the same size and thickness made so as to fit on to the air pump so that the air in it could be condensed. It seemed to rise about ¼° on forcing in 1 atmos. and to sink as much on letting it out. If this thermom. was placed in a cup of oil it alterd not more than ⅕° by the same operation, so that it seemd as if it alterd rather more than it would have done by the simple pressure on its ball, but is very doubtful.

A cylind. glass bottle about 5 inches in diam. and 7½ high in the cylind. part with a brass cap to fasten it to the air pump was prepared, the glass being much thicker than usual. A receiver also was prepared. A receiver also about 6½ in. diam. and 10½ high was prepared.

June, 1783. This bottle with the small thermometer used in the last experiment in it was screwed on to air pump and the air condensed 1 atmos. The thermometer rose 5° on condensing and sunk 2°½ on letting out the air: the event being the same in 2 trials.

It was then exhausted 2^ce^ running, to within 3 inch. when it sunk 3°½ in 1st trial and 4° in 2nd and rose 5°¼ on letting in in both trials.

The receiver was then placed over the bottle, the bottle with the thermometer in it remaining as before, so that the air in the receiver was not altered. The thermometer then rose 5°½ on condensing and sunk 2°¼ on letting out the air. It also sunk 3°½ on exhausting, and rose 5° on letting in.

The thermometer was then placed between inner and outer glass, the bottle as before so that the air between the glasses still remaind unaffected. It was not at all alterd by exhausting or condensing the air in the bottle or letting the air in or out. The thermom. was then placed in bottle left open and inclosed within receiver, the air in both being condensed and exhausted. It then sunk 6° on exhausting and rose 5°¾ on letting in. It also rose 5° on condensing and rose [sunk] 9°¾ on letting out.

The bottle with the thermometer in it was then shut, so that air in it was unalterd; the air in outer receiver being exhausted or condensed the thermom. was not at all alterd thereby.

The thermom. was then placed between the two glasses where the air was condensed and exhausted, the bottle being closed. It then sunk 1°¼ on exhausting and rose 2° on letting in air. It also rose 2°½ on condensing and sunk 1°½ on letting out the air.

The tube of the thermometer being broke in this experiment and consequently the tube remaining open, it was inclosed in bottle and that screwed on on plate. It then rose 5°·6 on condensing and sunk 2° on letting out. It also sunk 2°·1 on exhausting and rose 5°·6 on letting in, which is pretty nearly the same as it did before it was broke.

TABLE SHOWING DETAILS OF FOREGOING OBSERVATIONS

Bottle screwed on with thermometer in it.

Time		Heat
9.51		65°
52	condensed 30 inches	70
56		66
57	let out	63½
10. 1		64¾
2	condensed 30 inches	69¾
6		66
7	let out	63½
14		64½
16	exhausted to 27	61
23		63½
24	let in	68¾
28		65
30	exhausted to 27	61
35		63½
36	let in	68¾

The receiver was then placed over it.

45		65¼
46	condensed	70½

Time		Heat
10.50		67°
	let out	64$\frac{3}{4}$
55		65$\frac{1}{4}$
	exhausted	62$\frac{1}{4}$
11.3		65$\frac{1}{2}$
	let in	70$\frac{1}{2}$

Thermometer placed between inner and outer glass.

4.31		66
33	inner exhausted	do.
33		do.
	air let in	do.
	condensed	do.
	let out	do.

Thermometer placed in inner bottle included in receiver in which last the air is condensed or exhausted.

5.19	bottle open	66
22	exhausted	60
27		65
	let out	70$\frac{3}{4}$
35		66$\frac{3}{4}$
37	condensed	71$\frac{3}{4}$
	let out	62
	bottle shut	
8.35		63$\frac{3}{4}$
45		64
47	condensed	do.
50		64
	let out	do.
9.12		64$\frac{1}{3}$
16	exhausted	64$\frac{1}{4}$
	let in	64

Thermometer placed between inner and outer glass, air between them condensed and exhausted: inner glass closed.

10.18		67$\frac{1}{3}$
29		67$\frac{1}{2}$
31	exhausted	66$\frac{1}{4}$
43		67$\frac{3}{4}$
	air let in	69$\frac{3}{4}$
45		69
52		68$\frac{1}{3}$
	condensed	70$\frac{3}{4}$
57		69
	air let out	67$\frac{1}{2}$

The tube of thermometer was then broke and consequently remained open: it was then enclosed in bottle and screwed on on plate.

Time		Heat
4·35		11·7°
	condensed	17·3
4·43		12
	let out	10
4·58		10·4
	exhausted	8·3
5·2		11
	let in	16·6

TO TRY WHETHER DAMP AIR IS OF SAME SPEC. GRAV. AS DRY

[Suggested experiment.]

In room in which air is very dry fill a flask with the air of a room by blowing air through it with bellows and weigh it carefully. Then fill it with damp air by blowing air into it in same room through a cylinder filled either with fresh water or water more or less impregnated with salt, or filled with wet paper and weigh it again.

N.B. For fear the water, etc. in the cylinder should be cooled by the evap. it will be more accurate if the cyl. is bent so as to be immersed in water.

TO SEE WHETHER BULK OF PERFECTLY DRY AIR IS INCREASED BY SATURATING WITH MOISTURE IN RATIO OF DEPRESSION AS THAT HEAT BY WATER TO PRESSURE OF ATMOSPHERE ON IT [*sic*]

[Suggested experiment.]

Fill barometer carefully and in cold weather. Let in air through cyl. of wet paper by app. in figure. Try it carefully in different heats. Then let in a little water and repeat the experiments. It may be calculated how much the depression on the addition of water should be according to this hypothesis and consequently whether it agrees with hypoth.

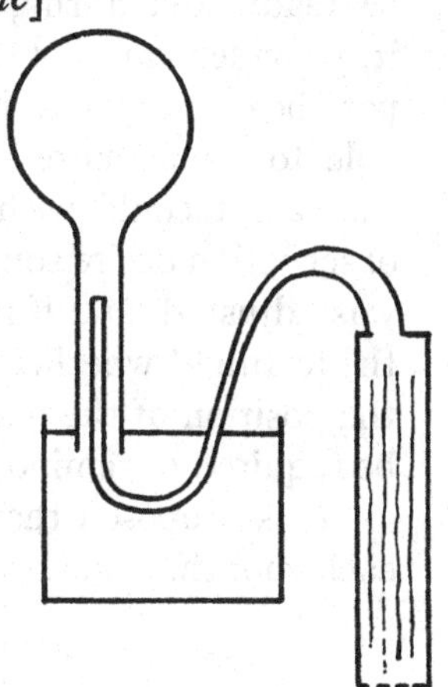

INSTRUCTIONS TO THE CLERK OF THE ROYAL SOCIETY CONCERNING THE METEOROLOGICAL OBSERVATIONS TO BE MADE ON BEHALF OF THE SOCIETY AT THE SOCIETY'S HOUSE IN CRANE COURT[1]

That he [Mr Robertson] observe the heat of the thermometer and barometer as early as he conveniently can in the morning not later than 7 o'clock in summer and 8 in the winter, and also about 2 in the afternoon which is supposed to be about the hottest part of the day.

That he measure the rain every morning about the same time as the thermometer except when the quantity is trifling, and in particular on the first morning of each month, and that he set down the sum of all the quantities measured in each month except on the 1st day together with that measured the 1st day of the succeeding month as the whole quantity fallen that month, and that he also set down the quantity fallen each year.

That he set down each day the direction of the wind and its strength together with any other observations relating to the weather that he may think fit. It will be sufficient to set down the wind once each day, namely at the time of the afternoon observation of the thermometer, and to distinguish its strength into 3 degrees, namely, calm, brisk and violent.

That during one fortnight of the year he observe the horizontal and dipping needle 5 times a day, namely, as early in the morning as convenient, about 2 in the afternoon, a little before he goes to bed, and about ½ way between each of those times and also to set down the mean variation and dip during that time. As to the dipping needle it must be observed that it will be necessary always to observe whether the needle is truly ballanced before he begins his course of observations, but if it is then found to be truly adjusted, I imagine there can be no danger of its requiring to be readjusted during that fortnight in which it is proposed to observe it, provided care is taken not to handle it, or take it out of the case compass box or to move it about during that time. But as to this, I shall be able to speak more positively in a month or 2. If this is found to be the case I could wish the Council would always before the fortnight of observation desire some one of [the] Society to examine whether the needle was adjusted, and if not to make it so, and also to examine at the end of the fortnight whether it continued so. At the same time he might examine the position of the horizontal needle. I would propose also that the Clerk be required to remind the Council of it every year in proper time.

It is proposed that the greatest and least heat of the thermometer in each month, together with the mean morning and midday heat of each

[1] [Cf. p. 53 and p. 112.]

month and each year; the greatest and least height of the barometer in each month; the quantity of rain in each month and each year; and the mean variation and dip of the needle in each year be printed at the end of the last part of the Transactions for each year.

The thermometer is proposed to be placed out of one of the windows of the Norfolk Library[1] and as in this situation the sun will shine upon it in the morning in summer it is proposed to fitt a piece of board as a screen. It could not be placed so as to be defended from the sun in summer unless it was placed out of one of the windows of the lower story where it is imagined the situation would be too confined to show the heat truly. It is supposed to be unnecessary observing the thermometer indoors.

The rain gage is proposed to be placed on the leads of the house not very far from the middle and raised $3\frac{1}{2}$ feet above them, that is on a level with the wooden rails. There is no chimney which will be elevated above it in an angle of more than   degrees.

A weathercock is proposed to be placed either on one of the chimneys of the house, or on the wooden rails which surround the leads, which may conveniently be seen from the garden.

The barometer, variation compass and dipping needle, are proposed to be placed in the Norfolk Library and a mark is proposed to be placed on the South Wall of the Museum which by means of a telescope fixed to the compass will serve to place it in the proper position.

ATMOSPHERIC REFRACTION

Some time ago, in discoursing with you about the extraordinary refractions which Wales observed in Hudson's Bay, I said that if the radius of curvature of a ray by refraction was less than that of the earth, I did not see how it was possible for there to be any visible horizon; but on considering the matter further I find that it is possible.

To explain this, let AB represent the surface of the sea, and CD an arch of a concentric circle. Suppose that the radius of curvature of a ray of light passing nearly horizontally through the air, at a small height above the sea, is less than the radius of curvature of the earth; and suppose that

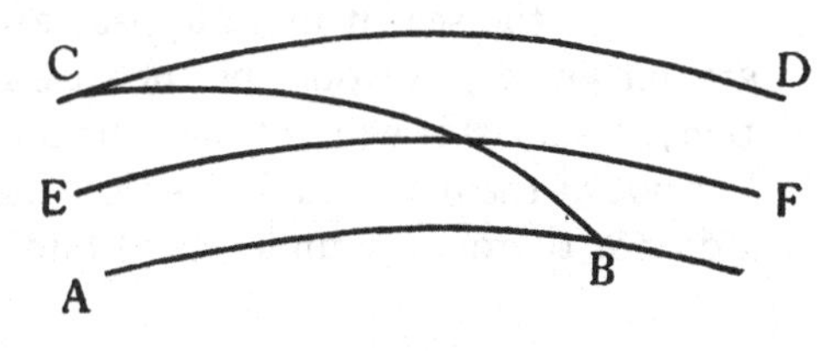

the greater the height of the ray of light above the sea, the greater

[1] [The Arundel or Norfolk Library, *Ex dono Henrici Howard Norfolciencis*, was presented to the Royal Society in 1666–7, as a New Year's gift, at the instigation of John Evelyn. It was collected by Thomas, Earl of Arundel, during his embassy at Vienna, and consists chiefly of first editions of books, published soon after the invention of printing (Maitland). It may be regarded as the nucleus of the Society's library. See Weld's *History of the Royal Society*, vol. I. 196.]

its radius of curvature, as it is natural to suppose; so that at the height of the line *CD* above the sea, its radius of curvature shall be exactly equal to that of the arch *CD*. It is plain that no ray of light passing horizontally at a greater height than that of *CD*, can ever fall on the line *AB*; but a ray passing horizontally ever so little below that height, will fall on it. Therefore let *CB* represent the path of a ray of light passing horizontally at an infinitely small distance below *CD*: it is plain that no object on the earth can appear elevated above the horizon in a greater angle than that of *CBA*; and all objects beyond a certain distance will appear elevated in very near that angle; and consequently the visible horizon will appear elevated by the angle *CBA*. Moreover if the height of the line *CD* (and the refracting power of the air below it) is such, that a ray of light coming (to the observer's eye) from an object at no very great distance (from him) may pass near *CD*, the horizon may appear distinct; but if it is such that no ray of light can pass near *CD* but what comes from a very distant object then the horizon must appear very indistinct; just as it does to an observer situated on a very high mountain. 2ndly, the higher the observer is above the sea, the less will the visible horizon appear elevated; for drawing the concentric circle *EF* between *AB* and *CD*, the line *CB* cuts *EF* in a less angle than it does *AB*. 3rdly, though the observer is near land the visible horizon will appear uniform, as in the open sea; for as objects of very different distances and colours will all be crowded into the same point, they will exhibit an uniform appearance; but objects at such a distance as to appear much below the visible horizon may appear distinct; therefore an observer at sea near the shore should see the objects on the shore distinct, and above them he should see the visible horizon appearing like that of a sea placed beyond the shore.

N.B. What is said in this article goes on a supposition that the height of the line *CD* above the level of the sea, is the same over the land as over the sea, which is not very likely to be the case. 4thly, if a distant object appears at the same height as a nearer object to an observer near the surface of the sea, it must appear above it if the observer is placed at a greater height, as would be the case if there was no extraordinary refraction; so that this will not account for Wales's seeing the fort on the deck, but not at the mast-head. As for what Wales says about the islands of ice I do not thoroughly understand him.

THE REFRACTION ON A MOUNTAIN SLOPE

The following paragraph is extracted from a paper by Sir Joseph Larmor, on "The Influence of Local Atmospheric Cooling on Astronomical Refraction," *Monthly Notices, R. Astronomical Soc.* vol. LXXV, 1915:

After the above was written, it was found in looking through Henry Cavendish's extensive MS. calculations relating to the planning by the

Royal Society, between 1772 and 1774, of the Schehallien experiment for the determination of the Earth's mean density, in which Cavendish took a very prominent part, that he had then considered the problem of accidental refraction-error due to the mountain, substantially as here. If the summit is colder than it is at the same level directly above the foot of the mountain, the strata of equal atmospheric density tilt upward towards the summit; and the apparent position of a star, chosen near the zenith to avoid ordinary refraction, will be deflected away from the mountain, thus making the observed effect of attraction of the vertical towards the mountain too small. Cavendish calculated roughly that for a defect of temperature of $12\frac{1}{2}°$ F. at the summit, which he took as an extreme estimate, and for a hill sloping at 21° to the horizon, the error would be about $0''\cdot6$ on each side of the hill, depending, as we have recognised, only on the temperature difference and not on the height. The formula (A) gives for a local temperature change of this total amount, from a mountain of that slope to the air above it, a value of about $0''\cdot5$ for z small, if the strata are taken parallel to the slope, whereas Cavendish made them inclined to the slope so that the temperature fell along it. In the Schehallien observations the effect of the attraction of the mountain came out as $11''\cdot6$, which Cavendish's extreme estimate for refraction would increase by 10 per cent., thus, as it happens, leading to consistency with modern determinations. There does not appear to have been any attempt to apply such a correction, either to Maskelyne's Schehallien observations, or to the later series by Sir Henry James at Arthur's Seat, where the result was 3 or 4 per cent. too small, while owing to flatter conditions the refraction error would be much less: see the account in Poynting's Adams Prize Essay of 1894 on *The Mean Density of the Earth*, pp. 15–22.

Note added by Sir Joseph Larmor. The Schehallien determination of the Earth's mean density was published in *Phil. Trans.* 1798, and the result of the Michell-Cavendish vibration experiments in *Phil. Trans.* 1798. They are fully described by Airy in his treatise on the "Figure of the Earth" in *Encyclopaedia Metropolitana*, 1830. He remarks on the "various contrivances of practical calculation" in Hutton's determination of the attraction of Schehallien, and he adds in a footnote, "Most of these, it appears, were suggested by Mr Cavendish. And it appears that nearly all the preliminary calculations of the attraction of Skiddaw, etc. [which had been suggested for the determination] were made by Mr Cavendish," as he had learned from inspection of the Cavendish Papers.

METEOROLOGICAL OBSERVATIONS AT MADRAS

(Extract from a letter by Cavendish.)

In England the heat of the water in deep wells or quick springs is very nearly equal to the mean heat of the air, and it seems well deserving inquiry whether it is the same in other countries; for if it is so, it would afford the readiest way of comparing the mean heat of different climates. As your correspondent's observations, if continued, will tell us the mean heat of the air at Madrass I should be very glad if he would also try the heat of the water in wells of the place, the deeper and less exposed to the air the better. It will be sufficient to try it once or twice at opposite seasons of the air [*sic*], but if he would be so good as to try it on 2 or 3 different wells it would be the better. If it is a draw well, it will be sufficient to draw up a bucket of water and try the heat with his thermometer. If it is a pump I would recommend to him to pump a few minutes before he tries the heat, as the water which comes first is what is contained in the body of the pump, which perhaps may be of a different heat from that in the well.

I am informed that the usual way of cooling their water at Madrass is to expose it to the open air in porous earthen vessels. I should be very glad if he would now and then try the heat of the water in them at different times of the day, and different seasons of the year, and also set down the heat of the air at the same time, so as to shew how much the water is cooled by the evaporation.

If there should happen to be any Taffoon (typhoon) while he is there, I could wish that he would observe the alterations of the barometer and wind and weather as closely as he conveniently can from the time of the first presages of it to the end, and that if he has an opportunity he would endeavour to collect how far it extended and at what time it began and ended at different places; and that he will set down any other circumstances which may occur to him that he thinks will tend to make us better acquainted with the nature of those extraordinary phenomena.

The exposition of your correspondent's thermometer without doors, which was placed under a shady tree, seems not quite unexceptionable, as I am afraid that the air under the tree may be cooled by the evaporation from the leaves. What confirms me in this opinion is that the thermometer without doors was seldom hotter than that within doors in the middle of the day, and was commonly considerably cooller in the morning, whereas if it had not been for that cause I imagine the thermometer without doors would commonly have been considerably hotter than that within doors in the middle of the day.

All the portable barometers of Ramsden which I have seen have a Vernier division by which we may observe the height to 100ths and 200ths of an inch, and have a screw at the bottom by which the quicksilver in the cistern may be adjusted to the proper height. If it would not be too much trouble to your correspondent, it would be better if he would set down the height of the barometer to hundredths of an inch, and it would be more satisfactory if he would mention whether he frequently adjusts the quicksilver in the cistern or whether he trusts to its remaining always the same. I need not say that if a person would be accurate in his observations he should examine the height of the quicksilver in the cistern frequently.

CAVENDISH'S REGISTERING THERMOMETER

In Wilson's *Life of Cavendish*, p. 477, is a description, with illustrations, of a Registering Thermometer contrived by Cavendish. The account is as follows:

"The above figures [p. 396] represent a front and back view of Cavendish's Register Thermometer. This instrument was presented by Sir Humphry Davy to Professor Brande, and is included in the collection of old apparatus in the Royal Institution.

This instrument consists essentially of a large glass tube containing alcohol, the expansion and contraction of which acts upon mercury contained in the recurved or inverted syphon-termination of the tube. A considerable portion of the alcohol tube is exposed to the atmosphere, but it is sheltered from the rain by a roof-like cover. The only opening in this tube is at the top of the left-hand limb of the syphon. The surface of the mercury here carries an ivory float, from the top of which proceeds a silken line, and this, passing twice round the periphery of a wheel grooved for the purpose, falls down and hangs loosely, with a small balance weight at its extremity.

The register is performed in the following manner: The axis of the wheel which carries the cord carries also a light index hand: this hand moves in a vertical plane some distance behind the graduated circle; but near the top of this hand projects a short horizontal piece carrying a vertical needle which in the right hand figure is now pointing at 50°. On either side of this index is a friction needle, which accompanies the index to the extreme limit of its range, and stops there. In the figure we may suppose the index hand to have advanced to nearly 80° on one side, and, having pushed the friction hand thus far, the alcohol began to contract, the index to recede, thus leaving the friction hand to record the highest temperature that had been attained. The alcohol continuing to contract, the index would recede until it came in contact with the friction hand on the other side: here the extreme limit appears to have been something below 40°; the temperature then beginning to rise, the index hand had reached 50° at the time of observation.

In order to make a new observation, a bent lever (which in the figure appears to be pointing near 30°) is turned round by means of a central thumb screw, first on one side and then on the other, so as to place the two friction hands in contact with the index hand at the point indicated at the time of setting. The instrument is then left to itself, and after

REGISTER THERMOMETER.

some hours, or next day, the friction hands will be found separated as before.

The two faces are glazed with plate glass, and a hole is made in the glass in the right hand figure for allowing the thumb-screw to pass out. The lower part of this glass is covered with tin-foil to the height of about 2 inches. Both faces are moreover provided with doors, which close with a spring. These doors, and the outer case, are of sheet and bar iron, and the whole is very heavy. The whole height, from the base to the ridge of the cover, is about 18 inches; the height of the glazed part is 11½ inches, and it is 6 inches across.

The brass back of the instrument (left-hand figure) is furnished with a number of projecting brass pegs: the top now consists of 4 pegs, the two outer ones of which have sharp points; all the others being blunt or rounded. I have not been able to discover the object of those pegs, but Professor Brande informs me, that when the instrument came first into his possession a number of pieces of bibulous were stuck upon them, the object of which was probably to keep the interior of the instrument dry.

The instrument is now quite out of order: the alcohol appears to have oozed out, probably by capillary action; the glass is also in some places corroded by the mercury, as we often see in old instruments in which impure mercury has been used."

Among the Cavendish MSS. preserved at Chatsworth is an account of the calibration of this instrument, which seems to have been devised in

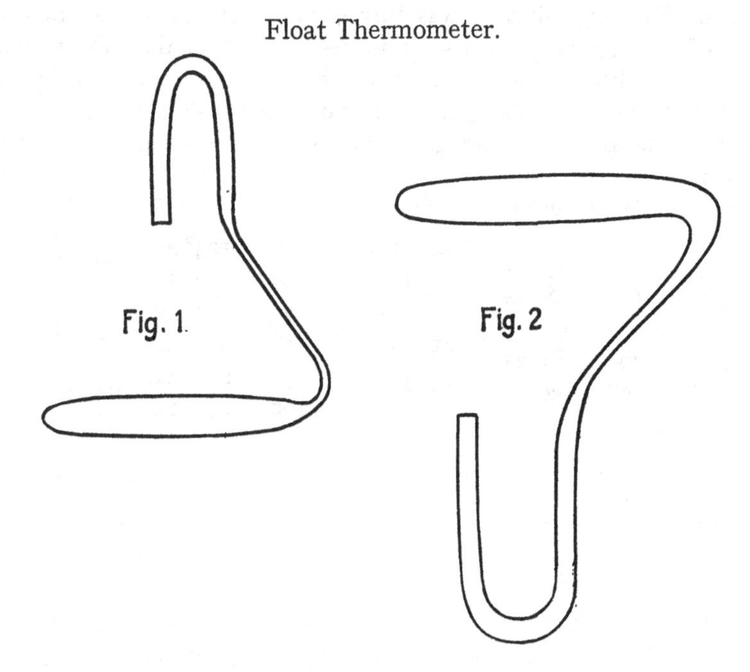

1779, whilst he was still living with his father at Great Marlborough Street. The account is, of course, only of historical interest, but it is here reproduced, with the accompanying figures in facsimile, as another illustration of the love of accuracy and scrupulous attention to detail which characterised all Cavendish's experimental work.

"June 12, 1779. The proportion of the bore of the 2 legs to each other tried by filling the th. with spts. with some ☿ in the legs with the brass frame fastend on and heating the cylinder in sand in the position of fig. 2."

[The heights of the mercury in the two limbs of the U tube were then read to 500ths of an inch as the position of the mercury was caused to fluctuate by the expansion and contraction of the spirit, whereby the alteration in the bore of the tube in which the float was to be placed could be ascertained. The observations extended over several days. On June 17, we read, "The ☿ was got out and a proper quantity of spt. put in and compaird in freezing mixt. [ice and salt?] with the standard [thermometer] *S* in position of fig. 1."

A large number of comparisons were made at different temperatures by the same thermometer and after certain corrections for expansion of the scale, etc. had been made a table was constructed showing the height of mercury in the leg corresponding to the true temperature.]

It was found by putting on the float and observing what division on the scale the bottom of the float answered to when the hand stood at a certain mark on the dial plate and when it was turned round one revolution that 1 revol. answered to 5·258 on scale or 5·516 inches. A piece of tin with two holes in it at a proper distance was fastend to the scale to enable the eye to look perpend. and avoid parallax; therefore 1 divis. on scale answers to 7·832 mins. Therefore the angles answering to different degrees on divided circle are as follows:

Deg.	Angle	Diff.
−10	0° 0′	
		19° 11′
0	19 11	
		19 43
10	38 54	
		20 14
20	59 8	
		21 16
30	80 24	
		21 41
40	102 5	

Deg.	Angle	Diff.
		21° 40′
50	123° 45′	
		22 42
60	146 27	
		22 51
70	169 18	
		23 14
80	192 32	
		24 9
90	216 41	
		24 40
100	241 21	

MATHEMATICAL AND DYNAMICAL MANUSCRIPTS

Arranged and Described by SIR JOSEPH LARMOR, F.R.S.

IT is difficult to advance any definite dates for this part of Cavendish's activity. One thing that emerges clear is that the elucidation of physical and astronomical phenomena by mathematical thought was one of his main permanent interests: and if he had no other claim to renown he would be entitled to rank high among the theoretical physicists of his period, though more deeply versed (as was then usual in England) in the physical significance than in the formal perfection of the analysis.

His more special occupation with Physical Astronomy and with the Dynamics and Figure of the Earth may fairly be held to centre around the year 1773; for at that time he was very actively engaged in forwarding the enterprise of the Royal Society for the determination of the mean density of the earth, and was indeed the guiding spirit in that project. This period is usually considered to have been barren as regards physical astronomy in Britain: but though the improvement of analysis and the conduct of surveys fell mainly to the share of France, these papers incidentally afford evidence that the Schehallien determination of the Earth's mean density was not an isolated episode. The dynamical principles guiding the rising science of Geodesy, and their physical outcome, as distinct from mathematical method—the practical discussion of the relation of the geometrical form of the Earth to gravity surveys by pendulum, the dynamical variation of latitudes, the investigation and control of atmospheric refraction—were elucidated with knowledge and insight worthy of the secular observations, involving the discovery of Precession and Nutation and of Aberration, organised by the previous generations of British workers. And it is to be noted that though the scientific discussions in the papers here described were never given to the world, and Britain seems to have been then largely isolated except in mechanical construction, yet they were in the hands of Maskelyne and Hutton and the other writers who compiled the published accounts, and of the committees with which they worked, and were in fact drawn up for their use.

The time and concentration required for the effective prosecution of such studies over so wide a range, especially when combined with the fundamental experimental determinations in other sciences that went on in the same years, were amply sufficient to account for the strengthening of Cavendish's habits of reserve and isolation. He was a natural philosopher so profound and universal as to have no time to be anything else.

FRAGMENTS ON MATHEMATICS

Cavendish seems to have spent much time perhaps at an early period on mathematical investigation: and like other great physicists he was equally at home in algebra and in arithmetical calculation. The manuscripts appear to be in no case mere copies from other sources, but reflect the writer's own thought. Though hampered in analysis by the fluxional method, it is clear that his mathematical powers were of a high order.

It will be sufficient as an indication of the range of his activity to give brief references to some of the more striking papers. He seems to have been making preparations at one time for a book on trigonometry. There exists a preliminary syllabus for a treatise on the principles of mechanics.

A fragment on the axioms of geometry: "If a line be of such a nature that no part of it can be made to pass through different points while the extremities of it remain fixed, it is called a straight line: otherwise it is called a curve line." "...A straight line is shorter than any other line which can be drawn between 2 points; for if this is denied it must be said that there may be drawn some curve line between the 2 points which is either as short or shorter than any line which can be drawn between those points, 'but that has been shown to be impossible.'"

Problems of geometrical maxima and minima, e.g. the circle has greatest area for given perimeter, developed at considerable length. "On the best proportion of the circles for the carpenter's oval."

"Surface of a cone cut by oblique plane" expressed by a series. "That sections of spheroids and parabolic cones are ellipses."

Dissection of a parallelogram so as to form another parallelogram.

"To find the shape of a parallelopiped from its appearance to the eye."

A summary of spherical trigonometry. "As to supplemental triangles in sph. trig."

A discussion of the form of triangles for which errors of measurement entail least results: "Fluxions of plane triangles."

A detailed discussion of the theory of deviation from a mean, in the form—Let A and B whose chances of winning are as $a : b$ play a very great number (n) of games together, to find the probability that A's gains neither exceed nor fall short of the probable estimate by more than $p\sqrt{n}$ stakes.

The ways of paying a given sum in coins of x, y, and z units.

A number of problems in algebra, sometimes worked out in extensive detail, such as "To find 3 or 4 numbers the sum, sum squares, sum cubes and sum fourth powers being given."

Newton's rule for interpolations: with numerical table headed "Interpolating by second differences, correction to be added or subst. to number formed by 1st diff."

OPTICAL FRAGMENTS

There is a discussion "of the figure of glasses necessary to bring rays to a focus and of the aberration of rays"; also on the error of deviation due to a prism being placed slightly out of the plane of symmetry and the position of symmetrical deviation.

There is a calculation "on the aberration in reflecting telescopes used in Herschel's manner"; and another with numerical values "on the aberration of rays passing through a spherical lens."

There is a discussion of the best conditions for vision of the wire micrometer.

He took an interest in the chromatic aberration of light, and made many experiments on the compensation of prisms of various materials solid and liquid, with mathematical calculations, between 1770 and 1790. It does not appear that any distinct conclusions emerged. There is a fragment on the condition for an achromatic object glass: Dollond, who perfected the achromatic telescope, died in 1761.

In other fragments the problem of astronomical refraction is discussed in general terms: especially he considered that the effects of moisture and temperature of the air required separate discussion and investigation. His estimate of the effect of temperature on observations on a mountain slope is reprinted below (p. 406). As regards local refraction in an observatory:

If the strata of air near the opening of the observatory and within the observatory were horizontal it would certainly be right to place the thermometer by which you estimate the heat as near as possible to the object glass: but in reality I imagine that they will be far from horizontal, and consequently there is so little dependance upon that part of the refraction which the ray suffers in passing from the outside of the observatory to the object glass that it seems indifferent whether the thermometer is placed within or on the outside of the observatory, from which I think the only way to be exact is to endeavour to make the air inside the observatory as nearly of the same heat as that outside as possible, though I imagine astronomers will not think this a very pleasant method in winter.

He discusses the correction of the refraction for variations of the thermometer and barometer.

An estimate is made of the amount of light stopped in passing across a shower of rain.

ATTRACTIONS AND GEODESY

There is a set of carefully prepared manuscripts mainly discussing the conditions of choice of a mountain for observation of deflection of the plumb-line, preparatory to the Schehallien experiment, probably for the use of the Committee of the Royal Society (1772–4). One of them is entitled in another hand "Mr Cavendish's rules for computing the attraction of mountains on plumb-lines."

There is a manuscript "On the choice of hills proper for observing attraction: given to Dr Franklin," which begins as follows: "The Royal Society are desirous of making some experiments to examine whether hills are capable of exerting any sensible attraction, and how great it is, as a means of determining the mean density of the Earth: and they are in search of hills proper for the purpose...."

Another packet is entitled "Computation for Skidda." Near the end of his "Figure of the Earth," Airy refers to Hutton's calculations (*Phil. Trans.* 1778) for the reduction of the Schehallien observations: "Indeed the various contrivances of calculation in this Paper will be found well worth the attention of any practical person." In a footnote he adds "Most of these it appears were suggested by Mr Cavendish. And it appears that nearly all the preliminary calculations of the attraction of Skiddaw, etc. were made by Mr Cavendish. This we have ascertained from an inspection of his papers, which we have had an opportunity of examining through the kind permission of his Grace the Duke of Devonshire."

There is a manuscript entitled "Paper given to Maskelyne relating to attraction and form of the Earth" to which the following letter refers. It is addressed

To the Honble Henry Cavendish
at Lord Charles Cavendish's
Great Malbro' Street.

Greenwich Jan. 5 1773.

Dear Sir,

Inclosed I return you your rules and directions for the choice of hills having a considerable attraction; which I have taken the liberty to take a copy of: I think them well calculated to procure us the information that is wanted. According to your Table, I should estimate, that the valley called Glent-Tilt, lying on the N.W. side of the mountain Ben-Glae in Scotland, should produce a defect of attraction on the two opposite sides of 36″, supposing the mean depth 1000 yards, the shape spheroidical and the length of the valley 8 miles, the breadth 4 miles (I believe it is less) and the angle which the direction of the valley makes with the meridian 50°. Col. Roy, from whose account these dimensions are taken, says it makes an angle of 50° or more with the true meridian. If the mean density of the earth exceed that at the surface 5 times[1], there will still remain 7″ attraction. I think the dimensions of this extraordinary valley deserve a more particular inquiry. I shall be obliged to you for a line to acquaint me, whether you found any thing material in those papers of the late Mr Robins, which you examined that have not been printed; as the proof of Mr Call's paper (making mention of them at the end of his account of the draught of the 12 signs in an Indian Pagoda) is now in my hands; and I would add a note about the papers. I am Sir,

Your very humble Servt,

N. Maskelyne.

[1] [As regards this excessive estimate, cf. p. 404.]

The deviation of the plumb-line by the tides was under discussion as the two following extracts show:

Suppose that the tide in the Bristol channel rises 50 feet that the channe is 10 mile broad (which I imagine must be its utmost breadth in any part where the tide rises to that height) and that it is extended infinitely both ways in length in a straight line: suppose too that at high water the sea touches the bottom of the cliff and retreats 50 feet from it at low water: the difference of attraction at high and low water on a particle of matter at the foot of the cliff is 797 feet, which would make the plumb line deviate $1\frac{3}{4}$ seconds if the mean density of the earth is the same as that of the surface.

If the sea retreats 400 feet from the cliff at low water everything else being the same the difference of attraction is 588 feet.

If the sea is 60 mile broad and the tide rises 40 feet the high water mark touching the cliff and the low water mark 50 feet distance the difference of attraction is 780 feet.

Since I saw you I have looked again into Boscovich's book (De litteraria expeditione &ª.) and find that what he says about the attraction of the tides does not differ so much from what I said in the paper I gave you as I imagined: he supposes the arm of the sea to be 100 mile broad in which case he says the plumb line will deviate 2″·38‴ if the mean density of the earth is the same as that of the sea [of the surface?]. I supposed the sea to be 10 miles broad and found according to one supposition that it should make the plumb line deviate $1''\frac{3}{4}$.

There is a paper entitled "Rules for computing the error caused in measuring degrees of latitude by the attraction of hilly countries." By calculating for a section of Italy perpendicular to the Apennines he estimates that on an arc of $2\frac{1}{6}$ degrees the length of a degree would be shortened by 210 toises, supposing the mean density of the Earth to be that of the surface.

He considers a section through the middle of the degree that had been measured in Pennsylvania, and perpendicular to the Alleghanies which are inclined at 37° to the meridian, and makes out that the attraction of the hills increased it by 24 toises. "The whole effect of want of attraction of the Atlantic = 1·376 which should diminish the length of a degree by 119 toises." An estimate for the degree measured at the Cape of Good Hope is a shortening of 174 toises due to the proximity of the ocean and a lengthening of 33 due to a range of hills. The attraction of the Cordillera, near Quito, is treated in detail on the basis of the diagrams in Bouguer's *Figure de la Terre*; as pendulum observations were available[1].

The following conclusions of a carefully prepared manuscript may be of

[1] [For historical detail as to all these surveys see Airy's *Figure of the Earth*, a very complete and important treatise written for the *Encyclopaedia Metropolitana* in 1830 before he left Cambridge for Greenwich.]

historical interest as indicating how far advanced the ideas of gravitational geodesy were in Cavendish's hands at this early date.

The attraction of the Cordillera on the pendulum at Quito will be much the same as if it was extended infinitely of the same level in all directions: only in order to allow for the attraction of Pinchincha and the other mountains rising above Quito which rather diminish the gravity let us suppose Pinchincha to be a segment of a sphere its altitude being 968 toises and the radius of its base 4800 toises. The height of Quito above the level of the sea is 1466. Whence the mean density of the earth should be 4,44 times that of the surface. If the observed difference of lengths had been $\frac{1}{100}$ of line less the mean density would have come out 3,83.

If there should be a greater quantity of matter under the Cordillera of less than the usual density of the surface of the earth (which is not unlikely considering that all the hills in have probably been volcanoes) the length of the pendulum at Quito or Pinchincha will be less than it would otherwise be which would make the mean density of the earth appear greater than it really is.

It is likely that the mean density of the earth should be several times greater than that of the surface though the internal parts of the earth are composed of the same materials as the surface, as the density of the internal parts may very likely [be] increased many times by the great pressure which they suffer.

These results, which he compares with estimates based on Canton's measure of the compressibility of water, are of course greatly in excess of the strikingly correct guess of Newton. A previous estimate led him to a ratio less than 3.

I know but 2 practicable ways[1] of finding the density of the earth first by the going of a pendulum in the foregoing manner and secondly by finding the deviation of the plumb line at the bottom of a mountain by taking the meridian altitudes of stars. The first way is by much the most easy but the latter seems much the most satisfactory.

If the mean density of the earth is the same as that of the surface the attraction of a conical hill will accelerate a pendulum at the top of it per day 32″,4 seconds into the versed sine of the angle which the side of the hill makes with the perpendicular. [Height not specified.]

The attraction of a hill in the form of a small segment of a sphere will make the plumb line deviate at its ft. 32,8 seconds. So that in this respect the latter method seems rather more exact. But the main point is that in the latter method the result seems much less affected by any irregularity in the density of the internal parts of the earth than in the former method.

Let us suppose for example that the earth consists of a nucleus *abd* covered with an outer crust of a considerably different density and let us suppose that

[1] [This was of course written long before Michell's plan by torsional vibration was known, which Cavendish carried out in 1798. The date of the Peru arcs is 1736, of Clairaut's *Figure de la Terre* is 1743, of Canton's experiments on compression of water is 1762, of the Schehallien experiment is 1778.]

this nucleus is of rather an irregular figure and consequently that the outer crust is much thinner in some places as for example in *A* than in others, it is plain that the force of gravity on the surface at *A* and at a height above the surface will not be to each other in the inverse duplicate ratio of their distances from the center as they would otherwise be and consequently no certain conclusion could be drawn from experiments of the pendulum at the top and bottom of a mountain in such place. Whereas in the latter method of finding the density of the earth I do not imagine that the result will be sensibly affected by any irregularity of this kind. That there is some such irregularity in the structure of the earth seems likely I think from observations of the pend. in different places, which differ more than I should think could be owing to the error of experiment particularly if you compare...

There follows a numerical discussion of the consistency of the observed variations of gravity with the latitude. Similar discrepancies appeared in a discussion by Bouguer: this subject is emphasized by Airy, *loc. cit.*

If the earth is a regular spheroid the length of the pendulum swinging seconds increases in going from the equator to the pole in proportion to the square of the sine of latitude or in proportion to the versed sine of 2^{ce} the latitude. The 4th column of the following table contains the length of the pendulum computed according to this rule on a supposition that the difference of length between the equator and pole is 2,3 lines, the length at the equator being chose such that the sum of the numbers in the 5th column (which are the excesses of the observed lengths above the foregoing) shall be nothing; the 6th column gives the mean of the 6 first and of the 4 last excesses in the foregoing column and the 7th, 8th and 9th columns are the same things on a supposition that the difference of the length between the equator and pole is 2,5 lines.

It appears from hence that in general if you suppose the difference of length between the equator and pole to be 2,3 lines the observed lengths fall short of the computed by more in the lower latitudes than in the higher, whereas if you suppose the difference to be 2,5 lines they fall short more in the higher latitudes than in the lower, as appears more plainly by the 6th and 9th columns in which are given the mean excess for the 6 first places (which may be looked upon as the lower latitudes as the versed sines of the doubled latitudes of those places are considerably less than the mean) and of the 4 last places which may be considered as the higher latitudes; whence it seems as if the true difference of length was between 2,3 and 2,5 lines or $\frac{1}{184}$ of the whole length.

It has been demonstrated by Clairaut that if the earth consists of spheroidical strata not much different from spheres and that the density is the same in all parts of the same stratum that the difference of the 2 axes divided by the whole axis exceeds $\frac{1}{230}$ as much as the difference of gravity divided by the whole force of gravity falls short of it, whence it should seem as if the difference of the axes was about $\frac{1}{306}$ which is a quantity which will agree much better with the precession of the equinoxes than a greater difference. The measures of a degree agree rather better with the supposition of $\frac{1}{230}$ than with that of $\frac{1}{306}$ but the difference is not great[1].

I think it would very well become the admiralty to send a vessel to observe the longitudes of such places as are most frequented by our ships as till then the method of finding the longitude at sea will be of very imperfect use. If they should do so it will tell us the length of the pendulum in many different climates without any addition trouble. I suppose you will think the most convenient way of finding the longitude will be by taking the difference in time of the transits of the moon and some fixed stars by a transit instrument. If they dont propose to stay long in any place I should think the best way would be not to endeavour to fix the instrument in the meridian but to point it to the pole star a little before the first observation and mark the time then to take the transit the moon and 2 or more stars one of which should if possible be before the moon and the other after it and the nearer to the same parallel of declination the better. This by an easy calculation will determine the going of the clock and the times of the transits so that the longitude might be determined with tolerable accuracy even in a single night: after the observ. is over you may see whether the instr. has altered its place by seeing how many revol. of the adj. screw it requires to make the instr. point again to the pole star.

It is needless saying that the instrument ought to stand on the ground and not to touch the walls of the observatory: there should also be a floor to the observatory which should be supported by the walls of the observatory and should by no means touch the instrument or rest on the ground near it as without these precautions it is impossible that the instrument should be steady.

Finally there is his estimate of the error from refraction on a sloping hill-side, probably in connexion with Schehallien, as follows (see *supra*, p. 393).

On the irregular refraction caused by the heat being different near the side of the hill from what it is at a distance from it.

Let the top of the hill be B feet above the place of observation, and let it be elevated above the horizon in an angle whose tangent is T; suppose that there is N degrees difference between the heat of the air at the top of the hill and at the same height perpendicularly over the place of observation. The alteration caused in the density of the air by N degrees of heat is the same as is caused by $N \times \frac{26800}{450} = 60N$ feet of altitude; 26800 feet being the height of an uniform atmosphere; Therefore if a line be drawn through the summit of the hill, in such manner that the density of the air shall be the same in all parts of it (o

[1] [See the section on Precession of the Equinoxes, under the heading Astronomy (p. 436), for further consideration of these questions, in relation to the internal constitution of the Earth. In connexion with the there following section on Tidal Friction and its argument which only needed the consideration of angular momentum (by Kelvin and Darwin) to complete it, it may be noted that Kant's theory on this subject dates from 1754, and Thomas Wright's views on cosmogony from fou[r] years earlier.]

a line of uniform density as I shall call it) the tangent of inclination of that line to the horizon will be $\frac{60 \times NT}{B}$ seconds. If we suppose that the heat of the air at any intermediate height on the side of the hill differs from the heat at the same height perpendicularly over the place of observation by a quantity which is to N degrees as the height of that place above the place of observation is to B, the line of uniform density will be equally inclined to the horizon at all heights less than that of the top of the hill; but at a little height above it I suppose the line of uniform density will be nearly horizontal. Now the refraction of a star at no very great distance from the zenith $= 57'' \times$ tan. zenith distance; and its refraction in passing through a portion of the atmosphere answering to a difference in the height of the barometer equal to D inches is $57'' \times \frac{D}{30} \times$ tan. zen. dist. whence we may conclude that the irregular refraction of a star near the zenith caused by this inclination of the line of uniform density to the horizon will be $57'' \times \frac{D}{30} \times \frac{60NT}{B} = \frac{114DNT}{B}$ which as $D = \frac{28B}{26800}$ nearly (the mean height of the barometer on the side of the hill being supposed 28 inches) equals $'',12 \times NT$.

If we suppose the inclination of the side of the hill to be $21^\circ.50''$ and therefore T to be $= ,4$ and $N = 12^\circ\frac{1}{2}$ which in all probability is on the outside of the truth the irregular refraction will be only $\frac{6}{10}$ of a second.

MECHANICS AND DYNAMICAL THEORY

There is a list of titles, possibly in Dean Peacock's writing, with a memorandum "Sent to Lord Brougham March 28, 1845, Theory of the Kite, and On Flying."

The early paper endorsed "Remarks relating to the Theory of Motion" has been printed *verbatim*, p. 415. It is very clearly written, with numerous precise erasures as the composition proceeded. It is in fact an argument for the conservation of mechanical and thermal energy, which the writer proposes to name "mechanical momentum" in distinction from "ordinary momentum" which depends on direction[1]. After a demonstration of the conservation of mechanical energy, and a specific introduction of the idea of potential energy, he infers with great distinctness that heat must be the "mechanical momentum" of the internal vibrations of bodies: but in proceeding to the energies and heats of chemical affinity he grasps the

[1] [Compare Young's discussion of the subject in his *Lectures* (1807), VIII, On Collision, in which the term energy was first introduced. The tacit avoidance by Cavendish of the term *vis viva*, made classical by Leibniz and D. Bernoulli, is noteworthy, and possibly arose from a connotation then more prominent than now: the absence of knowledge of the more analytical writings of d'Alembert would be natural in a British physicist of his time.]

new features and concludes that "this way of explaining it is insufficient[1]." Incidentally there is a discussion of a tinfoil crusher-gauge for transient pressures. The manuscript probably belongs to an early period: but the only clue to a date is a reference to *Phil. Trans.* No. 477, which is unavailing as that part was published in 1745. The controversy which he elucidates, as to whether force is measured properly by change of momentum or by "mechanical momentum," is supposed to have been composed at any rate from the point of view of the Continent where it was mainly carried on, by d'Alembert in 1743 in his *Traité de Dynamique*.

Various special problems relating to moving dynamical systems are solved in separate papers: but the work contains nothing remarkable.

A manuscript of considerable length developed on geometrical lines "On the motion of a solid of revolution from its centrifugal force."

A calculation entitled "Catenaria blown by the wind" with an estimate of "resistance of air on chain": also an investigation "On the form of the Catenaria in which the strength of the chain is everywhere proportional to the tension."

A mathematical and experimental investigation, with extensive calculations, of the strength under flexure of wooden bars of various sections sawn from logs, and their vibrations when one end is fixed, with a view

[1] [It is sometimes stated that Cavendish's views on the question whether heat is a substance were ambiguous. No such epithet can apply to the very remarkable corollaries at the end of this paper, which so clearly anticipate the modern doctrine of energy. It is unfortunate that there is no means of assigning a date to these expository essays; one is inclined to refer them mainly to the early period of study, before he joined the Royal Society in 1760 at 29 years of age. The unsystematic form of his special dynamical reasoning makes it unlikely that as regards it Cavendish was indebted to previous writers, and so is in favour of an early date. The earlier electrical manuscripts, largely experimental, of which "the style...leaves no doubt that they were intended to form a book" belong to the years 1771–3 (see Clerk Maxwell's very illuminating *Introduction*).

The heat as well as the pressure of air had been referred mathematically, in a general way, to the *vis viva* of the free motions of its particles or molecules by D. Bernoulli, *Hydrodynamica*, as early as 1738 (cf. Cor. 1 *infra*); and he extended widely the principle of Vis Viva, as Cavendish does in this paper, in *Berlin Memoirs* ten years later (cf. Lagrange, *Méc. Anal.*).

In Maxwell's *Theory of Heat*, p. 72, there is an interesting discussion of the implication in Black's term latent heat, that heat is an imponderable substance as distinguished from ponderable gases which Black was himself the first to recognize as distinct kinds of matter. "The analogy between the free and fixed states of carbonic acid and the sensible and latent states of heat encouraged the growth of materialistic phrases as applied to heat: and it is evident that the same way of thinking led electricians to the notion of disguised or dissimulated electricity, a notion which survives even yet, and which is not so easily stripped of its erroneous connotation as the phrase 'latent heat.' It is worthy of remark that Cavendish, though one of the greatest chemical discoverers of the age, would not accept the term 'latent heat.'" Maxwell goes on to quote a footnote (see p. 151 *supra*) of *Phil. Trans.* 1783 as from J. D. Forbes, *Encyc. Brit.*, Dissertation VI.]

to computation of the tapering lines for masts: also a discussion of the "vibrations of a straight uniform spring left to itself."

An investigation of the form of a solid of given volume so that its attraction at a point shall be maximum. [See Todhunter's *History*.]

A manuscript apparently intended to be sent to Dr Hutton entitled "Explanation of Mr Hutton's solution of Maseres' problem about vibrating string."

Form of arch for various modes of loading.

Fragments on the resistance of the air to projectiles, in connexion with Newton's *Principia*.

A geometrical and analytical paper on the "map pentagraph."

"Calcul. of force of engine turned by reaction of two jets of water for mechanical purposes."

"On the vibrations of pendulums whose centers of suspension move." Young's result is obtained that the effect is the same as if the pendulum were prolonged to a fixed centre. (For rolling motion, no correction is required.)

"Concerning the spinning of tops by Mr Mitchell." Doubtless his friend, Rev. John Michell[1] (1724–1793). "This point is what is known by the name of the center of percussion. How I came not to take notice of it I do not know."

"Pendulum." Records of swings on different days in case and out of case. "It is supposed that in this and the following pages the pendulum was compared with father's clock in new building."

"Pendulum with rolling motion by Nairne." A close scrutiny of two pendulums with reference to symmetry and slipping on edge and decrement of free swing. "Hence it should seem that the time was nearly equal to the difference of the logarithms of the 2 arches of vibration multiplied by 225: and therefore the time in which it changes from 2° to 1° should be $67\frac{1}{2}'$: by a mean of the observation in p. 7 it seemed to do the same with the former motion in $63'$."

A long series of experiments on moduli of bend and twist for glass tubes and iron and brass wires under load. The ratios seem to be very concordant. Further "experiments on twisting of wires of silver, iron and other metals tried by the time of a vibration."

In a manuscript "Concerning waves" there are general remarks on the character of wave-motion on water: and interesting guesses of the cause of the discrepancy between the velocity of sound and Newton's theoretical value, considering after Euler particles repelling according to the law r^{-1} or near thereto, finally rejecting all proposed explanations.

"Concerning the vibrations of pendulums suspended from the same horizontal bar." An unsatisfactory attempt at general explanation of the

[1] [See short *Memoir of John Michell*, by Sir Archibald Geikie, Cambridge University Press, 1918.]

fact that "If 2 clocks, the length of whose pendulums never differ by above a certain quantity, are fixed to the same horizontal bar, they will keep moving constantly together: one clock will never be before or behind the other by as much as one vibration." [See Ellicott, *Phil. Trans.* 1739.]

"On centrifugal pendulum" including a form of isochronous check.

A paper on the "Vena Contracta": nothing of note except a careful diagram of form near the orifice.

A series of carefully reduced experiments on flow of water through a glass tube ·2117 inches in diameter and 44·1 inches long: "the pressure required to overcome the friction in velocities of 34 and 19 in. per $'' = \frac{1}{388}$ of the length multiplied by the $\frac{5}{3}$ power of the vel. in in. per $''$ but in greater velocities seems to increase in a greater proportion."

A discussion of the mode of action of spiral springs.

"On error in pend. beating dead, supposing the accel. from the action of clockwork to be uniform while weight acts upon it, and that it is uniformly retarded during rest of time with such force as to keep the vibrations of same length." Inconclusive.

"Question about Tower of Babel." Triangular pyramid?

A paper "On the shape of the teeth in rack work": encloses a beautifully written manuscript ending with the following on a separate page:

Dr Young takes the liberty of sending for Mr Cavendish's inspection a copy of what he means, with Mr Cavendish's permission, to insert in his syllabus respecting the teeth of wheels. He believes that the point of contact G will seldom if ever fall between E and F.

Welbeck St. Thursday 3 Sept 1801.

In "A Syllabus of a course of lectures on Natural and Experimental Philosophy" printed at the press of the Royal Institution in 1802, this memorandum slightly changed occupies §§ 178–181: and in § 179 which contains the argument in small print there is the sentence "For the substance of this demonstration I am indebted to Mr Cavendish."

DYNAMICAL VARIATION OF LATITUDE

Let an oblate spheroid revolve round an axis not coinciding with the axis of the spheroid, and suppose this spheroid to be placed at such a distance from any other matter as not to be influenced by the attraction thereof. Let A be the pole of the spheroid, and P the pole on which it revolves, Ee the equator the motion being from E to e; and let the difference of the axes of the spheroid divided by the whole axis be called α. The point A endeavours by the centrifugal force to move towards P; by this compound motion the point P, or the point on which the spheroid actually revolves, is continually shifting its place on the surface of the spheroid, *id est* the spheroid revolves on a different part of its substance from what it did before; the motion of the point P being such as to describe a circle round A as a center in the direction from P to p (*id est* in the same direction that the spheroid revolves) in the same time that it makes $\frac{1}{\alpha \times \text{cs}\, AP}$ revolutions round its axis.

With respect to absolute space the point P describes a small circle round the point π: in the same time that the spheroid makes 1 revolution round its axis, and in the same direction; the radius $P\pi$ being to PA [strictly sin $P\pi$ to sin PA] as $\alpha \times$ cs PA to one.

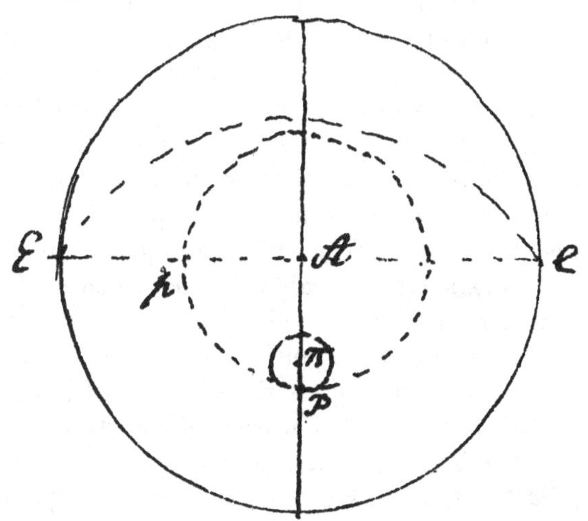

If the axis on which the earth revolves does not coincide with the axis of the spheroid, I imagine the precession of the equinoxes and nutation of the pole will not be sensibly affected thereby; but the pole will revolve on the surface of the earth round the axis of the spheroid in 230 days (supposing the difference of the axes to be $\frac{1}{230}$ part[1]) being sometimes on the European side thereof and sometimes on the American: whereby the latitudes of all the places on the earth will be continually altering, but the pole will constantly point to the same part of the heavens that it would otherwise do. Suppose for example that the pole of the earth was removed 1″ from the pole of the spheroid, at one time the latitude of London and all the places in the same meridian would be 1″ more northerly than the mean: consequently the sun and all the stars would cross the meridian 1″ more to the south than they would otherwise do: but in about

[1] [As regards this estimate, see p. 405. The correct Eulerian factor is $C/(C-A)$, where C, A are the moments of inertia, being 304 days. This supposes the earth to be rigid: if not, $C-A$ must be the value when the distortion produced by centrifugal force of diurnal rotation is supposed taken off, which should give the observed value 428 days. See *M. N. R. Astron. Soc.* Nov. 1906; *Proc. Cambridge Phil. Soc.* May, 1896. For a homogeneous spheroid the Eulerian period for small amplitude would be α^{-1}, as the text indicates: thus Cavendish was probably acquainted with Euler's general abstract result (*Mechanica*, 1736; *Theoria Motus*, 1765) which he here converts (possibly for the first time) into practical geometrical form, not far removed from the modern one in terms of Poinsot's rolling momental ellipsoid. Cf. *Proc. Camb. Phil. Soc.* 1896. In fact the rotational motion of the Earth is represented precisely by the circle pP of the diagram rolling on the much smaller fixed circle with centre π situated inside it.]

115 days afterwards the latitude of the same places would be 1″ more southerly than the mean, and then all the stars would cross the meridian 1″ too much to the north.

The historical interest of this subject has prompted renewed examination of the lengthy and complex computations in the manuscript, "On the motion of a solid of revolution from its centrifugal force." By powerful but hardly elegant use of spherical geometry he arrives at an expression for the centrifugal effect of the motion, giving a final result of the same type as in this fragment: but it would involve a tedious examination to find whether the value of his coefficient (α) agrees in general with the Eulerian one.

As a geometrical lemma he establishes the law for finding the angular velocity which is the resultant of three given ones round axes OP, Op, Om marked by points P, p, m on a spherical surface.

There is also a separate paper with the object probably of making the problem of free rotation manageable, in the absence of the resources of analysis, for the general form of solid without an axis of symmetry. He establishes for the case of a right solid the principle of kinetic equivalence: "the resistance of the whole solid is the same...as if $\frac{1}{6}$ of the whole quant. matter was placed in the centers of each of the six faces of the parallelepiped," which reduces that problem to one relating to particles.

In Euler, *De Motu*, 1765, Ch. XII there is only a passing reference to the effect on latitudes, which is the application that is of outstanding modern importance, and is Cavendish's main concern in the formulation here reproduced.

EFFICIENCY OF AN UNDERSHOT WATER WHEEL

Calculation of maximum and force of undershot wheels in which the float boards are of such a breadth that the water shall dash over them and in which little or no water can escape between the floats and the sides.

All the water which issues through the sluice loses the excess of its velocity above that of the wheel.

Therefore the pressure upon the wheel is such as would in a given time communicate to all the water which issues through the sluice in that time a velocity equal to the excess of the velocity of the issuing [entering?] water above that of the wheel and therefore (as the quantity of water which issues through the sluice in a given time is given) is proportional to that excess.

Therefore the quantity of work done by the wheel is proportional to the velocity of the wheel into the excess of velocity of the issuing water above that of the wheel and is a maximum when the velocity of the wheel is $\frac{1}{2}$ that of the issuing water.

The pressure acting upon the wheel when it moves with $\frac{1}{2}$ the velocity of the issuing water is $\frac{1}{2}$ the weight of the water which issues from the sluice in the time in which a heavy body will in falling by its own weight acquire the velocity of the issuing water.

The velocity of the wheel in the same case is such as would in the abovesaid time move through a space equal to the height from which a body must fall to acquire the velocity of the issuing water.

Therefore the greatest effect of an undershot wheel is such as would in a given time raise a weight equal to that of ½ the water which issues from the sluice in that time to the height from which a body must fall to acquire the velocity of the issuing water.

The fallacy of the common answer to this problem consists chiefly in supposing the water to act only upon that float nearest the sluice whereas in reality the

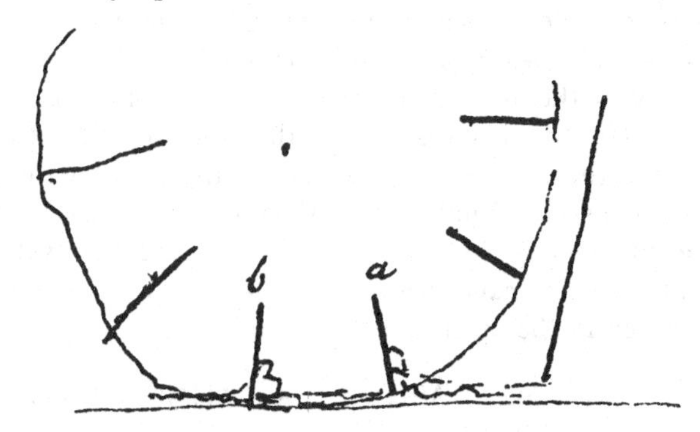

water contained between the float (*a*) and (*b*) continues to act upon the float (*b*) (after the float (*a*) is interposed between it and the sluice) till it comes to move with no greater velocity than the wheel.

In a manuscript "On the engine for raising water by a centrifugal force" it is shown that when the water spurts out from the end of a rotating horizontal pipe "The force [work] requisite to move this engine for any time is such as is sufficient to raise the quantity of water which issues from the horiz. pipe in that time to 2ce the height βD." The difference appears of course chiefly as kinetic energy of the issuing water.

ON THE MOTION OF SOUND

The method of proof in the carefully written manuscript, which he describes as *infra*, is to assume that when the piston vibrates *in any manner* the air along the tube takes up the same type of vibration but after a time represented by the velocity of propagation, and then to verify that this description fits with the motion of a portion of air as determined by the difference of pressure at its two ends, provided however the compression is small. There is a discussion of what may be expected when the amplitude of vibration is so great that this verification holds good only for the turning points of the vibration, remotely after the manner of Rayleigh's *Theory of Sound*, II, § 251. It is noted that the velocity of a wind adds to that of the sound.

The occasion of my writing this paper was as follows Sr I. N. [Newton] proved that if the vibrating body moved with a velocity regulated according to the same law as a pendulum vibrating in a cycloid, the particles of air would vibrate with a velocity regulated according to the same law, but does not prove that they will if the body vibrates in any other manner. Euler proved that Newtons demonstration was equally applicable if the body moved in the same manner as a body vibrating in a parabola and some other figures and if I mistake not supposed it an objection to the demonstration; this made me suspect that the case was that according to whatever law the body vibrated the air would do the same and set about trying to demonstrate it.

Besides making it general there are 2 or 3 other points in which this demonstration differs from Newtons first in reducing his 3 propositions into one which makes it a good deal clearer 2ndly it is drawn from the well known property that the density of the air is in proportion to the compression instead of the property which Newton deduces from it that the particles of air repel with a force inversely as the distance and 3rd that in the 1st part the air is supposed to be confined in a straight uniform canal because Newtons demonstration is in reality applicable only to that case. In the 2nd part however it is extended to the case of a conical canal which comes to much the same thing as if the sound is produced in the open air.

THE DISCHARGE OF A CANNON BALL

Internal ballistics. "Investigation of the loss of force owing to the inertia of the powder."

The density of the inflamed air will be greater at bottom of the barrel than near the bullet as having a greater quantity of matter to put in motion; the proportion which the density at bottom and at bullet near to each other remains pretty nearly the same in whatever part of the barrel the bullet is in after the space through which the bullet has moved bears any considerable proportion to the space occupied by the powder. The force with which the powder actually acts upon the ball is to the force with which it would act upon it if the powder had not inertia as the density of the inflamed air at the bullet to the mean density or the density which it would be of supposing it every where alike.

On this basis a calculation is made by fluxions which allows also for the escape of the inflamed air past the sides of the ball, on the basis "Elasticity of air gener. by powder before ball is moved = 1000 atmospheres": of course the cooling effect of expansion does not appear. Results and data from "Rob" (= Robins) are adopted as follows:

Height bullet discharged with vel. 1700 would rise in vacuo = 44930 feet = $8\frac{1}{2}$ miles therefore its greatest range would be 17 miles its actual range is supposed less than $\frac{1}{2}$ mile.

Vel. of 24l. cannon ball discharged from piece 10 feet long with full charge of powder id est 16 pounds = near 1650 feet per ″ the greatest range of such a shot in vacuo would be about 16 miles its real range is less than 3 miles its resistance at that vel. is near 23 times its weight.

REMARKS ON THE THEORY OF MOTION

If the force by which a body is accelerated is given the velocity acquired by it will be as the time during which it is accelerated. If the time is given the velocity acquired will be as the accelerating force.

Therefore the velocity acquired by any body is as the accelerating force multiplied into the time during which it acts.

The velocity acquired by a body acted on by a given accelerating force is as the square root of the space through which it is accelerated. If a body accelerated by a given force acquires a certain velocity v in falling through a given space s the same body acted upon by a force which is to the 1st as (a) to (1) will in falling through the same space acquire the velocity $v \times \sqrt{a}$ for the body will in the same time that it described the space s by the 1st force describe the space $s \times a$ by the 2nd and at the end of it acquire the velocity $v \times a$ therefore if in falling through the space $(s \times a)$ it will acquire the velocity $v \times a$ in falling through the space (s) it will acquire the velocity $v \times \sqrt{a}$.

Therefore the velocity acquired by a body falling through a given space is in the subduplicate ratios of the accelerating force.

Therefore the velocity acquired by a body is as the square root of the accelerating force multiplied into the square root of the space passed over.

From hence appears the nature of the dispute concerning the force of bodies in motion; for if you measure this force by the pressure multiplied into the time during which it acts the quantity of force which a moving body will overcome or the force requisite to put a body in motion or in other words the force of a body in motion will be as the velocity multiplied into the quantity of matter, but if you measure it by the pressure multiplied into the space through which it acts upon the body the force of a body in motion will be as the square of velocity multiplied into the quantity of matter. The 1st way of computing the force of bodies in motion is most convenient in most Philosophical enquiries but the other is also very often of use, as the total effect which a body in motion will have in any mechanical purposes is as the quantity of matter multiplied into the square of its velocity; for in all mechanical purposes the force must be measured by the weight or resistance to be overcome multiplied into the height to which it is raised or the space through which it is moved, thus it requires an equal force to raise the weight of one pound 2 yards as it does to raise 2 pound 1 yard for the same force which is employed to raise 1 pound to 2 yards will by a proper machine raise 2 pounds 1 yard height. What I have here said will appear plainer by examples.

If a body moving with 1 degree of velocity is able to compress 1 spring to a certain distance the same body moving with 2 degrees of velocity will compress 4 springs to the same distance.

If in any machine a weight by descending communicates any degree of motion to the parts of the machine as in fig. 1st Pl. 1st [p. 428] where the weight (a) is suspended by the string AB which is wound round the axis BC so that the weight cannot descend without putting in motion the fly dfg, then if the weight in descending any given height can make the fly revolve with any given velocity it will require 4 times that weight descending from the same height or the same weight descending from 4 times that height to make the fly revolve with 2^{ce} that velocity; when it is required to make the fly revolve with 2^{ce} the velocity by 4 times the weight descending from the same height this will not be exactly true, because as part of the weight is employed in accelerating itself there is then a greater weight to be accelerated than in the first case.

For like manner if one man can by applying his strength in the most advantageous [way] communicate a given velocity to any machine it will require 4 men to work during the same time to communicate 2^{ce} that motion to the engine, or it will require the force of one man for 4 times that time.

It must be observed here that as a man or any other cannot move with more than a certain velocity, and as the faster he moves the less force he is able to turn the engine with, therefore when it required to move the engine with 2 degrees of velocity the velocity with which the man moves should be only ½ as great in respect to that of the engine as when it is to be moved with only 1 degree of velocity.

The thickness of wall or timber which a cannon ball will force its way through is as the square of the velocity.

The work which an engine turn[ed] by a stream of water will do is as the quantity of water which strikes the wheel multiplied into the square of the velocity; for the pressure exerted upon the wheel is as the quantity of water × its velocity and the velocity with which the wheel may turn is as the velocity of the stream.

As this way of computing the forces of bodys in motion is very often of service, and as some theorems of considerable use in philosophy may be deduced from it, I would have some name by which to distinguish this way from the other; because it expresses the effect which a moving body will produce in mechanical purposes. I think it would not be amiss if it was called the mechanica force or *mechanical momentum* of bodies in motion.

The 1st Cor. It follows from the known property of the lever, namely tha in order to an equilibrium the power and weight must be to each other inversel as their respective velocitys, that in fig. 1st the resistance to the motion of th fly caused by the force spent in accelerating in part of it, as *g*, is the same a would be caused by a body placed in any other part of the fly (for instance i its center of gravity) whose momentum (computed according to the usua manner) multiplied into its velocity is equal to that of the body *g* multiplie into its velocity, that is whose mechanical momentum is the same; and as thi property obtains equally in all the other mechanical powers as in the lever w may conclude that in any mechanical engine whatever in which part of th weight is employed in putting the engine in motion, it being supposed to mov freely without friction, that the mechanical momentum acquired by the engin whilst its center of gravity descends through a given space is the [same] as i would acquire if the whole matter of the engine was collected on its center c gravity or is the same as the whole matter of the engine would acquire b falling through the space which its centre of gravity descends: in most cases th truth of this is very evident and from what will come after it will appear that it is equally true in the most complex cases.

Lemma fig. 1st, 2nd, 3rd. Plate 2nd. If any force *D* acts at *D* upon the crooked lever *DGC* whose center of motion is *G* in the direction *DB* perpendicular to *DG*, and another force *C* sufficient to keep the lever in equilibrio acts at *C* in the direction *Cc* perpendicular to *CG*, then the force with which the center of motion *G* is pressed in the direction *Gg* or in the same direction in which *D* acts is equal to the force *D* + that part of *C* which acts in the same direction as *D*; only it must be observed that as in 1st and 3rd figures the force *C* resolves it self into 2 the one acting in a direction perpendicular to that of *D* the other directly opposite to it, this last must be looked upon as negative and that in fig. 3rd the sum of *D* + the above mentioned part of *C* will be negative because in that case *G* is pressed in central direction to *D*. The force with which *G* is pressed in the direction *GD* = that part of the force *C* which acts in that direction. The truth of this Lemma is pretty evident by inspecting the figures.

When 2 bodies impinge on each other directly it is pretty evident that the center of gravity of the 2 bodies will move with the same velocity before and after the stroke, or the sum of their momenta taken in a given direction will be the same after the stroke as before, but it is not so plain that the same thing will take place when they impinge obliquely; this however may be shewn to be as constant a law as the other.

To do this no more is required than to prove Fig. 4th Pl. 2nd that the force which must be impressed upon the body *DNQ* whose center of gravity is *C* in the direction *BD* perpendicular to *DC* in order to give the body a given momentum in the line *BD* is the same which would be required to give it the same quantity of momentum provided the force was applied directly on its center of gravity.

The abovementioned force impressed at *D* will be spent partly in giving the center of gravity of the body a motion in the direction *BD* and partly in making the body revolve round its center of gravity; now it must be observed that as the body revolves round its center of gravity at the same time that the above mentioned center moves in the direction *BD* there will be a certain point in it which will be at rest immediately after the impression of the force, whilst all the other parts of the body have exactly the same motion as if they revolved round that point as a field center just as when a wheel rolls along that point which touches the ground stands still during the instant of its touching it. Let *G* taken somewhere in the line *DC* produced represent that point (for it will be somewhere in that line) and suppose the whole quantity of matter in the body to be collected in any number of points *LMH* and *a*. The velocity with which any point *L* moves is as *GL* and the direction of its motion in the line *lL* perpendicular to *GL*, the reaction of the point *L* or the force with which it resists being put into motion by drawing *L* from *L* towards *l* endeavours to make *G* move from *G* towards *O* were it not hinderd by the reaction of the rest of the body. By the preceding Lemma the force with which *G* is drawn towards *O* by the reaction of *L* is equal to that part of the force impressed at *D* which is spent in giving *L* its momentum + that part of the reaction of *L* which acts in the same direction as the force impressed at *D* (or since the reaction of a body put in motion is in a directly contrary direction to the motion of the body) —that force which is required to give the point L its velocity in the same direction as the force impressed at D. The case is the same in respect to any other point of the body. Therefore the force with which G is pressed towards O by the united reaction of all the parts of the body is equal to the whole force impressed at D minus the force which would be spent in giving the whole body its velocity in the direction BD supposing the force applied directly on its center of gravity; but as the point G is at rest this is necessarily nothing, therefore the force impressed at D is the same which would be required to give the body the same quantity of momentum if the force was applied directly on its center of gravity[1].

It appears also from the latter part of the preceding lemma that (as the velocity of M is proportional to MG) the force with which G is pressed by the resistance of M along the line DG either to or from G according to which side of DG the point M lies is proportional to $MG \times \frac{sr}{sm} = Mr$; hence appears the truth of what I before took for granted that G must be somewhere in the line passing through D and the center of gravity (for if it was not it must have a motion in the direction DG) and also that the center of gravity must move in a line parallel to BD.

When 2 perfectly elastick bodies of whatever shape strike each other in any manner whatever the sum of their mechanical momenta, not computed in one given direction but in any direction whatsoever, will be the same after the stroke as before.

When you say that the momenta of 2 bodies computed by the mass multiplied in the velocity are the same after the stroke as before you consider it only as made in a given direction, therefore if a body moves in a contrary direction you look upon its momentum as negative, if at right angles to it as nothing; but in this case whatever is the direction of a bodys motion I still look upon it as positive and the same as if it was made in any other direction.

There is no need here of running into disquisitions concerning the nature of absolute and relative motion; for suppose any given point to be at rest and compute the motion your system of bodies as they are in respect to that point, the truth of the proposition will be the same whether that point is really at rest or moving uniformly forwards in a right line[2].

1st case. Let the 2 bodies impinge directly on each other; suppose the quantity of matter in the 1st body as A and in the 2nd as B and let the velocity of the 1st body before the stroke be equal to $x + B$ and that of the 2nd to

[1] [This involved argument is to prove that the centre of gravity moves as if all the forces were transferred to that point and the mass were all collected there. The general theorem of conservation of the motion of the centre of gravity where there are no extraneous forces is established by Newton in the Introduction to the *Principia*: the present extension is ascribed by Lagrange to d'Alembert, and directly follows from his analytical method. Though Cavendish's method resembles Newton's in its geometrical character, he threshes the subject out for himself as usual, without any dependence on authorities.]

[2] [That is, interchanges of energy in impacts are not affected by a uniform motion of translation of the whole system: and this could not be so, unless the momentum is conserved.]

$x - A$; it is plain that by altering the value of x in respect of A, $x + B$ may be made to bear any assignd proportion to $x - A$ and therefore this is an universal expression for the velocity of the 2 bodies[1]; after the stroke the velocity of the 1st body $= x - B$ and that of the 2nd $x + A$; before the stroke the sum of the mechanical momenta of the bodies

$$= \overline{x + B}^2 \times A + \overline{x - A}^2 \times B = x^2A + x^2B + 2ABx - 2ABx + A^2B + B^2A$$

the momenta after the stroke $= \overline{x - B}^2 \times A + \overline{x + A}^2 \times B$ which is likewise equal to $x^2A + x^2B - 2ABx + 2ABx + A^2B + B^2A$ [whatever x may be].

Case 2nd. If any 2 bodies whose centers of gravity are A & B and the directions of whose motion are expressed by the lines FA and BC strike each other obliquely, supposing however that the point in which they strike one another and that they are void of friction by which means they will acquire no revolving

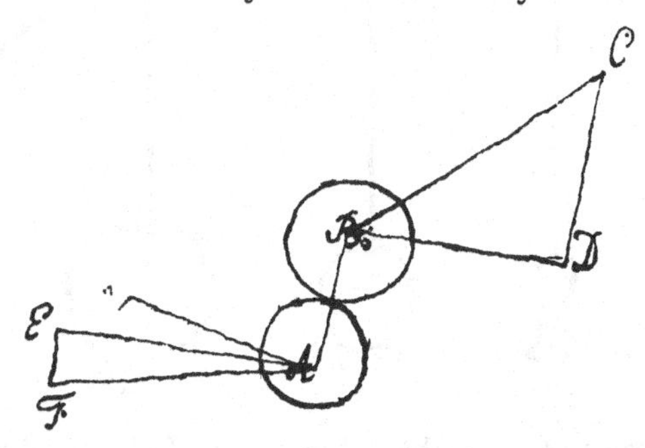

motion by the stroke, they will retain the same quantity of mechanical momentum after the stroke as before; for by resolving the motions FA and BC into FE and EA, CD and DB whereof FE and CD are parallel to the line AB joining their centers of gravity and EA and DB perpendicular to it, by the last proposition the sum of their mechanical momenta along the line FE or CD is the same before and after the stroke whilst their motions along EA and DB are not at all alter'd; therefore as whole mechanical momentum of it is equal to the sum of its mechanical momenta along FE and EA because $FA^2 = FE^2 + EA^2$ the sum of the entire mechanical momenta of A and B are the same after the stroke as before.

To prove that the same thing will take place computing the mechanical momentum of a body by the sum of the mechanical momenta of all its parts, though the bodies acquire a revolving motion from the stroke, we need only prove that supposing a body at rest whose center of gravity is C to be acted upon at A by any force in the direction BA perpendicular to AC, that the body will react at A with the same force as another body acted upon in the direction of its center of gravity of such a bigness that it should receive the same quantity of mechanical momentum.

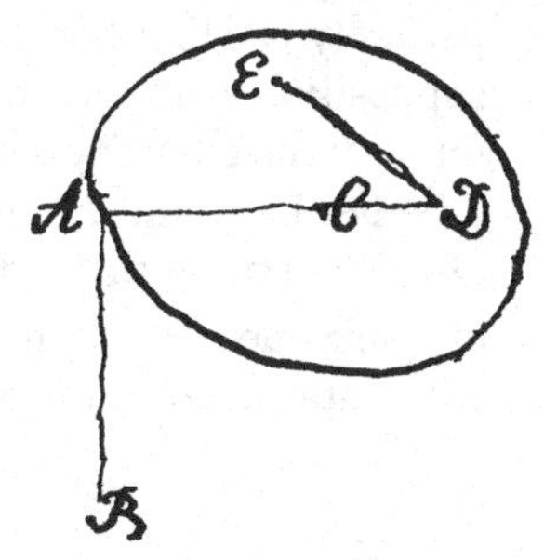

For by the preceding prop. there will be a certain point D which will be at rest immediately after the stroke therefore the velocity communicated to any particle

[1] [It would be simpler to write $x + kB$ and $x - kA$, where k is arbitrary; and so on.]

E will be as DE its distance from that point and the force with which it reac
upon it will be to the force with which another particle of the same size place
at A would react upon it[1] as $DE^2 : DA^2$ therefore the point E reacts with the sam
force as a point placed at A whose size should be to that of E as DE^2 to D
and whose mechanical momentum would consequently be the same.

Because in the collision of elastick bodies there is very often an encrease
momentum as usually computed some people have thought there might be
perpetual motion made from it, and indeed it at first seems very likely that
might be employed in mechanical performances to encrease the force.

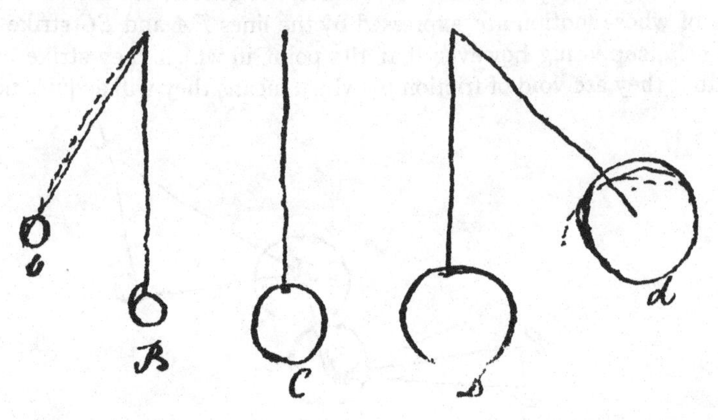

Thus suppose B, C, D to be any number of elastick balls encreasing in siz
from B to D let the ball B be drawn to any distance b and let fall again it wi
strike C and then D and the body D will move after the stroke with a muc
greater momentum than what B acquired by the fall[2]; it should seem therefo
that if when D was arrived at its greatest hight d it should be stopt by a catc
in such manner that its weight might be employed in moving the engine, that
would move it with more force than what was required to raise the ball B. But o
the contrary it appears by the foregoing proposition that if the motion of all th
other balls after the stroke could be in like manner applied to move the engine tha
their effect would be but rarely equal to what was required to raise the ball B

It appears also from this proposition that whenever 2 nonelastick or im
perfectly elastick bodies strike there is a loss of mechanical momentum; upo
supposition that the bodies are intirely void of elasticity and strike each othe
directly the quantity which is lost may thus be calculated. Let the weight of th
2 bodies be A and B and their respective velocities $x + B$ and $x - A$ as before[3]
after the stroke they will move both together with the velocity x and the sur
of their mechanical momenta will be $x^2 \times \overline{A + B}$, subtract that from the
momenta before the stroke and there will remain

$$A^2B + B^2A = \frac{AB}{A + B} \times \overline{A + B}^2$$

[1] [That is, the acceleration of a particle E is as DE, therefore the *work* absorbe
by its kinetic reaction is as DE^2.]

[2] [Because B has now acquired reversed momentum by its rebound.]

[3] [See footnote on previous page.]

equal to the momentum of a body whose weight is equal to the product of the weights of the 2 bodies divided by their sum and whose velocity is equal to that with which the 2 bodies meet.

If one body is supposed to stand still before the stroke the quantity of mechanical momentum lost by the stroke will be to the whole mechanical momentum before the stroke as the weight of the body struck to the sum of the weights of both bodies.

From hence it appears how much force is lost when you drive a nail or wedge &[a] with a hammer unless the weight of the wedge is very small in proportion to the hammer: if the hammer and wedge are of the same weight $\frac{1}{2}$ of the force will be lost.

From hence it follows that if a man holds a board in his hand and a bullet whose weight is small in comparison of that of the board is shot against it so as to pierce through it, the shock which the man will receive will be small in respect of what might be expected from the force requisite to pierce through it. For the same reason though light has a very sensible effect in putting the small particles of bodies in motion id est in causing heat, yet it is only by the nicest experiments that it can be found to have any effect in pushing bodies out of their places.

If a bullet moving with a given velocity pierces through the board in a certain time this same bullet moving with 10 times its velocity will pierce through it in less than $\frac{1}{10}$ part of that time; therefore as the force with which the parts of the wood resist being displaced is much the same whether they are struck fast or slow, the momentum communicating to the board in this latter case will be less than $\frac{1}{10}$ of what was communicated to it in the other, and the mechanical momentum less than $\frac{1}{100}$, consequently the mans hand will receive $\frac{1}{100}$ of the shock in this case than it would in the first; for the same reason a bullet moving very briskly will pass through the board without shattering it whereas a bullet which is almost spent will shatter it greatly.

When 2 soft or elastick bodies strike against each other the pressure exerted by one body on the other during the stroke is inversly as the space through which the bodies yield or are compressed by the stroke; therefore if the space through which they are compressed becomes nothing, or which is the same thing if both the bodies are perfectly hard, the force exerted by one body upon the other during the instant of the stroke is infinite. This property I believe was first taken notice of by Borelli.

Therefore in driving a wedge or any such thing by a hammer, if both the hammer and wedge were perfectly hard and the body to which the wedge was to be drove into was perfectly void of elasticity, every stroke would have some effect however small the stroke was; but if either the hammer or wedge yield to above a certain degree it is plain that pressure exerted on the wedge by the stroke of the hammer may be less than what is sufficient to move the wedge; moreover though the stroke be sufficient to move the wedge yet there will still be some force lost in compressing the hammer and wedge before they are overmuch compressed that their resistance shall be sufficient to move the wedge, and as the quantity of mechanical momentum lost this way will be the same whether the stroke is great or small, provided it is not too small as to be unable to move the wedge, there will be less force lost in proportion the greater the stroke is with which the wedge is drove.

This property will also serve to explain an odd experiment upon unannealed glasses related in *Phil. Trans.* No. 477 namely that if you drop into one of those glasses a small piece of flint or diamond or hard temper'd steel though not bigger than a small pea the glasses would brake, though you might drop into it a large piece of iron or lead, wood, or ivory &[a] without breaking; unluckily most of the bodies which he dropt into it of those kinds which broke the glasses were rough and sharp such as flints which broke the glasses the easiest of anything, he however says that he dropt a pearl of something less $\frac{1}{12}$ of an inch into a glass which broke it and says that he has often dropt musquet balls into them without breaking. This may pretty easily be accounted for by supposing [un]annealed glass to be extreamly little compressible; thus suppose the degree of compression of unanneald glass or the distance to which it may be compressed by any given force to be as 1 that of the pearl as 2 and lead as 4000, then the pressure exerted on the glass by the pearl is as great as that by a piece of lead of 1300 times its weight which is a greater disproportion than that of the weight of the pearl to the musquet ball, though there is no need of supposing it so great because the musquet ball, by being larger bears upon a greater part of the glass than the pearl; but if the compression of the glass had been 100 instead of 1 then the pressure exerted by the pearl would equal to that by a piece of lead of no more than 40 times its weight. There are some parts of this experiment, particularly the glasses sometimes not breaking till $\frac{1}{2}$ an hour after the stroke, which seem very difficult to account for.

It perhaps may not be impossible to contrive a method upon this principle of finding the degrees in which different bodies are compressible by the same force. Let the body *A* composed of the materials whose compressibility you have a mind to try be suspended as a pendulum by the line *CA* with a thin piece of tin foil or gold leaf stuck against its side, and let another body *B* of the same materials hung by the line *DB* be drawn to any distance from the perpendicular *E* and let fall against it so as to compress the tinfoil; after this has been repeated

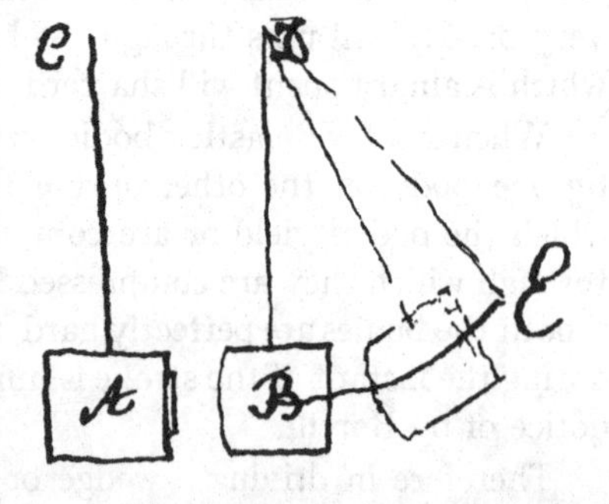

a sufficient number of times the resistance of the tinfoil will be so much increase that the stroke of the body *B* will not be sufficient to compress it any furthe which may easily be seen by observing whether the breadth of it increases; the to compare the compressibility of any other kind of matter with it exchang the bodies *A* and *B* for others of the same size and shape with the former, c the matter required, fix the same piece tinfoil used before to one of them an increase the height from which you let the body fall till you find that it is jus able to compress the tinfoil and no more. There are 3 principal causes whic disturb the accuracy of this experiment: the 1st is that the tin foil is not perfect unelastick, therefore when the resistance of the tinfoil is so great that its thickness cannot be decreased by the stroke it does not follow that it is not at all compressed by the stroke, but only that it restores itself to its former position, consequently the proportion of the compression found by this experiment does not express the real proportion of the compression of the bodies themselves but of the sum of the compressions of the body and tinfoil together—the best way I know of getting over this difficulty is by making the tinfoil so thin that its compression shall be small in respect to that of the bodies: the 2nd is that as the resistance of all bodies increases the more they are compressed we cannot be sure that the law in which it increases is the same in all bodies, but however that will not be of much signification in such an experiment as this; but the most formidable objection I think is this as it is impossible to make the body B constantly to keep exactly the same situation when it strikes A, the inside of it of which strikes A must be made a little convex to avoid striking it with its edge and consequently the part of its surface with which it touches the tinfoil will vary according as the body is more or less compressible; this must be avoided as much as possible either by making the radius of curvature reciprocally as the compression found by experiment, or by observing in how great a space the body and tinfoil touch one another by marking one of them by some colour which will easily come off or by engraving fine strokes in the body and seeing how much of the tinfoil receives the impression; but whatever way you use it can hardly be done with any great exactness.

Fig. 1st Pl. 3rd. Let the body A be attracted by or repelled from the points B, C, D, E, and a severally with forces which are always equal at equal distances however unequal at unequal distances, and let the body A by their united attractions or repulsions be carried to a, take Bb equal to Ba and Cc equal to Ca, Dd equal to Da &a then the quantity of mechanical momentum which the body A will acquire or lose in falling from A to a will be equal to the sum of the mechanical momenta which it would acquire or lose by falling from A to b by the attraction or repulsion of B singly and from A to c by the attraction or repulsion of C singly and from A to d by the attraction or repulsion of D singly &a: thus suppose that the body A by falling from A to b by the attraction of B would acquire the mechanical momentum B that in moving from A to c against the repulsion of C it would lose the momentum C and that in moving from A to d against the attraction of D it would lose the momentum D then the mechanical momentum acquired by A in falling through $Aa = B - C - D$.

For assume any 2 points in Aa, a and A infinitely near to each other and assume $B\beta$ equal to BA, and $Bb = B$a and in like manner assume

$$C\kappa = C\text{A and } C\text{k} = C\text{a}, \quad D\delta = D\text{A} \quad \text{and} \quad D\text{d} = D\text{a} \text{ \&}^{a}$$

then will the body A supposing it to fall through aA by the attraction of B alone receive the same increase of mechanical momentum as it would by falling through bβ by the same force, by Newt. Prin. Prop. 40th: in like manner it will lose the same quantity of momentum in moving through aA against the repulsion of C as in moving through kκ against the same repulsion, and it will be the same with D, E &a consequently as the mechanical momentum given to a body by any number of forces acting upon it together is equal to the sum of the momenta which they could give it separately, and as the body will receive the same encrease of Mechanical momentum in falling through aA whatever velocity it enters that space with, it will in falling through aA by the united attractions or repulsions of all the bodies receive an encrease of Mechanical momentum equal to the sum of those it would receive in falling through bβ by the attraction of *B* through kκ by the repulsion of *C* &ª: therefore the quantity of mechanical momentum acquired in falling through *Aa* [under] the united attractions or repulsions of all the bodies equals the sum those it would in moving through *Ab*, *Ac*, *Ad* &ª by their proper attractions or repulsions singly. Q.E.D.

If any point *B* instead of being considerd as immoveable is in motion then the encrease of mechanical momentum produced in *A* by the attraction of *B* above what it would acquire by the attraction or repulsion of the other bodies added to that produced in *B* by the same attraction is equal to that produced in a body in falling by the same attraction through the space by which those 2 bodies approach one another.

Therefore if there is a system of bodies *A*, *B*, *C*, *D*, *E* &ª attracting or repelling each other in the abovemention'd manner compute the mechanical momentum which it could produce in or take away from *B* in falling from *B* to *A* and also the momentum which it could produce in *C* in falling from *C* to *A* as also what it could produce by the same means in all the other bodies *D*, *E* &ª: in like manner compute the momentum which the attraction of *B* could produce in all the bodies which it attracts except *A* (as the momentum produced by their mutual attraction was before computed) compute also the momentum which *C* could produce in all the bodies it acts upon except *A* and *B* and do the same thing by all the other bodies, then the sum of these additional mechanical momenta added to the real momenta[1] with which the bodies are moving will remain constantly the same and will not be alterd by their actions upon one another.

Cor 1st. If any number of perfectly elastick bodies or particles mutually repellent in the manner above described are included in any space and put in motion, the sum of their mechanical momenta though it will not be constantly strictly equal because the sum of those additional momenta which are necessary to be added in order to make it so is not always the same, will yet be nearly so if the number of bodies be very great, but whether the number of bodies be great the sum of their mechanical momenta can never continue either encreasing

[1] [Namely, the *kinetic energy* and the *potential energy*, the latter being here proved to subsist in a field of attractions provided the forces depend only on the distances Compare the same argument and reservation, expressed more analytically, in Helmholtz's *Erhaltung der Kraft*, 1847. The potential energy of extended springs is explicitly introduced in Cor. 5 *infra*: it is the same as Daniel Bernoulli's *vis viva potentialis* mentioned by Euler, *De Curvis Elasticis*, § 1, 1744. The theorem of *vis viva* was extended to bodies moving under mutual forces of attraction by D. Bernoulli in *Berlin Memoirs*, 1748. In *Hydrodynamica*, Sec. 10 (1738), D. Bernoulli had proposed the air thermometer as the natural standard of "heat"; also he calculates the potential energy of compression of air (at uniform temperature), and speaks of the definite amount of "vis viva quae in carbonum pede cubico latet" being liberated by combustion, and (referring to Amontons) he speaks of its mechanical utilization.]

or diminishing for ever (because the bodies cannot continue either approaching nearer to or receding further from each other for ever) but must sometimes encrease sometimes diminish and their mechanical momenta taken at a medium will remain always the same.

If you suppose them to be elastick bodies there is no need that they should be spherical since the only reason why I supposed the repulsive or attractive force of each body to be equal at equal distances is this, that if I supposed the repulsive or attractive force of the body to be as is represented in fig. 2nd pl. 3rd where *A* represents the central body and whose central force is supposed equal in each part of the same line *BCD*, *bcd* &a which are driven further distant from each other and from the center at *B* and *b* than at the opposite part *D* then a body moving along *BbecC* would be more acted upon by the central body in moving from *B* to *e* than from *e* to *C*. But if you suppose the attractive or repulsive force of the body to be as represented in fig. 3rd Pl. 3rd where *D* represents the body and where the lines *ABC*, *abc*, $\alpha\beta\kappa$ &a in all parts of which the attraction is supposed of equal strength[1] are drawn in such manner that if you draw lines as $B\beta$ perpendicular to *ABC* or *abc* the distances of those lines from one another measured along $B\beta$ shall be everywhere the same, then it will be impossible for a body to come within the action of the body in such a manner but what it must be as much attracted or repelled in descending towards the body as in rising from it and the case will be the same as to preserving the same quantity of mechanical momentum as if the force was everywhere equal at equal distances from the center. In this case as well as that of elastick bodies the bodies will by their actions upon one another acquire a revolving motion[2] but this will make no difference by what has been said before.

The truth of this corollary is greatly confirmed by a preceding passsage where I shewed that the mechanical momentum of 2 perfectly elastick bodies was not alterd by their striking against each other.

Cor 2nd. Heat most likely is the vibrating of the particles of which bodies are composed backwards and forwards amongst themselves; therefore if bodies are composed of particles attracting or repelling one another in the manner above described their heat must remain constantly the same except as far as it is alterd by receiving from or communicating heat to other bodies, and whenever 2 bodies of different heats are mixed together or otherwise placed so that one may receive heat from the other, one will receive as much mechanical momentum or in other words as great an encrease of heat multiplied into its quantity of matter as the other loses so that the sum of their mechanical momenta may remain unalterd. But there is plainly both an encrease and loss

[1] [He is here struggling without success towards the modern idea of curves of constant potential: but the conclusion is correctly established at the top of the preceding page. The figures are on p. 430.]

[2] [Compare in connexion with this and the end of the previous paragraph the modern dynamical theory of gases; but Cavendish is intent mainly on the balance of energy. The nature of pressure and elasticity had been present to Huygens long before. Compare also more especially the theory of heat expounded in the next paragraph, which involves the steady state of internal motion, above deduced, with a definite amount of internal energy, for each temperature or state of "heat."]

of heat without receiving it from or communicating it to other bodies, as appears from the fermentations and dissolutions of various substances in which there is sometimes an encrease sometimes a loss of heat as well as from the burning of bodies in which there is a vast encrease of heat above what can reasoneably be supposed to be produced by the action of emitting light; and as this I think cannot with the least probability be supposed to arise from the attracting or repelling particles approaching nearer or receding further from one another, by which means the sum of those abovementiond additional momenta may be alterd, the particles must either not attract or repel equally at equal distances or must act stronger when placed in some particular situations than others or something else of that nature[1]. One would be apt at first to explain this by supposing them to attract or repel some kind of bodies stronger than others; but then it should seem as if there should always ensue an encrease heat whenever 2 bodies are mixed which mix together with any degree of force whereas there is often produced a great degree of cold thereby as in mixing salt and water. There are other reasons too which seem to shew that this way of explaining it is insufficient.

Cor 3rd. When any number of rays of light strike any body so as to be reflected backwards and forwards within it without ever emerging from it they will communicate their whole momentum to it, but when they are reflected from the body immediately as they will be returned with nearly the same swiftness with which they struck it they will communicate a very small part of their momenta to it; which is the reason why black bodies heat so much faster in the sun than white.

Cor 4th. When a vibrating body causes any pulse or sound in the air supposing it to be a perfectly elastick fluid, as soon as the body has performed a vibration compute the quantity of mechanical momentum communicated to

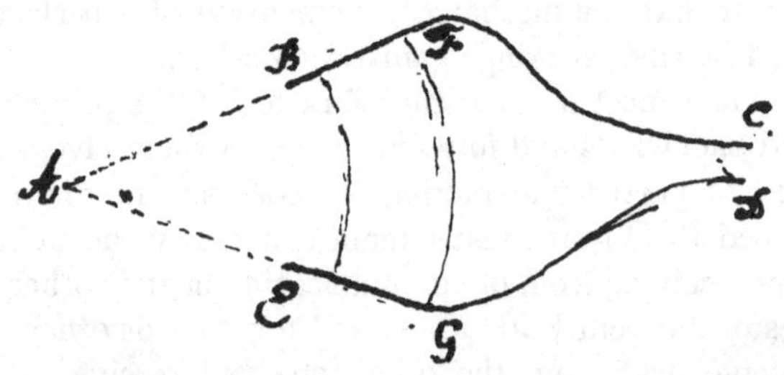

the air together with the momentum which would be produced by the compressed or rarefyed air restoring itself to its natural situation; that quantity will be neither encreased or diminished as the pulse moves forwards, and i there is any hollow vessel *BCDE* whose length *BC* is greater than the length o

[1] [Alternatively, some source of intrinsic or structural potential energy (e.g. o orbital motions in an electrical molecule) must be drawn upon when chemical change occur. This well-founded reservation may afford the reason for some remarks o Cavendish in other connexions, which have been interpreted as leaning toward the caloric theory. See also footnote, p. 408.

It is noteworthy that a very significant aphorism of Newton, *calor est agitati partium quaquaversum*, heat is uncoordinated internal motion, occurs abruptly in a chemical memorandum in Latin, printed in Horsley's Edition, vol. iv, pp. 397–400.

a pulse with its large end *BE* open and turn'd towards the sound as soon as the pulse is intirely within the vessel or which is the same thing when the hindermost part of the pulse is within the line *BE* compute as before the quantity of momentum of the air within the vessel; that quantity will not be alterd as the pulse proceeds on towards the narrower part of the vessel therefore the velocity of the vibrating fluid together with the compression and rarefaction of the air and the length of the pulse will be greater in the narrow part of the vessel than the other. This is the reason why deaf people hear better by applying those kind of funnels which they sometimes make use of to their ears.

[1][For like manner when the tide runs up any branch arm of the sea which is broader and deeper at the mouth than further in, the quantity of momentum of the water within the arm encreased by that which might be produced by all the particles of water which are raised above their natural level falling down to their level will remain still the same as the tide proceeds to the narrower part of the arm, and consequently the height to which the tide rises together with the velocity of the water and perhaps length of the tide or pulse will be encreased, supposing the water to be perfectly fluid or to yield perfectly easily to any motion impressed upon it and to be intirely void of cohesion and friction, for in that case as no force could be lost friction or the impinging of one body against the other the case would hold equally good as in a perfectly elastick fluid.

N.B. I here consider the motion of tides as exactly analogous to the pulses of air in sound.

Any kind of waves or such like motions either in air or water would upon the same supposing the fluid as before to be void of friction &ª would continue their motions in it for ever.]

Cor. 5th, fig. 1st. Suppose any number of bodies as *A*, *B*, *C*, *D* &ª connected together in any manner by perfectly elastick springs and suspended from one or more fixed points as *E*, *F* &ª to be put in motion, the sum of their me-

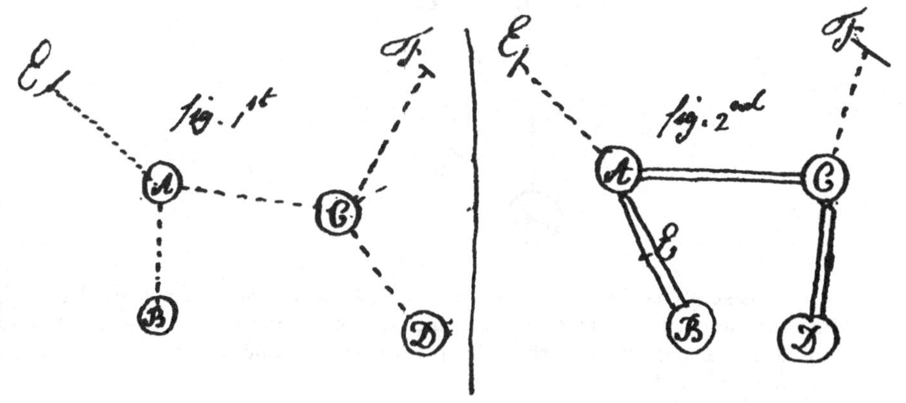

chanical momenta added to the momenta which might be produced by the restoring of the springs to any given degree of tension and the falling of the bodies to their proper level will remain constantly the same.

Fig. 2nd. The same thing will hold good if the bodies instead of being connected

[1] [The paragraphs within brackets are marked to be deleted in the manuscript.]

by springs are connected by solid inflexible rods only made so that the joints shall be intirely void of friction or shaking and that the parts of the engine are so constructed that they cannot impinge against one another; for suppose that the rod *AB* instead of being absolutely inflexible were perfectly elastick but of such a nature that it would require an infinite force to compress or bend it, in that case it would certainly hold good, but as long as this rod is acted upon by only finite powers it is of no signification whether it is inflexible or of such an elasticity here mention'd; but if 2 of these rods were to strike against each other in which case the force will be infinite then there will be a total difference between the 2 cases and there will be a great loss of force if the rod is incompressible; and it is plain from the nature of the lever and all mechanical powers and from what was said concerning bodies striking each other obliquely that the parts of the rod *AB* not being acted on immediately by the springs but by the mediation of the lever *AB* can make no alteration.

And in general we may conclude that whenever any system of bodies is in motion in such a manner that there can be no force lost by friction imperfect elasticity or the impinging of unelastick bodies, that then the sum of the mechanical momenta of the moving bodies added to the sum of the abovementioned additional momenta will remain constantly the same[1].

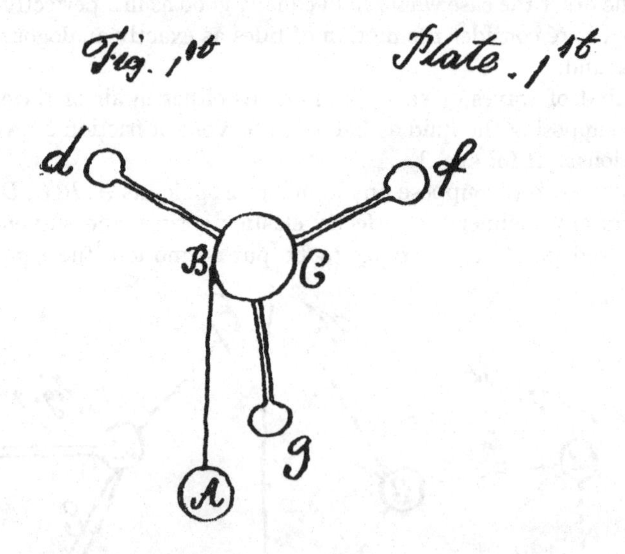

[1] [This surely is the earliest precise enunciation of the principle of the conser vation of energy, kinetic and potential, including enumeration of the causes that lea to its degradation, which on the principles of Cor. 2 would be into heat of precisel equivalent amount.]

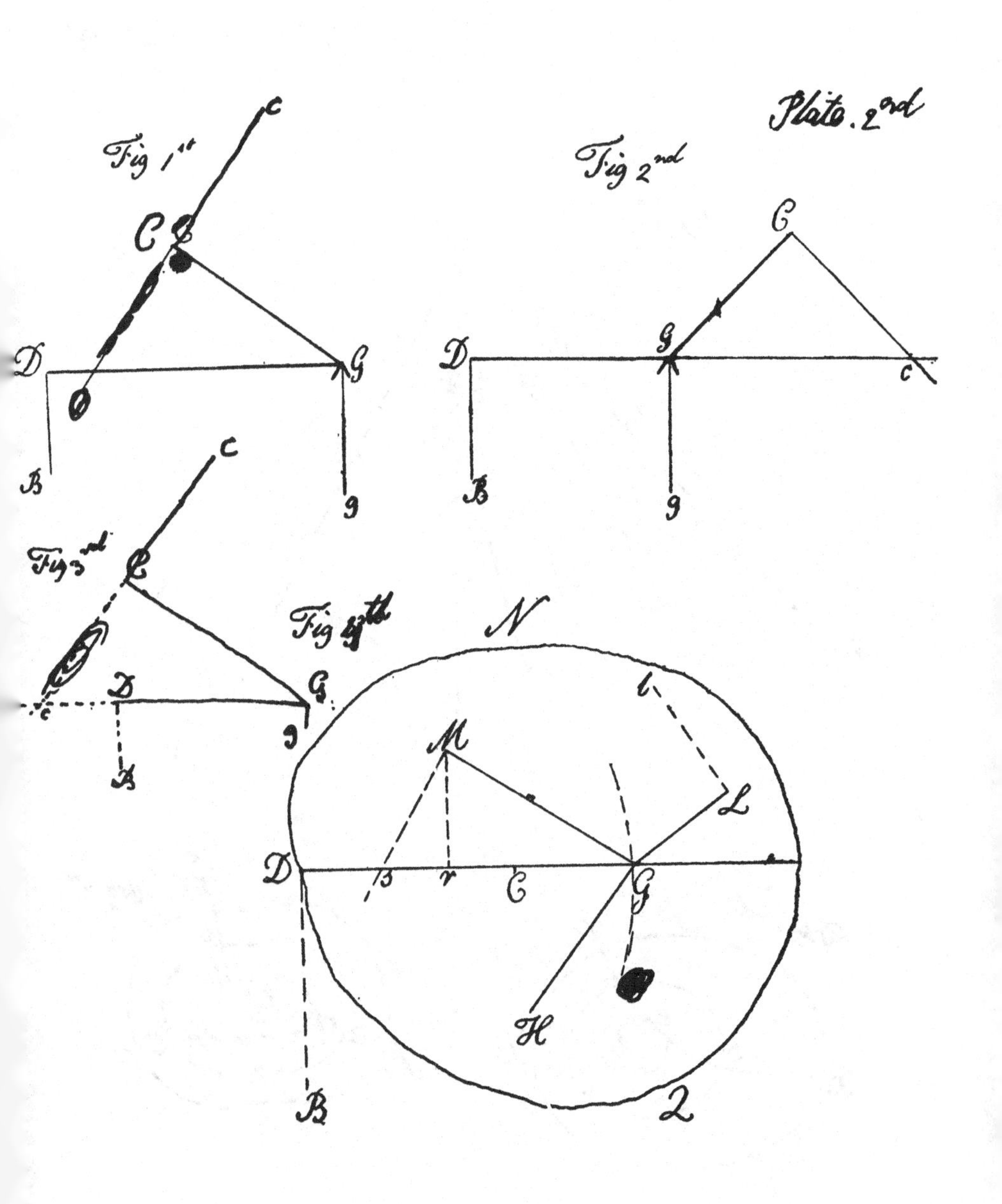
Plate. 2nd
Fig 1st
Fig 2nd
Fig 3rd
C
D
G
B
g
c
N
M
l
L
s
r
H

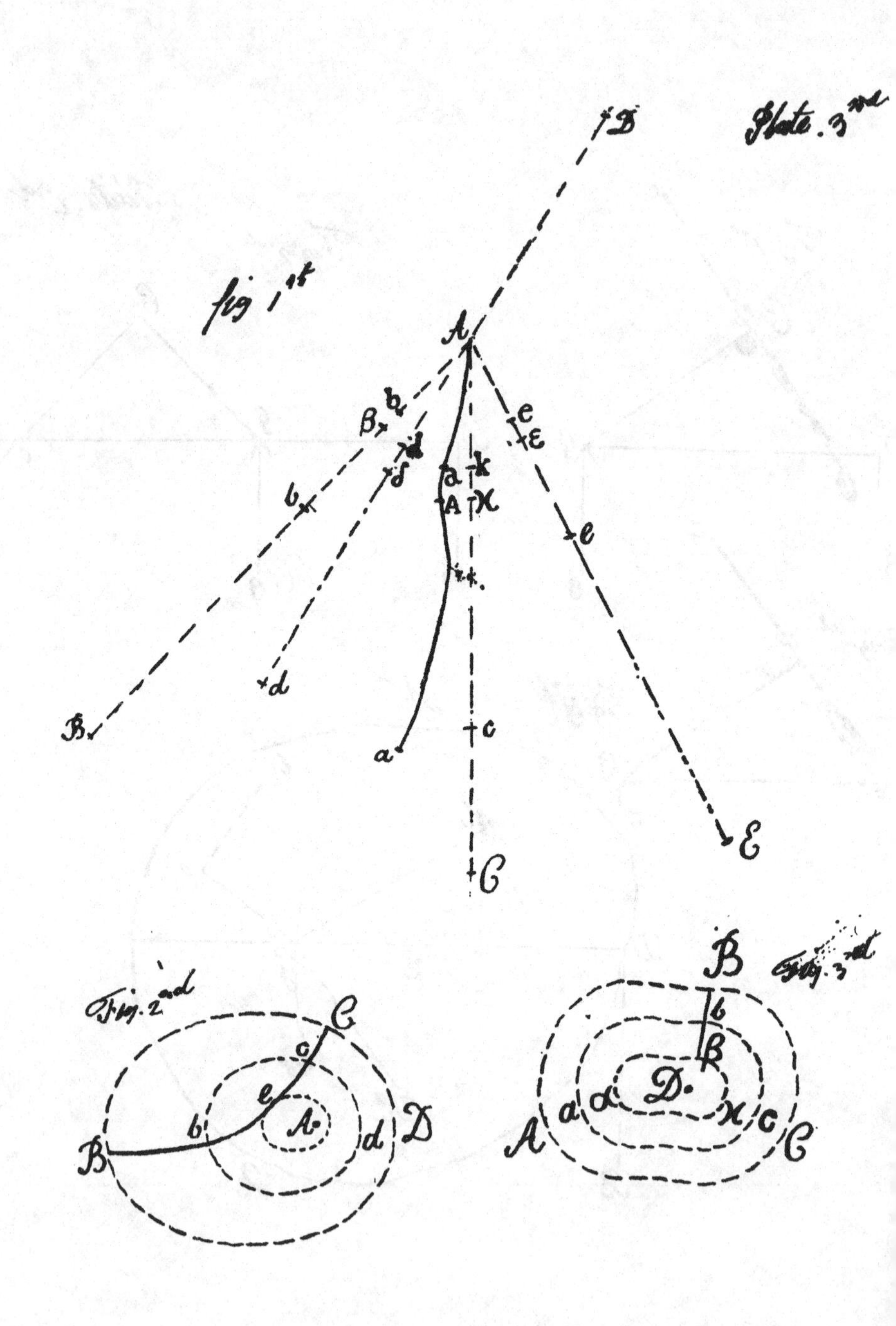
Plate. 3me
fig 1st
Fig. 2nd
Fig. 3rd

CAVENDISH AS A GEOLOGIST

CAVENDISH throughout the course of his long life rarely left London or thought to give himself a holiday. His habits were so fixed and regular that any interruption to the routine of his daily existence was irksome to him and was resented with an almost peevish impatience. Whatever rest he needed he seemed to find in change of occupation. The charms of the country had apparently few attractions for him. But at one period, viz. from 1785 to 1793, he was led to take up the study of Geology, then in its infancy as a science, possibly through the influence of his friend the Rev. John Michell, who occasionally dined with him at the Royal Society Club, and in whose conversation and correspondence he took great interest. In the summer and autumn of 1785, 1786, 1787 and 1793 he made a series of driving tours through portions of central and southern England and Wales in company with Sir Charles Blagden, at one time Secretary of the Royal Society, partly with a view of observing their geological and mineralogical features, and partly to familiarise himself with technical and manufacturing operations depending upon the applications of chemistry and physics. Accounts of these journeys are to be found among the Chatsworth Manuscripts. They are written in part by Cavendish himself, in part, apparently, by Blagden or by an amanuensis. As narratives of travel they have no particular value, nor do they afford any special information concerning such manufacturing processes as he was able to witness, as, for example, dyeing in the west of England, iron manufacture and mechanical engineering at Birmingham and its vicinity when he visited Watt, alum manufacture at Whitby, etc. He nowhere expressed any feeling for natural scenery and he seemed wholly unaffected by it. But, as may be surmised, whatever could be quantitatively ascertained was religiously noted. Thus he regularly observed by means of a way-wiser the distance travelled each day. He constantly read the height of the barometer as indicating the rise and fall of the roads he traversed. He observed the angles which distant hills and ridges subtended and noted the temperatures of the deep wells he came across.

Such scientific interest as these itineraries may possess is mainly confined to the geological observations they record, which are summarised in a special paper in Cavendish's handwriting. The manuscripts were accordingly submitted to Sir Archibald Geikie who was so good as to look through them and to contribute the following statement of his impressions of Cavendish's merits as a geologist.

NOTE ON CAVENDISH AS A GEOLOGIST

By Sir Archibald Geikie, O.M., K.C.B., F.R.S.

Cavendish evidently had a keen interest in tracing out the distribution of some of the more conspicuous geological formations across central and southern England. But he identified them, as he confessed, only by their superficial characters. These characters however are often deceptive. One of them on which he seems to have laid considerable stress was their colour. When he found a *yellow* limestone he concluded that it lay immediately below the Clay [Gault] which underlies the Chalk. Michell pointed out to him that another yellow limestone, below the "Lyas," was extensively developed from Leicestershire to Yorkshire and further north. Cavendish accepted this statement, calling Michell's formation the *Ancient Yellow Limestone.* There can be no doubt that this was what we know as the Magnesian Limestone which overlies the Coal-measures, as Michell pointed out.

Cavendish recognised that a succession of formations could be traced from the Chalk downwards through Clay (Gault) then Sands (Greensand) to his "yellow limestone" which seems to have been any portion of the calcareous zones in the Oolitic series, below which he saw that the "Lyas" lies. He seems to have formed a general notion of the distribution and trend of these formations in the wide extent of country which he had traversed. He perceived that they are less inclined to the horizon than the older rocks which lie to the west of them, and he correctly inferred that "for the most part the farther we go to the W. or N.W., the lower the strata we meet with."

He gives a curious proof that, as he confesses, "a very imperfect knowledge of the strata is to be acquired by merely looking at the surface, as is my case." He took the clay which lies below the gravels of London to be the same as that which lies below the Chalk, and he supposed that the Chalk has been entirely washed away from the London plain. He did not know that this thick formation lies still intact *below* the London Clay, protected from denudation by the mass of overlying Tertiary deposits.

Cavendish entirely missed the meaning of the Derbyshire Toadstone, preferring Michell's idea that it is "clay which was heated in its place," instead of Whitehurst's, who recognised the volcanic origin of the rock.

It is curious to reflect that while Cavendish was perambulating England in geological excursions, William Smith was busy with those observations and inferences among the very same rocks which gave him the true key to the stratigraphical succession, and laid a sound foundation for Stratigraphical Geology.

I do not think that Cavendish's paper should be published. It is of no geological importance, and it would add absolutely nothing of any consequence to his scientific renown.

CAVENDISH'S ASTRONOMICAL MANUSCRIPTS

Among the Cavendish Manuscripts preserved at Chatsworth are a number of disconnected papers relating to Astronomical Subjects. Many of them are simply notes or memoranda on matters of passing interest, or connected with the work of Committees of the Royal Society. Others are concerned with the mathematical treatment of observations. A few have been written out in detail as if for preservation or future reference. The great majority are without titles, and in many cases it is very difficult, if not impossible, to discover their meaning. Sir Frank Dyson was so good as to undertake to examine them. He writes as follows: "I have been through the Cavendish papers....They were extremely interesting, and one cannot help regretting that he did not publish more....I have indicated what they are about and made a few extracts."

ON CAVENDISH'S ASTRONOMICAL PAPERS

By Sir Frank Watson Dyson, LL.D., F.R.S.,
Astronomer Royal.

These papers, many of them very scrappy and not intended for publication, show the wide range of Cavendish's interests in all subjects connected with Astronomy. This is indicated by the titles:

The Moon's Atmosphere.
Table of Tides at various places.
On the Moon's parallax.
The Attraction of an Elliptic Wedge.
On the light of the full moon.
Alteration of shape in planets' orbits due to resisting Medium.
Method of adjusting the equatorial sector.
Herschel's planet.

In the last paper he concludes "On the supposition of a $\left\{\begin{matrix}\text{parabolic}\\ \text{circular}\end{matrix}\right\}$ orbit its distance from the sun during the time of its having been seen is $\left\{\begin{matrix}20\\ 19\end{matrix}\right\}$ times the distance of the Sun."

In another paper he deals with the difficulty caused by the custom of using the true equinox, and recommends the use of the Mean Equinox (i.e. corrected for nutation). The complete change of astronomical practice in this matter was made later by Bessel and Airy.

Evidently Cavendish was in close touch with Maskelyne, and two of Maskelyne's letters are found in this packet of papers.

Among a number of observations at different places to determine the height of Blanchard's balloon are a series made by Cavendish at Greenwich under the heading "Observations of the Altitude of Blanchard's balloon 16 Oct., 1784 with Bird's Astronomical Quadr. at the Royal Observatory. 4 ft. radius."

He was very much interested in the transit of Venus, and there are computations and correspondence as to the best places for its observation. Among the places mentioned in the letter to Dr Morton, Wardhus was occupied by Father Hell, and Tahiti by Captain Cook. In connection with the transit he considered the effect which would be produced by an atmosphere on Venus, and foresaw to some extent the difficulties which might arise in the observations. Incidentally he made some experiments on the *Minimum visible* under different conditions. He found 45″ for a small notch and 3″ for the diameter of a wire seen against skylight.

Cavendish maintained for a considerable time an interest in the difficult problem of the determination of the orbit of a comet from three observations. Among the papers there is a list of all the Comets whose orbits had been computed, and there are a number of scrappy papers of computation. These are undated. A more finished essay, apparently intended for publication, was sent to Maskelyne, who, in his reply dated April 16, 1788, suggests that Cavendish should compute the orbit of the Comet discovered in that year by Miss Herschel, and supplies observations. There is another letter of Maskelyne's to Cavendish dated Oct. 9, 1799, in which he says that Sir Henry Englefield to whom he had communicated observations of the Comet of 1799 at the same time as to Cavendish, had forwarded him a preliminary orbit. "I send these to save you unnecessary trouble or that you may direct it with more advantage by commencing with the rough elements here given."

The orbit of the Comet of 1799 is "computed by the table of Boscovich's sagitta." Besides using this Newtonian method he also made the computation by "a fluxional process." There are a few short notes on Laplace's method. At the conclusion of one of these notes he remarks

For La Place's method we only find whether the supposition agrees with observation in one respect, but the greatest fault is that if the angle subtended by the Earth and Sun at the Comet at either observation is nearly right, a small alteration in the radius vector makes a great error in the heliocentric place and therefore a small error in that observation will make a great alteration in the parabola.

None of this work appears to have been published, and no reference is made to Cavendish in Sir Henry Englefield's book (1793) which gives an account of Boscovich's and Laplace's methods, or in Ivory's paper in the *Phil. Trans.* 1814.

There are short notes on planetary perturbations of Comets, which probably had their origin in discussions about the near approach of Lexell's Comet to Jupiter in 1779. One of these is drawn up in the form of precepts for a computer and is entitled "Written for person thought of for calculating perturbations of expected Comet."

I have made four short extracts from the papers:

(1) Letter to Dr Morton.

(2) The last page of a paper on the Precession of the Equinoxes.

(3) A part of a very short note on the influence of the tides on the earth's rotation.

(4) An isolated scrap on the bending of a ray of light by gravitation, which is of interest, as the possibility of the bending of a ray of light by a gravitational field is at present engaging attention, though Cavendish was working on a corpuscular theory. This may have been suggested by Query 1 of Newton's *Opticks* "Do not Bodies act upon Light at a distance, and by their action bend its Rays, and is not this action (*coeteris paribus*) strongest at the least distance?"

(1) *Letter to Dr Morton.*

The best way of finding the parallax of the sun from the transit of Venus in 1769 is by a comparison of the duration of the transit in different places for which purpose it should be observed in such places where the difference of duration is the greatest. The duration is greatest about Tornea and Wardhus in Lapland. It is to be hoped that Swedes and Danes will send observers to these places. The place where the duration is least is in some of the islands supposed to be in the South Sea to the South of the Equator, and which would consequently be the best place to compare with Tornea and Wardhus. If observers could be sent there, which I imagine there is no probability of, the next best places are Cape Corrientes in Mexico, where the duration is above 16 minutes less than at Tornea, and California, where the duration is from 16 to 15 minutes less than at Tornea. The Royal Society proposes to send observers to California which is a better place than Cape Corrientes, as at this latter place the transit will end so little before sun set that there is great danger of the sun being hid in clouds.

It is very desirable that the transit should be observed also in some place where the duration is of an intermediate length between that at California and the North as it would serve as a check in case of error in either of other observations and besides that would be particularly useful in case the observation should fail at either of the other places. I know of no place so proper for this purpose as Kamptschatka. In the southern part of the peninsula of Kamptschatka the duration is near 8 minutes less than at Tornea and rather more than 8 minutes greater than at the south point at California, so that the difference is sufficient to deduce the parallax very well by comparing the duration at Kamptschatka with that at either of the foregoing places, supposing the observation to fail at one of them; I know of no other place proper for this purpose except the north western parts of Hudson Bay, where I believe it would be extremely difficult to send observers as ships can hardly get there early enough in the summer to land observers there before the transit, and it would be hardly practicable for observers to winter there. Besides that, if it was practicable to send observers there, it would not be so proper a place as Kamptschatka as the sun is much lower and consequently there is more danger of its being hid in clouds.

(2) *On Precession of the Equinoxes.*

The French Academicians by the mensuration of a degree determine the difference of the axes to be $\frac{1}{178}$ of the whole, but it has been demonstrated that if the earth be supposed to consist of spheroidical strata and is of the same density in each part of the same stratum, however different in different strata, or however different the degree of ellipticity of those strata, that then the difference of the 2 axes divided by the axis + the difference of gravity at the equator and pole divided by gravity = $\frac{5}{2}$ of the centrifugal force at the equator divided by gravity; the Academicians have also determined the difference of gravity to be $\frac{1}{180}$, whence the difference of the axes should come out $\frac{1}{318}$, therefore, as the difference of axes as observed by the Academy can not take place without assuming some very improbable hypothesis of the density of the earth, or by denying the theory (which seems too well founded to be shaken by this) and as the different measures of degrees agree so little with one another, as well as because the irregularity of the surface of the earth (particularly in the high mountains of Peru) may cause an alteration in the direction of gravity, and by that means disturb the accuracy of the experiment, I think we may fairly reject this mensuration and assume that difference of axes which agrees with the difference of gravity or $\frac{1}{318}$; this, if you suppose the annual precession caused by the sun to be 15″·4, answers very well, but if you suppose it to be 12″ or 8″·7 it may be reconciled with experiment by supposing that the degree of ellipticity of the spheroidical strata diminish as they approach nearer to the center. Thus, if you suppose the earth to consist of a spherical nucleus of an uniform density covered with a spheroidical shell of less density, and if you suppose the diameter of the nucleus to be $\frac{10}{11}$ of the diameter of the earth and its density to that of the outer shell as 13 : 9, the difference of axes of the earth and difference of gravity ought to be the same as I have here supposed and the precession caused by the sun would be 12″ or the mean quantity, but I know no way of accounting for a precession 8″·7 unless you suppose the density of the earth at first to increase and then diminish again as you approach towards the center; if you will grant this, you may account both for the difference of axes and precession by supposing the earth to consist of an hollow sphere whose outer diameter is $\frac{19}{20}$ and inner diameter $\frac{1}{10}$ of that of the earth and filled with a matter of less density and covered with a spheroidical shell and that the density of the outer shell hollow sphere and inclosed matter are to each other in the proportion of 1, 6 and $\frac{42}{100}$.

(3) *On the diminution of the diurnal motion of the earth in consequence of the tides.*

If there was no loss of force by friction the sum of the *vis viva* of the ☽ in its orbit round ⊕, the rotatory *vis-viva* of ⊕ considered as one Mass (or the visible) and the *vis viva* of the Water in respect of ⊕ (or the invisible *vis viva*) should remain unaltered except by the attracting parts approaching nearer. But if the invisible *vis viva* is diminished by friction, and the loss is continually supplied by the attract. ☽ then the sum of the *vis viva* of ☽ and of the visible *vis viva* of ⊕ must be as much diminished as to compensate that. But the only way by which the loss of invisible *vis viva* can be compensated is by the water being at a medium raised higher on that side of ⊕ which has left ☽ than on the other, and this will diminish the visible *vis viva* of ⊕ and increase that of ☽, and the increase of *vis viva* of ☽ is to the diminution of visible *vis viva* of ⊕ directly as their angular velocities, or as 1 : 13, and conseq. the diminution of visible *vis viva* of ⊕ is to the diminution of the invisible *vis viva* by friction as 13 to 12, and therefore may be considered as equal. It must be observed, however, that this increase of *vis viva* of ☽ will increase its distance from ⊕, and therefore will actually diminish its *vis viva*, but this does not affect the justness of the foregoing conclusion.

(4) *To find the bending of a ray of light which passes near the surface of any body by the attraction of that body.*

Let s be the centre of body and a a point of surface. Let the velocity of body revolving in a circle at a distance as from the body be to the velocity of light as $1 : u$, then will the sine of half bending of the ray be equal to $\frac{1}{1 + u^2}$.
[This deflection is half the amount given by Einstein's law of gravitation.]

CAVENDISH'S MAGNETIC WORK

CAVENDISH'S interest in Magnetism, and especially Terrestrial Magnetism, would appear to have originated through his association, as an investigator, with his father. Lord Charles Cavendish possessed instruments for observing magnetic declination and dip and seems to have occupied himself in making systematic observations in the garden of his house at Great Marlborough Street in which he was assisted by his son. It was probably at their instigation that the Royal Society included measurements of these elements in the scheme of meteorological observations which they instituted at their house in Crane Court, on which Cavendish, by desire of the Society, reported in 1766 (see p. 112).

The Chatsworth Manuscripts contain a considerable number of papers by Cavendish on Magnetism and related subjects. These have been very carefully examined by Dr Charles Chree, F.R.S., the Superintendent of the Kew Observatory, who writes as follows concerning them.

ON THE CAVENDISH MSS. RELATING TO MAGNETISM AND ASSOCIATED SUBJECTS

By CHARLES CHREE, Sc.D., LL.D., F.R.S.,
Superintendent of the Kew Observatory.

The Cavendish MSS. relating to magnetism consist of (1) a bundle of small octavo pages about 6½″ × 4″, usually two leaves, sometimes only one. Two leaves are usually paged as four pages. The pages run from 2 to 256, but numbers 167, 168, 209, 210, 223, 224, 225, 240, 241, 242, 243, 254, 255 and 256 each include two pages. There is an additional leaf 200 A, numbered on one side only. These all relate to declination observations. The observations on pp. 2—10 seem experimental. The year to which they refer is not given. On pp. 2, 3, 4 there are no dates, only the days of the week. On pp. 5—10 there are dates ranging from July 3 to December 30, but no year. Pp. 11, 12, 13 relating to observations in May and June 1782 seem similarly preliminary. The observations recorded on pp. 14 *et seq.* are mostly systematic and cover declination observations taken at Hampstead and Clapham between 1782 and 1809. The usual particulars given are the date, the hours of observation, the readings of the two ends of the needle and their mean. In 1805 and subsequent years, as a rule one reading only is given with a correction. Also some explanation is given, especially in p. 200 A, of how the corrections were arrived at. In addition to the actual declination observations, there are for each year the calculations from which the several mean declinations were derived.

Six pages (unnumbered) of the same size as above, and two slips summarise the mean declinations from 1782 to 1803. The pages and slips to a considerable extent are duplicates of one another.

The above form the basis of the tables giving declinations, mean values, ranges, etc. at Hampstead and Clapham.

A separate group of pages of similar size numbered 1 to 23, having subject "Effect of heat on magnets" indicated on one page. The results have been summarized.

(2) A bundle of small octavo pages about $6\frac{1}{2}'' \times 4''$ marked "Horizontal needle" on outside page. These are variously paged and include dip as well as declination observations. The first part, numbered pp. 1 to 61, deals with declination and dip observations made in 1773, 1774 and 1775. A good many of the observations are experimental. Others compare different magnets or aim at determining errors due to peculiarities of instruments, presence of magnetic matter, etc. A good many refer to observations made at the Royal Society House, or with the Royal Society needles. The observational data referring to the Royal Society House have been set out subsequently.

(3) There follow pages numbered 1 to 6, the first headed "Trial of dipping needles for R.S.," and five unnumbered leaves with a scrap of paper relating apparently to needles by Ramsden and Nairne.

Then pages numbered 1 to 4, the first headed "Sissons dipping needle."

Then four leaves describing some dipping needle.

Then four leaves unnumbered headed "springing of needles" giving some values of Young's modulus. These have been made use of in what follows.

Then four leaves, two blank, headed "Trial of long bars by compass."

Then eleven leaves, some blank, numbered 1 to 15, describing results of experiments on strength of variously shaped needles which are subsequently dealt with.

Then a series of pages numbered 1 to 58 dealing with dip observations. Some of the observations seem experimental. There is a synopsis of the experiments recorded on pp. 11—16, but their object is not explained. On pp. 17, 18, 19, 25, 26, 27, 28, 33, 36, 37, 38, 39, 40, 41, 42, 43 there are results of dip observations taken several times a day throughout a number of months, presumably during several years of which 1775 was certainly one (cf. p. 39). [There seems no attempt to deduce from them the nature of the diurnal variation, but it is difficult to imagine what other purpose Cavendish can have had in view. Presumably when he started observing the probable extent of the diurnal range was unknown to him *v. infra.*] Pp. 45—54 deal with dip observations made in August 1778 in London and various places in England, the results of which have been reproduced. Pp. 55—58 deal with dip observations made in 1791 which have also been utilised. In the same bundle are two sheets with dip results obtained with Nairne's needle inside a house, and also with its plane out of the magnetic meridian. There follow eight pages, some of them blank, giving some declination results. The place of observation is not stated. One page containing a summary of results for 1788, 1789, 1790 is headed "Gilpin's observing needle." The declinations it gives are fully 10′ in excess of those obtained in the same years at Clapham. A second set of data for 1789 give lower values for the declination than those at Clapham. At the end of the packet are a few pages having apparently nothing to do with magnetism. One contains some information as to radii of gyration; a second refers to an elasticity experiment in which a modulus of elasticity was derived for crown glass; a third refers to some astronomical calculation.

There are finally a bundle of miscellaneous papers with sheets of various sizes. These comprise: a paper, paged 1 to 13, entitled "Bending of tapering needle by its weight." This includes the mathematical solution of the problem and some numerical results which are referred to subsequently.

A MS. of two sheets containing declination results obtained in Cecil Street and Pall Mall between 1759 and 1775 (*v. infra*).

Three sheets dealing with declination and dip results obtained in June and July 1776 and 1778 at the Royal Society's House (see p. 465 *et seq.*).

One sheet dealing apparently with error in (Royal Society?) compass in 1777.

One sheet and small scrap of paper dealing with error of (Royal Society's?) compass in 1779 and 1780.

A sheet giving some declination results in Pall Mall in May 1787.

A sheet giving results of hourly declination observations with "Mr Gilpin's observing needle" in June 1788 (see p. 472 *et seq.*).

A sheet summarising declination results in Pall Mall from 1782 to 1791.

A sheet containing apparently the results of half hourly observations of declination from 6 a.m. to 7 p.m. on Oct. 24th (?) of an unspecified year, at an unspecified place. The range shown is 12′.

A sheet giving some comparative declination results for Clapham and Pall Mall, and some yearly means at Hampstead and Clapham.

A sheet giving "Variation at Royal Society" in June, July and September 1802 and 1803.

Two sheets (six pages) describing some experiments with different suspensions in a declination needle, referring apparently to effects of moisture.

Eight pages (three blank) apparently notes from some work on Terrestrial Magnetism, including dip results obtained at a number of places in England in 1720.

Sheet referring to old declination observations, copied apparently from Gellibrand's work on Magnetism.

Sheet containing some unexplained data headed 1774 and 1775.

Large sheet folded in four dealing with some unexplained experiments apparently with dip needles near a disturbing magnet in 1778 and 1779.

Sheet folded in two giving results of deflection experiments made in 1776 with a poker and the cast iron "cheeks" of a stove.

Sheet folded in two giving results of deflection experiments made in 1776 with an iron and a steel bar supplied by Elwell.

Three large sheets folded in four dealing with deflection experiments made with blistered steel, cast iron and forged iron bars supplied by Elwell (see p. 444).

Sheet with some unexplained diagrams of parallelograms having apparently something to do with prospects obstructed and not obstructed by trees at some unspecified place.

One sheet folded in two, one side being the solution of a problem in spherical trigonometry, the other side inscribed "Computation how great the dip must be that the error caused by moving needle a given distance from magnet(ic) meridian shall be a maximum and how great the error is in that case."

Ten large and ten small pages (some blank) giving calculations in spherical trigonometry. One page is inscribed "Examination whether direction of horizontal needle (i.e. declination needle) is in the small circle passing through the two points of surface in which dip = 90°, the two magnetic poles being supposed to be at a great and equal depth below surface and the distance of the two above mentioned points being near 180°."

MS. eight large pages inscribed "The method of balancing the needle after it is made magnetical." This refers to dip needles furnished with balancing screws.

MS. paged 1 to 5 inscribed "To find the true dip from observations made in the four different ways when the difference between those ways is considerable." This refers to the case when there are considerable differences between the dips obtained with the circle facing east and west and with the two ends dipping.

Four pages, three blank, inscribed "On the different forms of constructing a dipping needle," refers to cases where the axle rolls on horizontal planes and on friction wheels.

Four pages dealing with effect of ship's iron on compass (see p. 463).

MS. paged 1 to 26, and sheet with two figures, inscribed "On the different construction of dipping needles." This deals with various sources of error including some numerical results for bending of different shaped needles (see p. 453).

A single sheet, folded in two, inscribed "For Captain Pickersgill."

A MS. paged 1 to 7 inscribed "For Cook and Bayley."

A MS. paged 1 to 17 inscribed "Directions for using the dipping needle for Dalrymple."

These three MSS. contain instructions to travellers and are dealt with subsequently (see p. 462).

Three scraps of paper unintelligible by themselves.

The substance of these papers may be conveniently arranged and dealt with as follows:

§ 1. Introduction.
§ 2. Experiments on "fixed" (permanent) and "moveable" (temporary) magnetism.
§ 3. Effect of Heat on Magnets.
§ 4. Strengths of Magnets of various cross-sections.
§ 5. "Springing" (Elastic Bending) of Needles.
§ 6. Errors in Observed Dip due to bending of Dip Needles.
§ 7. Sources of error in Dip Observations.
§ 8. Instructions to Observers and General Notes.
§ 9. Dip Observations.
§ 10. Diurnal Variation of Dip.
§ 11. Declination Observations.
§ 12. Secular Change of Declination.
§ 13. Diurnal Variation of Declination.
§ 14. Disturbed Days.

INTRODUCTION

§ 1. To facilitate the comprehension of the experimental work on magnets which Cavendish executed, it is desirable to consider first how he measured the strength of magnets.

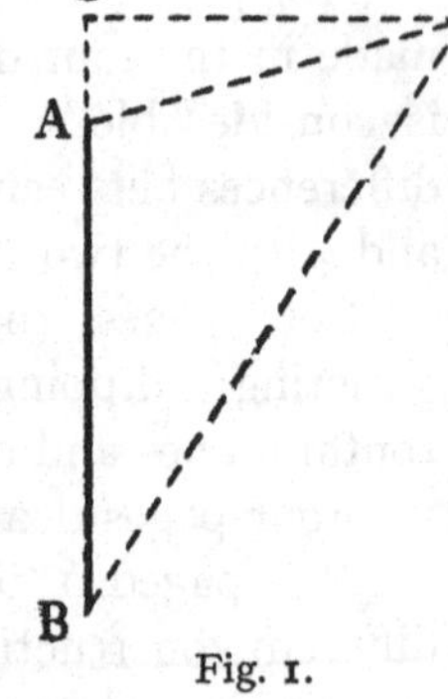

Fig. 1.

Suppose A and B to represent the poles of a magnet in a vertical position, and P a point at a horizontal distance $PO = d$ from the vertical line BA. Let $+\mu$ and $-\mu$ represent the equal strengths of the two poles. Let l denote the length AB, and let ψ_1 and ψ_2 denote the angles APO and BPO. If $m \equiv \mu l$ be the magnetic moment of the magnet, it is easily proved that the horizontal component F of the resultant magnetic force at P is directed along OP and is given by

$$F = (m/ld^2)(\cos^3\psi_1 - \cos^3\psi_2) \quad\text{..................(1).}$$

Suppose now that a small compass needle NS, capable of motion only in a horizontal plane, has its centre at the point P, and that Fig. 2 represents this needle as deflected out of the magnetic meridian MM'. We may suppose the needle so short compared with the horizontal distance d of its centre from the deflecting magnet that the magnetic force may be regarded as the same at the positions occupied by the two poles N and S. Suppose this force F to make an angle α with the perpendicular to the magnetic meridian, and ϕ to be the inclination of the deflected needle NS to the magnetic meridian. Then we obviously have

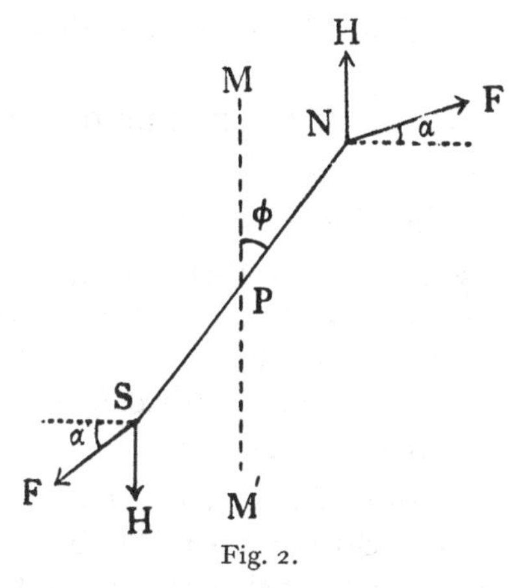

Fig. 2.

$$F \cos \alpha = (H + F \sin \alpha) \tan \phi.$$

If α be small, and ϕ not too large, a close approximation is

$$F = H \tan \phi \quad \text{.................................(2).}$$

If the same magnet in different magnetic conditions—or a series of magnets in succession, identical in pole distance—occupies the fixed position AB, we get for the magnetic moment by combining (1) and (2)

$$m = CH \tan \phi \quad \text{..............................(3),}$$

where C is a constant determined by the values of l and d and the consequent values of ψ_1 and ψ_2.

For very exact work (2) would have to be replaced by a more complicated expression which allowed for the finite length of the deflected compass needle, and allowance would have to be made for variations in H during the time of the observations. There were no magnetographs in Cavendish's time, and he naturally was obliged to treat H as a constant. Also, instead of taking changes in m as measured by changes in $\tan \phi$, he regarded them as measured by the changes in ϕ itself. As ϕ was not in all cases small, this cannot be regarded as altogether satisfactory. Still the results suffice to give a good general idea of the nature of the phenomena, and this seems to have been all Cavendish really aimed at.

The objects he seemed to have had principally in view were to ascertain what kinds of iron or steel were most suitable for temporary and permanent magnets; how magnets were affected by changes of temperature; and what shapes should be given to magnets to secure high strength for a minimum of weight.

In his experiments Cavendish seems invariably to have got one end of the deflecting magnet—either the upper or the lower end—level with the auxiliary compass needle. He may have supposed the poles of bar magnets to be situated at the extreme ends, but there is nothing to show that he did. The position is obviously one which could be recovered with considerable accuracy without any elaborate machinery.

EXPERIMENTS ON "FIXED" (PERMANENT) AND "MOVEABLE" (TEMPORARY) MAGNETISM

§ 2. The first experiments we shall consider were made in 1776. They seem to have been led up to by some experiments, to which there are references in several notes, which employed various iron objects that happened to be at hand, including part of a stove, a poker and an iron bar from an electrical machine. After this preliminary stage, to quote Cavendish's own words,

> Some bars were got from Elwell 31¾ inch long, 2·1 (inch) broad and about 0·5 (inch) thick. On May 29, 1776, one of each (kind) was made magnetical, the marked end being the South pole. In trying the experiment the bars were placed perpendicularly (i.e. in the vertical position) against a wall 25 inches distant from the centre of the (auxiliary compass) needle, 91°¼ to west of usual magnetic north, either the top or bottom of the bar being always on a level with the needle. They were kept constantly with the marked end upwards till after the observation of June 30, after which they were kept with the mark downwards.

These remarks apply to four only of the Elwell bars, including an unnumbered "blistered steel" bar, and three others, all called No. 1, distinguished as "blistered steel hammered," "cast iron" and "forged iron" respectively. It is obvious that what the statement means is that, except for the short time when experiments were in actual operation, the magnets up to June 30 were kept with the south (marked) pole upwards, and so in the position in which the earth's vertical field tended to strengthen them; but after June 30 they were kept south-pole down, and so with the earth's field tending to demagnetise them.

The plan of the experiments was as follows:

With the marked, i.e. south, pole up, the lower or north pole was brought level with the compass needle, the deflection of whose north pole ϕ_1 to the east of the magnetic meridian was noted; then, still with south pole uppermost, the upper or south pole was brought level with the compass and the deflection ϕ_2 to west of magnetic north was noted. The algebraic difference of the two angles, i.e. $\phi_1 + \phi_2$ taken numerically was called A. The magnet was then inverted so that the marked pole was downmost, and readings were again taken of the compass needle, ϕ_3 counted + to east, when the lower, i.e. marked pole, was level with the compass, and ϕ_4, counted + to west, when the upper pole was level with the compass. The numerical sum $\phi_3 + \phi_4$ of these two angles was called B. To see the significance of these angles, let m_p represent the magnetimoment arising from the permanent, or as Cavendish called it the "fixed" magnetism, and m_t the magnetic moment arising from the temporary magnetism, induced by the earth's vertical field, or as Cavendish calls it the "moveable" magnetism. When the marked pole was up, the magnetic moment was $m_t + m_p$; when it was down—supposing for mathematical convenience the temporary magnetism to be the stronger—the magnetic moment was $m_t - m_p$.

If now we suppose with Cavendish that (3) is replaceable by

$$m = (1/C')\,\phi \qquad (4),$$

where C' is a constant, we have

$$\left.\begin{aligned} A &= 2C'\,(m_t + m_p) \\ B &= 2C'\,(m_t - m_p) \end{aligned}\right\} \qquad (5),$$

whence

$$\left.\begin{aligned} m_p &= (A - B)/4C' \\ m_t &= (A + B)/4C' \end{aligned}\right\} \qquad (6).$$

Thus the variations in $A - B$ and in $A + B$ tell us the changes in the permanent and temporary magnetic moments. These variations are shown in the following table.

TABLE I. "Fixed" and "moveable" magnetism.

Date		Blistered Steel		Blistered Steel (hammered), No. 1		Cast Iron, No. 1		Forged Iron, No. 1	
		$A-B$	$A+B$	$A-B$	$A+B$	$A-B$	$A+B$	$A-B$	$A+B$
1776		° ′	° ′	° ′	° ′	° ′	° ′	° ′	° ′
May	29	27 30	8 30	—	—	—	—	—	—
"	30	27 20	8 24	90 42	4 8	68 32	3 48	26 18	12 42
"	31	—	—	89 50	3 50	68 14	3 26	24 12	11 22
June	2	27 32	8 28	90 42	3 52	69 0	3 40	24 19	11 35
"	6	27 26	8 40	90 48	4 4	69 11	3 35	24 1	11 35
"	19	27 21	8 41	90 48	3 52	69 13	3 37	23 46	11 28
"	22	27 5	8 39	90 51	3 45	68 55	3 41	23 40	11 24
"	30	27 11	8 35	90 54	4 0	69 12	3 32	23 27	11 49
July	2	26 51	8 39	90 43	4 3	69 1	3 37	22 51	11 35
"	29	26 40	8 46	90 15	3 55	68 55	3 49	22 53	11 33
August	10	26 32	8 36	89 30	3 56	68 15	3 41	22 40	11 36
October	9	25 30	8 34	89 38	3 42	68 14	3 36	22 25	11 19
"	27	25 34	8 34	89 34	3 50	68 2	3 32	22 28	11 22
1777									
June	2	24 20	8 44	86 43	4 3	66 2	3 48	21 30	11 34
"	13	24 25	8 31	86 35	3 51	66 0	3 36	—	—

In considering the significance of these figures it must be remembered that C' in (4) is not an absolute constant, but varies with H, that the induced magnetism necessarily varies with changes in the vertical force, and that the magnetic moment of a magnet is a quantity having a temperature coefficient. Thus some irregular fluctuation was inevitable in the figures in Table I, even if the apparatus had possessed every modern refinement. That the fluctuations should be so regular as they are, considering the rough character of the apparatus, is a remarkable tribute to Cavendish's care and skill as an observer.

When we look at the large size of $A - B$ in the case of the second and third bars, the criticism that the use of the deflection angles instead of their tangents was under the circumstances very inexact naturally suggests itself. But it should be remembered that $A - B$ has to be divided by 4 to get the real size of the angle. If we take as an example $22°\cdot5$, i.e. $\pi/8$, we have

$$(\tan 22°\cdot5) \div (\pi/8) = \cdot414/\cdot393 = 1\cdot05,$$

so that even in this case the departure from unity is not very serious.

The first observation with the forged iron bar gives an outstandingly large value for both $A + B$ and $A - B$ which it is difficult to account for satisfactorily. With this exception the fluctuations in the values in the several $A + B$ columns appear to be fortuitous. They supply no evidence of any change in the induction coefficient in the earth's vertical field. With the $A - B$ columns, i.e. the permanent magnetic moment, it is quite otherwise. In the case of the blistered steel bars, whether hammered or not, there was evidently little if any change in the permanent moment so long as the bars were kept south pole up; but subsequent to June 30, 1776, when the earth's vertical field tended to demagnetise the bars, the loss of moment is clearly apparent. The cast iron bar seemed if anything to increase in moment up to June 30, 1776, but subsequently there was an undoubted fall. The forged iron bar seems to have lost moment from the start. Its percentage loss of moment subsequent to June 30, 1776, was similar to that in the other bars.

The differences between the individual bars—which, it will be remembered, were identical in size—are very striking. If the two blistered steel bars were of the same material, differing only in the hammering to which the one was subjected, the effect of the treatment was surprisingly large. Roughly speaking, it halved the temporary, and trebled the permanent magnetism, both excellent things for a horizontal force collimator magnet. The cast iron bar had apparently a slightly lower temporary moment than the hammered steel, and a permanent moment some 25 per cent. less. The forged iron bar was evidently the softest, its temporary induction being three or four times as large as that of the hammered steel and cast iron bars.

Of the remaining bars obtained from Elwell, Cavendish writes:

The three bars marked No. 2 were kept constantly with the mark upwards. In the first trials with them the compass was placed in the same position as with the other bars (i.e. the horizontal distance of the compass from the wall agains which the magnets stood was 25 inches); in the others (i.e. in the later experiments) it was placed....17·8 (inches) from the wall.

The three No. 2 bars represented, like the No. 1 bars, blistered steel hammered, cast iron and forged iron. Apparently these bars were not stroked, as the No. 1 bars were, but had acquired a small amount of permanent magnetism before the readings commenced, presumably simply from continued exposure in one position to the earth's vertical field. The observations on them commenced on May 30, 1776. On May 31 after readings had been taken with the compass 25 inches from the wall, they were repeated with the distance reduced to 17·8 inches, and all the subsequent readings were taken at this reduced distance. As to the subsequent treatment of the bars, Cavendish writes of the blistered steel No. 2 bar that after the readings taken on June 10, 1777, it was "struck 100 times on anvil, falling 1·6 inches by its weight. This was repeated thrice, giving 300 blows in all." After the reading on June 11, 100 blows were repeated twice. Then 100 blows were repeated each day until June 26. It was then "struck almost every day" until July 12, "struck several times" between July 12 and August 21, and finally "struck

TABLE II. "Fixed" and "moveable" magnetism.

Date		Blistered Steel (hammered), No. 2		Cast Iron, No. 2		Forged Iron, No. 2	
		$A-B$	$A+B$	$A-B$	$A+B$	$A-B$	$A+B$
1776		° ′	° ′	° ′	° ′	° ′	° ′
May	30	1 3	5 3	1 30	3 44	0 45	10 15
,,	31	0 52	5 2	1 35	3 41	0 18	10 12
,,	31	2 20	11 0	3 15	8 15	1 4	22 10
June	2	2 25	11 15	3 12	8 18	1 29	22 45
,,	19	4 10	11 20	3 24	8 30	1 25	22 25
,,	25	3 57	11 13	3 24	8 20	1 40	22 34
July	28	4 38	11 18	3 33	8 37	1 2	22 58
August	10	4 38	11 18	3 35	8 29	1 30	22 50
October	10	5 35	11 15	3 40	8 34	2 0	23 6
,,	29	5 2	10 56	3 37	8 23	1 47	22 37
1777							
June	8	6 53	10 47	4 5	8 15	1 57	22 37
,,	10	6 52	10 52	—	—	—	—
,,	11	—	—	4 14	8 30	1 51	22 5
,,	11	8 4	11 10	—	—	—	—
,,	12	—	—	4 32	8 22	9 57	23 49
,,	26	8 35	11 5	4 44	8 30	10 37	23 59
July	12	8 38	10 58	4 49	8 23	10 50	23 38
August	21	8 50	11 4	4 43	8 17	10 52	23 32
October	1	9 22	11 2	5 2	8 32	11 30	22 54

sometimes" between August 21 and October 1. The cast iron No. 2 bar was struck 250 times on June 11, 1777, and 100 times on June 12 before the readings taken on that day. The forged iron No. 2 bar was struck 300 times on June 11, and 200 times on June 12 before the reading was taken. Apparently after July 12 the cast iron and forged iron bars were treated identically with the blistered steel bar. The results of the observations appear in Table II, the significance of $A - B$ and $A + B$ being the same as in Table I.

The first of the horizontal lines in the table marks the reduction in the distance between the magnet and the compass. Only the readings above this line are immediately comparable with those in Table I. The second horizontal line marks the introduction of the experiments intended to show the effect of mechanical agitation on the magnetic moment.

Comparing the values of $A + B$, i.e. the temporary magnetic moments, observed on May 30 and 31, 1776 before the reduction of distance, with those given for the corresponding No. 1 bars in Table I, we see that the differences are small. This means that the presence or absence of a large permanent magnetic moment makes little difference to the temporary magnetic moment.

The reduction effected on May 31, 1776, in the distance of the compass needle was doubtless intended to secure greater sensitiveness, the permanent magnetic moments being so small. Between May 31, 1776 and June 11, 1777 any change in the permanent moments must be regarded as arising from simple exposure to the earth's vertical field, acting steadily in the direction to increase the magnetism. The bars all show some increase of moment, but it is decidedly larger in the blistered steel bar than in the others. All this time, it will be observed, the temporary magnetic moment remained the same, or very nearly so, just as in the case of the No. 1 bars in Table I.

The introduction of mechanical agitation of the bars on June 11, 1777, produced at once a remarkable increase in the permanent magnetism of the forged iron bar. It had also a considerable effect on the blistered steel bar, but not much on the cast iron bar. The successive applications of the treatment tended apparently to further increase the permanent moment, but only to a comparatively minor extent. There is nothing to suggest that the mechanical agitation had any effect on the induction coefficient for temporary magnetism of the blistered steel or cast iron bars, but there is at least a suspicion of an increased induction coefficient in the forged iron bar.

If we take the formula (1) and simplify it by assuming the poles at the ends of the magnets—the distance being probably about $\frac{1}{12}$ of the magnet's length—we find that increasing the distance d from 17·8 to 25 inches would reduce the force F to about 0·43 of its value at the shorter distance. If we compare the sum of the $A + B$ deflection angles on May 31 before and after the increase of distance we obtain a ratio of 0·46. We shall thus certainly not be far wrong if we regard the angles observed on October 1, 1777, in Table II when multiplied by 0·45 as fairly comparable with the corresponding angles in Table I. We thus infer that mechanical agitation alone sufficed in the case of the forged iron to secure a permanent magnetic moment about a quarter the size of that obtained by stroking in the usual way. This agrees with the conclusions we should derive if we assumed the temporary magnetic moments equal in the bars Nos. 1 and 2.

Effect of Heat on Magnets

§ 3. A small packet of papers entitled "Effect of heat on magnets" deals with experiments on the effects produced by immersing magnets in water at about 115° F., the previous temperature being that of the air, about 65° F. usually. The experiments were made in July, but the year is not stated.

The effects of a sudden rise of temperature on a magnet or on an unmagnetised bar of iron or steel are somewhat complex. There is first of all a shock effect, apparently similar to that caused by mechanical agitation. If an ordinary magnet, especially one recently magnetised, has a sudden rise of temperature, such as occurred in Cavendish's experiments where the bar was lowered into a glass of hot water, an immediate loss of magnetic moment occurs. The loss may not be altogether permanent, i.e. if the magnet is kept for some days after the incident at its original temperature, a partial recovery of moment may take place. But a considerable part at least of the loss seems to be permanent. If the experiment be repeated a second time, there is usually still further loss, but the loss is less. After some repetitions there is at least an approach to a cyclic state of matters, in which the lower magnetic moment is associated with the higher temperature. The cyclic change, which alone represents the effect of a true temperature coefficient, is usually small, and unless there are a very large number of observations accidental changes in the earth's field affecting the readings of the auxiliary compass may introduce an undesirably large element of uncertainty, which can be eliminated only when magnetograph records are available. This will I hope explain why only a short summary of the results is given.

The experiments were confined to three bars. One of these, No. 4, had apparently been magnetised for some time. It was 10 inches long; its other dimensions are not recorded. Its marked end was a north pole. The observations were made exactly in the same way as in the case of the experiments summarised in Tables I and II. The double deflection angle, i.e. the algebraic difference of the readings of the auxiliary compass when level first with the one then with the other of the two poles of No. 4, is recorded in all cases, the marked end of No. 4 being sometimes up sometimes down. The differences between the results with mark up and down were very small, showing that No. 4 had a very small temporary induction coefficient. Its temperature coefficient was apparently also small. The shock effects are however apparent. On the first occasion the marked, i.e. north, pole was down when the sudden rise of temperature, amounting to 63·5° F., occurred. The double deflection angle fell from 44° 0′ to 43° 14′. On the next occasion, with the marked end up, a rise of 56° F. was accompanied by a fall of 48′ in the deflection angle. After this experiment No. 4 was kept for about a fortnight marked pole up, and so suffering the demagnetising effect of the earth's vertical field. This reduced the double deflection angle by about $2^\circ\frac{3}{4}$. Subsequently two more sudden changes of temperature similar to the two previous were made. The first took place with the marked end up, the double deflection angle falling 36′. The second also took place with the marked end up, the double deflection angle falling only 8′. Some hours thereafter observations were made at normal temperature and a rise of 5′ was observed, as compared with the reading taken at the high temperature. This suggests that a close approach had been made to the cyclic state.

The two other bars had not been stroked prior to the commencement of the experiments, and possessed originally only a small amount of permanent magnetism, acquired under exposure to the earth's vertical field. They were both $19\frac{1}{2}$ inches long, the cross-section being a square of side 0·75 inch. No. 1 was of steel, No. 2 of iron.

When experiments began on No. 1 the lower end, whether the marked or the unmarked end, was a north pole, but the moment was somewhat larger when the marked pole was up. Thus the marked end was the south pole of the permanent magnetism, but the temporary considerably exceeded the permanent magnetism. When the first sudden rise of temperature, amounting to 41° F., was applied, the marked pole was up, and the double deflection angle rose 34′. Thus the heating acted apparently in the main as hammering would have done, facilitating the action of the earth's vertical field. On the second occasion a sudden rise of 48° F. of temperature occurred with the marked pole down. There was now a small fall in the double deflection angle with marked pole up, but a considerable increase in the angle with marked pole down. Also the deflection angle was now greater with the marked end down than with it up. Thus apparently the shock enabled the earth's field to reverse the magnetism. After this the bar was stroked so as to make the marked the north pole, and in the intervals between the experiments it was kept marked pole up, and so suffering demagnetisation from the earth's field. The cycle: normal, hot, normal temperature was applied on each of two days, one day intervening, the range of temperature being 47° F. on the first occasion and 52° F. on the second. On the first occasion the double deflection angle on return to the normal temperature showed a fall of 2° 58′; on the second occasion there was again a fall, but only of 25′

The change from the intermediate hot to the final normal temperature was accompanied on both occasions by a rise in the double deflection angle, amounting to 41′ on the first occasion and 67′ on the second. The double deflection angle was about 37° on the second occasion. The passage from the high to the normal temperature on that occasion would certainly not be accompanied by more than a small fraction, if any, of the permanent loss sustained during the cycle. Taking the moment as proportional to the deflection angle, we get as a rough measure of the temperature coefficient per 1° F:

$$q = 67 \div (37 \times 60 \times 52) = \cdot 0006.$$

This is at least of the right order, but high for good magnet steel.

The No. 2 or iron bar was treated in a similar way to No. 1. Originally it showed little if any permanent moment, the double deflection angle being mark up 10° 7′, mark down 9° 50′. But on being suddenly raised 44° F. in temperature, the marked end being down, the readings were: mark down 11° 35′, mark up 9° 55′. Thus heating had the effect of a mechanical shock, the bar tending to have the pole that was down at the time made the north pole of a permanent magnet. Two days later the effect seemed to have largely disappeared, the bar appearing again to be nearly neutral. The temperature was again suddenly raised, the mark being this time up, and again the bar assumed a small permanent moment, the marked end being this time the south pole. The bar was then stroked, converting it into a permanent magnet with the marked end a south pole. The permanent and temporary moments in this bar were now so nearly equal that when the marked pole was down the deflection angle was almost nil. The temperature cycle: normal, hot, normal was applied on two days, one day intervening. On the first occasion the deflection angle fell both with the rise and the subsequent fall of temperature, the combined effect being a reduction in the double deflection angle from 19° 38′ to 18° 54′. On the second occasion the apparent differences between the readings were too small to possess any significance.

Strengths of Magnets of various cross-sections

§ 4. During May of some unspecified year Cavendish made a number of experiments on two needles made by Nairne and nine made by Elwell. Of the latter five were old and four new. He exposed them to a variety of treatment and tested their strengths, employing an auxiliary compass in the way already described. The deflection angles were of the order of 10°, so that replacing the angle by its tangent would have made little difference to the numerical results. After concluding the experiments Cavendish tried whether any formula could be found which represented satisfactorily the relation between the magnetic moment and the dimensions of the cross-section. The only calculations carried to a conclusion related to five magnets all by Elwell, but supplied at two different dates.

The information given as to their weights and dimensions in different places is not in all respects consistent, but the following data accord closely with the values employed by Cavendish for log bt and log $(b + t)$, where b denotes the breadth and t the thickness, and they also fit fairly with the recorded values of the weights. The dimensions are in inches. The descriptive terms and letters are those applied by Cavendish.

TABLE III. Dimensions of various magnets.

Bar	Length	Breadth	Thickness
"straight"	12·03	0·154	0·096
"square"	11·95	0·164	0·164
"flat"	12·03	0·600	0·041
B	12·03	0·50	0·073
E	11·90	0·26	0·067

The first three of these represented the earlier, the two last the later consignment. A remark by Cavendish seems to imply that the two lots were supposed to be of the same steel, but possibly it was intended to apply not to the bars B and E but only to one bar A of the later set, the results for which are not included. In any case the treatment of the older and newer bars as regards annealing may have been quite different. It will be seen that for practical purposes the lengths were all the same, 12 inches.

Cavendish tried a formula of the type

$$m = C\,(b + t)^p\,(bt)^q \quad \text{.........................(7)},$$

where C was regarded as constant for the bars of the same consignment. After trying various values of p and q he drew the following conclusion: "Therefore if we suppose the force of the bars to draw (the compass) needle aside to be as $(b + t)^{\frac{1}{4}} \times (bt)^{\frac{5}{8}}$ it will agree as well with observation as any proportion." In other words the best values of p and q in (7) are $p = \frac{1}{4}$, $q = \frac{5}{8}$.

The formula makes the results for bars "straight," "square" and "flat" agree pretty closely, and likewise the results for B and E, but the values of C in the two cases differ sensibly.

Cavendish also writes the formula he approved in the form

$$\{(b + t)^{0\cdot25}/(bt)^{0\cdot125}\} \times (bt)^{0\cdot75} \quad \text{.....................(8)},$$

presumably with the object of showing that in bars of the same length and similar form of section the magnetic moment varies as $D^{1\cdot5}$, where D represents the linear dimension of the cross-section. The weight of course varies as D^2, so that the magnetic moment increases less rapidly with the area of the cross-section than does the weight. Another way of regarding the result is that it gives moment $\propto$ (weight)$^{\frac{3}{4}}$.

It will be noticed that the formulæ tried were of the type

$$C' s^p S^q \quad \text{...............................(9)},$$

where s denotes the perimeter, S the area of the cross-section, and C' is a constant.

Cavendish proceeded to try whether any results could be obtained calculated to throw light on the best shape to give to dip needles, whether they should be of uniform width throughout, a common shape apparently in older needles, or should taper towards the ends. He thus carried out experiments in the way already described on the magnetic moments of three needles B, C, D of the same length and approximately the same thickness, B being of uniform width, C tapering from 0·5 inch at the middle of the length to about $\frac{1}{4}$ inch at the ends, and D still more tapering. The needles were apparently of the same steel. Particulars of the experimental results appear in the following table. The dimensions are in inches, but I have converted the weights, w, from pennyweights to grammes. The values in needle B of the quantities in the two last lines have been taken as unity for comparative purposes.

TABLE IV. Strength of dip needles of various shapes.

Needle	B	C	D
w	56·7	42·2	33·4
Length	12·03	12·03	12·03
Breadth at mid-length	0·5	0·5	0·5
Breadth at ends	0·5	0·24	0·06
Thickness (approx.)	0·07	0·07	0·07
Moment/w	1	1·14	1·19
Moment/$w^{\frac{3}{4}}$	1	1·07	1·04

One would rather infer that Cavendish regarded the formula

$$\text{moment} \propto w^{\frac{3}{4}}$$

as at least approximately satisfied by the results. From the remarks he made later in connection with the shape of dip needles, he evidently decided that the desirability of a taper in the needles was proved. We learn incidently that when fitted as dip needles the above would have an axle weighing about 11·7 grammes.

"SPRINGING" (ELASTIC BENDING) OF NEEDLES

§ 5. Cavendish also made experiments on what he calls the "springing of needles," meaning the elastic deflection under a load. In this case "needle" meant a rectangle of uniform breadth and small thickness. It was supported at the ends, in the flat position, and loaded at the middle. The deflection immediately under the load was measured, exactly how is not stated, but evidently with high precision. The "needles" seem to have been in general about 12 inches long, varying in width from 0·60 to 0·16 inches. A few particulars are given in each case.

For instance, weights of 2 oz., 8 oz. and 16 oz. applied successively to one of the "needles" caused it to "spring" 0·04, 0·15 and 0·305 inches. This presumably was intended partly as a check on the application of Hooke's law, and partly as some guidance to the choice of suitable weights. If we denote the length, breadth and thickness of the "needle" by $2a$, b and t, the observed "spring" or deflection by y, and the applied weight by w, the formula applied by Cavendish for the determination of Young's modulus E may be written

$$E = 2wa^3/ybt^3 \qquad \text{(10)}.$$

Cavendish himself uses A as the symbol for Young's modulus, and he expresses it in what is now rather an unusual way, viz. as a length modulus with the inch as unit of length. In other words, E is to be regarded as the length in inches of a bar of the material which if attached to the end of a sample bar of the same section would double its length, supposing what of course is not the case, that Hooke's law applied in that extreme case.

The following values are quoted by Cavendish for certain "needles" to which his descriptive titles are assigned. For comparison with modern data I have added a column showing the equivalent values of E in gramme weight per cm.², taking with Cavendish 7·8 as the specific gravity in each case.

TABLE V. Young's modulus in iron and steel needles.

"Needle"	Thickness in inches	E: Length modulus in inches	E: Grammes weight per cm.²
Flat	0·0417	$106{\cdot}9 \times 10^6$	$21{\cdot}2 \times 10^8$
Straight	0·0953	84·7	16·8
Iron	0·0667	106·2	21·0
Iron hammered	0·0566	104·0	20·6
B	0·0732	103·6	20·5
C	0·0669	109·3	21·6

In another place Cavendish comments on the fact that the difference in elastic properties between iron and steel is not to any great extent due to the size of their elastic moduli, but to the weight which they can stand without permanent deformation. The values obtained above accord well with modern results. The thicknesses assigned to the bars called "flat," "straight" and "*B*" are sufficiently close to those given for the bars similarly designated in the experiments made in connection with formula (7) to render it almost certain that they were the same bars.

ERROR IN OBSERVED DIP DUE TO BENDING OF DIP NEEDLES

§ 6. Cavendish's experiments on the elasticity of needles were presumably suggested by his interest in a subject to which there are many references in his papers, the bending of dip needles. That the bending of a dip needle will introduce an error into the dip observed with it is one of those propositions which though really true are apt to be accepted fo erroneous reasons. If one's attention is exclusively directed to the dipping end, the fact that bending of the needle under its own weight will bring the end nearer the vertical than it otherwise would be, and so increase the dip, appears sufficiently obvious; while if one's attention happens to be concentrated on the other end, the exactly opposite conclusion is acceptable. On reflection one sees that when one reads both ends of the needle, as is invariably done, the two effects neutralise one another, at least to a first approximation. It is true notwithstanding that the bending of needles does introduce an error, and that for a reason which Cavendish correctly apprehended. But if his views on the subject ever reached the notice of his contemporaries, they do not seem to have produced any permanent impression, and it was left for Sir Arthur Schuster[1], less than 30 years ago, to call attention to the fact in a convincing way.

In the figure suppose AB to represent diagrammatically the dip needle as it would be if weightless, and $A'B'$ to represent its actual position when supported at its centre C. If AB makes an angle ϕ with the horizontal, gravity g may be resolved into $g \sin \phi$ along and $g \cos \phi$ perpendicular to AB. Only the latter component tends to bend the needle. Thus G the centre of gravity of the bent needle is on the perpendicular to AB

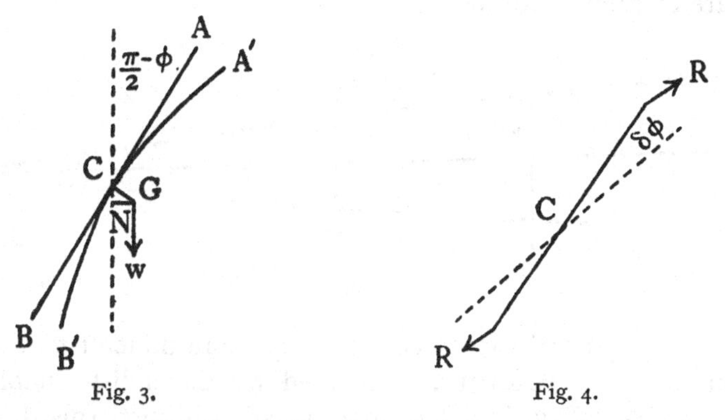

Fig. 3. Fig. 4.

through C, and its distance GC from $AB = \bar{y} \cos \phi$, where $\bar{y}$ represents the reduction in the height of the centre of gravity due to bending when the needle is horizontal. If $GN = GC \sin \phi$ be the perpendicular from G on the vertical through C, the gravitational couple tending to turn the needle is wGN, where w is the weight of the needle, or substituting for GN in terms of $\bar{y}$ and ϕ, $w\bar{y} \sin \phi \cos \phi$. If the consequent deflection from the position, which the needle if weightless would assume be $\delta\phi$, the magnetic couple is obviously $mR \sin \delta\phi$, where m is the magnetic moment of the needle and R the total magnetic force.

Thus $$mR \sin \delta\phi = w\bar{y} \sin \phi \cos \phi,$$

or $$\sin \delta\phi = w\bar{y} \sin \phi \cos \phi / mR \quad \text{.......................(11).}$$

[1] *Philosophical Magazine*, March 1891, p. 275.

Cavendish did not actually demonstrate this formula, but contented himself with announcing it in the following terms:

Error caused in dipping needle by its bending is ang(le) whose sine is to the rad(ius) as mot(ion) cent(re) grav(ity) × weight needle × sine × cosine of dip to ½ length needle × force applied at end needle sufficient to draw it 90 degrees out of true direction.

A figure shows at once that the somewhat circuitous phrase "½ length needle × force applied at end sufficient to draw it 90 degrees out of true direction" merely represents mR. The angle $\delta\phi$ is so small that $\sin \delta\phi$ may be replaced by $\delta\phi$. It represents in all cases a reduction in the angle of dip. There is no doubt as to what "motion centre gravity" means because it is the title Cavendish actually attaches to $\bar{y}$ as defined above.

Cavendish calculated the value of $\bar{y}$ in three cases, viz. first when the needle is of uniform width throughout, second when it tapers uniformly from the centre to the sharp ends, and third when it tapers uniformly from a width $2b$ at the centre to a width $2f$ at either end. The last case (cf. Fig. 5, showing the half needle $BFF'B'$) is more general than and includes the first two as particular cases. The particular cases are the identical cases treated by Sir Arthur Schuster, and the results obtained for them are identical with his.

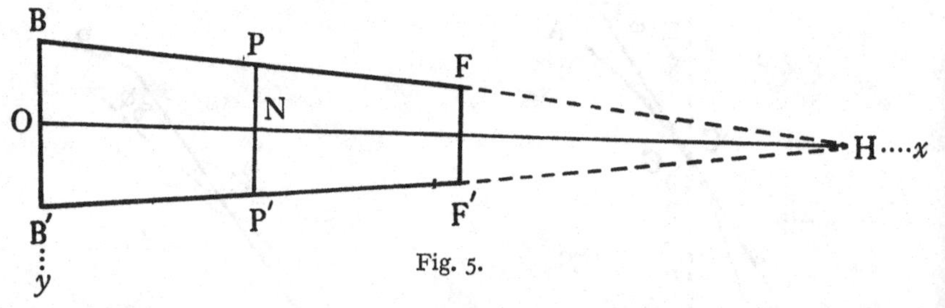

Fig. 5.

The more general case is considerably more difficult mathematically but an ingenious substitution adopted by Cavendish enabled him to surmount the difficulties. His analysis, slightly modernised, is substantially as follows: If Ox and Oy be taken along and perpendicular to the central line ON of the needle, the ordinary elastic equation on the Bernoulli-Euler theory is $E\omega\kappa^2 \dfrac{d^2y}{dx^2}$ = bending moment arising from the weight to the right of the section PP', at distance $ON = x$. Here $\omega\kappa$ represents the moment of inertia of the cross-section through N about the perpendicular to PP' through N. In Fig. 5, $BFF'B'$ is to be regarded as the outline of the half needle in the mid-plane of its thickness, this mid plane being vertical and containing the horizontal axis Ox and the vertical axis Oy, the positive direction of y being downwards. If t be the thickness and η represents PN, then for the cross-section through N

$$\omega\kappa^2 = 2t\eta^3/3.$$

Suppose the two edges BF, $B'F'$ of the needle to be prolonged and meet in a point H, which lies by symmetry on the median line ON, and let α denote the angle BHO. Then by regarding the portion $PFF'P'$ of the needle as the difference of the two triangles PHP' and FHF', it is easily seen that the bending moment of the mass to the right of the section PNP' is

$$\eta . \eta \cot \alpha . \tfrac{1}{3}\eta \cot \alpha . g\rho t - f . f \cot \alpha \,(\eta \cot \alpha - \tfrac{2}{3} f \cot \alpha)\, g\rho t,$$

where ρ is the density of the material and g gravity.

This reduces to $\quad g\rho t \cot^2 \alpha \,(\tfrac{1}{3}\eta^3 - f^2\eta + \tfrac{2}{3} f^3)$.

Again $\quad x = ON = OH - HN = \cot \alpha \,(b - \eta)$,

and so $\quad \dfrac{dx}{d\eta} = -\cot \alpha$,

and $\quad \dfrac{d^2y}{dx^2} = \tan^2 \alpha \dfrac{d^2y}{d\eta^2}$.

Thus the bending equation becomes

$$E\,(\tfrac{2}{3})\, t\eta^3 \tan^2 \alpha \frac{d^2y}{d\eta^2} = g\rho t \cot^2 \alpha \,(\tfrac{1}{3}\eta^3 - f^2\eta + \tfrac{2}{3} f^3),$$

or

$$\frac{d^2y}{d\eta^2} = \frac{3g\rho \cot^4 \alpha}{2E}\left(\frac{1}{3} - \frac{f^2}{\eta^2} + \frac{2}{3}\frac{f^3}{\eta^3}\right) \qquad \text{......................(12).}$$

Integrating this in two steps, noticing that $\dfrac{dy}{d\eta}$ and y both vanish when $\eta = b$, we find

$$y = \frac{3}{2}\frac{g\rho \cot^4 \alpha}{E}\left\{\frac{1}{6}b^2 + f^2 - \frac{2}{3}\frac{f^3}{b} - f^2 \log b - \frac{1}{3}b\eta - \frac{f^2}{b}\eta \right.$$
$$\left. + \frac{1}{3}\frac{f^3}{b^2}\eta + \frac{1}{6}\eta^2 + \frac{f^3}{3\eta} + f^2 \log \eta \right\} \text{...(13).}$$

The logarithm is to the Napierian base.

If $\bar{y}$ denote the distance of the centre of gravity of the bent needle below ON, a being the half length of the needle, we have

$$\bar{y} = \int_0^a y\, 2\eta\, dx \div \text{area of } BFF'B'.$$

Noticing that area $BFF'B' = (b^2 - f^2) \cot \alpha$, and changing from x to η as variable, we have

$$\bar{y} = \frac{3g\rho \cot^4 \alpha}{E\,(b^2 - f^2)} \int_f^b \{\eta\,(\tfrac{1}{6}b^2 + f^2 - \tfrac{2}{3}f^3 b^{-1} - f^2 \log b)$$
$$+ \eta^2\,(-\tfrac{1}{3}b - f^2 b^{-1} + \tfrac{1}{3}f^3 b^{-2}) + \tfrac{1}{6}\eta^3 + \tfrac{1}{3}f^3 + f^2 \eta \log \eta\}\, d\eta.$$

Carrying out the integration, and replacing $\cot \alpha$ by $a/(b - f)$, we find eventually, in exact agreement with Cavendish,

$$\bar{y} = \frac{g\rho a^4}{E\,(b - f)^5\,(b + f)}\left\{\tfrac{1}{24}b^4 - \tfrac{1}{2}b^2 f^2 + \tfrac{2}{3}b f^3 - \tfrac{15}{8}f^4 + 2b^{-1}f^5 - \tfrac{1}{3}b^{-2}f^6 + \tfrac{3}{2}f^4 \log \frac{b}{f}\right\}$$
$$\text{......(14).}$$

For the special case of the pointed needle putting $f = 0$ we find

$$\bar{y} = g\rho a^4/24Eb^2 \quad\ldots\ldots\ldots\ldots(15),$$

which is identical with Prof. Schuster's equation (3), allowing for the difference of notation, Prof. Schuster's l, a, Y answering to our a, $2b$, E respectively.

The other special case treated by Prof. Schuster, $f = b$, does not follow conveniently from the above, because the method of proof tacitly assumes that a and *therefore* $(b - f)/a$ is finite, and the expression (14) assumes the form 0/0 if we put $f = b$. It was treated independently by Cavendish, who found

$$\bar{y} = 3g\rho a^4/20Eb^2 \quad\ldots\ldots\ldots\ldots(16),$$

which is in agreement with Prof. Schuster's equation (4).

Cavendish evaluated (14) for a series of values of b/f, finding for

b/f	= 1	1·5	2	3	∞
$\bar{y} \div (g\rho a^4/Eb^2)$	= 0·150	0·124	0·108	0·090	0·047.

The needle he was particularly interested in had

$$a = 6,\ b = \tfrac{1}{4},\ f = \tfrac{1}{8}, \text{ in inches, and so } b/f = 2.$$

For its length modulus he took 107×10^6 inches, in other words he replaced E by $g\rho \times 107 \times 10^6$. Thus he had

$$\bar{y} = \frac{6^4 \times 4^2 \times 0{\cdot}108}{107 \times 10^6} = 0{\cdot}0000209 \text{ inches.}$$

Cavendish writes down, however, 0·0000265 inches.

The numerical factor given for the case $b/f = 2$ is at least very approximately correct. I find 0·1085, while to bring the result for $\bar{y}$ up to Cavendish's value we want the factor to be 0·136 approximately. A possible explanation is suggested by the series of values quoted in § 5 for Young's modulus. The value which appears above, 107×10^6, obviously answers to that of the first "needle" on the list described as "flat." If, however, we replace it in the calculation by 85×10^6, the value of the next "needle," which is described as "straight," employing the more exact value 0·1085 of the factor we agree exactly with Cavendish's figure 0·0000265 inches.

He proceeds

> I find also that the force which must be applied to the end of the needle in order to draw it aside 90 degrees is 1/600 of the weight of the needle, consequently the error caused by the bending of the needle in this climate when the dip is 72°½ is about 2½ minutes, but in a place where the dip was 45° would be about 4½ minutes.

This is easily shown by means of equation (11) if we notice that in modern language Cavendish's statement is equivalent to $mR = \frac{1}{600}wa$ where a is 6 inches.

Substituting in (11) we have

$$\sin \delta\phi = 100 \times 10^{-7} \times 265 \times 0{\cdot}5 \sin (145^\circ) = 0{\cdot}00076,$$

or $$\delta\phi = 2'{\cdot}6 \text{ approx.}$$

For $\phi = 45^\circ$, $$\sin \delta\phi = 100 \times 10^{-5} \times 265 \times 0{\cdot}5 = 0{\cdot}001325,$$

or $$\delta\phi = 4'{\cdot}6 \text{ approx.}$$

The phraseology adopted by Cavendish naturally suggests that the magnetic moment was actually found by balancing the magnetic couple against a gravitational couple. This would be quite practicable. One might, for instance, determine the weight required to be suspended from one end to bring the needle into the horizontal position, or into the position at right angles to the natural dip. But, as a matter of fact, a different procedure seems to have been adopted. This is what Cavendish writes on the subject:

Nairne's needle is 12 inches long $\frac{1}{2}$ inch broad in middle and $\frac{1}{4}$ at end, therefore force applied at end sufficient to turn it on centre is 10/36 of what it would be if all the matter was collected at end, therefore if the needle vibrates in 5″, force required to draw it aside 90° $= \dfrac{6 \times 10}{39{\cdot}12 \times 25 \times 36} = \dfrac{1}{600}$ of weight nearly.

From a small calculation on the same page, deducing the ratio 10/36 as quoted above, it is clear that what Cavendish means by "force applied at end sufficient to turn it on centre" is simply the moment of inertia I of the needle about its axis of rotation. He is obviously employing the ordinary formula for the time T of swing from rest to rest, viz.

$$T = \pi \, (I/mR)^{\frac{1}{2}}.$$

This gives $$mR = \pi^2 I/T^2.$$

His first statement really means

$$I = \frac{10}{36} \frac{w}{g} a^2,$$

whence $$mR = a \left\{ \frac{10}{36} \frac{wa}{(g/\pi^2)} \frac{I}{T^2} \right\}.$$

Taking the inch as unit of length,

$$a = 6,$$
$$g = 32{\cdot}2 \times 12 = 386{\cdot}4,$$
$$g/\pi^2 = 39{\cdot}12 \text{ very nearly.}$$

Thus $$mR = a \times 10 \times 6w \div (39{\cdot}12 \times 36 \times 25)$$
$$= aw/600 \text{ very nearly.}$$

The ratio 10/36 is deduced as follows:

force necessary to turn this needle on its centre is to that necessary to move the same quantity of matter placed at the end

$$:: \int x^2 (b - x \tan \alpha)\, dx \;:\; \tfrac{1}{2}(b+f)\, a^3$$
$$:: \tfrac{1}{3} b a^3 - \tfrac{1}{4} a^4 \tan \alpha \;:\; \tfrac{1}{2}(b+f)\, a^3$$
$$:: 4b - 3a \tan \alpha \, [\equiv (4b - 3b + 3f)/12] \;:\; \tfrac{1}{2}(b+f)$$
$$:: (b + 3f)/6 \;:\; b + f.$$

As $b = 2f$ the ratio is 5 : 18, or 10 : 36 as stated.

The introduction of $\frac{1}{2}(b+f)\,a^3$, i.e. $\frac{1}{2}(b+f)\,a \times a^2$, is alone sufficient to make it quite clear that what Cavendish had in view was the moment of inertia. The integral given, however, is not an absolutely complete expression for the moment of inertia of the needle—or rather the quarter needle. The complete expression is

$$\int_0^a \{x^2(b - x\tan\alpha) + \tfrac{1}{3}(b - x\tan\alpha)^3\}\,dx.$$

For the present case in which $b = 2f = a/12$, the ratio of the term neglected to that retained proves to be $3\,(f/a)^2 : 1$, or $1 : 192$. Thus the neglect of the term, whether intentional or otherwise, is quite immaterial.

Sources of error in Dip Observations

§ 7. Cavendish discusses several other sources of error in dip observations. If the needle be in a vertical plane inclined at an angle α to the magnetic meridian, the observed dip is necessarily too great. If ϕ be the true and $\phi + \delta\phi$ the observed dip, we have obviously

$$\tan(\phi + \delta\phi) = \tan\phi\sec\alpha \quad \text{......................}(17).$$

This is the most convenient form for exact logarithmic determination of $\phi + \delta\phi$, and so of $\delta\phi$, when ϕ and α are known.

When α is not large the approximate formula

$$\delta\phi = \tfrac{1}{4}\,\alpha^2\sin 2\phi \quad \text{..............................}(18)$$

is convenient. It shows at once, as remarked by Cavendish, that for a given error in the meridian setting of the dip circle, the consequent error in the dip is greatest when the true dip is 45°.

In Cavendish's time, in London, ϕ was about $72°\frac{1}{2}$, and so $\sin 2\phi = 0{\cdot}574$. This reduction factor had consequently to be applied to deduce from the errors calculated for $\phi = 45°$ the corresponding errors in the case of observations made in London near the end of the eighteenth century. For $\phi = 45°$ Cavendish quotes for $\alpha = 2°$, and for $\alpha = 5°\frac{1}{8}$ the respective values 1′ and 7′. In the latter case, I suspect, he inadvertently took $5°\frac{1}{8}$ instead of $5°\frac{5}{8}$ (i.e. half a point). For 2° and $5°\frac{5}{8}$ more exact values are 1′·05 and 8′·35. Consequently in Cavendish's time an error of half a point in the setting of the dip circle in London would have entailed an error of only 4′·8 in the observed dip.

Another source of error considered by Cavendish is the existence of a protuberance on the axle of a dip needle, resulting from imperfect polishing. In his time dip needles were either carried on friction wheels or with their ends rolling on flat planes, so that the conditions were not the same as now exist. He concludes that if the height of the protuberance be regarded as constant, the consequent maximum error in the dip will vary as the square root of the diameter of the axle, while if the height of the protuberance be supposed to vary as the diameter of the axle, the error also will vary as this diameter. He uses these results as an argument for having the axles of as small diameter as is compatible with their being sufficiently strong to bear the weight of the needle satisfactorily. I do not on this point follow Cavendish's reasoning, which must, I think, make some assumptions not fully disclosed. Something will depend in practice on how the observations are actually taken. In modern practice the difficulty presents itself usually from the occurrence of rust on the axle. It is now usual to set the microscope wire to the estimated mid position of the point of a vibrating needle, and the error consequent on the presence of a speck of rust depends a good deal on whether one observes with a large or small arc of vibration. Cavendish seems to assume the needle to be at rest when read. The calculations he made as to the effect of a protuberance, of what he considered a probable size, evidently somewhat startled him. He remarks

it seems surprising how it should be possible (to make) the axis so true as that the needle should not be liable to a greater error, and indeed the only way by which I can account for it is by supposing that the axis (axle) and plane on which it rolls do not actually touch but are kept from one (another) by a repulsive force.

This remark is taken from a MS. inscribed "On the different construction of dipping needles." In it Cavendish enumerated the following four principal sources of error: (i) imperfections of the axle, (ii) departure of the axle from horizontality, (iii) observing out of the magnetic meridian, (iv) bending of the needle. As regards (i) his opinion that the axle should be as fine as the weight of the needle allows has been already mentioned. No sensible error he says should arise from (ii) assuming ordinary care is exercised in levelling the dip circle, provided readings be always taken as he advocates with the instrument facing both east and west. As regards error (iii) the figure he obtained, viz. 1′, as the maximum error of dip for an error of 2° in the setting of the circle, shows that with reasonable care in determining the magnetic meridian no sensible error should arise from this cause in observations on land. At sea (iii) is a more serious source of error, but Cavendish had been assured that unless it is very rough it is rare for a ship to deviate as much as half a point from the direction it is intended to steer in, so that the average departure from the magnetic meridian during the taking of a dip observation ought with proper care to be but a small fraction of half a point, and the consequent error in the dip should thus be much under 7′ (or more exactly 8′). As regards (iv) he had found in the way already explained that with a needle such as he himself used the effect of bending was to reduce the dip observed in London about 2′½.

The following is a summary of the practical conclusions reached:

1°. The less the diameter of the ends of the axle (where it rolls), supposing them to be equally well ground, the less the error arising from irregularities of shape.

2°. The slenderer the needle, the less will be the error arising from defects in the axle, because the magnetic moment is larger in proportion to the weight in a slender needle. Also the more slender the needle, the less need be the diameter of the axle. A very slender needle, however, is liable to increased error from bending.

3°. The longer the needle, the greater the errors to which it is liable. The longer the needle the greater the error due to bending, also the greater the weight, and so necessarily the thicker the axle. There is, however, a practical limit to the reduction in length because the pointing of a long needle can be read more accurately than that of a short one.

4°. It is better that the needle should taper than have a uniform breadth from centre to ends (as seems to have been the case with many of the older needles). Experiments he had made showed the ratio of the magnetic moment to the weight to be greater in a tapering needle than in one of uniform width. By tapering the needle the weight is reduced, and so the necessary thickness of axle. The time of vibration is also reduced, which is advantageous from the observational point of view. Finally he ventures on the following anticipation of the shape now generally adopted in dip needles: "If the weight of the axis (axle) is very small in proportion to that of the needle, I should think it would not be worse if it was made still more tapering, or even brought almost to a point." A note indicates that he thought the ratio of the weight of the blade of the needle to that of the axle should not be less than 3 : 1.

It is interesting to find that Cavendish anticipated the modern idea that small magnets have many theoretical advantages. In others of his MSS. he remarks on the difficulties and errors associated with the use of the large dip needles—some of them about four feet in length—employed by certain of the old English observers.

Instructions to Observers and General Notes

§ 8. Several short manuscripts deal with observations at sea. They are entitled "For Cook and Bayley. Directions for the use of the dipping needle," "Directions for using the dipping needle for Dalrymple," "On the different forms of constructing a dipping needle," etc. Mr Dalrymple it seems, was to sail to Madras round the Cape of Good Hope, while Captain Cook was about to sail presumably on one of his three great voyages of exploration which began respectively in 1768, 1772 and 1776. There is naturally a good deal that is common in the several MSS., and much of what is said is by way of instruction as to the use of instruments now obsolete. There are, however, various remarks of interest. Referring to the subject of ships magnetism, he writes:

If there are no large iron bars in the ship except such as stand upright the these bars will be equally magnetised and the direction of magnetism in the will be the same whatever tack the ship goes on. Consequently if they draw the (compass) needle out of its true direction towards the west when the ship sails on one course, they will draw it as much to the east when the ship sails on the direct contrary course. But if there are any large horizontal bars in the ship this will not be the case, for the direction of magnetism in these bars will be reversed by the ship's turning round. The guns are large horizontal bars, but as they are of cast iron I believe they will not easily acquire magnetism, and when they have acquired it, its direction will not easily be changed.

Cavendish wrote of course in the days of wooden ships, when such comparatively small amount of iron as there was, except in the shape of cannon, was likely to be soft iron. His remarks on cast iron were presumably based on the experiments summarised in Tables I and II, which show it to have a small induction coefficient for temporary magnetism. He proceeds:

It would be of great use if a way could be discovered of finding by an easy experiment at sea how much the needle is drawn out of its true direction in different positions of the ship. I say of doing it at sea, because in all probability the quantity by which it is affected will be very different in different parts of the world. If there are no horizontal bars in the ship this may be done in this manner. First find the variation in the usual way with the ship's head to the north by the compass, then turn round 4 points and observe as before, and proceed in that manner till you have got all round the compass. The mean of these 8 observations will be the true variation, whence you may find how much the variation is affected by the iron work in each position of the ship.

After commenting on the practical difficulties, he proceeds

If either from the quantity of horizontal iron or from other causes this method of finding the error of the compass is impracticable, it still might be possible doing it in a harbour in this manner: Let the ship be brought in a line between 2 objects on shore, and take the bearing of those objects by the compass, with the ship's head in different directions, while another person places himself on shore, also in a line between those 2 objects and takes their bearing by another compass.

A note attached to this adds

An easier way will be for the person on shore to place himself in any situation and to take the bearing by the compass of the observer on board the ship, at the same time that the observer on board the ship takes the bearing of the person on shore.

In another place the suggestion is made that the observers on shore and aboard should interchange compasses and repeat the observations.

"Perhaps," Cavendish adds, "by making experiments in different parts of the world rules might be found out by which a person who knows how much his compass is affected by iron work in one part of the world may find how much it would be in another."

These remarks, it must be remembered, were written long before the day of Poisson and Archibald Smith.

In the instructions to Captain Cook and Dalrymple, Cavendish empha sises the fact that not only must dips be taken with the circle facing bot east and west, but further that the result is imperfect unless the pole be reversed. He allows, however, that time for taking the full experimen with the poles reversed may not be always available. In this event h suggests that on the occasions when time allows, the process of reversin the poles should be gone through several times, so as to get a reliabl value for the difference between the readings with marked and unmarke poles dipping, and so for the correction to be applied when readings ar taken with only one pole dipping. He thought it specially important tha this should be done before crossing the magnetic equator. He mention how best allowance may be made for the error resulting when reading are taken with one pole only dipping. Suppose, for example, that th marked pole gives the larger dip. Suppose the excess on one occasion t be found to be $\delta\phi$, the true dip—i.e. the mean of the dips obtained on tha occasion first with the marked end and then with the unmarked en dipping—being ϕ_1; and suppose on the next similar occasion the exces to be $\delta\phi_2$, the true dip being ϕ_2. Then on any intermediate occasion, whe ϕ was the observed dip from observations with the marked pole onl dipping, the true dip may be taken as

$$\phi - \tfrac{1}{4}(\sec\phi_1\,\delta\phi_1 + \sec\phi_2\,\delta\phi_2)\cos\phi.$$

Cavendish merely states the result, but the proof is easily supplied. Th difference between the two ends arises from the centre of gravity bein on one side or the other of the centre of the axle. It is obviously neare the end which when dipping gives the bigger dip, and unless the needl is lop sided may be supposed to be in the median line at a distance c fror the axis. With the marked end dipping, the gravity couple $wc\cos\phi$ where w is the weight of the needle, pulls the needle through the angle $\tfrac{1}{2}\delta$ out of the direction of the true dip. Thus if m be the magnetic momen and R the total force

$$mR\sin(\delta\phi/2) = wc\cos\phi.$$

The angle $\delta\phi$ in any reasonably well made needle is so small that the sin may be replaced by the angle, and so

$$\delta\phi = 2\,(wc/mR)\cos\phi.$$

Thus we have

$$\tfrac{1}{4}(\sec\phi_1\,\delta\phi_1 + \sec\phi_2\,\delta\phi_2) = \frac{wc}{2}\left(\frac{1}{m_1R_1} + \frac{1}{m_2R_2}\right).$$

And if we may suppose $\dfrac{1}{m_1R_1} + \dfrac{1}{m_2R_2} = \dfrac{2}{mR}$, we obviously have

$$\phi - \tfrac{1}{4}(\sec\phi_1\,\delta\phi_1 + \sec\phi_2\,\delta\phi_2) = \phi - \tfrac{1}{2}\delta\phi,$$

as it ought to be.

If a needle is always magnetised in a uniform way, with the same bar magnets, the moment acquired will naturally be nearly uniform, and the variations of total force with latitude or longitude are much less rapid than those of vertical or horizontal force. The correction of course is put forward not as a perfect one, but only as the best that is forthcoming. w really varies slightly with latitude, but any such variation would be negligible in view of the several uncertainties. Under the circumstances supposed, the needle was presumably freshly magnetised only when the poles were reversed. If this were the case, m would naturally tend to fall, and $\delta\phi$ would correspondingly increase. The possibility of an indirect effect of m upon the dip—the moment not being exactly the same when the marked and unmarked poles dip—is in fact one of the weak points in dip observations.

It should be added that Cavendish himself seems generally if not always to have observed the period of oscillation of the dip needle, or rather the time occupied by a given number of oscillations. At a fixed station this should afford a very good check on the uniformity of the results obtained when stroking the needle on different occasions. Unless a needle is unusually thick or of exceptionally hard steel, it does not require very powerful bar magnets to practically "saturate" it. Thus if any of Cavendish's original needles whose period he has recorded remains in working order, and could be identified, it might possibly serve to give an approximate estimate of the intensity of the total (i.e. resultant) force in his day.

Amongst the instructions to explorers are two which have nothing to do with magnetism, but may perhaps be mentioned for their general interest. Cavendish expresses a wish that observations should be made at frequent intervals on sea temperatures, the water being apparently secured with a bucket lowered overboard. An officer who had made previous observations of the kind had concluded that in all cases the sea temperature closely approached that of the air, but Cavendish thought confirmation desirable. His second suggestion was that in any land explorations measurements should be made of the temperature of any deep well, or natural spring having a considerable flow. He thought in this way, judging by his experience in England, that a good guess might be made at the mean annual temperature of the place.

Dip Observations

§ 9. There are amongst the magnetic papers a good many notes of dip observations. They mostly refer, however, to comparisons made of different needles—e.g. "Royal Society's needle" and "father's (Lord Charles Cavendish's) inverting needle," which are hardly of general interest. Certain of the comparisons seem to have been made at the "Royal Society's House," which as a place of observation seems to have suffered from the presence of "Dr Knight's magnets[1]." One of the series of comparisons seems worth recording, as it was on an elaborate scale, and was carried out in the garden of Lord Charles Cavendish's house in Great Marlborough Street, a presumably undisturbed place. The date was 1775. The details are of interest as showing the accuracy reached at the time in the construction of dip needles.

TABLE VI. Dip observations in London in 1775, with various needles.

Description of needle	Mark upwards, Instrument facing East	Mark upwards, Instrument facing West	Mark dipping, Instrument facing East	Mark dipping, Instrument facing West	True dip
	° ′	° ′	° ′	° ′	° ′
Royal Society's	72 46	71 59	72 8	72 40	72 23
Nairne's	72 54	72 28	71 45	73 21	72 37
"My new needle"	72 34	72 20	71 41	73 23	72 30
"My old needle"	71 40	73 53	72 19	72 19	72 33
Nairne's new needle	73 8	72 0	73 15	71 57	72 35
Sisson's	73 1	71 49	71 57	73 0	72 27
Means	72 40	72 25	72 11	72 47	72 31

In most cases several observations seem to have been taken with each needle, the days of observation including October 10, 11, 13 and 14, 1775. But the last needle, by Sisson, was tried at a different time, April 15, 1775, and only the final means are given in its case.

The observations being in the garden were doubtless taken by daylight, and so presumably, considering the season of the year, not very far from noon. The total range of the regular diurnal variation in October is, however, on the average only about 1′½, so the precise hour is of minor importance. At the same time there is always the possibility of magnetic disturbance. Thus the differences between the different needles shown by the table are more likely to be over-estimates than under-estimates. There is no explicit statement as to whether the needles were all tried in the same dip circle. If they were tried in the same circle, the differences between the results obtained with the circle facing east and west were in considerable measure due to the needles themselves. They are larger, but still not so very much larger, than the differences one gets with modern instruments.

The MSS. include particulars of observations made apparently in the Royal Society's House in 1775, 1776 and 1778. Observations in all three cases were taken at 7 a.m., noon, 2 p.m. and 10 or 11 p.m., on a number of successive days. On each occasion the needle was read only with the one end dipping and with the circle in only one position. But on different occasions the position of the circle was varied, and the pole which dipped

[1] Mr Harrison, when Assistant Secretary Royal Society, informed me that these magnets after being kept in the Royal Society's custody for about 100 years were eventually transferred to the South Kensington Museum.

was altered, so that the earlier and later observations of the series were made with different poles dipping. It was thus possible to obtain from the series as a whole a final mean value for the dip fairly representative of what would have been got if complete observations with both poles dipping and with the circle facing both east and west had been made on each occasion. The observational mean thus found in 1775, 72° 30′·2, was brought up to 72° 31′½ by the application of a correction of + 1′·3 to allow for the disturbing effect apparently of Dr Knight's magnets. The observations extended from June 19 to July 4, and included in all 64 separate readings, 16 at each hour. The mean from the four hours combined ought to be nearly free from accidental results, but the means for the separate hours seem affected by considerable probable errors. A mean derived from the four hours of the day selected might be expected to exceed the true mean for the day, but only by about 0′·1.

The data for 1776 were similar in character, the only difference being that the last hour of observation was 11 p.m. instead of 10 p.m. The observations in this case extended from June 21 to July 7. There were at least two observers, one of whom described as "Young Rob(erton)" was apparently regarded as the more reliable. The final mean derived from his observations alone was 72° 30′. The mean when all the observations were included seems to be the same. Cavendish says 72° 31′, but the difference was apparently due to an arithmetical error which he subsequently corrected.

In 1778 the last hour of observation was sometimes 10 and sometimes 11 p.m. In other respects the procedure was apparently the same as in 1775 and 1776. The sheet itself gives only the four mean results derived from the east and west positions of the circle and the two poles dipping. The final mean from the four combined is 72° 26′. In the case of the 1776 and 1778 observations nothing is said as to any correction.

About this time it is known from other sources that Cavendish was in general charge of the magnetic and meteorological observations carried on at the Royal Society's House. The results of these observations were published from time to time in the *Philosophical Transactions*, from which I have extracted the following particulars, including all I could find for the inclination:

TABLE VII. Dip in London at Royal Society's House.

Year	Mean dip ° ′
1775	72 30
1776	72 30
1777	72 25
1778	72 26
1779	72 21
1780	72 17

The results given here for 1776 and 1778 are identical with those given above, and the same is also true of 1775 if the correction made on account of Dr Knight's magnets is omitted. It will be noticed that the result obtained for 1775 in the Royal Society's House agrees closely with the mean which Cavendish himself obtained from observations with six needles in the garden of the house in Great Marlborough Street during the same year. This encourages the hope that the results obtained at the Royal Society's House give a fairly reliable indication of the true dip in London, and of its secular change from 1775 to 1780. It would seem to have been decreasing at the rate of about 2′½ per annum.

The only later systematic results which I have observed in the MSS. were obtained in 1791 with two needles by Nairne, described as "thick" and "thin" respectively, and one by Sisson. The observations with the two Nairne needles were made on August 4 and 7, those with the Sisson needle on October 2. Judging by the differences in the several positions, the Sisson needle was the same as that similarly described in Table VI. The Nairne needles may also have been the same as figured in Table VI, but that seems more doubtful. The mean dips given by the three needles were respectively 71° 45′½, 71° 43′ and 71° 50′, the final mean from the three being 71° 46′. If we combine this with the mean derived for 1775 from all the needles in Table VI we deduce a fall of 45′ in 16 years, or an average fall of 2′·8 per annum.

Other dip observations possessing a present day interest are some taken on a limited magnetic survey which Cavendish conducted during 1778. He observed in London before and after his journey, in the course of which he took dip observations at Oxford, Birmingham, Towcester, St Ives and Ely. The observations in London were taken on August 8, 10, 19 and 22, in a garden (probably that of the house in Great Marlborough Street). At Oxford the observations were made on August 14 "in garden of Observatory." The observations in Birmingham followed on the 15th "in bowling green." The observations at Towcester and St Ives were apparently made both on the 17th, the former in a garden, the latter in a room. The observations at Ely were made on the 18th in a garden. The place of observation at St Ives was not altogether satisfactory, owing to the proximity of iron. Observations made, however, in two positions, one considerably nearer the disturbing object than the other, differed by only 2′, so the disturbing effect was presumably very small. Two needles described respectively as "new" and "old" were employed; but on return to London "the balancing screws were found to be loose" in the "old" needle, and results obtained with it were discarded. Cavendish gives the dip observed at each station, its difference from the London dip, and the difference of the geographical coordinates of the place from London. His figures, with information as to the apparent change in dip between 1778 and 1891, obtained by comparing Cavendish's figures with those obtained by Rücker and Thorpe, appear in the following table:

TABLE VIII. Observations of dip in London and elsewhere in 1778.

Place	Dip	Difference from London			Excess above dip in 1891
		Dip	N. Latitude	W. Longitude	
	° ′	′	′	′	° ′
London	72 19	0	0	0	4 52
Oxford	72 48	29	13	72	4 56
Towcester	73 1	42	36	60	4 58
St Ives	72 35	16	48	4	—
Ely	72 41	22	54	15	4 43
Birmingham	73 5	46	58	112	4 50

In comparing the results for 1778 and 1891 it must be remembered that the actual sites occupied by Cavendish and by the observers in Rücker and Thorpe's survey were not the same. Even a small distance between the sites occupied might mean several minutes in the dip. The differences between the latitudes and those of London given by Cavendish agree fairly well with those assigned by Rücker and Thorpe to the places of the same name, if we suppose "London" to be at about 51° 31′, or about 3′ north of the mean latitude of Greenwich and Kew. Thus in taking a mean value from Kew and Greenwich as the "London" value for 1891 we should do fairly well. The longitude differences however given by Cavendish would seem to require "London" to be very nearly in the meridian of Greenwich, whereas the supposed site was about 10′ west.

Taking a mean from the five places for which secular change data are available we find 4° 52′, a result identical with that obtained from London alone. This gives 2′·6 for the average annual fall from 1778 to 1891, but we know that between 1860 and 1891 the average rate in London was only about two-thirds of this, so for part of the time it must have been considerably more rapid.

Cavendish did not content himself with merely obtaining the observational results, but drew the following inference as to the direction of the isoclinal lines (i.e. lines of equal dip):

By comparing London, Ely and Oxford dip should increase

30′ by going 1° to north,
18′ ,, ,, ,, ,, west;

according to which supposition

dip at Birmingham should be 62′½ greater than at London,
,, ,, Towcester ,, ,, 36′ ,, ,, ,, ,,

Therefore lines of equal dip should seem to run about 44° to south of west and dip should increase about 42′ by going 1° to N.W.

Accepting Cavendish's figures for the longitude and latitude differences, and employing the method of least squares, I get the following results:

Employing all the data, change of 24′ per 1° Lat. and of 16′ per 1° Long.

Omitting St Ives, change of 27′ per 1° Lat. and of 15′ per 1° Long.

If, confining ourselves to London, Ely and Oxford, we suppose the longitude of London 10′ in excess of the value assigned by Cavendish we obtain larger changes, viz. 34′ per 1° Lat. and 21′ per 1° Long. As it so happens, however, the differences between the directions deduced for the isoclinals from these several sets of figures amount to only a few degrees. All make the isoclinals run approximately N.E. and S.W.

If we take Rücker and Thorpe's figures for their districts III and IV, which seem the nearest comparable, we find their isoclinals running only about 16° north of east.

The number of Cavendish's stations would not warrant any great confidence in the accuracy of his result. The dip and declination results, however, appropriate to 1778 are at least consistent with a considerably more southern position of the north magnetic pole than that obtaining in 1891, so we should expect a difference between the isoclinals at the two epochs in the direction which Cavendish's figures show.

Diurnal Variation of Dip

§ 10. According to E. Walker's *Terrestrial and Cosmical Magnetism*, p. 170, the existence of a regular diurnal variation in the dip was first announced by Arago in 1827. Using a modern dip circle, an ordinary observation of dip with a single needle occupies some 20 minutes, and consists usually of 16 or more readings of each end. The accuracy expected is at the best of the order ± 1′. With a circle and needle in one position, two independent readings taken in immediate succession differ more often than not, the difference not infrequently exceeding 3′. Unless an observer were to remain almost steadily at work, the determination of the diurnal variation by complete sets of observations would hardly be feasible. Single readings taken in one position of the circle, with one end only of the needle dipping, will not give the true dip; but the differences between them are obviously capable of giving the diurnal variation, if a sufficiently high standard of accuracy is attainable. Accident, however, is likely to play too large a part, unless the observations extend over a large number of days.

It is difficult to imagine any object in taking a large number of observations at stated hours of the day, other than the investigation of the diurnal variation. A good many pages of Cavendish's manuscript contain such observations, and it is thus reasonable to suppose that he definitely wished to find out whether a diurnal variation did or did not exist. We know that the subject had been mooted in his day, because

Walker, *loc. cit.*, p. 169, tells us that Gilpin, a contemporary of Cavendish's, had definitely pronounced against a diurnal variation.

In some of the earlier pages of the MS. there are notes of dip observations, sometimes as many as eight or nine, made at intervals throughout one and the same day. At first these showed differences amongst themselves of 15′ or 20′, the differences occurring in too erratic a way to be natural phenomena. Some improvements of instrument or method must have been introduced, because in the later pages of the MS. the differences are less, and the latest observations of the kind, which alone are allotted to a definite year, 1775, show remarkably small differences. The observations were taken with a "new" needle in the months of August, September, October and November. There was usually one morning observation between $9\frac{1}{2}$ h. and $11\frac{1}{2}$ h., and an evening observation between 19 h. and 20 h., or between 23 h. and 24 h. Sometimes there were evening observations at both these times, and more than one morning observation. There were also occasional observations between $14\frac{1}{2}$ h. and $16\frac{1}{2}$ h. The choice of hours varied from day to day.

We know now that in London the dip in the months August to October is at a maximum about 10 h., and the change between 9 h. and 11 h. is small. Again the minimum occurs in the late evening, and there is little variation between 19 h. and 24 h. Thus the hours selected by Cavendish for his observations suggest that he had recognised the existence of a diurnal variation, and knew approximately the usual times of maximum and minimum. There is, however, no direct evidence of this, and all we can now do is to show that his observations of 1775 constitute practically a demonstration of the existence of a sensible diurnal variation.

Taking all the days in which there were both morning and evening observations, and forming the difference between each morning and evening observation of the same day, I got 14 such differences in August, 13 in September, 6 in October and 2 in November. The sum of the excesses of the morning over the evening reading in these 35 cases amounted to + 66′, giving a mean excess of + 1′·89. The difference was negative in only four cases. The largest single difference was only + 7′, and only three other differences exceeded + 4′. If we take the ordinary day dip results for Kew, from the mean of the 11 years 1890 to 1900, and calculate a morning-afternoon difference, or as we shall call it a "range," from

$$\tfrac{1}{2}\{(10) + (11)\} - \tfrac{1}{4}\{(19) + (20) + (23) + (24)\},$$

where (10), for instance, signifies the dip at 10 h., we get + 1′·92 for August, + 1′·73 for September, + 1′·48 for October and + 0′·68 for November. If we derive a final mean range from these four results, by weighting the August, September, October and November values in the ratio 14 : 13 : 6 : 2, corresponding with the numbers of Cavendish's observations, we get + 1′·71. Results vary in reality from year to year, according to sun-spot frequency, the differences between successive years being often much larger than 0′·18. Thus the range given by Cavendish's observations has not merely the right sign, but is also of the right order of magnitude. His observations in October and November, when the true range is reduced, gave very small differences, which practically neutralised one another. If we had confined ourselves to his August and September observations, we should have got for the mean excess of the morning reading + 2′·56, and in no single case would the evening reading have exceeded the morning reading on the same day.

Observations of Magnetic Declination

§ 11. Cavendish's work on magnetic declination includes a certain amount of discussion of instrumental details, and a certain amount of instrumental comparison, but it is mainly concerned with the results of systematic observations. With regard to instruments, it seems only necessary to remark on the importance he attached to inverting the needle, so as to eliminate any collimation error (i.e. error arising from non-coincidence of the magnetic axis and the sighted line). Several detached sheets of MS. give particulars of mean annual values of declination deduced from the observations taken. For 1782 and later years there are in addition full particulars of the individual observations. There do not seem, however, to be any such particulars for the earliest series of observations included in the following table:

Table IX. Magnetic Declination in Cecil Street.

Year	Period of observation	Declination (West) ° ′
1759	June 25 to August 5	18 53·9
1766	August	20 0·0
1767	October	20 22·0
1768	August	20 34·5
1769	October	20 44·0

In the sheet containing these results Cavendish says with reference t the year 1759:

The variation (i.e. declination) by a mean of about 300 observations mad in Cecil Street from June 25 to August 5 was 19° 20′·9. By two observation made at the same hours on different days by Dr Bradley at Greenwich an myself in Cecil Street it appeared that my meridian was not correct, and tha I made the variation 27′ too westwardly. The true variation therefore wa 18° 53′·9.

This latter is the value given in Cavendish's table from which Table I is copied. It is added that the result for 1766 was "by a correct meridian. Cavendish further mentions that for 1767 and 1768 Mr Canton got 20° 3 and 20° 50′ respectively—values in excess of Cavendish's by about 16′-

but, referring apparently to 1769, he adds "Mr Canton told me that he had just found that his meridian had been incorrect ever since the year 1765, and that he had now corrected it, and that he made the variation exactly the same this year as mine."

One cannot but feel some doubt whether at this early date Cavendish fully realised that declination may be expected to vary somewhat at places only a few miles apart. He leaves it uncertain whether the simultaneous observations taken by himself and Dr Bradley were astronomical or magnetic. It is only in the former case that reliance could be placed on the deduction he drew as to an error in his assumed meridian. If the difference between the declination at Cecil Street and Greenwich was half as large in 1759 as that between Kew and Greenwich now is, it would amount to about 12′. Without knowing what manner of man Mr Canton was, one cannot be certain that his discovery that his meridian was in error by the precise amount which brought his results into agreement with Cavendish's, was wholly uninfluenced by his recognition of Cavendish's ability[1]. In any case, presumably, we may assume that Cavendish adhered throughout to the meridian he finally adopted in 1759, so that the result his figures give for the secular change, viz. + 1° 50′·1 in ten years, or an average of + 11′·0 per annum, is probably pretty exact. This, it need hardly be said, is an unusually high value.

Judging by the number of observations, some 300, made in 1759 in about six weeks, Cavendish must then have observed much more frequently throughout the day than he did in later years. His earlier work probably disclosed to him the general character of the diurnal variation, and guided him in the selection subsequently made of observation hours.

The same sheet includes certain later data for Pall Mall, viz. 21° 13′ for August and September 1774 and 21° 24′·5 for July 1775. These were presumably taken not by Cavendish himself, but probably by Dr Heberden,

[1] Mr John Canton, the son of a broad-cloth weaver, was born at Stroud in 1718. Coming up to London in 1737, he became an assistant in the well-known Academy in Spital Square of which he ultimately became head-master. He died in 1772 in the 54th year of his age. He is described as "a man of very genteel and modest behaviour" who "gained the respect and acquaintance of the most eminent philosophers of his time." He was elected into the Royal Society in 1751, and received the Copley Medal the same year for his paper (*Phil. Trans.* Vol. 47, p. 31) on "A Method of making Artificial Magnets." He afterwards served on the Council for which he was again chosen on two subsequent occasions. He was the first in England to repeat, in 1752, Franklin's experiment of "drawing electric fire from the clouds during a thunder-storm." In 1762 he was again awarded the Copley Medal for his work on the Compressibility of Water (*Phil. Trans.* Vol. 52, p. 640). He contributed several papers on Electricity and Magnetism to the *Phil. Trans.* He was a member of a Committee appointed by the Royal Society to consider the best means of protecting St Paul's Cathedral from lightning, and his name is associated with the phosphorescent substance he prepared by calcining oyster shells with sulphur. [EDITOR.]

to whom Cavendish later refers as observing in Pall Mall[1]. One would also infer that Cavendish was inclined to apply a correction of + 15′·5 to the Pall Mall results.

The sheets already referred to in § 9 containing particulars of dip results during 1775, 1776 and 1778 in the Royal Society House also contain particulars of corresponding declination observations. These were taken at the same hours as the dip observations, viz. at 7 a.m., noon, 2 p.m. and 10 or 11 p.m. The mean from these four hours in the months of June and July would at the present epoch in the average year be about 1′·4 in excess of (i.e. more westerly than) the true mean declination for the day.

In 1775 observations were made on all days from June 18 to July 4 inclusive. The mean resulting declination was 21° 42′·9. It is added that the instrument had no error, but that a correction of about 2′·8 was necessary to allow for the effect of Dr Knight's magnets, bringing the declination up to 21° 45′·7.

In 1776 the observations extended from June 21 to July 7, and the resulting mean was 21° 47′. Nothing is said as to the necessity for any correction.

In 1778 the observations extended from June 29 to July 13. The mean for the uncorrected readings was 22° 20′, but on a separate sheet a correction of − 9′ on account of instrumental error is applied, bringing the value down to 22° 11′.

The *Philosophical Transactions* gives the mean values observed at the Royal Society's House for the years 1774 to 1780. The values it gives for 1775 and 1776 are identical with the above, if we omit the correction of 2′·8 on account of Dr Knight's magnets in 1775. But the value assigned to 1778 is only 21° 55′·5, while the values assigned to 1777, 1779 and 1780 are respectively 22° 12′, 22° 4′·5 and 22° 41′. These fluctuations are such as to preclude the possibility of high accuracy in the results.

From 1782 to 1809 Cavendish made a practice of taking daily declination observations throughout one or more of the summer months. During the earlier years 1782 to 1785 these observations were taken at his residence in Hampstead, but in 1786 and later years at Clapham, where he had gone to live. In the earlier years the number of the daily observations varied considerably, being sometimes only two, but frequently from five to eight. One of the observations was usually at an hour nor far from 8 a.m., and

[1] Dr William Heberden, eminent as a physician, was a Fellow of St John's College, Cambridge. He practised in the University for some years and subsequently repaired to London when he was elected into the Royal Society. He is best known as a medical writer, and was the first to give "a clear and satisfactory account of that painful thoracic disease called Angina pectoris." He possessed "a liberal and enlightened mind, a refined and classical taste, and an uniform complacency of disposition." He died in 1801 at the age of 91. [EDITOR.]

another between noon and 2 p.m., and there was frequently an observation late at night near 10 or 11 p.m. In the later years there were usually only two observations, and not infrequently only one each day, and the hours became much more uniform. At the same time observations tended to be slightly earlier in the day in some years than in others.

In the earlier years Cavendish took the largest and least of the daily readings, irrespective of when these occurred, and formed them into two separate groups. From the mean of the greatest and the mean of the least he derived a mean which he regarded as a measure of the declination, except for instrumental error. During part of the time he observed with the marked face of his needle up, and during the remainder of the time with the marked face down. Combining the mean reading mark up (representing the mean of the corresponding greatest and least readings) with the mean mark down, he obtained his final value for the declination. The greatest reading practically always was a reading taken between noon and 2 p.m., and the least reading was in the great majority of instances observed between 7 a.m. and 9 a.m. Again in the earlier years the needle was inverted at a purely arbitrary date, or dates, no distinction being drawn between days in different months. For instance in 1782 observations commenced on June 26 with mark upwards. The needle was inverted on July 4, and again on July 16. The observations from June 26 to July 4, and from July 16 to 26, were combined in one group representing mark up, while those from intermediate days represented mark down. One final mean was then obtained representing the whole series of observations.

After 1788 it was usual to form more than one mean, but still during a good many years without regard to calendar months. Thus in 1789 the days of observation formed three groups, the first including April 23 to May 15, the second May 15 to June 5 and the third June 5 to June 27. By that time Cavendish seemed to have assured himself that in his own instrument the difference between mark up and mark down was insignificant, and whilst in every year the needle was inverted at least once, he did not apparently consider it essential to have mark up and mark down readings represented in every mean. Thus in 1789 all the readings between April 23 and May 15, and again all those between June 5 and June 27 were taken with mark down, while those taken between May 15 and June 5 were taken with mark up. Thus two of his three groups represented mark down exclusively and only one mark up, but in forming a mean for the year he gave equal weight to each of the three groups.

For 1800 and later years Cavendish in forming his groups recognised the calendar months, and it is not clear whether for a considerable number of years previously he had made a practice of inverting his needle or not.

In the latter part of 1803 Cavendish got a new instrument, and later he seems to have come to the conclusion that the results he obtained with it in 1804 and 1805 were unreliable. At all events he does not include mean results from those years in his tables. From 1806–9 he adopted a considerably more elaborate method of reduction, which experiments made on his new compass had shown to be expedient. This involved readings with the mark—or as he then called it "knob"—both up and down.

The ordinary procedure became eventually a reading about 8 a.m. called a "morning" reading, and a reading about 1 p.m. called an "afternoon" or "evening" reading. Means were derived from morning and evening readings grouped separately, and then a final mean from the morning and evening means. The change of terminology from "greatest" and "least" to "morning" and "evening" seems to have occurred in 1799, but the approach to a uniform practice of morning and afternoon readings was made gradually, anterior to that date.

The time of the "least" or "morning" reading tended on the whole to become later, but fluctuated from year to year. It averaged in 1782 about 7 h. 45 m., in 1786 about 8 h. 10 m., in 1790 about 8 h. 40 m., in 1795 about 8 h. 10 m., in 1800 about 8 h. 20 m., and in 1805 about 8 h. 30 m. During the subsequent years it was about 8 h. 30 m., but in 1809 was 8 h. 50 m.

The time of the "greatest" or "evening" reading varied a good deal on individual days in the earlier years, but the average time did not fluctuate very widely. It was in 1782 about 1.30 p.m., the same in 1786, in 1790 about 1 p.m., in 1795 about 1.30 p.m., in 1800 about 1.40 p.m., but in 1801 and 1802 about 1 p.m. In 1805 and later years the time was very uniform at about 1 p.m.

The maximum, i.e. extreme westerly value, in the regular diurnal variation of declination occurs in London between 1 and 2 p.m. G.M.T., the whole year round. Near the hour of maximum the rate of change is very small, and thus the fluctuation in the time of Cavendish's afternoon reading would have little effect.

The time of the minimum in the diurnal variation is more variable. Near midwinter the forenoon minimum is as late as 9 a.m., but from May to September it occurs at 7 a.m., or at least nearer 7 than 8 a.m. Thus even in the years when Cavendish's morning observations were earliest the minimum was already past on the average day, and in the years when his observations were latest the rise since the minimum was quite appreciable.

Another feature of the diurnal variation is that it is not symmetrical. The maximum departs more from the mean for the day than does the minimum. The consequence of this will be more clearly apparent on considering the following results representing an average from Kew data of the eleven years 1890 to 1900. They may be regarded as representative of the average year in London at the end of last century, and there is no reason to suspect any very large change in such a phenomenon as th general character of the regular diurnal variation in the course of a single century.

TABLE X. Influence of hours of observation on value of Declination.

	Excess above true mean for the day				
	May	June	July	August	September
	′	′	′	′	′
Mean from 7 a.m. and 1 p.m.	0·9	0·5	0·7	1·3	1·7
,, ,, 8 ,, ,, ,,	1·0	0·6	0·9	1·5	1·7
,, ,, 9 ,, ,, ,,	1·7	1·2	1·4	2·2	2·3

It is clear that the mean obtained by Cavendish may be expected to be in every case slightly in excess of the true mean, i.e. the mean that would be obtained from hourly readings extending over the whole day. Also the excess will naturally be larger the later the average time of the morning reading. It will also be noticed that the excess may be expected to be a minimum in June and to increase from July to September. It is greater, it may be added, in years of many sun-spots, when the diurnal range is specially large, than in years of few sun-spots.

Supposing declination to be increasing at the time, as was the case during the years of Cavendish's observations, the true value would naturally increase from June to September. We should thus have the natural effect and the observational effect conspiring to raise the apparent value of the declination from one month to the next.

One peculiarity of Cavendish's treatment remains to be mentioned. In a minority, but still not a negligible minority of cases, at least in some months and years, he had only one observation, either morning or evening, in the day. These single observations were used when forming the morning and evening means, i.e. certain days were represented only in one of the two categories. It may seem at first sight that this is immaterial. It would be so if every day of the month were identical with every other day, but this is not the case. In the first place the type of the regular diurnal variation is constantly varying. Thus if we had a majority of morning observations from say the first half of the month, and a majority of evening observations from the second half, it would clearly be unsatisfactory. In the second place, there is the uncertain influence of disturbance. Only five hours intervene between an 8 a.m. and a 1 p.m. observation taken on the same day, as compared with 19 hours between a 1 p.m. and an 8 a.m. observation on consecutive days. The influence of disturbance is more to be apprehended in the second case than in the first.

These remarks will, I hope, explain why it seemed expedient to derive a new set of means from Cavendish's observations, treating the calendar months separately, and including only those days when there were both morning and afternoon readings. Table XI gives Cavendish's own results, and Table XII the new results. The figures in Table XI are not in every case absolutely identical with those Cavendish himself gives. In some cases I came across arithmetical slips, which it seemed well to correct. But these were all comparatively trifling, and the differences from Cavendish's figures were seldom as large as 0′·2.

In Table XI, when Cavendish treated all the data of the year together, obtaining only one mean, that mean appears only in the column devoted to the mean for the year. In all other cases the mean for the year is simply the arithmetic mean of the means for the two or more sub-periods under which Cavendish grouped his observations. These yearly means appeared in most years in a MS. table which Cavendish had drawn up. When they were lacking I have taken the arithmetic mean of the mean values for the sub-periods, as being the method approved by Cavendish. The tendency for the means for the later periods to exceed those from the earlier periods of the same year, which we were led to anticipate, will be readily recognised. There are in fact few years where it does not manifest itself.

TABLE XI. Declination Results at Hampstead and Clapham.

Year	Date					Mean declination from	
						Groups or months ° ′	Years ° ′
1782	June	26	to	July	24		22 36·8
1783	June	24	,,	July	15		46·9
1784	May	16	,,	June	23		54·1
1785	May	21	,,	July	10		23 3·3
1786	July	10	,,	August	6		12·5
1787	May	4	,,	June	6		15·3
1788	May	2	,,	May	29		23·1
1789	April	23	,,	May	15	23 28·5	
,,	May	15	,,	June	5	31·5	
,,	June	5	,,	June	27	30·3	30·1
1790	May	18	,,	June	28		36·6
1791	May	20	,,	May	31	39·3	
,,	June	2	,,	June	11	39·1	
,,	July	2	,,	August	3	41·0	39·8
1792	June	16	,,	July	8	48·4	
,,	July	8	,,	August	3	49·7	
,,	August	3	,,	August	29	50·3	49·5
1793	July	5	,,	July	21	52·1	
,,	July	21	,,	August	9	52·8	52·5
1794	June	18	,,	July	9	55·8	
,,	July	10	,,	August	12	57·7	
,,	August	12	,,	August	27	58·6	57·4

TABLE XI. Declination Results at Hampstead and Clapham—*contd.*

Year	Date	Mean declination from Groups or months °	'	Years °	'
1795	July 7 to July 25		56·2		
,,	July 26 ,, August 26		59·0		
,,	August 27 ,, October 1	24	0·7		58·6
1796	June 28 ,, July 27	23	59·3		
,,	July 27 ,, August 20	24	0·6		
,,	August 20 ,, September 21		2·2	24	0·7
1797	July 11 ,, August 12		0·8		
,,	August 12 ,, September 12	24	1·6	24	1·2
1798	May 29 ,, June 23	23	59·3		
,,	June 24 ,, July 31	24	1·0		
,,	July 31 ,, August 31		2·6		
,,	September 1 ,, October 1		3·7	24	1·6
1799	July 12 ,, August 24				0·6
1800	July		2·6		
,,	August		4·1		
,,	September		4·5		3·7
1801	June		4·5		
,,	July		6·6		
,,	August		7·7		
,,	September		8·6		6·8
1802	June 23 to July 31		7·0		
,,	August		9·0		
,,	September		10·1		8·7
1803	June		8·1		
,,	July		9·2		
,,	August		12·0		
,,	September		11·6		10·2
1806	August		15·3		
,,	September		16·1		15·7
1807	June		16·0		
,,	July		16·5		
,,	August		17·3		16·6
1808	June	24	15·4		
,,	July		16·2		
,,	August		17·6		
,,	September		18·7	24	17·0
1809	June		16·1		
,,	July		19·4		
,,	August	24	19·6	24	18·4

TABLE XII. Declination Results at Hampstead and Clapham.

Year	Month	Number of days of observation	Mean declination For month	Mean declination For year
			° ′	° ′
1782	June	4	22 37·3	
,,	July	20	36·7	22 36·8
1783	June	6	46·9	
,,	July	13	46·8	46·9
1784	May	4	54·3	
,,	June	5	54·8	54·6
1785	May	5	23 2·8	
,,	June	3	4·2	23 3·3
1786	July	7	12·5	
,,	August	2	17·2	13·5
1787	May	23	15·4	
,,	June	3	16·0	15·5
1788	May	22		22·7
1789	May	23	30·0	
,,	June	19	30·3	30·0
1790	May	9	37·1	
,,	June	21	36·5	36·7
1791	May	8	39·5	
,,	June	3	39·0	
,,	July	8	40·5	40·0
1792	June	8	48·3	
,,	July	9	49·4	
,,	August	3	50·2	49·1
1793	July	18	52·1	
,,	August	8	54·0	52·7
1794	June	5	55·4	
,,	July	18	56·3	
,,	August	8	57·9	56·9
1795	July	16	56·7	
,,	August	14	59·8	
,,	September	25	24 0·7	59·3
1796	June	2	23 57·8	
,,	July	20	59·6	
,,	August	21	24 1·3	
,,	September	12	2·0	24 0·7
1797	July	7	0·5	
,,	August	18	1·8	
,,	September	6	24 1·2	24 1·3
1798	June	24	23 59·6	
,,	July	18	24 1·2	

TABLE XII. Declination Results at Hampstead and Clapham—*contd.*

Year	Month	Number of days of observation	Mean declination For month		Mean declination For year	
			°	′	°	′
1798	August	26	24	2·7		
,,	September	19		3·5	24	1·7
1799	July	6		0·3		
,,	August	4		0·7		0·5
1800	July	22		2·8		
,,	August	21		4·2		
,,	September	10		4·8		3·7
1801	June	16		4·4		
,,	July	17		6·4		
,,	August	22		7·6		
,,	September	24	24	8·5	24	7·0
1802	June	7	24	5·5		
,,	July	20		7·3		
,,	August	27		9·1		
,,	September	21		10·2	24	8·7
1803	June	13		8·4		
,,	July	22		9·0		
,,	August	25		11·9		
,,	September	24		11·5		10·5
1804	September	26				12·6
1805	June	21		9·9		
,,	July	19		10·5		
,,	August	23		9·9		
,,	September	23		8·6		9·7
1806	August	24		15·3		
,,	September	24		16·1		15·7
1807	June	22		16·0		
,,	July	24		16·5		
,,	August	24		17·3		16·7
1808	June	28		15·4		
,,	July	18		16·2		
,,	August	24		17·6		
,,	September	25		18·7		17·0
1809	June	22		16·1		
,,	July	15		19·4		
,,	August	21	24	19·6	24	18·2

Table XII gives the number of days on which each monthly mean is based. In it the mean value for the year was arrived at by allowing equal weight to each individual day. The days, in fact, were put in a single group irrespective of the months they belonged to. In some years, e.g. 1789, there were months containing only one or two days of observation, from which satisfactory monthly means could not have been derived. These solitary days were used, however, in forming the annual means, so that these means were based in some cases on a slightly larger number of days than contributed to the monthly means. As already explained, days with only one observation were excluded, so that the number of days on which Cavendish observed exceeded that shown in all years. In most cases, especially in the later years, the differences between the yearly means in Tables XI and XII are very small; in a good many cases there is absolute identity. The absolutely largest difference, occurring in 1795, is only 0′·7. In that year Cavendish's third group included about twice as many days as his first, and the mean values he obtained from the two differed by 4′·5. Thus it naturally makes a considerable difference whether one allots equal weight to individual groups or to individual days.

As already stated, the place of observation was changed from Hampstead to Clapham between the 1785 and 1786 observations, and the change of site might of course have entailed a very decided discontinuity in the figures. The mean annual values, whether in Table XI or Table XII, do in fact at first sight rather suggest a small discontinuity. Taking Table XI, it will be seen that the apparent secular change from 1785 to 1786 was 9′·2, while the mean annual change from 1782 to 1785 was only 8′·8. The secular change, however, from 1786 to 1787 was only 2′·8 as compared with 7′·8 in the following year. Thus the most probable explanation seems to be an unduly high value for 1786 unconnected with the change of site, or else an abnormally low secular change between 1786 and 1787. It happens fortunately that some additional light on this point is derivable from data given in one of Cavendish's MS. They represent the mean annual values obtained by an independent observer, Dr Heberden, who observed from 1782 to 1791 in Pall Mall, presumably on one spot.

In the event of any substantial difference between the values of the

TABLE XIII. Variation (Declination) in Pall Mall by Dr Heberden.

Year	From	To	Heberden	Cavendish less Heberden
			° ′	′
1782	May 20	June 24	22 22·5	14·3
1783	June 10	June 25	34·7	12·2
1784	May 12	May 28	40·4	13·7
1785	May 7	May 23	49·2	14·1
1786	May 18	May 27	23 0·1	12·4
1787	May 3	May 21	2·0	13·3
1788	April 22	May 7	7·0	16·1
1789	May 8	May 15	19·8	10·3
1790	May 12	May 22	32·1	4·5
1791	May 12	May 26	25·5	14·3

declination at Cavendish's places of observation at Hampstead and Clapham, the difference between his annual means and Heberden's should have exhibited a decided change after 1785. On the contrary, if we compare the mean difference from the preceding years with that from the succeeding years, we reach different conclusions as to the sign of the difference between Hampstead and Clapham according as we derive means in each case from three or from four years. Heberden's results, equally with Cavendish's, show an exceptionally large change between 1785 and 1786, and an exceptionally small one between 1786 and 1787. Considering the nature of the observations, the uniformity of the difference between Cavendish and Heberden, with the exception of the one year 1790, is really surprising. In that year presumably some error entered into Heberden's observations, as the reversion in an isolated year, 1790–1791, of the secular change is a highly improbable event at an epoch when that change is substantial.

Reference has already been made to declination data from Pall Mall for 1774 and 1775, and to the fact that Cavendish seemed to regard them as requiring a correction of + 15′·5 to make them comparable with his own earlier observations in Cecil Street. Unless Heberden changed his instrument or methods between 1775 and 1782, the natural inference would be that the Cecil Street, Hampstead and Clapham results were all fairly comparable, the difference of site representing not more than 2′ or 3′ of declination.

Secular Change of Declination

§ 12. Table XIV presents the secular change data derivable from Tables XI and XII. The figures opposite any year represent the increase of declination during the previous twelve months. A minus sign indicates a fall. The first five columns of secular change data give the results obtained by comparing the mean values of the declination for months of the same name in consecutive years as given in Table XII. The sixth column gives the mean obtained by allowing equal weight to each of the monthly results. The seventh column gives the secular change derived from the mean results for the year in Table XII, and the last column the corresponding results derived from Cavendish's yearly values in Table XI.

A comparison of the corresponding figures in the last three columns of Table XIV will give an idea of the order of the uncertainties arising from variations in the method of handling the observational results.

Cavendish himself formed no annual means for 1804 or 1805. Thus no figure is entered under either of these years in the last column of Table XIV and the figure assigned to 1806 really represents ⅓ of the apparent rise between 1803 and 1806.

While I have given in Table XII the values which Cavendish's observations, treated in the same manner as in other years, give for the declination

TABLE XIV. Secular change of Declination in London, 1782–1809.

Year	From monthly mean values: May	June	July	Aug.	Sept.	Mean	From all observations of year	From Cavendish's yearly means
	′	′	′	′	′	′	′	′
1783		9·6	10·1			9·9	10·1	10·1
1784		7·9				7·9	7·7	7·2
1785	8·5	9·4				9·0	8·7	9·2
1786							10·2	9·2
1787							2·0	2·8
1788	7·3					7·3	7·2	7·8
1789	7·3					7·3	7·3	7·0
1790	7·1	6·2				6·6	6·7	6·5
1791	2·4	2·5				2·4	3·3	3·2
1792		9·3	8·9			9·1	9·1	9·7
1793			2·7	3·8		3·2	3·6	3·0
1794			4·2	3·9		4·1	4·2	4·9
1795			0·4	1·9		1·2	2·4	1·2
1796			2·9	1·5	1·3	1·9	1·4	2·1
1797			0·9	0·5	− 0·8	0·2	0·6	0·5
1798			0·7	0·9	2·3	1·3	0·4	0·4
1799			− 0·9	− 2·0		− 1·4	− 1·2	− 1·0
1800			2·5	3·5		3·0	3·2	3·1
1801			3·6	3·4	3·7	3·6	3·3	3·1
1802		1·1	0·9	1·5	1·7	1·3	1·7	1·9
1803		2·9	1·7	2·8	1·3	2·2	1·8	1·5
1804					1·1	1·1		
1806				1·1	1·5	1·3	1·7	1·8
1807				2·0		2·0	1·0	0·9
1808		− 0·6	− 0·3	0·3		− 0·2	0·3	0·4
1809		0·7	3·2	2·0		2·0	1·2	1·4
1782 to 1791		6·9	7·1				7·0	7·0
1791 to 1800			2·5				2·6	2·7
1800 to 1809			1·8	1·7			1·6	1·6

in 1804 and 1805, I am inclined to regard the data for 1805 as untrust worthy. The data for 1804 have been used in obtaining the secular chang assigned to that year in Table XIV, but in deducing from Table XII th secular change for 1806 I discarded the results for 1804, employing as in the case of Cavendish's own results the data for 1803.

The last three lines give mean results for the three 9-year periods into which the 27 years, 1782 to 1809, can be subdivided.

Even at a modern first-rate observatory it is undoubtedly an optimistic view to regard mean annual values of declination as reliable to 0′·1. Thus to look for that degree of accuracy in the secular change data derivable from Cavendish's observations would be utterly unreasonable.

It will be remembered (see p. 473) that Cavendish's observations in Cecil Street gave +11′·0 as the mean value of the secular change between 1759 and 1769. If we combine his mean annual values for 1769 in Cecil Street and for 1782 in Hampstead we obtain as the mean secular change during these 13 years +8′·7. Heberden's results in Pall Mall from 1782 to 1791 give the same mean value +7′·0 as Cavendish's corresponding figures given in Table XIV. Thus all the results available point to a decline, comparatively slow at first and then more rapid, in the secular change. The rate of decline however fell off after 1800, and we know from other sources that the actual turning point when the declination reached its extreme westerly value was not attained near London until about 1818.

Diurnal Variation of Declination

§ 13. Table XV gives particulars of the ranges obtained from Cavendish's "greatest" and "least" or "morning" and "evening" readings, making use only of those days in which both morning and evening readings were available. The results for the five months May to September are given separately. The column headed "all days" gives the results obtained by combining all the observation days of the year in one group. In the average year there is not very much difference between the ranges of the regular diurnal variation in May, June, July and August. There is a decided though not very large falling off in September. The last column gives the sun-spot frequency for the year taken from Wolfer's great table.

In the later years when the observation hours tended to become stereotyped the range in Table XV should represent pretty nearly that obtained from the values assigned to the two hours 8 a.m. and 1 p.m. or 2 p.m. in the regular diurnal inequality, and should thus be just a shade under the range of the regular diurnal inequality itself. In the earlier years the larger number of daily observations would tend somewhat to increase the range, because the hours at which the extreme values appear fluctuate from day to day. On the other hand, the hours at which the extreme values were most likely to occur were often not the actual hours of observation chosen.

For comparison, I have given in the last three lines the ranges of the diurnal inequality at Kew in May and June, and the means derived from these two months, for the three years 1890, 1893 and 1870. Of these 1890 represents, like 1784, 1798 and 1809, sun-spot minimum, while 1893 and 1870 represent, like 1787 and 1804, sun-spot maximum.

As is well known, the fact that the range of the regular diurnal variation tends to be small in years of few sun-spots, and large in years of many sun-spots, was discovered about the middle of last century by Lamont and Wolf. It is obvious from Table XV that if he had had sun-spot data before him, Cavendish might well have discovered the result prior to the end of the eighteenth century. It would be very interesting to know what

TABLE XV. Daily Ranges from observations at Hampstead and Clapham.

Year	May	June	July	August	September	All days	Sun-spot frequency
	′	′	′	′	′	′	
1782		13·5	13·5			13·5	38·5
1783		13·2	14·0			13·7	22·8
1784	13·5	11·6				12·4	10·2
1785	14·0	12·3				13·4	24·1
1786			18·3	10·5		16·6	82·9
1787	20·1	15·5				19·6	132·0
1788	19·8					19·8	130·9
1789	14·5	12·9				13·8	118·1
1790	13·3	12·0				12·4	89·9
1791	12·8	13·0	12·4			12·8	66·6
1792		12·1	13·8	12·0		12·8	60·0
1793			12·4	12·4		12·4	46·9
1794		9·5	11·5	10·1		10·8	41·0
1795			7·8	9·0	9·1	8·7	21·3
1796		5·5	7·7	8·6	10·0	8·5	16·0
1797			7·5	8·9	9·2	8·6	6·4
1798		8·7	8·9	9·3	9·8	9·1	4·1
1799			9·3	8·8		9·1	6·8
1800			8·7	9·4	6·7	8·6	14·5
1801		8·5	9·0	11·3	10·5	10·0	34·0
1802		8·7	10·2	10·4	10·0	10·1	45·0
1803*		11·0	10·8	10·8	10·1	10·6	43·1
1804					10·1	10·1	47·5
1805		9·9	10·5	9·9	8·6	9·7	42·2
1806		11·4		10·7	9·2	10·5	28·1
1807		8·9	7·4	9·6		8·6	10·1
1808		9·6	9·7	10·3	9·4	9·7	8·1
1809		7·4	8·0	6·2		7·1	2·5
1890	8·6	9·1				8·8	7·1
1893	13·3	13·9				13·6	84·9
1870	16·7	15·1				15·9	139·1

* The mean range from 17 October days in 1803 was 8′·0.

impression the enormous size of the ranges which he was recording near sun-spot maximum in 1787 and 1788 made upon his mind.

It will be generally admitted that if Wolfer's table can be at all relied on very large ranges were to be expected in 1787 and 1788, but the fact that the ranges then observed by Cavendish are so largely in excess of those for 1870, a year apparently of even larger sun-spot frequency, may rouse suspicion. According to observations of recent years the range of the diurnal inequality of declination bears a linear relationship to sun-spot frequency, so that for equal increments in the one we expect at least approximately equal increments in the other. This refers, however, to the mean diurnal inequality for the whole year. Sun-spot frequency varies not merely from month to month but from day to day. The relation between the magnetic diurnal range and sun-spot frequency is doubtfully shown on individual days, and imperfectly shown in individual months. In a year of sun-spot maximum we confidently expect that the range of the diurnal magnetic inequality will be well above the average in every month of the year, but we expect this excess to be considerably larger in some months than in others. This is apparent even in the few recent data in Table XV which are based on magnetograms. If we calculated the range for 1870 from those for 1890 and 1893 on the simple basis of sun-spot frequency we should get a range fully 1′ in excess of that actually found. Still the excesses of the ranges in 1787 and 1788 are undoubtedly difficult to account for in this way. There is, however, a very obvious explanation of the difficulty, though it cannot claim to be more than a probability. The relation established between declination range and sun-spot frequency is based on a comparison of years of many and few sun-spots having approximately the same mean epoch. It thus relates to years which are closely alike if not identical so far as the absolute values of the magnetic elements at the place are concerned. But it is a very different matter when we come to comparing years of sun-spot maximum such as 1893 and 1787 separated by over a century. The most generally accepted theory as to the ultimate cause of the regular diurnal magnetic variation is that it consists of electrical currents in the upper atmosphere, assisted probably by earth currents. But if we suppose the diurnal variation to be due to currents situated either in the air or the earth, then if the currents remain the same, the range will vary inversely as the value of H, the horizontal force. The accurate measurement of horizontal force dates only from the epoch of Gauss and Lamont, so we have no direct knowledge of the value of H in London in, say, 1787. If, however, we consult the Kew records we find that the average yearly rise in H from 1860 to 1865 was identical with the average yearly rise from 1860 to 1900. If we suppose the force to have had the same mean annual change since 1787, we find that while the value of H in 1893 was 0·18238 C.G.S. units, in 1787 it was only 0·15906. Even this would seem too high an estimate, for if we combine it with a dip of 72°—the approximate value deduced for 1787 from Cavendish's observations of 1791 and the secular change given by it and his other observations—we obtain for the vertical force in 1787 a value which would require its mean annual fall from 1787 to 1860 to have been about three times that observed for the average year between 1860 and 1865. If, on the other hand, we assume that the rate of fall of the vertical force was the same from 1787 to 1860 as it was from 1860–1865, we find for the vertical force in 1787 a value which combined with a dip of 72° makes H as low as 0·147. As the rate of fall of V diminished very decidedly after 1865, this latter estimate of H deserves considerably less weight than the first. If we suppose the value of H to have been 0·155 C.G.S. units in 1787 we shall probably not be very far wrong. But 0·155 is only 0·85 of the value of H in 1893. Thus if the declination range varies inversely as the value of H at the time, a range of 19′·6 in 1787 would, under the conditions as to force existing in 1893, have been only 19′·6 × 0·85, or 16′·7. This seems quite reasonable for a sun-spot frequency of 132.

According to the Schuster-Balfour Stewart theory the intensity of the electrical currents in the upper atmosphere varies as the value of the vertical force. If this be true, it would naturally mean larger currents in 1787 than in 1893, and so tend equally with the reduced value of H to enhanced ranges at the earlier date.

As the criticism was most likely to originate with theorists, I have thought it well to treat this aspect of the case in some detail. Probably, however, the plain man will derive more confidence from the very strong confirmatory evidence of the accuracy of Cavendish's range figures which is fortunately derivable from data recorded in his MSS. He gives in the first place the morning and afternoon declination readings taken in Pall Mall, presumably by Dr Heberden, from May 4 to May 21, 1787. Taking all Dr Heberden's 18 days we obtain for the mean range 19′·7, Cavendish's own mean result for the same month as given in Table XV being 20′·1. Fifteen of the 18 days on which Heberden observed were also days on which Cavendish had both morning and afternoon readings. The mean ranges for these 15 days were Heberden 20′·1, Cavendish 21′·7. Cavendish's observations were not at absolutely fixed hours, and very probably the same was true of Heberden's, so absolute identity is not to be expected. Considering the nature of the instruments we should not expect any very high degree of accuracy in individual readings, and high precision is not to be expected in the ranges for individual days. But errors in defect would be as likely as errors in excess, thus some conclusions can be drawn from the individual data. A large mean range might signify that the majority of individual ranges were large, or that there were a few very large ranges. The latter case would mean the existence of days of large disturbance, the former that the regular diurnal variation was specially developed. Of the 18 ranges observed by Heberden the least was 1[illegible] and the largest 29′, while ten lay between 17′ and 23′. Of the 23 ranges observed by Cavendish, in the same month, the least was 13′·5 and the largest 27′·5, and 13 lay between 16′·5 and 23′·5. Thus the phenomenon was clearly not due to the occurrence of a small number of highly disturbed days, but to a persistently large range in the regular diurnal variation.

Further evidence tending in the same direction is derivable from the readings taken four times a day at the Royal Society's House. One of the years included, 1778, is the year of absolutely largest sun-spot frequency in the whole of Wolfer's list. Again Cavendish's MSS. include particulars of observations taken in 1788, a year nearly equal to 1778 in sun-spot frequency, during 12 hours on every single day of June. Where and by whom these observations were taken is not stated; the only information respecting them is the remark "Mr Gilpin's obs(erving) needle" on the back of the single sheet of paper on which they are entered. The presumption is that they were taken by Cavendish himself, or under his supervision, somewhere in London. Table XVI embodies the information derivable from these various sources as to the nature of the regular diurnal variation. The results are given as excesses above the value at 7 a.m., the natural hour of the daily minimum in summer. For comparison, corresponding data are given for 1890 and 1893 at Kew, and the mean sun-spot frequencies given in Wolfer's table for the several years are added. The results for 1775, 1776 and 1778 include days from July as well as June, the results for the other years are from June only.

TABLE XVI. Diurnal Variation of Declination in London.

	Forenoon				Afternoon							
Year	7	8	10	Noon	1	2	4	6	8	10	11	Sun-spot frequency
	′		′	′	′	′	′	′	′	′	′	
1775	0·0			9·1		10·6				2·0		7·0
1776	0·0			7·5		8·0					2·0	19·8
1778	0·0			16·7		19·0				8·5		154·4
1788	0·0	0·6	8·3	17·3	19·2	19·2	13·3	7·4	6·2	5·8	5·7	130·9
1890	0·0	0·3	3·3	8·0	8·9	9·1	7·1	5·1	4·2	3·8	3·7	7·1
1893	0·0	0·4	5·0	12·2	13·4	13·9	10·8	7·5	6·0	5·5	5·2	84·9

In 1776 individual readings at the Royal Society's House were more than usually erratic. The diurnal inequality got out by Cavendish from all the observations comes

7 a.m	Noon	2 p.m.	11 p.m.
0	4′	4′	− 6′

Evidently Cavendish was not satisfied, as he got out a second set of figures, confining himself to observations made by "Young Rob(erton)." It is from these that the results in Table XVI were derived. There were however only five such readings at 7 a.m., and one of them must either have been in error or taken at a considerably disturbed time, as it made the declination 10′ *higher* than at noon and 9′ *higher* than at 2 p.m. If this reading were omitted—it is 19′ higher than any other reading taken by Roberton at the same hour—the figures for 1776 at noon, 2 p.m. and 11 p.m. would each be increased by 4′. This would obviously fit in much better with the other results. If this emendation is accepted, the phenomena exhibited in the older years are so similar to those derived from magnetographs in recent years that their substantial accuracy can hardly be questioned. Whatever the true explanation may be, the large size of the regular diurnal variation in 1778 and 1788 can hardly be doubted.

The hours of observation in 1788 included 6 a.m., and the mean value for that hour was 0′·1 lower than that for 7 a.m., so that the range of the regular diurnal inequality in June 1788 was actually 19′·3. The mean absolute range in that month, i.e. the mean of the ranges derived from the highest and lowest reading of each day, irrespective of the hour of occurrence was 20′·4. It was again unquestionably a case not of a few highly disturbed days with abnormally large ranges, but of persistently large regular diurnal variation throughout the month. The 30 ranges varied only from 14′ to 27′, and 23 of them lay between 17′ and 23′.

Disturbed Days

§ 14. In England, 4 a.m. to 2 p.m. is the time of the day when magnetic disturbance is least common, thus when daily observations are confined to 8 a.m. and 1 p.m., or hours approximating thereto, as was the case in many years with Cavendish's observations, many days may appear quiet which were in reality considerably disturbed. But when observations are taken at 10 or 11 p.m. as well, as was the case in 1776 and 1778, and still more so when they are taken twelve times a day as in June 1788, the chance of any considerable disturbance failing to show itself is not very large. It is a pretty common belief, though not a very well-grounded one—1893 for instance was a conspicuous example to the contrary—that years of exceptionally large sun-spot frequency are years when magnetic disturbances are specially large and numerous. Thus the presence or absence of specially disturbed conditions in years of such abnormal sun-spot development as 1778, 1787 and 1788 is of considerable interest. As regards 1788, the observations with Gilpin's needle establish beyond a doubt that the month of June was a distinctly quiet month, and there are no special indications of disturbance in the observations available for 1778 and 1787. It cannot of course be safely inferred from the character of one or two months what was the character of the whole year. But it can at least be said that while there is conclusive evidence that the regular diurnal variation was abnormally large in 1778, 1787 and 1788, the evidence so far as it goes is against any special development of magnetic disturbance. This does not mean of course that disturbance was non-existent. That is practically never the case, a certain amount of disturbance being the rule rather than the exception. There were undoubtedly some days of considerable disturbance in these years, and one such case, May 24, 1788, is clearly indicated by Cavendish's observations. He observed five times that day. The readings he got at 11.20 a.m. and 1.10 p.m. were identical—in itself a sign of disturbance—and they each exceeded the reading at 7.50 a.m. by 40′. It was a time of large regular diurnal variation, still this represents a considerable disturbance. Again, on the 13th of the same month declination was lower at 5.20 p.m. and at 7.50 p.m. than at 8.25 a.m.—by 5′ and 10′ respectively—whereas on a normal day the difference should have been considerably the other way.

The following are the principal other occasions of disturbance which I have noticed:

On July 20, 1782 declination fell 24′ between 2.45 p.m. and 10.20 p.m. and rose 6′ between 10.20 p.m. and 11 p.m. The observer enters "auror(a)" against the 10.20 and 11 p.m. observations. Such casual notes are very rare, but whether Cavendish associated the disturbance with the aurora, or merely recorded the presence of aurora as a very rare event in London in July, we cannot be certain. The disturbance continued until the 21st, declination *falling* 4′ between 7.40 a.m. and 9.20 a.m., and being 6′ higher at 7.40 a.m. than at 7.30 p.m.

On May 23, 1789, the reading at 9 a.m. was 3′ *higher* than at 4.45 p.m., and only 4′ lower than at 1.30 p.m.

On June 21, 1790, the reading at 12.40 p.m. was only 2′ higher than at 8.50 a.m. instead of 12′ as usual.

On July 5, 1794, the readings at 8.10 a.m. and 1.40 p.m. differed by only 0′·5, instead of 11′½ as usual.

On August 30, 1795, the readings at 8 a.m. and 1.10 p.m. differed by only 1′·5, instead of 9′ as usual.

On August 20, 1796, the reading at 12.30 p.m. was about 10′ higher than usual, making the range for the day fully double the normal.

On July 19, 1800, the reading at 8.30 a.m. exceeded that at 12.40 p.m. by 4′·5, but Cavendish has drawn his pen through it, so it may have been considered doubtful.

On August 23, 1800, the reading at 2.30 p.m. was only 2′ higher than that at 8.30 a.m., instead of 9′½ as usual.

On July 3, 1801, the reading at 1 p.m. was only 1′ higher than that at 8.30 a.m., instead of 9′ as usual.

In 1804 the excesses of the afternoon over the morning readings on September 6 and 20 were respectively only 0′·6 and 2′·3, instead of 10′ as usual.

On September 21, 1805, the afternoon reading was only 0′·5 higher than the morning reading, instead of 9′ as usual.

On August 23, 1806, declination was 6′·8 *lower* at 1 p.m. than at 8.10 a.m., and on July 5, 1808, it was 0′·2 *lower* at 1 p.m. than at 8.50 a.m.

On August 25, 1809, declination was only 1′·4 higher at 1 p.m. than at 9 a.m., instead of 6′ as usual.

The above includes all the cases where I observed the range for the day to fall below 3′, but ranges under 6′ in summer months may certainly be regarded as abnormal, and the number of ranges between 3′ and 6′ was considerably greater than that of ranges below 3′. Allowance must, however, be made for the fact that errors of 1′ or more must frequently have occurred with an instrument such as Cavendish used, even in the most skilled hands, so that the possibility of not infrequent errors of ± 2′ in the daily range must be borne in mind.

INDEX TO VOL. II

CAMBRIDGE: PRINTED BY J. B. PEACE, M.A., AT THE UNIVERSITY PRESS

PLATE I

[TAB. VII]

Fig. 1.

Fig. 2.

Fig 3.

Fig. 4.

Fig. 5.

Fig. 6.

Plate II

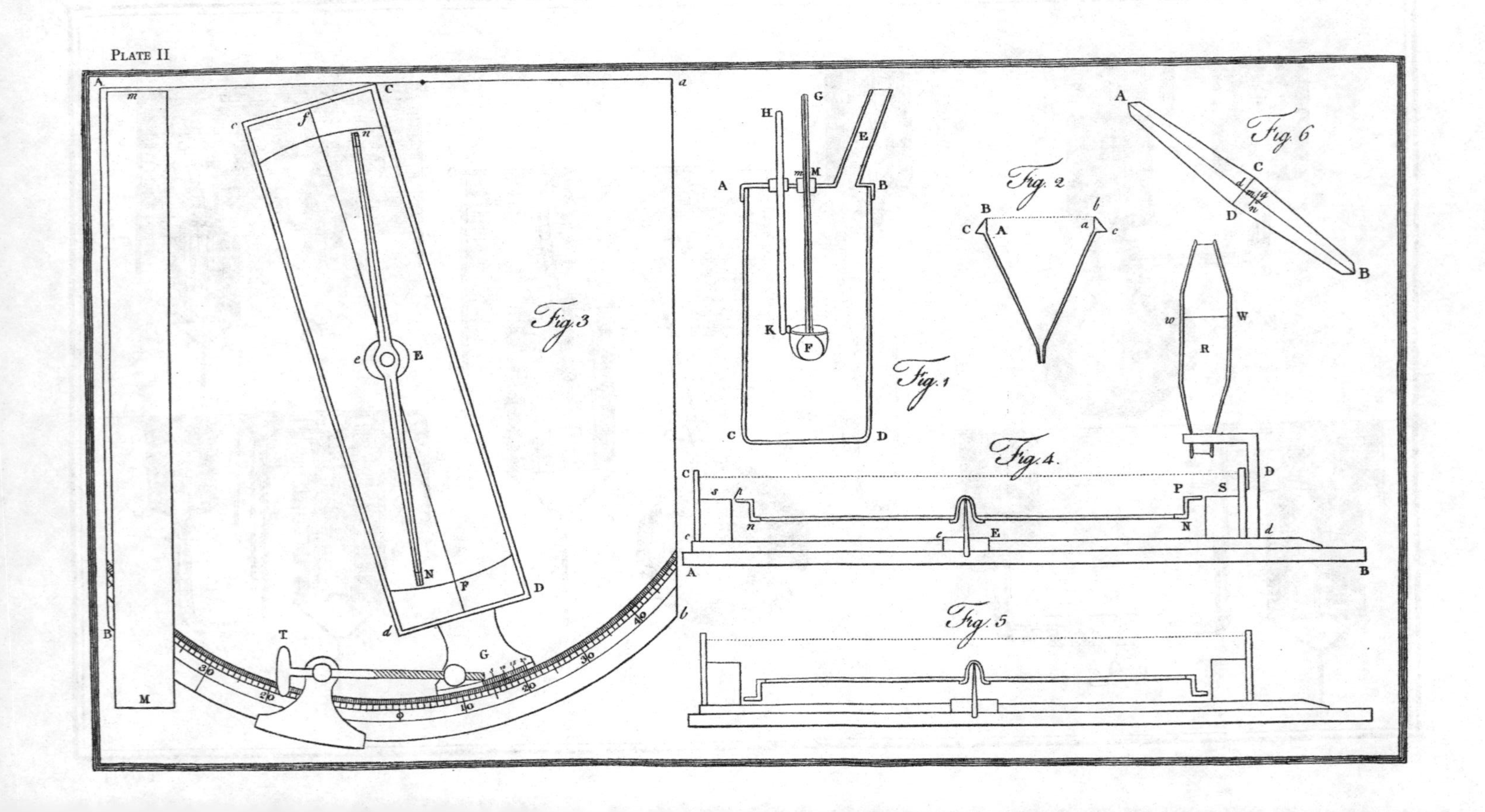

PLATE III

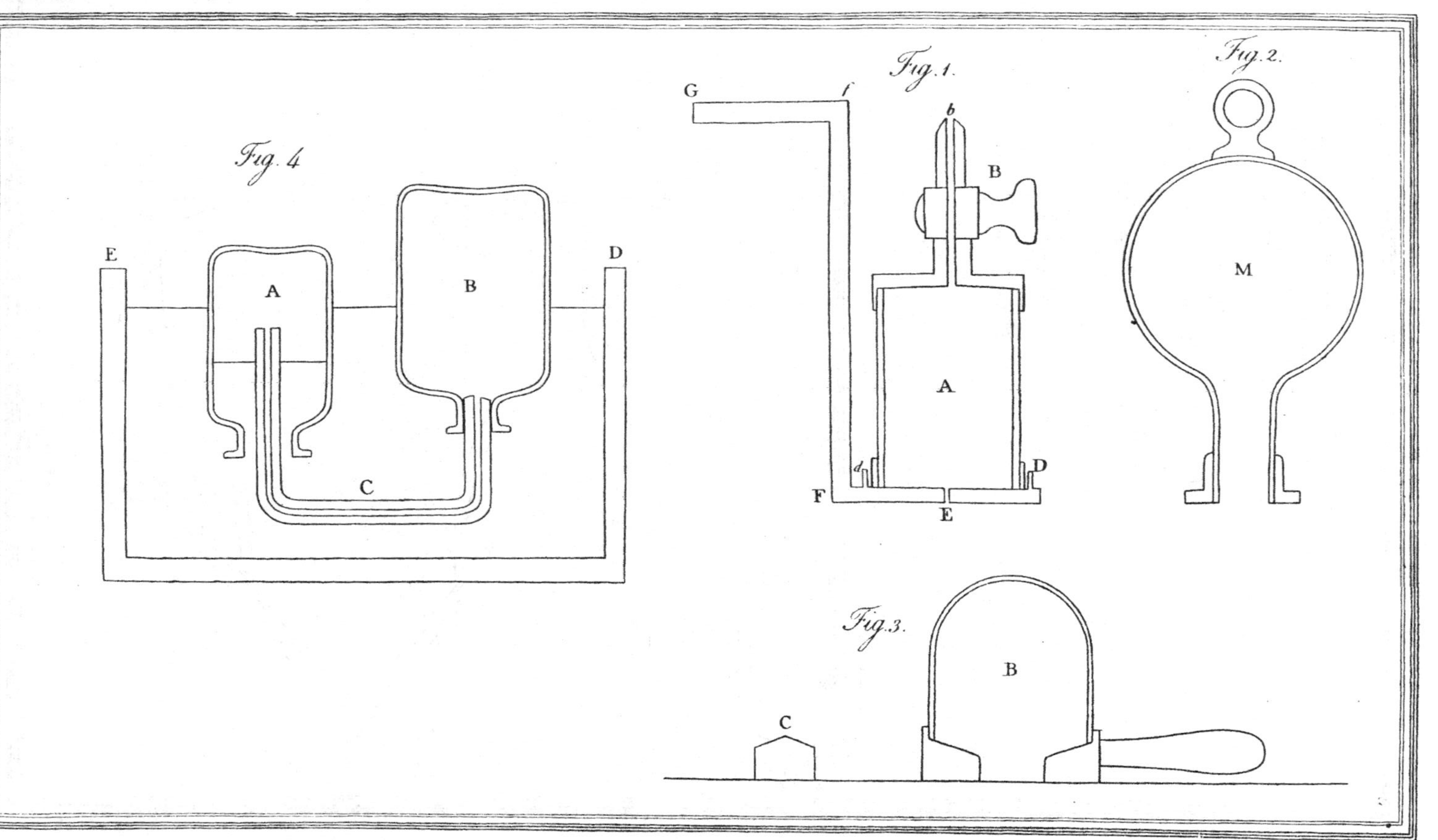

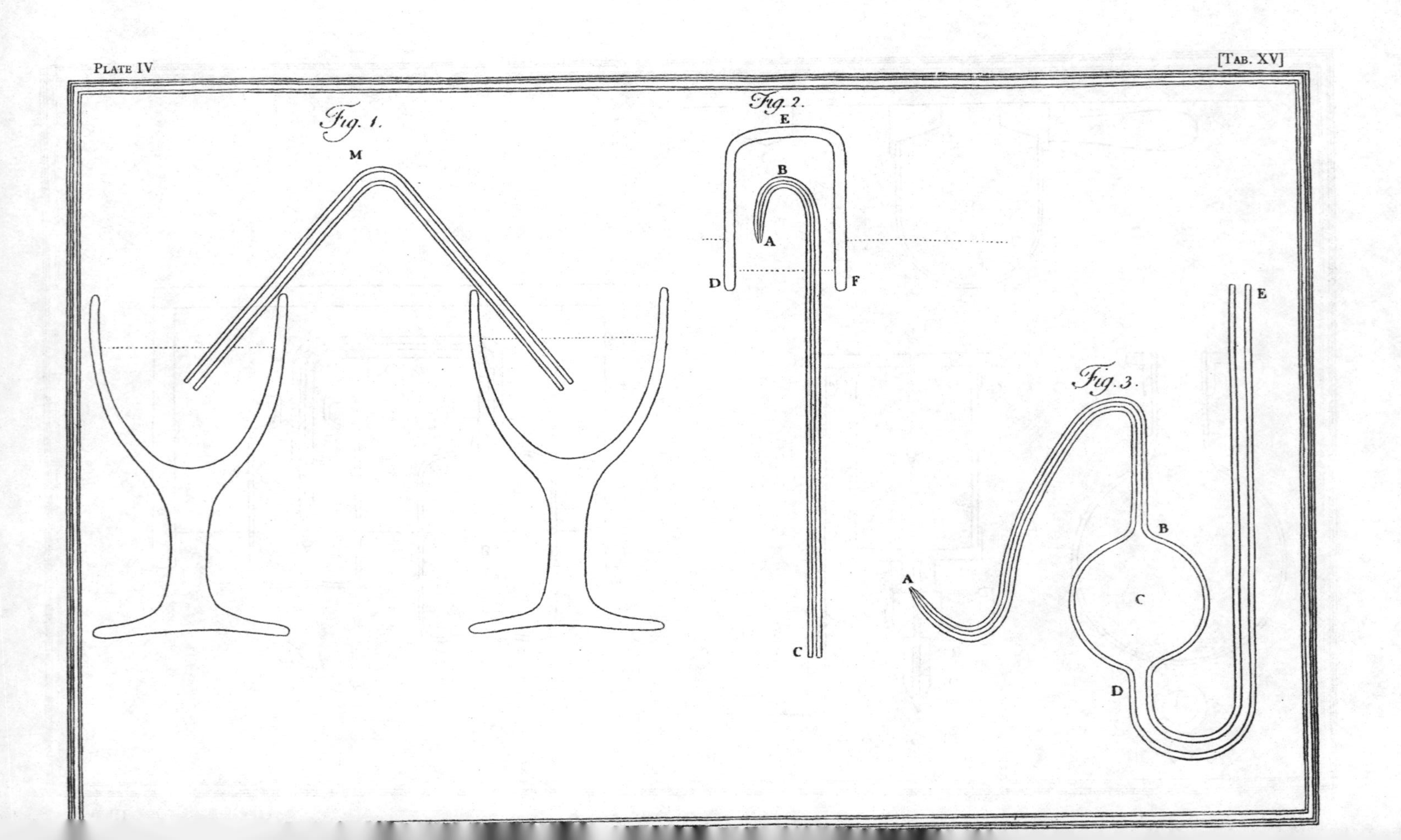
Fig. 1.
M
Fig. 2.
E
B
A
D
F
C
Fig. 3.
E
B
A
C
D

Fig. 1

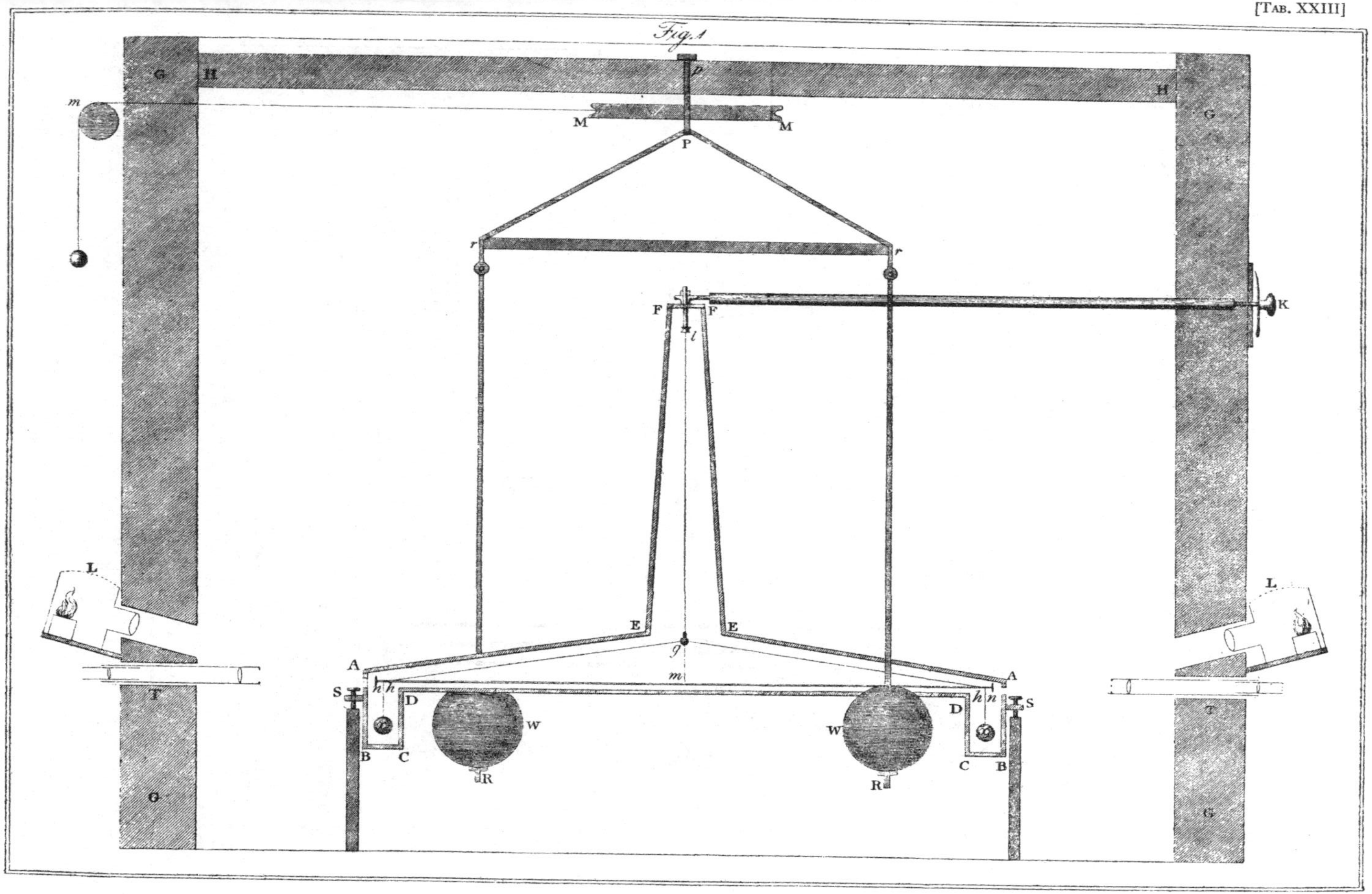

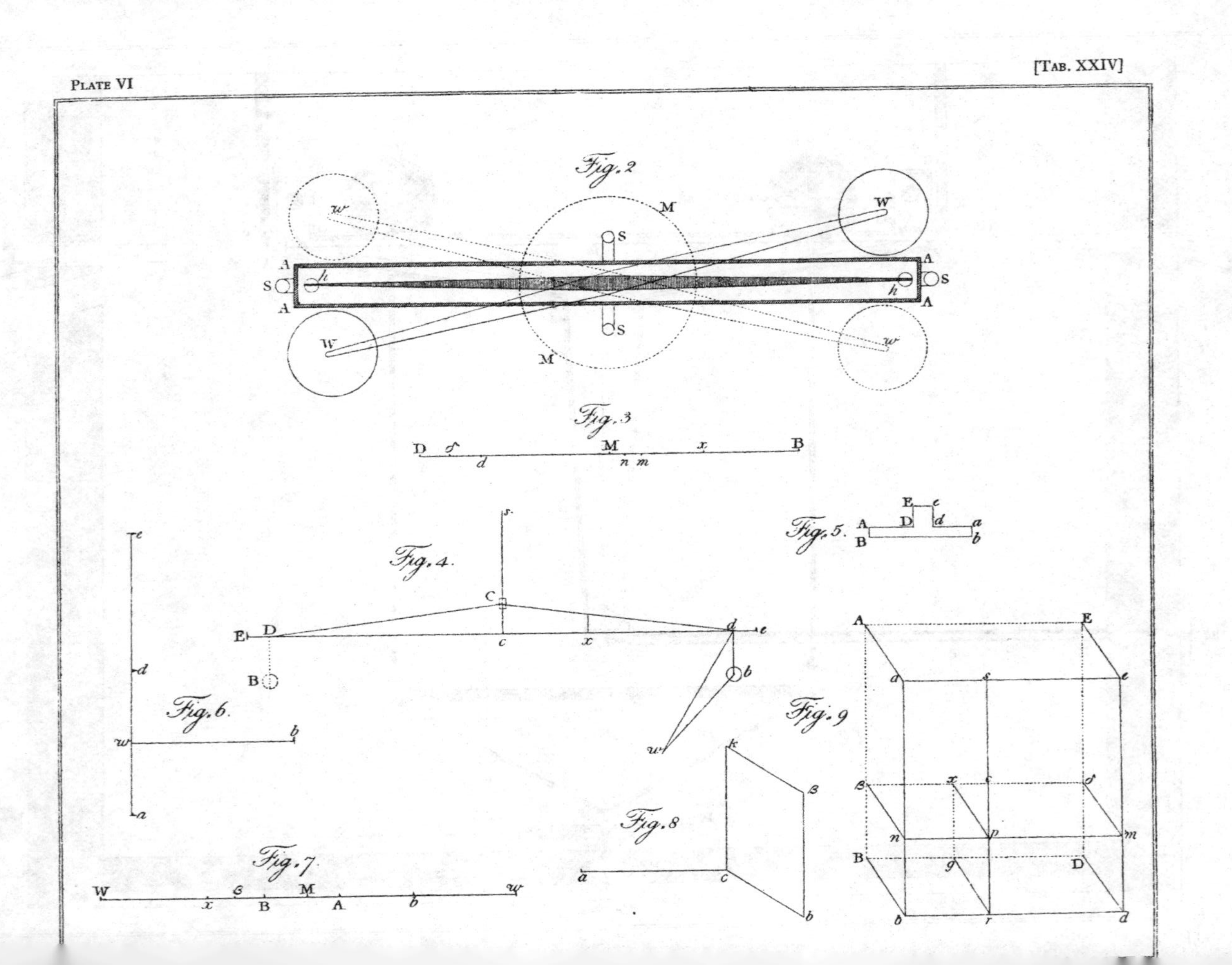
Fig. 2
W
w
M
S
A
h
Fig. 3
D
δ
d
M
n m
x
B
Fig. 4.
s.
C
E
D
c
x
d
e
b
B
w
Fig. 5.
E
e
D
d
A
a
B
b
Fig. 6.
e
d
w
b
a
Fig. 7.
W
x
β
B
M
A
b
w
Fig. 8
k
β
a
c
b
Fig. 9
A
E
a
s
e
β
x
δ
n
p
m
B
g
D
b
r
d

9 781108 01822

Printed in the United States
By Bookmasters